Reflections on Experimental Science

World Scientific Series in 20th Century Physics

Published

Vol. 1   Gauge Theories — Past and Future
         edited by R. Akhoury, B. de Wit, and P. van Nieuwenhuizen

Vol. 2   Scientific Highlights in Memory of Léon van Hove
         edited by F. Nicodemi

Vol. 3   Selected Papers, with Commentary, of T. H. R. Skyrme
         edited by G. E. Brown

Vol. 4   Salamfestschrift
         edited by A. Ali, J. Ellis and S. Randjbar-Daemi

Vol. 5   Selected Papers of Abdus Salam (with Commentary)
         edited by A. Ali, C. Isham, T. Kibble and Riazuddin

Vol. 6   Research on Particle Imaging Detectors
         by G. Charpak

Vol. 7   A Career in Theoretical Physics
         by P. W. Anderson

Vol. 8   Lepton Physics at CERN and Frascati
         edited by N. Cabibbo

Vol. 9   Quantum Mechanics, High Energy Physics and Accelerators: Selected Papers
         of J. S. Bell (with Commentary)
         edited by M. Bell, K. Gottfried and M. Veltman

Vol. 10 How We Learn; How We Remember: Toward an Understanding of Brain
         and Neural Systems — Selected Papers of Leon N. Cooper
         edited by Leon N. Cooper

Vol. 12 Sir Nevill Mott — 65 Years in Physics
         edited by N. Mott and A. S. Alexandrov

Vol. 13 Broken Symmetry — Selected Papers of Y. Nambu
         edited by T. Eguchi and K. Nishijima

Forthcoming

Vol. 11 30 Years of the Landau Institute — Selected Papers
         edited by I. M. Khalatnikov and V. P. Mineev

Vol. 15 Encounters in Magnetic Resonances — Selected Papers of Nicolaas Bloembergen
         (with Commentary)
         edited by N. Bloembergen

Vol. 16 Encounters in Nonlinear Optics — Selected Papers of Nicolaas Bloembergen
         (with Commentary)
         edited by N. Bloembergen

Vol. 17 The Collected Works of Lars Onsager (with Commentary)
         edited by P. C. Hemmer, H. Holden and S. K. Ratkje

World Scientific Series in 20th Century Physics — Vol. 14

REFLECTIONS ON EXPERIMENTAL SCIENCE

Martin L. Perl

Stanford Linear Accelerator Center
Stanford University

World Scientific
Singapore • New Jersey • London • Hong Kong

Published by

World Scientific Publishing Co. Pte. Ltd.
P O Box 128, Farrer Road, Singapore 912805
USA office: Suite 1B, 1060 Main Street, River Edge, NJ 07661
UK office: 57 Shelton Street, Covent Garden, London WC2H 9HE

The editor and publisher would like to thank the following publishers for their assistance and their permission to reproduce the articles found in this volume:

American Physical Society, Elsevier Science Publications B. V., Editions Frontieres, American Institute of Physics, John Wiley & Sons Inc., Institute of Physics, Deutsches Elektronen-Synchrotron DESY, and The Institute of Particle Physics.

Library of Congress Cataloging-in-Publication
Perl, Martin L., 1927–
    Reflections on experimental science / Martin L. Perl.
      p. cm. -- (World Scientific series in 20th century physics ; vol. 14)
    Includes bibliographical references and index.
    ISBN 981022429X -- ISBN 9810225741 (pbk)
    1. Particles (Nuclear physics) -- Experiments. I. Title. II. Series.
QC793.412.P47   1996
539.7/2--dc20   95-48852
                                                                                         CIP

British Library Cataloguing-in-Publication Data
A catalogue record for this book is available from the British Library.

Copyright © 1996 by World Scientific Publishing Co. Pte. Ltd.

All rights reserved. This book, or parts thereof, may not be reproduced in any form or by any means, electronic or mechanical, including photocopying, recording or any information storage and retrieval system now known or to be invented, without written permission from the Publisher.

For photocopying of material in this volume, please pay a copying fee through the Copyright Clearance Center, Inc., 222 Rosewood Drive, Danvers, Massachusetts 01923, USA.

Printed in Singapore.

CONTENTS

Preface ix

PART A. THE DISCOVERY OF THE TAU LEPTON

A1. A Memoir on the Discovery of the Tau Lepton and Commentaries on Early Lepton Papers — 3

A2. Search for New Particles Produced by High-Energy Photons — 30
A. Barna *et al.*, *Phys. Rev.* **173**, 1391 (1968)

A3. Comparison of Muon–Proton and Electron–Proton Deep Inelastic Scattering — 42
W. T. Toner *et al.*, *Phys. Lett.* **36B**, 251 (1971)

A4. The Search for Heavy Leptons and Muon–Electron Differences — 48
M. L. Perl and P. A. Rapidis, SLAC-PUB-1496, (1974), unpublished

A5. Lectures on Electron–Positron Annihilation:
Part II. Anomalous Lepton Production — 139
M. L. Perl, *Proc. Canadian Inst. Particle Physics Summer School* (McGill Univ., Montreal, 1975), eds. R. Heinz and B. Margolis, p. 435

A6. Evidence for Anomalous Lepton Production in $e^+ - e^-$ Annihilation — 193
M. L. Perl, *et al.*, *Phys. Rev. Lett.* **35**, 1489 (1975)

A7. Review of Heavy Lepton Production in e^+e^- Annihilation — 197
M. L. Perl, *Proc. 1977 Int. Symp. Lepton and Photon Interactions at High Energies*, (Hamburg, 1977), ed. F. Gutbrod, p. 145

PART B. THE PHYSICS OF THE TAU LEPTON AND TAU NEUTRINO

B1. The Physics of the Tau Lepton and Tau Neutrino in 1995 — 219

B2. Comments on B2
The Future of Tau Physics and Tau-Charm Detector and Factory Design — 266
M. L. Perl, *Proc. Sixth Lake Louise Winter Institute* (World Scientific, Singapore, 1991), eds. B. A. Campbell, A. N. Kamal, P. Kitching, F. C. Khanna, p. 182

B3. Comments on B3
Beyond the Tau: Other Directions in Tau Physics — 308
M. L. Perl, *Proc. 2nd Workshop on Tau Lepton Physics* (World Scientific, Singapore, 1993), ed. K. K. Gan, p. 483

B4. Comments on B4
Tau Physics at Future Facilities 327
M. L. Perl, *Proc. Third Workshop Tau Lepton Physics, Nucl. Phys. B
(Proc. Suppl.)* **40**, (1955), ed. L. Rolandi, p. 541

PART C. INNOVATIONS IN EXPERIMENTAL METHODS AND NEW DIRECTIONS IN PHYSICS

C1. Comments on C1
Nuclear Electric Quadrupole Moment of Na^{23} 345
M. L. Perl, I. I. Rabi and B. Senitzky, *Phys. Rev.* **97**, 838 (1955)

C2. Comments on C2
Scattering of K^+ Mesons by Protons 349
D. I. Meyer, M. L. Perl and D. A. Glaser, *Phys. Rev.* **107**, 279 (1957)

C3. Comments on C3 and C4
The Use of a Sodium Iodide Luminescent Chamber
to Study Elastic and Inelastic Scattering of Pions in Hydrogen 355
M. L. Perl, L. W. Jones, and K. Lai, *Proc. 1960 Int. Conf. Instrumentation
High-Energy Phys.* (Interscience, New York, 1961), p. 186.

C4. Negative Pion–Proton Elastic Scattering at 1.51, 2.01, and
2.53 Bev/c outside the Diffraction Peak 363
K. W. Lai, L. W. Jones, and M. L. Perl, *Phys. Rev. Lett.* **7**, 125 (1961)

C5. Comments on C5
Pion–Proton Elastic Scattering at 2.00 GeV/c 365
D. E. Damouth, L. W. Jones, and M. L. Perl, *Phys. Rev. Lett.*
11, 287 (1963)

C6. Comments on C6
Neutron–Proton Elastic Scattering from 1 to 6 GeV 370
M. N. Kreisler *et al.*, *Phys. Rev. Lett.* **16**, 1217 (1966)

C7. Comments on C7
A High Energy, Small Phase-Space Volume Muon Beam 376
J. Cox *et al.*, *Nucl. Instrum. Methods* **69**, 77 (1969)

C8. Comments on C8
Measurement of the Inclusive Electroproduction of Hadrons 389
J. T. Dakin *et al.*, *Phys. Rev. Lett.* **29**, 746 (1972)

C9. Comments on C9
Small Multiplicity Events in $e^+ + e^- \to Z^0$ and Unconventional
Phenomena 395

M. L. Perl, *Proc. 22nd Rencontre de Moriond*, Vol. 1 (Editions Frontieres, Gif Sur Yvette, 1987), ed. J. Tran Thanh Van, p. 451

C10. Comments on C10
Rotor electrometer: New instrument for bulk matter quark
search experiments   418
J. C. Price *et al.*, *Rev. Sci. Instrum.* **57**, 2691 (1986)

C11. Comments on C11
Electron–Positron Collision Physics: 1 MeV to 2 TeV   427
M. L. Perl, *Proc. Intersections between Particle and Nuclear Physics*,
(Am. Inst. Phys., New York, 1988) ed. G. M. Bunce, p. 1185

C12. Comments on C12
Exploration of the limits on Charged-Lepton-Specific forces   450
C. A. Hawkins and M. L. Perl, *Phys. Rev.* **D40**, 823 (1989)

C13. Comments on C13
Notes on the Landau, Pomeranchuk, Migdal Effect: Experiment
and Theory   462
M. L. Perl, *Proc. Rencontres de Physique de la Vallee D'Aoste*
(Editions Frontieres, Gif Sur Yvette, 1994), ed. M. Greco, p. 567

C14. Comments on C14
Efficient bulk search for fractional charge with multiplexed
Millikan chambers   478
C. D. Hendricks *et. al.*, *Meas. Sci. Technol* **5**, 337 (1994)

PART D. ESSAYS IN PHYSICS

D1. Comments on D1
Popular and unpopular ideas in particle physics   493
M. L. Perl, *Physics Today*, **Dec**, p. 27 (1986)

D2. Comments on D2
Science in the Age of Accelerators   500
M. L. Perl, *Proc. Physics of Particle Accelerators*, (Am. Inst. Phys.,
New York, 1989) eds. M. Month and M. Dienes, p. 2098

PART E. REFLECTIONS ON EXPERIMENTAL SCIENCE
Reflections on Experimental Science   525

PREFACE

This book on experimental science, made up of my comments, scientific reprints, reflections, and a memoir, has been written for many types of readers. First and very important to me is the reader who has an interest in the realities of scientific work but is not interested in the technical details. For this reader the main parts of the book will be the comments before each scientific reprint, the final part of the book on "Reflections on Experimental Science" and perhaps the memoir on the discovery of the tau lepton which begins the book. This reader may occasionally want to glance at a reprint.

The second type of reader is one who is interested in the history of particle physics or of big science since the 1960's. I hope I have provided some insights into that history. I think the reprints will be useful to this reader because they describe the physics knowledge and theories and speculations at the time of the experiment.

I hope that the third type of reader will be some of my colleagues who will forgive me for some of my more judgmental statements. They will perhaps be amused by some of the reprints, recalling how ignorant we were and realizing how far we have come in elementary particle physics.

Most of all I hope that some readers will be students who are entering the experimental sciences or are young experimental scientists. I wish that a book like this had been available to me, I would not have made less mistakes but I would have felt less bad about my mistakes.

Experiments are the heart of science. Speculation is needed, calculations are needed, theories are needed; but in the end speculation, calculation and theory must all be tested by experiment. Are they confirmed by experiment? If not, the speculation is a dream, the calculations are wrong, the theory must be altered or discarded.

Much has been written on the history of great experiments and the lives of famous experimental scientists. There are many books on the philosophy of the scientific method and on the interaction between theory and experiment. There are specialized books on how to do experiments. This book "Reflections on Experimental Science" has a different purpose. I examine the ways in which experiments are conceived and carried out from the point of view of working experimental scientists, the experimenters themselves! I write about the emotional and other personal qualities of the experimenter which influence the day-to-day operation of experiments. I write about the effects of an experimenter's scientific community on the research. I tell something about fashion in scientific research, how a scientific community's interest in a subfield builds and then decays. And very important, I reflect on the subject of success and failure in experimental science.

Of course a great deal has been written about the very successful experiments, the great experiments that changed a science. Little has been written about the failed experiments, about experiments which didn't work or which didn't get the right answer. I don't mean the extravagant failures such as the search for cold fusion; I mean the small failures which occur in the work of every experimenter. It is a delicate task to write about many of these subjects: emotion in scientific work, pressures from the scientific community, fashions in research, experiments that fail. And so I have chosen to write primarily about my own experimental work in elementary particle physics.

Elementary particle physics is the part of the physical sciences which deals with the basic nature of matter, energy and time. The origins of elementary particle physics lie in the late nineteenth century when the first identified elementary particle, the electron, was discovered. The electron is the simplest example of an elementary particle; it is a very small piece of matter which cannot, to the best of our experimental knowledge, be broken up into simpler pieces of matter. That is, we have not been able to do experiments which break up the electron. Therefore the electron is one of the basic and elementary parts of matter. By experiment more then a dozen elementary particles have been discovered. These include two other particles which are like the electron but have much greater mass, the muon and the tau, and these particles along with the neutrinos are called leptons. However most of the weight of ordinary matter is made up of another family of six elementary particles! called quarks with the peculiar n ames of up, down, strange, charm, bottom, and top.

Elementary particle physics is also called high energy physics because the examination of the properties of elementary particles and the search for new elementary particles usually requires the production and collision of particles at very high energies. This in turn requires large experimental facilities at laboratories which may employ a thousand or more people. These experimental facilities consist of particle accelerators, particle colliders, and particle detectors which may require hundreds of scientists and engineers for their construction and operation. Thus my field is often called big science as opposed to small science; high energy physics along with astronomy and nuclear physics were the first of the big sciences. My research has often used the big science facilities in my field, but not always, as the reader will learn I have often obtained the most pleasure with my small science experiments.

In the comments on papers written after 1980 the reader will come across phrases such as 'new elementary particle physics', 'new phenomena in elementary particle physics', or simply 'new physics'. By 1980 experiment and theory had established a broad, quantitative, and mathematical picture of the elementary particles and fundamental forces, usually called the 'standard model' of elementary particle physics. However this model contains many parameters which cannot be calculated, and so must be measured. In addition there are many unexplained aspects of this model, for example we do not know why there are only four fundamental forces. The emphasis of the past fifteen years on the search for some new phenomena in elementary particle physics comes from the strong almost desperate desire of the particle physics community to break out of the standard model, to find a more basic theory in which more is explained and less has to measured. I feel this desperation and so the reader! will see many references to 'new physics'.

The development of scientific theories is of course a crucial part of the scientific method, but I write very little about theory because I don't do theoretical research work and I have no first hand knowledge of the day-to-day practice of a theoretical physicist. This book is not about the scientific method, it is about what experimenters do.

To set the background for this book I will write a little about my research life in physics. In 1955 I received a doctorate in atomic physics from Columbia University and went to the University of Michigan to begin working in elementary particle physics. I was an engineer turned experimental physicist and I am still an experimental physicist with an engineering flavor in my research. It has been an astonishing and fortunate forty years for me.

Astonishing because in those forty years we have learned so much about the fundamental nature of particles and forces. Astonishing because the technology of elementary particle physics experimentation has changed so much. In 1955 we used vacuum tube electronics. We used small main frame computers for calculations and some data analysis, but not for data acquisition since our interface to the computer was paper tape and punch cards. My first high energy physics research used the 3 GeV Cosmotron proton accelerator at Brookhaven National Laboratory and a non-magnetic propane bubble chamber with a volume of about 3 liters.

It has been a fortunate forty years for me in several ways. Starting with the experimental technology of 1955 I have been able to use in my research the new technology as it was invented and developed: optical spark chambers, transistor circuits, large main frame computers interfaced through terminals and electronic tape, high energy proton accelerators, electron linear accelerators, electron-positron colliders, drift chambers, electromagnetic shower calorimeters, and powerful desktop computers. I have been able to experiment in just about all parts of elementary particle physics ranging from the strong interaction process of pion-proton elastic scattering to the electromagnetic and weak interaction processes involved in the production and decay of tau leptons and charm quarks. I have been particularly fortunate that my thinking in the 1960's about the electron-muon problem led in the 1970's to the discovery of the tau lepton, using the then newly developed technologies of t! he electron-positron collider and the large-solid-angle, electronic, particle detector.

In this volume I describe my research over these forty years using two intertwined streams. One stream is made up of the reprints of articles and reviews which I authored alone or with my colleagues. I have selected the reprints with several goals in mind. I want to give a broad picture of the research I have done over the years and thus indirectly a partial history of high energy physics research. I want to illustrate various sizes of experiments: small experiments carried out with a simple apparatus and a few people such as the mid-1960's neutron-proton scattering experiment described in Reprint C6 or my present quark search (Reprint C14); and large apparatus, large group experiments such as the present day tau lepton research described in the reprints in Part B. Finally by the selection of reprints I want to illustrate the various flavors of experiments. There are the "bread-and-butter" experiments such as the study of the production of hadrons in inelastic electron sca! ttering (Reprint C8). There are th e speculative experiments such as the search for a new force between charged leptons (Reprint C12). There are the successful experiments on the physics of the tau lepton described in Part B. There is the once-in-a-lifetime discovery experiment, the finding of the tau lepton described in Part A. There are the unsuccessful experiments such as my first quark search using a rotor electrometer (Reprint C10). And finally there are the experiments I thought about but never carried out, for example the experiments I wanted to do at a tau-charm factory (Reprint B2).

However journal papers, proceedings papers, and review articles give a formal and desiccated picture of the experiment, they are the skeleton of the research. The pleasures, the anxieties, the lucky breaks, the bad breaks, are all left out. Often the real reasons for doing the experiment are left unstated. Therefore in this book there is a stream of comments which goes along with, and intertwines with, the stream of reprints. These comments put the flesh back on the skeleton of the reprint, they bring back the motivations, the emotions, the daily reality of the experiment. The memoir in Part A on the discovery of the tau lepton is part of the stream of comments.

This book has five parts. Part A is a history of the discovery of the tau lepton and includes six reprints of articles which were important in that history. Part B has a newly revised review of tau lepton and tau neutrino physics and three reprints of articles on other aspects of tau lepton physics. Part C, with fourteen reprints on non-tau physics, gives a broad picture of the flesh as well as the skeleton of experimentation in elementary particle physics. Part D contains reprints of two essays on particle physics.

Part E entitled "Reflections on Experimental Science" is a summary of what I have learned about the reality of experimental work. Some of these reflections developed over many years, some came to me as I assembled this book and wrote the comments. I don't claim uniqueness for these reflections, I am sure that experimenters in every science have had similar experiences and thoughts. But it has given me pleasure to write about the personal and human aspects of experimental work. I hope it gives pleasure to the reader.

Part A

The Discovery of the Tau Lepton

A MEMOIR ON THE DISCOVERY OF THE TAU LEPTON AND COMMENTARIES ON EARLY LEPTON PAPERS

CONTENTS

1. Introduction	3
2. SLAC, Leptons, and Heavy Leptons	4
3. Heavy Lepton Searches in the 1960's	6
4. Photoproduction Searches for New Charged Leptons	7
5. Studies of Muon–Proton Inelastic Scattering	7
6. Electron–Positron Colliding Beams and Sequential Leptons	8
7. The SLAC–LBL Proposal	11
8. Lepton Searches at ADONE	13
9. Discovery of the Tau in the Mark I Experiment: 1974–76	13
10. The 1975 Canadian Talks	14
11. First Publication: "We have no conventional explanation for these events"	17
12. Is it a Lepton? 1976–1978	19
13. Anomalous Muon Events	20
14. Anomalous Electron Events	22
15. Semileptonic Decay Modes and the Search for $\tau^- \to \nu_\tau \pi^-$ and $\tau^- \to \nu_\tau \rho^-$	24
16. The Tau Mass	26
17. The Tau Lifetime	27
References	28

1. Introduction

In the first part of this memoir I describe the earlier thoughts and work of myself and my colleagues at SLAC in the 1960's and early 1970's which led to the discovery of the tau lepton. I also describe the theoretical and experimental events in particle physics in the 1960's in which our work was immersed. I will also try to describe for the younger generations of particle physicists, the atmosphere in the 1960's. That was before the elucidation of the quark model of hadrons, before the development of the concept of particle generations. The experimental paths to progress were not as clear as they are today and we had to cast a wide experimental net.

In the second part I describe (a) the experiments which led to our 1975 publication of the first evidence for the existence of the τ, (b) the subsequent experiments which confirmed the existence of the τ, and (c) the experiments which elucidated the major properties of the τ.

As I travel this history I take side trips to comment on the papers reprinted in this section, papers which illustrate the knowledge and experimental techniques of those times.

Other histories of the discovery of the tau and related physics are Tsai (1992), Feldman (1992) and Harari (1992).

2. SLAC, Leptons, and Heavy Leptons

At the start of the 1960's, I was at the University of Michigan; our experiments were carried out at the Brookhaven Cosmotron and the Berkeley Bevatron, experiments in strong interaction physics. But I was becoming interested in lepton physics for a number of reasons. I liked experiments in which the results could be summarized in a few numbers or a few graphs. Thus I worked primarily in elastic scattering and other two-body reactions. I also liked experiments where the theory was relatively simple, and it was clear that strong interaction theory was not becoming simpler. On the other hand, the physics of leptons seemed a simpler world.

In the lepton world I was intrigued by the careful measurements being made on the (g-2) of the muon by Charpak *et al.* (1962), and on the (g-2) of the electron by Wilkinson and Crane (1963) at my University. I was also interested in the precision studies of positronium and muonium then in progress as well as other precision atomic physics experiments. (Indeed as a graduate student at Columbia University in the years 1950 to 1955, I worked under I. I. Rabi on an atomic beam experiment. And it was there that I first learned about positronium from Vernon Hughes.) These low energy studies of the charged leptons were in very capable hands, and I thought that it would be most useful for me to consider high energy experiments on charged leptons, experiments which might clarify the nature of the lepton or explain the electron-muon problem.

The opportunity appeared to think seriously about such experiments in 1962 when W. K. H. Panofsky offered me a position at the yet-to-be built Stanford Linear Accelerator

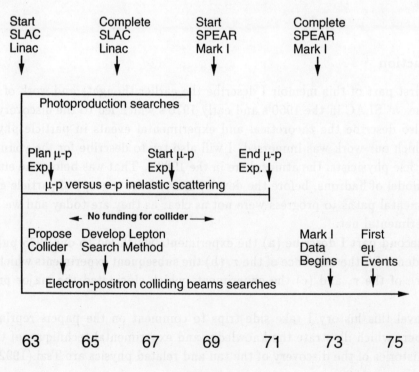

Fig. 1

Center. Here was a laboratory which would have primary electron beams, a laboratory at which one could easily obtain a good muon beam, a laboratory in which one could easily obtain a good photon beam for production of particle pairs. And on the same campus at the High Energy Physics Laboratory, the Princeton-Stanford e^-e^- storage ring was operating. (O'Neil et al. 1958, Barber et al. 1966).

From the time that the SLAC linear accelerator began operation in 1966 until the discovery of the τ in 1975, my colleagues and I cast a wide experimental net in our studies of leptons. These studies fell into three classes which I shall describe in turn: photoproduction searches for new charged leptons, studies of muon-proton inelastic scattering to seek $e - \mu$ differences, and e^+e^- colliding beam searches for new charged leptons. Figure 1 shows schematically the history of our three classes of lepton studies set against the construction history of the SLAC linear accelerator and the SPEAR e^+e^- storage ring.

Before turning to these studies, I describe the general thinking in the 1960's in the lepton world about the possible existence and types of new leptons. Since the 1950's a great deal of thought had been given to the concept of lepton number and lepton number conservation. This is not the place to record that intricate history. It is sufficient to note that by the beginning of the 1960's these concepts were well developed, although there was disagreement on how the leptons should be classified. And by the beginning of the 1960's there were papers on the possibility of the existence of charged leptons more massive than the e and μ, heavy leptons. I remember reading the 1963–1964 papers of Zel'dovich (1963) and of Lipmanov (1964). But since the particle generation concept was not yet an axiom of our field, older models of particle relationships were used. For example, if one thought (Low, 1965) that there might be an electromagnetic excited state e^* of the e then the proper search method was

$$e^- + \text{nucleon} \to e^{-*} + \cdots$$
$$e^{-*} \to e^- + \gamma$$

Or, if one thought (Lipmanov 1964) that there was a μ' which was a member of a μ, ν_μ, μ' triplet then the proper search method was

$$\nu_\mu + \text{nucleon} \to \mu^{-\prime} + \cdots$$

It is interesting to note in view of the decade later search for $\tau^- \to \nu_\tau \pi^-$ (Sec. 15) that Lipmanov (1964) calculated the branching fraction for this decay mode.

By the second half of the 1960's the concept had been developed of a heavy lepton L and its neutrino ν_L forming an L, ν_L pair. Thus in a paper written in 1968, Rothe and Wolsky (1969) discuss the lower mass limit on such a lepton set by its absence in K decays. They also discuss the decay of such a lepton into the modes

$$L \to e\bar{\nu}_e\nu_L, \mu\bar{\nu}_\mu\nu_L, \pi\nu_L.$$

Incidentally, in our 1971 proposal (Larsen et al. 1971) to SLAC to study e^+e^- annihilation physics using the SPEAR collider then under construction, we reference Rothe and Wolsky (1969) as indicative of the thinking on heavy leptons in the second half of the 1960's.

3. Heavy Lepton Searches in the 1960's

In 1971 I gave a paper at the Muon Physics Conference at Colorado State University entitled "The Search For Leptons and Muon–Electron Difference". I gave a revised review at a conference in Moscow in the Soviet Union the next year and revised it again in 1974 with Petros Rapidis as co-author. (Perl and Rapidis 1994). It was one of those papers that was never published because neither the Colorado nor the Moscow proceedings were published. But I have included the 1974 SLAC preprint as Reprint A4 in this book because it provides a good portrait of heavy lepton searches in the 1960's.

The paper begins with a review of the possible types of heavy leptons. It begins with heavy sequential leptons

$$e \ \mu \ \mu' \ \mu'' \cdots$$

and their associated neutrinos

$$\nu_e, \ \nu_\mu, \ \nu_{\mu'}, \ \nu_{\mu''} \cdots$$

(I proposed the term "sequential leptons" in the first version of this paper). The paper goes on with discussions of excited leptons

$$e^{*-} \to e^- + \gamma,$$

special pairs of leptons, stable and metastable leptons. We did not know if there were more leptons, and if there were more leptons, we did not know what direction should be taken in an extension of the lepton class of particles.

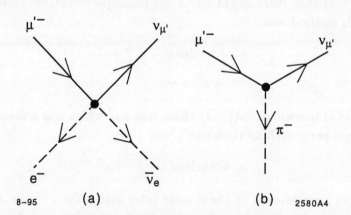

Fig. 2

Section 3 of this paper on decay properties shows that there was a transition period between the four-fermion contact interaction model of the weak interactions and the W and Z boson model. We knew about the latter hypothesis but there was no proof and so as illustrated by Fig. 2 taken from this paper, we used the four-fermion contact interaction to calculate decays.

The long Secs. 4 and 5 of this paper show that we also did not know the best way to search for new leptons. Today it is obvious to use

$$e^+ + e^- \to \ell^+ + \ell^-$$

$$e^+ + e^- \to \nu_\ell + \bar{\nu}_\ell;$$

as I write we are waiting for Lep 200 to begin operation so that the search can be continued for $m_\ell > m_Z/2$. In the 1960's lepton search methods included π and K decays, photoproduction, production in electron and proton beam decays, production in lepton-proton inelastic scattering; as well indirect searches in elastic lepton-proton scattering and in the static projections of the e and μ.

Our very ignorance made it a very exciting time. We were indeed naive about leptons in the 1960's. But it was wonderful to be able to hope that a small short experiment might turn up a new lepton. In the next section I give an example.

4. Photoproduction Searches for New Charged Leptons

Soon after the Stanford linear accelerator began operation, Fig. 1, we made one cast of our net (Barna *et al.* 1968) looking for a new charged lepton. We were looking for any new charged particle x from the reactions

$$e^- + \text{nucleus} \to + \cdots$$

$$\gamma + \text{nucleus} \to x^+ + x^- + \cdots$$

The search used the pair production calculations of Tsai an Whitis (1966); this experiment was the beginning of a long and fruitful collaboration between my colleague Y.-S.(Paul) Tsai and myself.

I have reprinted Barna *et al.* (1968) entitled "Search for New Particles Produced by High-Energy Photons" as Reprint A2. Certainly the photoproduction of pairs is a very general way to produce charged particles. This sort of search is bounded by three constraints. First the upper limit on the mass search range is set by the photon energy. With the 17.5 GeV/c incident e^- beam we could search up to 2 GeV/c^2. The second constraint is the magnitude of the photoproduction cross section. The cross section was known for point particle, but for an extended particles such as a hadron the form factor could only be guessed. The third and most serious constraint was that the particle had to be long lived, a lifetime greater than 10^{-8} s.

We were worried about this lifetime constraint and as the reader will see in Sec. II of Barna *et al.* (1968) we argued around the idea that the more massive the particle the shorter the lifetime. Perhaps it was a bit whistling in the dark, but we could have been right. We found no new particles and indeed there were no new particles to find.

5. Studies of Muon–Proton Inelastic Scattering

As SLAC was being built, Fig. 1, we were preparing to study muon proton inelastic scattering

$$\mu^- + p \to \mu^- + \text{anything}$$

to compare it with

$$e^- + p \to e^- + \text{anything}.$$

Extensive studies of $e-p$ inelastic scattering were planned at SLAC. Indeed, some of those studies led to the Nobel Prize being awarded to J. Friedman, H. Kendall, and R. Taylor. My hope was that we would find a difference between the μ and e other than the differences of mass and lepton number. In particular, I hoped that we would find a difference at large momentum transfers. Some of our hopes, or at least my hopes, were naive by today's standards of knowledge of particle physics. For example, I speculated (Perl 1971) that the muon might have a special interaction with hadrons not possessed by the electron.

Therefore, beginning in the late 1960's, we measured the differential cross sections for inelastic scattering of muons on protons, and then compared (Toner et al. 1972, Braunstein et al. 1972) the $\mu-p$ cross sections with the corresponding $e-p$ cross sections. Reprint A3 of this book is Toner et al. (1972) entitled "Comparison of Muon–Proton and Electron–Proton Deep Inelastic Scattering". The cross section $d\sigma/dq^2\,d\nu$ for charged lepton proton inelastic scattering depends upon the four-momentum q^2 transfered to from the lepton to the proton and on ν, the laboratory energy of the lepton. In this experiment we measured $d\sigma/dq^2\,d\nu$ for the $\mu-p$ system and compared it to $d\sigma/dq^2\,d\nu$ for the $e-p$ system by examining the ratio

$$\rho(q^2,\nu) = \frac{(d\sigma/dq^2\,d\nu)_{\mu p}}{(d\sigma/dq^2\,d\nu)_{ep}} = \frac{N^2}{(1+|q^2|/\Lambda^2)^2}$$

We were looking for a difference in magnitude, that is $N \neq 1$, or a difference in q^2 behavior of the cross sections. As discussed in Perl and Rapidis (1974), Reprint A4, these differences could come from a new non-electromagnetic interaction between the μ and hadrons or from the μ not being a point particle. However as summarized in Toner et al. (1972) we found no significant deviation of N from 1 and only a lower limit on Λ.

Other experimenters studied the differential cross section for $\mu-p$ elastic scattering and compared it with $e-p$ elastic scattering (Ellsworth et al. 1960, Camilleri et al. 1969, Kostoulas et al. 1974). But statistically significant differences between $\mu-p$ and $e-p$ cross sections could not be found in either the elastic or inelastic case. Furthermore there were systematic errors of the order of 5 or 10% in comparing $\mu-p$ and $e-p$ cross sections because the techniques were so different.

Thus it became clear that this was not a fruitful direction and I turned to the third cast of our net, the use of e^+e^- colliding beams to search for heavy leptons.

Before going on to e^+e^- colliding beams searches, the major subject of this memoir, a remark on electron-nucleon and muon-nucleon inelastic scattering. Since the 1960's the study of these processes and the related neutrino-nucleon inelastic scattering has become a major area in which the measurements are analyzed in terms of electroweak coupling constants and the F_1, F_2, F_3 structure functions. The $\ell^- - \gamma - \ell^-$, $\nu - Z^0 - \nu$ and the $\nu - W^+ - \ell^-$ vertices are fully explained by conventional electronweak theory. The new physics we sought in Toner al. (1972) have not been found although experiment have been carried out at much higher energies and much larger q^2.

6. Electron–Positron Colliding Beams and Sequential Leptons

The construction and operation of electron–positron colliders began in the 1960's (Voss 1994). By September 1967 at the Sixth International Conference on High Energy

PROPOSAL FOR A HIGH-ENERGY

ELECTRON-POSITRON COLLIDING-BEAM STORAGE RING

AT THE

STANFORD LINEAR ACCELERATOR CENTER

March 1964

It is proposed that the Atomic Energy Commission support the construction at Stanford University of a Colliding-Beam Facility (storage ring) for high-energy electrons and positrons. This facility would be located at the Stanford Linear Accelerator Center, and it would make use of the SLAC accelerator as an injector.

This proposal was prepared by the following persons:

Stanford Physics Department

D. Ritson

Stanford Linear Accelerator Center

S. Berman
A. Boyarski
F. Bulos
E. L. Garwin
W. Kirk
B. Richter
M. Sands

Fig. 3

Accelerators, Howard (1967) was able to list quite a few electron–positron colliders. There was the pioneer 500 MeV ADA collider already operated at Frascati in the early 1960's and, also at Frascati, ADONE was under construction. The 1 GeV ACO at Orsay and 1.4 GeV VEPP-2 at Novosibirsk were in operation. The 6 GeV CEA Collider at Cambridge was being tested. And, colliders had been proposed at DESY and SLAC (Ritson *et al.* 1964).

The 1964 SLAC proposal (Ritson *et al.* 1964), Fig. 3, already discussed the reaction

$$e^+ + e^- \to x^+ + x^-$$

and gave the total production cross section as

$$\sigma = \frac{\pi}{3} r_e^2 \left(\frac{m_e}{E}\right)^2 \beta \left[1 + \frac{(1-\beta^2)}{2}\right]$$

where r_e is the classical electron radius. This proposal did not directly lead to the construction of an e^+e^- collider at SLAC because we could not get the funding. About 5 years later with the steadfast support of the SLAC director, Wolfgang Panofsky, and with a design and construction team led by Burton Richter, construction of the SPEAR e^+e^- collider was begun at SLAC, Fig. 1.

It was this 1964 proposal and the 1961 seminal paper of Cabibbo and Gatto (1961) entitled "Electron–Positron Colliding Beam Experiments" which focussed my thinking on new charged lepton searches using an e^+e^- collider. As we carried out the experiments described in Sections 4 and 5, I kept looking for a model for new leptons, a model which would lead to definitive colliding beam searches while remaining reasonably general. Helped by discussions with my colleagues such as Paul Tsai and Gary Feldman, I came to what I later called the sequential lepton model.

I thought of a sequence of pairs

$$\begin{array}{cc} e^- & \nu_e \\ \mu^- & \nu_\mu \\ \ell^- & \nu_\ell \\ \ell'^- & \nu_{\ell'} \\ \vdots & \vdots \end{array}$$

each pair having a unique lepton number. I also usually thought about the leptons as being point Dirac particles. Of course, the assumptions of unique lepton number and point particle nature were not crucial, but I liked the simplicity. After all, I had turned to lepton physics in the early 1960's partly in a search for simple physics.

The idea was to look for

$$e^+ + e^- \to \ell^+ + \ell^-$$

with

$$\begin{aligned} \ell^+ &\to e^+ + \text{undetected neutrinos carrying off energy} \\ \ell^- &\to \mu^- + \text{undetected neutrinos carrying off energy} \end{aligned} \tag{1}$$

or

$$\begin{aligned} \ell^+ &\to \mu^+ + \text{undetected neutrinos carrying off energy} \\ \ell^- &\to e^- + \text{undetected neutrinos carrying off energy}. \end{aligned}$$

This search method had many attractive features:

- If the ℓ was a point particle, we could search up to an ℓ mass (m_ℓ) almost equal to the beam energy, given enough luminosity.

- The appearance of an $e^+\mu^-$ or $e^-\mu^+$ event with missing energy would be dramatic.

- The apparatus we proposed to use to detect the reactions in Eq. (1) would be very poor in identifying types of charged particles (certainly by today's standards) but the easiest particles to identify were the e and the μ.

- There was little theory involved in predicting that the ℓ would have the weak decays

$$\ell^- \to \nu_\ell + e^- + \bar{\nu}_e$$

$$\ell^- \to \nu_\ell + \mu^- + \bar{\nu}_\mu$$

with corresponding decays for the ℓ^+. One simply could argue by analogy from the known decay

$$\mu^- \to \nu_\mu + e^- + \bar{\nu}_e.$$

I incorporated the e^+e^- search method summarized by Eq. (1) in our 1971 Mark I proposal to use the not-yet-competed SPEAR e^+e^- storage ring.

My thinking about sequential leptons and the use of the method of Eq. (1) to search for them was greatly helped and influenced by two seminal papers of Paul Tsai. In 1965 he published with Anthony Hearn (Tsai and Hearn 1965) the paper "Differential Cross Section for $e^+ + e^- \to W^+ + W^- \to e^- + \bar{\nu}_e + \mu^+ + \nu_\mu$". This work discussed finding vector boson pairs W^+W^- by their $e\mu$ decay mode. it was thus closely related to my thinking, described above, of finding $\ell^+\ell^-$ pairs by their $e\mu$ decay mode. Tsai's 1971 paper (Tsai 1971) entitled "Decay Correlations of Heavy Leptons in $e + e \to \ell^+ + \ell^-$" provided the detailed theory for the applications of the sequential lepton model to our actual searches. The reader might look back at Table 2 from Tsai's paper. This table gives the decay modes and their branching ratios for various lepton masses, branching ratios which we are still trying to precisely measure today. Tsai's work was incorporated in the heavy lepton search part of the Mark I detector proposal.

In 1971 Thacker and Sakurai (1971) also published a paper on the theory of sequential lepton decays but it is not as comprehensive as the work of Tsai. The 1971 paper of Tsai was the bible for my work on sequential heavy leptons, and in many ways it still is my bible in heavy lepton physics. Recently, I read this paper to learn about τ polarization effects in τ decays. A more general paper "Spontaneously Broken Gauge Theories of Weak Interactions and Heavy Leptons" by James Bjorken and Chri Llewellyn Smith (1973) was also very important in keeping my thinking general.

7. The SLAC–LBL Proposal

After numerous funding delays, a group led by Burton Richter and John Rees of SLAC Group C began to build the SPEAR e^+e^- collider at the end of the 1960's. Gary Feldman and I, and our Group E, joined with their Group C and a Lawrence Berkeley Laboratory Group led by William Chinowsky, Gerson Goldhaber, and George Trilling to build the

Mark I detector. In 1971 we submitted the SLAC–LBL Proposal (Larsen *et al.* 1971) for the experiment using the Mark I detector at SPEAR. (The detector was originally called the SLAC–LBL detector and only called the Mark I detector when we began to build the Mark II detector. For the sake of simplicity, I refer to it as the Mark I detector.) The contents of the proposal consisted of five sections and a supplement as follows:

A.	Introduction	Page 1
B.	Boson Form Factors	Page 2
C.	Baryon Form Factors	Page 6
D.	Inelastic Reactions	Page 12
E.	Search for Heavy Leptons	Page 16
	Figure Captions	Page 19
	References	Page 20
	Supplement	

Thus the heavy lepton search was left for last and allotted just three pages because to most others it seemed a remote dream. But the three pages contained the essential idea of searching for heavy leptons using $e\mu$ events, Eq. (1).

I wanted to include a lot more about heavy leptons and the $e - \mu$ problem but my colleagues thought that would unbalance the proposal. We compromised on a 10 page supplement entitled "Supplement to Proposal SP-2 on Searches for Heavy Leptons and Anomalous Lepton-Hadron Interactions". The supplement began as follow.

"**1. Introduction**

While the detector is being used to study hadronic production processes it is possible to simultaneously collect data relevant to the following questions:

(1) Are there charged leptons with masses greater than that of the muon?

We normally think of the charged heavy leptons as having spin $\frac{1}{2}$ but the search method is not sensitive to the spin of the particle. This search for charged heavy leptons automatically includes a search for the intermediate vector boson which has been postulated to explain the weak interactions. This is discussed in Section 8.

(2) Are there anomalous interactions between the charged leptons and the hadrons?

In this part of the proposal we show that using the detector we can gather definitive information on the first question within the available mass range. We can obtain preliminary information on the second question — information which will be very valuable in designing further experiments relative to that question. We can gather all this information while the detector is being used to study hadronic production processes. Additional running will be requested if the existence of a heavy lepton, found in this search, needs to be confirmed. This is discussed in Section 5."

8. Lepton Searches at ADONE

While SPEAR and the Mark I detector were being built, lepton searches were being carried out at the ADONE e^+e^- storage ring by two groups of experimenters in electron–positron annihilation physics: One group reported in 1970 and 1973 (Alles-Borelli *et al.* 1970, Bernardini *et al.* 1973). In the later paper they searched up a mass of about 1 GeV for a conventional heavy lepton and up to about 1.4 GeV for a heavy lepton with decays restricted to leptonic modes.

The other group of experimenters in electron–positron annihilation physics was led by Shuji Orito and Marcello Conversi. Their search region (Orito *et al.* 1974) also extended to masses of about 1 GeV.

9. Discovery of the Tau in the Mark I Experiment: 1974–76

SPEAR and the Mark I Detector

The SPEAR e^+e^- collider began operation in 1973. Eventually SPEAR obtained a total energy of about 8 GeV, but in the first few years the maximum energy with useful luminosity was 4.8 GeV.

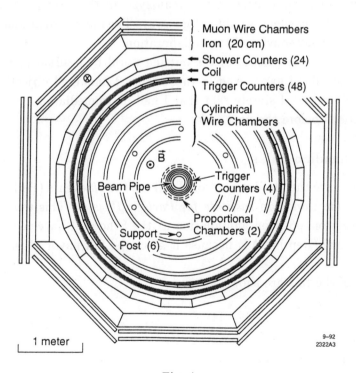

Fig. 4

We also began operating the Mark I experiment in 1973 in the form shown in Fig. 4. The Mark I was one of the first large-solid-angle, general purpose detectors built for colliding beams. The use of large-solid-angle particle tracking and the use of large-solid-angle particle identification systems is obvious now, but it was not obvious twenty years ago. The electron detection system used lead-scintillator sandwich counters built by our Berkeley colleagues. The muon detection system was also crude using the iron flux return which was only 1.7 absorption lengths thick.

Discovery of the $e-\mu$ events

Both detection systems worked just well enough, so in 1974 I began to find $e\mu$ events, that is events with an e, an opposite sign μ, no other charged particles, and no visible photons.

By early 1975 we had seen dozens of $e\mu$ events, but those of us who believed we had found a heavy lepton faced two problems: how to convince the rest of our collaboration and how to convince the physics world. The main focus of this early skepticism was the γ, e and μ identification systems: Had we underestimated hadron misidentification into leptons? Since our γ and e system only covered about half of 4π, what about undetected photons? What about inefficiencies and cracks in these systems?

The questions inside our Mark I Collaboration were answered by George Trilling, Gerson Goldhaber and Burton Richter putting together an independent team of collaboration members. The charge to that team was to reanalyze all the data to try to make the $e\mu$ signal go away. But the $e\mu$ signal would not go away. The independent analysis agreed with my work and that is what convinced the collaboration.

I worked through the skepticism of the outside world by gradually expanding the geographic range of the talks I gave. And in those talks, I answered objections if I could. If new objections were raised, I simply said that I had no immediate answer. I then worked on the new objections before the next talk.

10. The 1975 Canadian Talks

In June, 1975 I gave the first international talk on the $e\mu$ events (Perl 1975) at the *1975 Summer School of the Canadian Institute for Particle Physics*. This was the second of my two lectures on electron–positron annihilation at the School. I have reprinted this second lecture in this volume (Reprint A5) because it gives the full flavor of the first year of the discovery of the τ.

The contents of the talk are shown below

1. Introduction
    A. Heavy Leptons
    B. Heavy Mesons
    C. Intermediate Boson
    D. Other Elementary Bosons
    E. Other Interpretations
2. Experimental Method

3. Search Method and Event Selection
    A. The 4.8 GeV Sample
    B. Event Selection
4. Backgrounds
    A. External Determination
    B. Internal Determination
5. Properties of $e\mu$ Events
6. Cross Sections of $e\mu$ Events
7. Hypothesis Tests and Remarks
    A. Moments Spectra
    B. θ_{coll} Distribution
    C. Cross Sections and Decay Ratios
8. Compatibility of e^+e^- and μe Events
9. Conclusions

The talk had two purposes. First, to discuss possible sources of $e\mu$ events: heavy leptons, heavy mesons and intermediate bosons. And second, to demonstrate that we had some good evidence for $e\mu$ events. The largest single energy data sample, Table 1, was at 4.8 GeV, the highest energy at which we could then run SPEAR. The 24 $e\mu$ events in the total charge = 0, number photons = 0 column was our strongest claim.

One of the cornerstones of this claim was an informal analysis carried out by Jasper Kirkby who was then at Stanford University and SLAC. He showed me that just using the numbers in the 0 charge, 0 photons columns of Table 1, we could calculate the probabilities for hadron misidentification in this class of events. There were not enough eh, μh, and hh events to explain away the 24 $e\mu$ events.

The misidentification probabilities determined from three-or-more prong hadronic events and other considerations are given in Table 2. Compared to present experimental techniques the $P_{h\to e}$ and $P_{h\to\mu}$ misidentification probabilities of about 0.2 are enormous, but I could still show that the 24 $e\mu$ events could not be explained away.

And so the evidence for a new phenomena was quite strong, not incontrovertible, but still strong. What was the new phenomena: a sequential heavy lepton, a new heavy muon with the decays

$$M^- \to e^- + \bar{\nu}_e$$

$$M^- \to \mu^- + \bar{\nu}_\mu,$$

the intermediate boson W, or some weird resonance R with

$$e^+ + e^- \to R \to e^+ + \nu_e + \mu^- + \bar{\nu}_\mu?$$

Table 1. From Perl (1975). A table of 2-charged-particle events collected at 4.8 GeV in the Mark I detector. The table, containing 24 $e\mu$ events with zero total charge and no photons, was the strongest evidence at that time for the τ. The caption read:

"Distribution of 513, 4.8 GeV, 2-prong, events which meet the criteria: $p_e > 0.65$ GeV/c, $p_\mu > 0.65$ GeV/c, $\theta_{\text{copl}} > 20°$."

	Total Charge = 0			Total Charge = ±2		
Number photons =	0	1	>1	0	1	>1
ee	40	111	55	0	1	0
$e\mu$	24	8	8	0	0	3
$\mu\mu$	16	15	6	0	0	0
eh	18	23	32	2	3	3
μh	15	16	31	4	0	5
hh	13	11	30	10	4	6
Sum	126	184	162	16	8	17

Table 2. From Perl (1975). The caption read:

"Misidentification probabilities for 4.8 GeV sample"

Momentum range (GeV/c)	$P_{h \to e}$	$P_{h \to \mu}$	$P_{h \to h}$
0.6–0.9	.130 ± .005	.161 ± .006	.709 ± .012
0.9–1.2	.160 ± .009	.213 ± .011	.627 ± .020
1.2–1.6	.206 ± .016	.216 ± .017	.578 ± .029
1.6–2.4	.269 ± .031	.211 ± .027	.520 ± .043
weighted average using hh, μh, and $e\mu$ events	.183 ± .007	.198 ± .007	.619 ± .012

Sections 5 and 7 of the reprinted lecture were an attempt to identify the new phenomena. Figure 6 of the lecture shows the observed cross section, $\sigma_{e\mu,\text{obs}}$, to be non-zero above 4 GeV total energy and to have a maximum size of about 2×10^{-35} cm^2. Certainly too large for the higher order weak interaction process

$$e^+ + e^- \to e^+ + \nu_e + \mu^- + \bar{\nu}_\mu.$$

This cross section when corrected for acceptance gave a leptonic branching fraction of $(17 \pm 2)\%$, compatible with the sequential heavy lepton explanation. On the other hand, as discussed on page 24 of the lecture, the heavy meson explanation could not be ruled out.

And then there were some disquieting aspects of the date. Defining Θ_{coll} the

$$\cos \Theta_{\text{coll}} = -\vec{p}_e \cdot \vec{p}_\mu / (|\vec{p}_e||\vec{p}_\mu|)$$

so that $\Theta_{coll} = 0$ when the e and μ are moving in exactly opposite directions, we expect as shown in Fig. 19 of the lecture that some event will have large Θ_{coll}. But there were no events with $\Theta_{coll} > 80^0$, in contradiction to both heavy lepton and heavy meson hypotheses. This provides a lesson to young experimenters. Data with low statistics obtained from a crude apparatus should not agree with the correct hypothesis in all aspects. Indeed perfect agreement is suspicious.

The Canadian lecture ends with these conclusions:

> "(1) No conventional explanation for the signature $e\mu$ events has been found.
>
> (2) The hypothesis that the signature $e\mu$ events come from the production of a pair of new particles - each of mass about 2 GeV - fits almost all the data. Only the θ_{coll} distribution is somewhat puzzling.
>
> (3) The assumption that we are also detecting ee and $\mu\mu$ events coming from these new particles is still being tested."

I was still not able to specify the source of the μe events: leptons, mesons or bosons. But I remember that I felt strongly that the source was heavy leptons. It would take two more years to prove that.

11. First Publication: "We have no conventional explanation for these events"

As 1974 passed we acquired e^+e^- annihilation data at more and more energies, and at each of these energies there was an anomalous $e\mu$ event signal, Fig. 5. Thus, I and my colleagues in the Mark I experiment became more and more convinced of the reality of the $e\mu$ events and the absence of a conventional explanation.

An important factor in this growing conviction was the addition of a special muon detection system to the detector, Fig. 6a, called the muon tower. This addition was conceived and built by Gary Feldman. Although we did not use events such as that in Fig. 6b in our first publication, seeing a few events like this was enormously comforting.

Finally in December 1975, the Mark I experimenters published Perl *et al.* (1975) entitled "Evidence for Anomalous Lepton Production in $e^+ - e^-$ Annihilation". (This is Reprint A6.)

The final paragraph reads

> "We conclude that the signature $e - \mu$ events cannot be explained either by the production and decay of any presently known particles or as coming from any of the well-understood interactions which can conventionally lead to an e and a μ in the final state. A possible explanation for these events is the production and decay of a pair of new particles, each having a mass in the range of 1.6 to 2.0 GeV/c^2."

We were not yet prepared to claim that we had found a new charged lepton, but we were prepared to claim that we had found something new. To accentuate our uncertainty

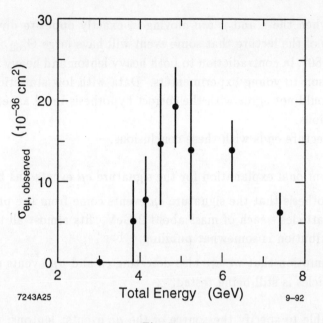

Fig. 5

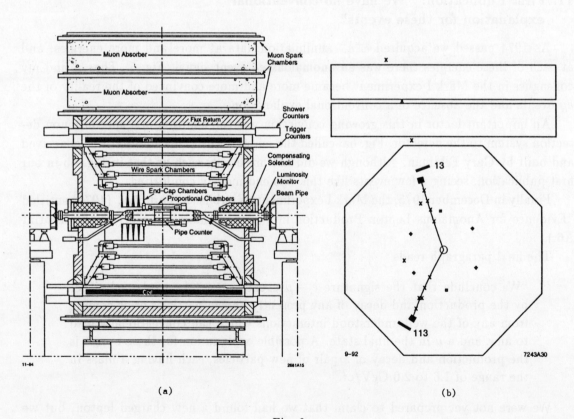

Fig. 6

I denoted the new particle by U for unknown in some of our 1975–1977 papers. The name τ came later. Incidentally, τ was suggested to me by Petros Rapidis who was then a graduate student and worked with me in the early 1970's on the $e - \mu$ problem (Perl and Rapidis 1975). The letter τ is from Greek τριτον for third — the third charged lepton.

Thus in 1975, twelve years after we began our lepton physics studies at SLAC, these studies finally bore fruit. But we still had to convince the world that the $e\mu$ events were significant and we had to convince ourselves that the $e\mu$ events came from the decay of a pair of heavy leptons.

12. Is it a Lepton? 1976–1978

Our first publication was followed by several years of confusion and uncertainty about the validity of our data and its interpretation. It is hard to explain this confusion a decade later when we know that τ pair production is 20% of the e^+e^- annihilation cross section below the Z^0, and when the τ pair events stand out so clearly at the Z^0.

There were several reasons for the uncertainties of that period. It was hard to believe that both a new quark, charm, and a new lepton, tau, would be found in the same narrow range of energies. And, while the existence of a fourth quark was required by theory, there was no such requirement for a third charged lepton. So there were claims that the other predicted decay modes of tau pairs such as e-hadron and μ-hadron events could not be found. Indeed finding such events was just at the limit of the particle identification capability of the detectors of the mid-1970's.

Perhaps the greatest impediment to the acceptance of the τ as the third charged lepton was that there was *no* other evidence for a third particle generation. Two sets of particles u, d, e^-, ν_e and c, s, μ^-, ν_μ seemed acceptable, a kind of doubling of particles. But why three sets? A question which to this day has no answer.

It was a difficult time. Rumors kept arriving of definitive evidence against the τ: $e\mu$ events *not* seen, the $\tau \to \pi\nu$ decay *not* seen, theoretical problems with momentum spectra or angular distribution. With colleagues such as Gary Feldman I kept going over our data again and again. Had we gone wrong somewhere in our data analysis?

Clearly other tau pair decay modes had to be found. Assuming the τ to be a charged lepton with conventional weak interactions, simple and very general theory predicted the branching fractions:

$$B(\tau^- \to \nu_\tau + e^- + \bar{\nu}_e) \approx 20\%$$
$$B(\tau^- \to \nu_\tau + \mu^- + \bar{\nu}_\mu) \approx 20\% \qquad (2)$$
$$B(\tau^- \to \nu_\tau + \text{hadrons}) \approx 60\%$$

Therefore experimenters should be able to find the decay sequences.

$$e^+ + e^- \to \tau^+ + \tau^-$$
$$\tau^+ \to \bar{\nu}_\tau + \mu^+ + \nu_\mu \qquad (3)$$
$$\tau^- \to \nu_\tau + \text{hadrons}$$

and

$$e^+ + e^- \to \tau^+ + \tau^-$$
$$\tau^+ \to \bar{\nu}_\tau + e^+ + \nu_e \qquad (4)$$
$$\tau^- \to \nu_\tau + \text{hadrons}$$

The first sequence, Eqs. (3), would lead to *anomalous muon events*.

$$e^+ + e^- \to \mu^\pm + \text{hadrons} + \text{missing energy} \qquad (5)$$

and the second, Eqs. (4), would lead to *anomalous electron events*

$$e^+ + e^- \to e^\pm + \text{hadrons} + \text{missing energy} \qquad (6)$$

One might also look for the sequence

$$e^+ + e^- \to \tau^+ + \tau^-$$
$$\tau^+ \to \bar{\nu}_\tau + e^+ + \nu_e \qquad (7)$$
$$\tau^- \to \nu^\tau + e^- + \bar{\nu}_e$$

leading to

$$\tau^+ + \tau^- \to e^+ + e^- + \text{missing energy},$$

and an analogous sequence for the μ decay modes. A student of mine, Frank Heile (Heile *et al.*, 1978) did find some weak evidence for the process in Eq. (7), but the background from radiative Bhabha pairs was a severe problem. Incidentally, the great improvement in detectors in 15 years is illustrated by contrasting this measurement with the beautiful determination of $B(\tau^- \to \nu_\tau + e^- + \bar{\nu}_e)$ by Akerib *et al.* (1992) using the CLEO II detector and the process in Eq. (7).

13. Anomalous Muon Events

The first advance beyond the $e\mu$ events came with three different demonstrations of the existence of anomalous μ-hadron events

$$e^+ + e^- \to \mu^\pm + \text{hadrons} + \text{missing energy}.$$

The first and very welcome outside confirmation for anomalous muon events came in 1976 from another SPEAR experiment by Cavalli-Sforza *et al.* (1976). This paper was entitled "Anomalous Production of High-Energy Muons in e^+e^- Collisions at 4.8 GeV".

I have in my files a June 3, 1976 Mark I note by Gary Feldman discussing μ events using the muon identification tower of the Mark I detector, Fig. 6a. For data acquired above 5.8 GeV he found the following:

"Correcting for particle misidentification, this data sample contains 8 μe events and 17 μ-hadron events. Thus, if the acceptance for hadrons is about the same as the acceptance for electrons, and these two anomalous signals come from the same source, then with large errors, the branching ratio into one observed charged hadron is about twice the branching ratio into an electron. This is almost exactly what one would expect for the decay of a heavy lepton."

This conclusion was published, Feldman *et al.* (1977), in a paper entitled "Inclusive Anomalous Muon Production in e^+e^- Annihilation".

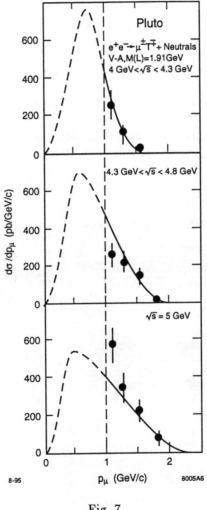

Fig. 7

The most welcomed confirmation, because it came from an experiment at the DORIS e^+e^- storage ring, was from the PLUTO experiment. In 1977 the PLUTO Collaboration, Burmester *et al.* (1977), published "Anomalous Muon Production in e^+e^- Annihilation as Evidence for Heavy Leptons". Figure 7 is from this paper. PLUTO was also a large-solid-angle detector and so for the first time we could fully discuss the art and technology of τ

research with an independent set of experimenters, with our friends Hinrich Meyer and Eric Lohrman of the PLUTO Collaboration.

With the finding of μ-hadron events I was convinced I was right about the existence of the τ as a sequential heavy lepton. Yet there was much to disentangle: it was still difficult to demonstrate the existence of anomalous e-hadron events and the major hadronic decay modes

$$\begin{aligned} \tau^- &\to \nu_\tau + \rho^- \\ \tau^- &\to \nu_\tau + \pi^- \end{aligned} \tag{8}$$

had to be found.

14. Anomalous Electron Events

The demonstration of the existence of anomalous electron events

$$e^+ + e^- \to e^\pm + \text{hadrons} + \text{missing energy}$$

required improved electron identification in the detectors. A substantial step forward was made by the new DELCO detector, Fig. 8, at SPEAR (Kirkby 1977, Bacino *et al.* 1978).

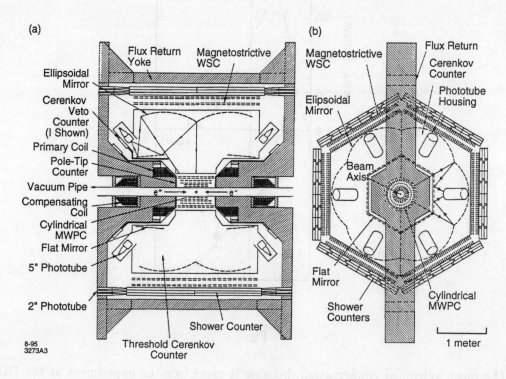

Fig. 8

In Kirkby's talk at the 1977 Hamburg Photon-Lepton Conference, "Direct Electron Production Measurement by DELCO at SPEAR", he stated

"A comparison of the events having only two visible prongs (of which only one is an electron) with the heavy lepton hypothesis shows no disagreement. Alternative hypotheses have not yet been investigated."

The Mark I detector was also improved by Group E from SLAC and a Lawrence Berkeley laboratory Group led by Angela Barbaro-Galtieri; some of the original Mark I experimenters had gone off to begin to build the Mark II detector. We installed a wall of lead glass

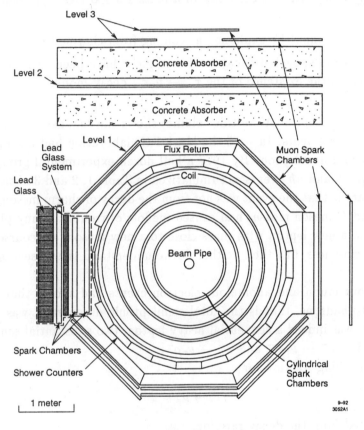

Fig. 9

electromagnetic shower detectors in the Mark I, Fig. 9. This led to the important paper (Barbaro-Galtieri *et al.* 1977)) entitled "Electron–Muon and Electron–Hadron Production in e^+e^- Collisions". The abstract read:

> "We observe anomalous $e\mu$ and e-hadron events in e^+e^- collisions at S-PEAR in an experiment that uses a lead-glass counter system to identify electrons. The anomalous events are observed in the two-charged-prong topology. Their properties are consistent with the production of a pair of heavy leptons in the reaction $e^+e^- \to \tau^+\tau^-$ with subsequent decays of $\tau^\pm$ into leptons and hadrons. Under the assumption that they come only from this source, we measure the branching ratios $B(\tau \to e\nu_e\nu_\tau) = (22.4 \pm 5.5)\%$ and $B(\tau \to h + \text{neutrals}) = (45 \pm 19)\%$."

15. Semileptonic Decay Modes and the Search for $\tau^- \to \nu_\tau \pi^-$ and $\tau^- \to \nu_\tau \rho^-$

By the time of the 1977 Photon Lepton Conference at Hamburg, I was able to report (Perl 1977) in a "Review of Heavy Lepton Production in e^+e^- Annihilation" that

> "(a) All data on anomalous $e\mu$, ex, ee and $\mu\mu$ events produced in e^+e^- annihilation is <u>consistent</u> with the existence of a mass 1.9 ± 0.1 GeV/c^2 charged lepton, the τ.
>
> (b) This data <u>cannot</u> be explained as coming from charmed particle decays.
>
> (c) Many of the expected decay modes of the τ have been seen. A very important problem is the existence of the $\tau^- \to \nu_\tau \pi^-$ decay mode."

This 1977 review (Reprint A7 in this volume) shows the changes that had occurred in τ lepton research in three years. In 1975 the Mark I Collaboration had been alone in arguing for the existence of a new lepton. By 1977 a half dozen experimental groups had evidence for anomalous $e\mu$, eh and μh events as summarized in Table 1, 2 and 3 of the 1977 review. I was also able to argue that the event was not from the decay of charm mesons. The closeness of the τ mass and D meson masses had caused me lots of trouble. Many physicists did not want to believe in a new lepton with mass close to a new meson. Of course we still do not understand this closeness, but then neither do we understand the closeness of the μ and π masses.

The anomalous muon and anomalous electron events had shown that the total decay rate of the τ into hadrons, that is the total semileptonic decay rate, was about the right size. But if the τ was indeed a sequential heavy lepton, two substantial semileptonic decay modes had to exist: $\tau^- \to \nu_\tau \pi^-$ and $\tau^- \to \nu_\tau \rho^-$.

First, the branching fraction for

$$\tau^- \to \nu_\tau + \pi^- \tag{9a}$$

could be calculated from the decay rate for

$$\tau^- \to \mu^- + \bar{\nu}_\mu \tag{9b}$$

and was found to be

$$B(\tau^- \to \nu_\tau \pi^-) \approx 10\%. \tag{9c}$$

Second, the branching fraction for

$$\begin{aligned} \tau^- &\to \nu_\tau + \rho^- \to \nu_\tau + \pi^- + \pi^0 \\ &\to \nu_\tau + \pi^- + \gamma + \gamma \end{aligned} \tag{10a}$$

could be calculated from the cross section for

$$e^+ + e^- \to \rho^0 \tag{10b}$$

and was found to be

$$B(\tau^- \to \nu_\tau \rho^-) \approx 20\% \,. \tag{10c}$$

One of the problems in the years 1977–1979 in finding the modes in Eqs. (9a) and (10a) was the poor efficiency for photon detection in the early detectors. If the γ's in Eq. (10a) are not detected then the π and ρ modes are confused with each other. Probably the first separation of these modes was achieved using the Mark I-Lead Glass Wall detector. As reported at the Hamburg Conference by Angelina Barbaro-Galtieri (1977)

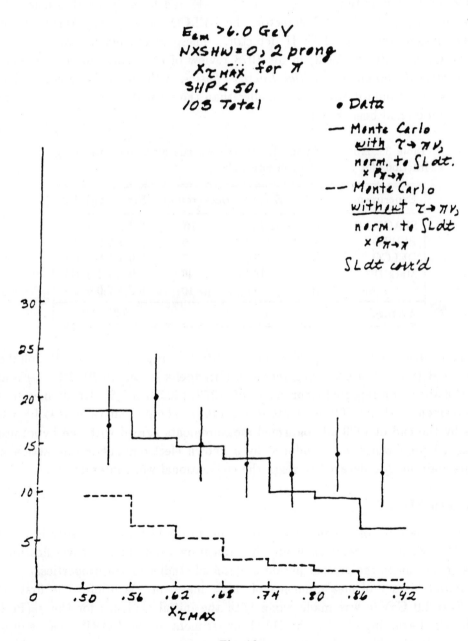

Fig. 10

$$B(\tau^- \to \nu_\tau \pi^-)/B(\tau^- \to \nu_\tau \rho^-) = 0.44 \pm 0.37\,.$$

Gradually the experimenters understood the photon detection efficiency of their experiments and in addition new detectors, such as the Mark II, with improved photon detection efficiency were put into operation.

In our collaboration the first demonstration that $B(\tau \to \nu_\tau \pi^-)$ was substantial came from Gail Hanson (1978) in an internal note dated March 7, 1978. She looked at a sample of 2-prong, 0-prong events with one high-momentum prong. Figure 10 taken from her internal note shows an excess of events, particularly at large x, if $B(\tau \to \nu_\tau \pi^-)$ is taken as zero.

Within about a year the $\tau \to \nu_\tau \pi^-$ decay mode had been detected and measured by experimenters using the PLUTO detector, the DELCO detector, the Mark I-Lead Glass Wall detector and the new Mark II detector. These measurements were summarized, Table 3 by Gary Feldman (1978) in his 1978 review of e^+e^- annihilation physics at the *XIX International Conference on High Energy Physics*. Although the average of the results in Table 3 is two standard deviations smaller than the present value of $(11.1 \pm 0.2)\%$, the $\tau^- \to \nu_\tau \pi^-$ mode had been found.

Table 3. From Feldman (1979), the various measured branching fractions for $\tau^- \to \pi^- \nu_\tau$ in late 1978.

Experiment	Mode	Events	Background	$B(\tau \to \pi\nu)$ (%)
SLACE-LBL	$x\pi$	≈ 200	≈ 70	$9.3 \pm 1.0 \pm 3.8$
PLUTO	$x\pi$	32	9	$9.0 \pm 2.9 \pm 2.5$
DELCO	$e\pi$	18	7	$8.0 \pm 3.2 \pm 1.3$
Mark II	$x\pi$	142	46	$8.0 \pm 1.1 \pm 1.5$
	$e\pi$	27	10	$8.2 \pm 2.0 \pm 1.5$
Average				8.3 ± 1.4

The year 1979 saw the first publications of $B(\tau^- \to \nu_\tau \rho^-)$. The DASP Collaboration using the DORIS e^+e^- storage ring reported (Brandelik *et al.* 1979) $(24 \pm 9)\%$ and the Mark II Collaboration reported (Abrams *et al.* 1979) $(20.5 \pm 4.1)\%$. Crude measurements, but in agreement with the 20% estimate in Eq. (10c). The present value is $(24.8 \pm 0.2)\%$.

Thus by the end of 1979 all confirmed measurements agreed with the hypothesis that the τ was a lepton which was produced by a known electromagnetic interaction and, at least in its main modes, decayed through the conventional weak interaction.

16. The Tau Mass

In the final sections of this paper I sketch some of the history of τ research in the years 1978 to 1982 when that research made the transition from the space verification of the existence of the tau to the present period of detailed studies of tau properties.

The initial history of measurements of the τ mass, m_τ, is brief. The first estimate $m_\tau = 1.6$ to 2.0 GeV/c was made along with the initial evidence for the τ (Perl *et al.* 1975). By the beginning of 1978 the DASP experiment at the DORIS e^+e^- storage ring showed $m_\tau = 1807 \pm 20$ MeV/c^2 (Brandelik *et al.* 1978).

By the middle of 1978 the DELCO experiment at SPEAR (Bacino *et al.* 1978) had made the best measurement $m_\tau = 1784^{+2}_{-7}$ MeV/c^2 as reported in a paper entitled "Measurement of the Threshold Behavior of $\tau^+\tau^-$ Production in e^+e^- Annihilation". This paper contained the classic measurement of the τ pair production cross section at low energy. (It was only in 1992, fourteen years later, that there was an improvement in the measurement of m_τ, the BES Collaboration using the BEPC e^+e^- collider reported (Bai *et al.* 1992) $m_\tau = 1775.9 \pm 0.5$ MeV/c^2.)

17. The Tau Lifetime

The last major property of the τ to be determined was the τ lifetime. Measurements of the τ lifetime, τ_τ, could not be made at the energies at which SPEAR and DORIS operated; the first measurement of τ_τ required the higher energies of PETRA and PEP. The best measurements required, in addition, secondary-vertex detectors. The first published measurement used a primitive secondary-vertex detector built by Walter Innes and myself to improve the triggering efficiency of the Mark II detector. (Feldman *et al.* 1982) Let by G. J. Feldman and G. H. Trilling we measured $\tau_\tau = (4.6 \pm 1.9) \times 10^{-13}$ sec.

Another early measurement was from the MAC experiment at PEP with $\tau_\tau = (4.9 \pm 2.0) \times 10^{-13}$ sec. (Ford *et al.* 1982).

Table 4. From Jaros (1982), the status of τ lifetime measurements in 1982.

Experiment	Number of Decays	Average Decay Length Error (mm)	$\tau_\tau (10^{-13}\,\text{s})$
TASSO	599	10	0.8 ± 2.2
MARK II	126	4	4.6 ± 1.9
MAC	280	4	$4.1 \pm 1.2 \pm 1.1$
CELLO	78	6	$4.7^{+3.9}_{-2.9}$
MARK II Vertex Detector	71	0.9	$3.31 \pm .57 \pm .60$

The modern era in τ lifetime measurements began with the pioneering work of John Jaros on precision vertex detectors (Jaros 1982). Table 4 taken from his paper show the status of τ lifetime measurements at the end of 1982. Theory predicts

$$\tau_\tau = \tau_\mu \left(\frac{m_\mu}{m_\tau}\right)^5 B(\tau^- \to \nu_\tau e^- \bar{\nu}_e),$$

which using modern values gives

$$\tau_\tau(\text{predicted}) = 2.9 \times 10^{-13} \text{ sec.}$$

Thus the 1982 measurement of τ_τ agreed with theory and the overall identification of the τ as a heavy lepton was complete.

This concludes this memoir on the discovery of the tau lepton. A search for new leptons and experiments on the old $e - \mu$ problem, research begun in the early 1960's, led twenty years later to the third charged lepton, the tau. The old $e - \mu$ problem became the new and still unsolved $e - \mu - \tau$ problem.

References

Abrams, G. S. et al. (1979). Phys. Rev. Lett. **43**, 1555.
Akerib, D. S. et al. (1992). Phys. Rev. Lett. **69**, 3610; Phys. Rev. Lett. **71**, 3395 (1993).
Alles-Borelli, V. et al. (1970). Lett. Nuovo Cimento **IV**, 1156.
Bacino, W. et al. (1978). Phys. Rev. Lett. **41**, 13.
Bai, J. Z. et al. (1992). Phys. Rev. Lett. **69**, 3021.
Barbaro-Galtieri, A. (1977a). Proc. 1977 Int. Symp. Lepton and Photon Interactions at High Energies (Hamburg, 1977), Ed. F. Gutbrod, p. 21.
Barbaro-Galtieri, A. et al. (1977b). Phys. Rev. Lett. **39**, 1058.
Barber, W. C. et al. (1966). Phys. Rev. Lett. **16**, 1127.
Barna, A. et al. (1968). Phys. Rev. **173**, 1391.
Bernardini M. et al. (1973). Nuovo Cimento **17A**, 383.
Bjorken J. D. and C. H. Llewellyn-Smith (1973). Phys. Rev. **D7**, 887.
Brandelik, R. et al. (1978). Phys. Lett. **73B**, 109.
Brandelik, R. et al. (1979). Z. Physik **C1**, 233.
Braunstein, T. et al. (1972). Phys. Rev. **D6**, 106.
Burmester, J. et al. (1977). Phys. Lett. **68B**, 297.
Cabibbo N. and R. Gatto (1961). Phys. Rev. **124**, 1577.
Camilleri, L. et al. (1969). Phys. Lett. **23**, 153.
Cavalli-Sforza, M. et al. (1976). Phys. Rev. Lett. **36**, 558.
Charpak, G. et al. (1962). Phys. Lett. **1**, 16.
Ellsworth, R. W. et al. (1960). Phys. Rev. **165**, 1449.
Feldman, G. J. (1978). Proc. XIX Int. Conf. High Energy Physics (Tokyo, 1978), Eds. S. Hounma, M. Kawaguchi, and H. Miyazawa.
Feldman, G. J. (1992). Proc. 20th SLAC Summer School on Particle Physics, SLAC-Report-412, Ed. L. Vassilian, p. 631.
Feldman, G. J. et al. (1977). Phys. Rev. Lett. **38**, 117.
Feldman, G. J. et al. (1982). Phys. Rev. Lett. **48**, 66.
Ford, W. T. et al. (1982). Phys. Rev. Lett. **49**, 106.
Hanson, G. (1978). SLAC–LBL Collaboration Internal Note, March 7, 1978.
Harari, H. (1992). Proc. 20th SLAC Summer School on Particle Physics, SLAC-Report-412, Ed. L. Vassilian, p. 647.
Heile F. B. et al. (1978). Nucl. Phys. **B138**, 189.
Howard, F. T. (1967). Proc. Sixth Int. Conf. High Energy Accelerators (Cambridge, 1967), Ed. R. A. Mack, p. B43.
Jaros, J. A. (1982). Proc. Paris High Energy Physics (Paris, 1982); J. Physique **43** Supp. C-3, 106.
Kirkby, J. (1977). Proc. 1977 Int. Symp. Lepton and Photon Interactios at High Energies (Hamburg, 1977), Ed. F. Gutbrod, p. 3.
Kostoulas, I. et al. (1974). Phys. Rev. Lett. **32**, 489.
Larsen, R. M. et al. (1971). SLAC Proposal SP-2.
Lipmanov, E. M. (1964). JETP **19**, 1291.
Low, F. E. (1965). Phys. Rev. Lett. **14**, 238.
O'Neil, G. K. et al. (1958). HEPL Report RX-1486.
Orito S. et al. (1974). Phys. Lett. **48B**, 165.
Perl, M. L. (1971). Physics Today, **July**, 34.
Perl, M. L. (1975a). Proc. Canadian Inst. Particle Physics Summer School (McGill Univ., Montreal, 1975), Eds. R. Heinzi and B. Margolis.

Perl, M. L. et al. (1975b). *Phys. Rev. Lett.* **35**, 1489.
Perl, M. L. (1977). *Proc. 1977 Int. Symp. Lepton and Photon Interactions at High Energies* (Hamburg, 1977), Ed. F. Gutbrod, p. 145.
Perl, M. L. and P. Rapidis (1974). SLAC-PUB-1496 (unpublished).
Ritson, D. S. Berman, A. Boyarski, F. Bulos, E. L. Garwin, W. Kirk, B. Richter, and M. Sands (1964). *Proposal for a High-Energy Electron–Positron Colliding-Beam Storage Ring at the Stanford Linear Accelerator Center.*
Rothe, K. W. and A. M. Wolsky (1969). *Euc. Phys.* **B10**, 241.
Thacker H. B. and J. J. Sakurai (1971). *Phys. Lett.* **36B**, 103.
Toner, W. T. et al. (1972) *Phys. Lett.* **36B**, 251.
Tsai, Y.-S. (1971). *Phys. Rev.* **D4**, 2821.
Tsai, Y.-S. (1992). *Proc. 20th SLAC Summer School on Particle Physics*, SLAC-Report-412, Ed. L. Vassilian, p. 623.
Tsai, Y.-S. and A. C. Hearn (1965). *Phys. Rev.* **140B**, 721.
Tsai, Y.-S. and V. Whitis (1966). *Phys. Rev.* **149**, 1348.
Voss, G. (1994). *Proc. Int. Conf. History of Original Ideas and Basic Discoveries in Particle Phys.* (Sicily, 1994), Ed. H. Newman.
Wikinson, D. and H. R. Crane (1963). *Phys. Rev.* **130**, 852.
Zel'dovich, Ya. B. (1963). *Soviet Phys. Uspekki* **5**, 931.

Search for New Particles Produced by High-Energy Photons*

A. BARNA, J. COX, F. MARTIN, M. L. PERL, T. H. TAN, W. T. TONER, AND T. F. ZIPF

Stanford Linear Accelerator Center, Stanford University, Stanford, California 94305

AND

E. H. BELLAMY†

High Energy Physics Laboratory, Stanford University, Stanford, California 94305

(Received 5 April 1968)

A search for new particles which might be produced by photons of energy up to 18 GeV is described. No new particles were found. Calculations of the Bethe-Heitler process are described which make it possible to state that this experiment would have detected non-strongly-interacting particles whose mass and lifetime lay in a definite range, did they exist.

I. INTRODUCTION

WE have used the new Stanford linear electron accelerator to search for hitherto unknown elementary particles, particularly for particles which do not have strong interactions. The basic idea behind this search was that through the photoproduction of particle pairs, any charged particle can be created provided it has an antiparticle and that there is sufficient energy in the incident photon. The Stanford linear electron accelerator provides for the first time an intense source of high-energy photons—up to 18 GeV in this experiment. The experiment consisted of a momentum-analyzed secondary beam and a pair of differential gas Čerenkov counters which allowed particles of various masses in that beam to be detected. We were particularly interested in looking for non-strongly-interacting particles, and provision was made separately to detect strongly- and non-strongly-interacting particles.

In any search for new particles, the method of search limits in some ways the properties of the particles that might be found. This experiment was sensitive to charged particles with lifetimes greater than 5×10^{-9} sec, and with a production cross section at least 10^{-7} times that of the muon. Within these limitations, we have not found any new particles. We have made calculations, described in this paper, of the electromagnetic pair production of particles of arbitrary mass and zero spin. The results of these calculations and those of Tsai and Whitis[1] for spin-$\frac{1}{2}$ particles enable us to make the positive statement that if such non-strongly-interacting particles existed with a mass less than that of the proton and a lifetime similar to that of the kaon, we would have detected them.

II. GENERAL CONSIDERATIONS ON THE EXISTENCE OF ELEMENTARY PARTICLES

In our mind, there are two basic problems in elementary-particle physics. One is to understand and to calculate how the particles interact. The other is to learn what particles exist and to formulate rules which limit the possible kinds of particles. The two problems are related. This can be seen most clearly in the case of the strongly-interacting particles. The mesons and the numerous short-lived particles which appear as resonances in the strong interaction seem to be an intimate part of the interaction itself, so that one can expect that a correct theory of the interaction would also explain and predict the multitude of particles.

In the case of the particles which do not interact strongly, the situation is very different. The only known particles are the photon, the electron, the muon, and the two types of neutrinos. There is no understanding of why these particles and no others should exist, although the electromagnetic and weak interactions can be calculated. In particular, there is the puzzle of the existence of both the electron and the muon, particles so dissimilar in mass yet alike in all other aspects. Because the interactions can be calculated, it is possible to postulate the existence of a new particle and to calculate its lifetime and its effect on known processes as a function of its mass. Many authors have done this.[2] However, all such calculations make the basic assumption that no radically new feature enters into the interaction which could alter the result by orders of magnitude. As an example only, consider the effect of strangeness on the strong interaction. The muon-electron problem seems so little understood that some new concept as unlikely as strangeness was, may be required for its solution. We therefore believe that experimental searches for new particles should not be inhibited by preconceived ideas that short lifetimes are to be expected for massive, weakly-interacting particles. These ideas are based on our current understanding of the physics involved. This is true also of estimates of the production of hypothetical particles in specific processes. For example, the fact that K mesons are not observed to decay into heavy muons[3] means that according to

---

* Work supported by the U. S. Atomic Energy Commission.
† On leave from Westfield College, University of London, London, England.
[1] Y. S. Tsai and V. Whitis, SLAC Users Handbook, Part D (unpublished) and (private communication).

[2] F. E. Low, Phys. Rev. Letters **14**, 238 (1965); F. J. M. Farley, Proc. Roy. Soc. (London) **A285**, 248 (1965); T. D. Lee and C. N. Yang, Phys. Rev. Letters **4**, 307 (1960); J. Schwinger, Ann. Phys. (N. Y.) **2**, 407 (1957); S. M. Berman *et al.*, Nuovo Cimento **25**, 685 (1962).
[3] A. M. Boyarski *et al.*, Phys. Rev. **128**, 2398 (1962); E. W. Beir, Ph.D. thesis, University of Illinois, 1966 (unpublished).

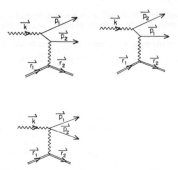

FIG. 1. The three Feynman diagrams upon which are based the calculation of photoproduction of a pair of spin-0 particles.

present understanding of the weak interaction, an exact analog of the muon does not exist with a mass intermediate between that of the muon and the kaon. This is a restricted and specific statement. In the spirit of the foregoing argument, one might then suppose that some additional selection rule or other restriction could exist which would prevent K-meson decay to the heavy muon. It is possible, in this spirit, to conceive of many other particles which might exist, but there is no need to list them here. The purpose of this discussion is to point out that considerations which are of great importance in predicting the effects of specific modifications of present knowledge are limited. Experiments should go beyond the range of such predictions, and be limited only by experimental considerations.

The least specific production process which we can imagine is electromagnetic pair production: It requires only that the particles exist in charged particle-antiparticle pairs. The production rate for this process can be calculated without making further restrictions other than the assumption that effects arising from the form factor of the particle produced can be neglected, so that an experiment using photoproduced particles has a known sensitivity for a general class of particles. A direct search for the particle itself does not introduce any assumption about specific decay modes.

III. OTHER PARTICLE SEARCHES

A. Experiments at Proton Accelerators and with Cosmic Rays

There have been many searches at proton accelerators, mostly unpublished, for new, long-lived, strongly-interacting particles either in beams or in bubble chambers. It is most unlikely that such particles exist in the mass range below $\frac{1}{5}$ GeV. In the case of particles with no strong interaction, the most likely production process is photoproduction by the photons from π^0 decay. However, the photon flux at such machines and in the cosmic radiation is low, and their path length in the target (measured in radiation lengths) is small, so that one can conclude that such searches do not place a useful limit on the existence of non-strongly-interacting particles. The neutrino experiments[4] have placed lower limits to the mass of the hypothetical intermediate boson, a particle of specific properties.

B. Experiments at Electron Machines

Coward et al.[5] have searched without success for particles with masses between m_e and $175m_e$. This search was basically similar to the present one in that the particles were photoproduced and detected directly. Definite limits were placed on the mass and lifetime of any particles which could be pair-produced. A short search has been carried out at the Cambridge electron accelerator (CEA) for particles of mass less than the kaon,[6] again without success. Also at CEA, a search has been made for a lepton which could be found in the process $e+p \rightarrow X+p$, in which the recoil proton was momentum-analyzed and a search made for a bump in its momentum spectrum.[7] Again, no new particles were found. These searches cover a limited mass range, or rely on special processes to produce the particles. The present search was intended to cover a wider mass range in a completely general manner.

IV. PARTICLE PRODUCTION BY ELECTRONS

The principal process by which electrons produce secondary charged particles in a thick target takes place in two steps. First, an electron radiates in the Coulomb field of a nucleus. The secondary particles are then photoproduced at another nucleus in the target by the bremsstrahlung. The direct electroproduction reaction $e^-+\text{nucleus} \rightarrow e^-+\text{nucleus}+X^++X^-$, where X^+ and X^- are the particles produced, can be described as photoproduction by virtual photons. It has been shown by Panofsky, Newton, and Yodh[8] that the spectrum of virtual photons associated with an electron is equivalent to the real bremsstrahlung spectrum which would be produced by the electron in a target of 0.02 radiation length. We can therefore neglect this process in a thick target. The photoproduction may be purely electromagnetic pair production, or it may involve the strong interaction. In this section, we describe some calculations of the electromagnetic pair production of spin-0 and spin-$\frac{1}{2}$ particles at 0° in a thick target. The yields to be expected are presented as a function of the mass of the particle produced. These yields represent lower limits for the production of possible new particles under the conditions of our experiment. We have made a number of approximations which we estimate will lead to an over-all error of the order of 20–30% in the yields.

[4] G. Bernardini et al., Nuovo Cimento 38, 608 (1965); R. Burns et al., Phys. Rev. Letters 15, 42 (1965).
[5] D. H. Coward et al., Phys. Rev. 131, 1782 (1963).
[6] J. S. Greenberg (private communication).
[7] C. Betourne et al., Phys. Letters 17, 70 (1965); H. J. Behrend et al., Phys. Rev. Letters 15, 900 (1965); R. Budnitz et al., Phys. Rev. 141, 1313 (1966).
[8] W. K. H. Panofsky, C. M. Newton, and G. B. Yodh, Phys. Rev. 98, 751 (1955).

TABLE I. The differential cross section at 0° for pair production of a spin-0 particle with 9 GeV/c momentum by a 15-GeV photon incident on a beryllium nucleus, or on a free proton. Values are given for various masses of the produced particle. $|t|_{\min}$ is the minimum value of the square of the four-momentum transfer to the target.

Particle mass (GeV)	Beryllium nucleus target			Free proton target	
	$|t|_{\min}$ (GeV/c)²	$\frac{1}{Z^2}\left(\frac{d^2\sigma}{d\Omega dp}\right)_{0°}$ $\frac{cm^2}{sr, GeV/c}$ No form factor	With form factor	$|t|_{\min}$ (GeV/c)²	$\left(\frac{d^2\sigma}{d\Omega dp}\right)_{0°}$ $\frac{cm^2}{sr, GeV/c}$ With form factor
0.105	2.3×10^{-6}	7.3×10^{-30}	7.0×10^{-30}	2.4×10^{-6}	6.7×10^{-30}
0.20	3.0×10^{-5}	4.2×10^{-31}	3.5×10^{-31}	3.1×10^{-5}	4.1×10^{-31}
0.40	5.0×10^{-4}	1.9×10^{-32}	1.1×10^{-32}	5.1×10^{-4}	1.7×10^{-32}
0.60	2.5×10^{-3}	3.0×10^{-33}	9.5×10^{-34}	2.7×10^{-3}	2.2×10^{-33}
0.80	8.1×10^{-3}	7.9×10^{-34}	1.4×10^{-34}	8.8×10^{-3}	4.4×10^{-34}
1.0	2.0×10^{-2}	2.6×10^{-34}	2.0×10^{-35}	2.3×10^{-2}	1.0×10^{-34}
1.5	0.105	3.1×10^{-35}	3.0×10^{-37}	0.157	2.7×10^{-36}
2.0	0.36	6.1×10^{-36}	7.2×10^{-39}	1.15	3.0×10^{-39}

However, since the yields are a rapidly decreasing function of mass, the effect of such an error is to change only slightly the upper mass limits of the experiment.

We shall now calculate the photoproduction of a pair of spinless particles of mass M and unit charge. Consider first the simplest case, coherent pair production, in which the target nucleus remains in its ground state. The reaction is calculated using the three Feynman diagrams shown in Fig. 1. k is the momentum of the incident photon. E_1, p_1 and E_2, p_2 are the energy and three-momentum of the produced particles, X_1 and X_2. r_1 and r_2 are the initial and final values of the three-momentum of the target nucleus, which has mass M and charge Z. In this experiment, we search for new particles produced at 0 ± 6 mrad to the direction of the incident electron beam. For particles above about 100-MeV mass, the cross section is sufficiently flat in the forward direction that we can use the 0° value. The other particle has spherical angles (θ_2,ϕ_2) with respect to the incident photon direction. The differential cross section[9] in the laboratory system has the form

$$\left(\frac{d^2\sigma}{d\Omega_1 dp_1}\right)_{0°} = \int_0^{2\pi} d\phi_2 \int_0^{\pi} \sin\theta_2\, d\theta_2$$
$$\times \left|\left(\frac{p_2 p_1^2}{2^{10}\pi^5 E_1 WM(E_2+W)(1-D\cos\theta_2)}\right)\langle|A_0|^2\rangle_{av}\right|. \quad (1)$$

The first term in the brackets is the phase-space factor and the second term is the square of the matrix element. W is the total energy of the recoil nucleus.

$$D = E_2(k-p_1)/p_2(E_2+W).$$

After averaging over the incident photon polarization,

[9] Equations (1) and (5) were directly derived from the Feynman rules given in J. D. Bjorken and S. D. Drell, *Relativistic Quantum Mechanics* (McGraw-Hill Book Co., New York, 1964), p. 285. Equation (1) when partially integrated agrees with the result of S. D. Drell, Stanford Linear Accelerator Report No. M-200-7A, 1960, p. 7 (unpublished). Equation (1) and the Drell result are exactly equal to one quarter of the cross section given for the coherent production of a pair of spin-0 particles by W. Pauli and V. F. Weisskopf, Helv. Phys. Acta 7, 709 (1934).

the matrix element squared has the form

$$\langle|A_0|^2\rangle_{av} = \frac{Z^2 8 M^2 E_1^2 e^6 p_2^2}{k^2 t^2 E_2^2}\left(\frac{\sin\theta_2}{1-\beta_2\cos\theta_2}\right)^2. \quad (2)$$

The electric charge is defined by $e^2 = 4\pi\alpha$, where α is the fine-structure constant. β_2 is the laboratory velocity of X_2. t is the square of the four-momentum transferred to the nucleus, defined by

$$t = (k-E_1-E_2)^2 - (\mathbf{k}-\mathbf{p}_1-\mathbf{p}_2)^2.$$

Inserting Eq. (2) into Eq. (1) and integrating over ϕ_2 yields

$$\left(\frac{d^2\sigma}{d\Omega_1 dp_1}\right)_{0°,\text{nucl,nff}}$$
$$= \int_0^{\pi} \sin\theta_2\, d\theta_2 \left[\left(\frac{Z^2\alpha^3}{\pi}\right)\frac{E_1 p_1^2 p_2^3 M}{k^3 E_2^2(E_2+W)(1-D\cos\theta_2)}\right.$$
$$\left.\times\left(\frac{1}{t^2}\right)\left(\frac{\sin\theta_2}{1-\beta_2\cos\theta_2}\right)^2\right] \quad (3)$$

for the production at 0° of a spin-0 particle. The subscript nff means that no form factor is included.

Consider the production of $m=0.2$ GeV particles by a 15-GeV photon. As θ_2 increases from 0°, the overwhelming variation in the integrand is in the last two terms. The first term, which is mostly the phase-space factor, decreases by 4% as θ_2 goes from 0.0 to 0.2 rad. But $|t|$ increases from 6×10^{-3} (GeV/c)² to about 1.2 (GeV/c)², which by itself leads to a 4×10^4 decrease in the integrand. The third term is 0 at 0° and reaches a maximum at $\cos\theta_2 = \beta_2$. In fact, the integrand is relatively large only in the region where the growth of the last term toward its maximum is not yet canceled by the $1/t^2$ term. Thus the cross section is due primarily to the small four-momentum transfer part of the reaction, 0.01 to 0.2 (GeV/c)² in this example.

In the production of much heavier particles, say, $m>0.5$ GeV, the major variation in the integrand of Eq. (3) is in the last two terms, as before. However, we find that we can no longer have very low values of $|t|$.

TABLE II. The numbers in the table are the ratio of the coherent photoproduction of spin-$\frac{1}{2}$ particles to that of spin-0 particles. The production is on beryllium and the particles are produced at 0° with 9-GeV/c momentum. Values are given for various incident photon energies and various masses for the produced particles.

Mass (GeV)	Incident photon energy (GeV)				
	18	16	14	12	10
0.105	2.3	2.3	2.7	3.9	5.4
0.20	2.2	2.3	2.7	3.8	10.7
0.40	2.2	2.3	2.6	3.8	11.4
0.60	2.2	2.3	2.6	3.7	12.9
0.80	2.2	2.3	2.6	3.9	16.2
1.0	2.2	2.3	2.7	4.0	
1.5	2.2	2.4	2.9	4.8	
2.0	2.3	2.5	3.3	6.3	

It is easy to see this by considering a very heavy target of mass much greater than the incident photon energy. Then the minimum four-momentum transfer squared is

$$|t|_{\min} = M^4 k^2 / 4 p_1^2 (k-p_1)^2$$

and $|t|_{\min}$ is proportional to the fourth power of the particle mass. Column 2 of Table I gives $|t|_{\min}$ when beryllium is the target and we have used a 15-GeV incident photon to produce a 9-GeV/c momentum particle As m goes from 0.105 to 2.0 GeV, $|t|_{\min}$ changes by a factor of 10^5.

Since t^2 enters in the denominator of Eq. (3) and most of the integrand comes from small $|t|$ values, we must expect a strong mass dependence in the cross section. This mass dependence is illustrated in column 3 of Table I. The cross section. divided by Z^2, calculated from Eq. (3) for production of a 0.105-GeV mass particle at 9 GeV/c and 0° by a 15-GeV photon on beryllium is 7.3×10^{-30} cm^2/sr (GeV/c). For a 1.0-GeV mass particle under the same conditions, the cross section, divided by Z^2, is 2.6×10^{-34} cm^2/sr (GeV/c). Thus, even in the simplest case of the production of spin-0 particles from a point nucleus, there is strong mass dependence. The finite size of the nucleus is taken into account by multiplying the cross section by the square of the nuclear form factor.[10] For beryllium, we obtain

$$\left(\frac{d^2\sigma}{d\Omega_1 dp_1}\right)_{0°,\text{nucl}} = \left(\frac{d^2\sigma}{d\Omega_1 dp_1}\right)_{0°,\text{nucl,nff}} \times \left(\frac{1}{1+26.7|t|}\right)^2. \quad (4)$$

Since $|t|_{\min}$ increases with the mass of the particles produced, the form factor reduces further the cross section for the production of massive particles. In column 4 of Table I, the cross section of column 3 is shown, but now with the effect of the nuclear form factor included. For a particle of 1-GeV mass and 9-GeV/c momentum, photoproduced by a 15-GeV photon, the form factor depresses the cross section by a further factor of 10.

This reduction of the coherent pair production requires the consideration of incoherent pair production which results from the interaction of a photon with an individual nucleon in the nucleus. In this process the form factor of the nucleon must be considered. But the nucleon form factor is less $|t|$-dependent than the nuclear form factor. Therefore, as the mass of the particle produced increases, the incoherent production becomes more important. Equation (5) gives a slightly simplified formula[9] for the production cross section of a pair of spin-0 particles on a free proton, in which some terms which are only important for $|t|>1$ (GeV/c)^2 have been neglected.

$$\left(\frac{d^2\sigma}{d\Omega_1 dp_1}\right)_{0°,\text{proton}}$$
$$= \int_0^\pi \sin\theta_2 \, d\theta_2 \left\{ \left(\frac{\alpha^3}{\pi} \frac{E_1 p_1^2 p_2^3 M}{k^3 E_2^2 (E_2+W)(1-D\cos\theta_2)}\right) \right.$$
$$\times \left(\frac{1}{t^2}\right) \left[\left(\frac{G_E^2 + |t|/4M^2|G_m^2}{1+|t|/4M^2|}\right) \left(\frac{\sin\theta_2}{1-\beta_2\cos\theta_2}\right)^2 \right.$$
$$\left. \left. + \frac{G_m^2 |t| k^2 E_2^2}{2M^2 E_1^2 p_2^2} \right] \right\}. \quad (5)$$

Comparison of Eq. (5) with Eq. (3) shows the following differences. Of course, Z^2 has been dropped. The last term is now more complicated and contains the nucleon form factors G_E and G_M. We replace G_E and G_M by the formula

$$G_E = G_M/2.793 = (1+1.41|t|)^{-2},$$

where $|t|$ is in (GeV/c)^2. With these substitutions, we have calculated the cross section for the production of spin-0 particles on a free proton. For the case of a 15-GeV incident photon producing a 9-GeV/c particle at 0°, we have given the results in Table I in column 6. For masses less than 0.5 GeV, the free proton cross section is almost the same as the nuclear cross section (divided by Z^2) without the nuclear form factor. As the mass increases, the nucleon form factor begins to reduce the proton cross section. But its effect is much less drastic than the effect of the nuclear form factor on the nuclear cross section. Therefore, for masses above 0.5 GeV, the incoherent cross section gains in importance over the coherent.

Incoherent pair production can take place upon neutrons as well as protons. To calculate this, we have used Eq. (5) with $G_E=0$ and G_M given by

$$G_M/1.913 = (1+1.41|t|)^{-2}. \quad (6)$$

The total production, coherent and incoherent, is calculated by Eq. (7).

$$\left(\frac{d^2\sigma}{d\Omega_1 dp_1}\right)_{0°} = \left(\frac{Z^2-Z}{Z^2}\right)\left(\frac{d^2\sigma}{d\Omega_1 dp_1}\right)_{0°,\text{nucl}}$$
$$+ Z\left(\frac{d^2\sigma}{d\Omega_1 dp_1}\right)_{0°,\text{proton}}$$
$$+ (A-Z)\left(\frac{d^2\sigma}{d\Omega_1 dp_1}\right)_{0°,\text{neutron}}. \quad (7)$$

[10] R. Hofstadter, Ann. Rev. Nucl. Sci. 7, 231 (1957).

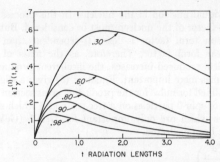

FIG. 2. Bremsstrahlung spectrum as a function of distance along the target t, in radiation lengths. The spectrum is given as the product of the photon energy k in GeV and the photon flux $I_\gamma^{(1)}(t,k)$, in number of photons per GeV. The numbers attached to the curves are the ratio of k to the incident electron energy.

The term $(Z^2-Z)/(Z^2)$, which multiplies the coherent cross section, is a rough way of taking into account the effect of the Pauli principle on the incoherent production.[11] We have used the elastic form factors of the nucleon for G_E and G_M. However, the breakup of the nucleon can also contribute to pair production. Unfortunately, there are as yet insufficient data to allow this to be calculated, and we have therefore neglected it.

All of the foregoing discussion applies to spin-0 particle production. To see the effect of the spins of the produced particles, we will consider the case of the coherent production of spin-$\frac{1}{2}$ particles at 0°. We obtain[12]

$$\left(\frac{d^2\sigma}{d\Omega_1 dp_1}\right)_{0°,\text{nucl},\text{nff}}$$
$$= \int_0^\pi \sin\theta_2 \, d\theta_2 \left[\left(\frac{\alpha^3 Z^2}{\pi} \frac{p_1^2 E_1 p_2^3 M}{k^3 E_2^2 (E_2+M)(1-D\cos\theta_2)}\right) \right.$$
$$\times \left(\frac{1}{t^2}\right)\left(\frac{\sin^2\theta_2}{1-\beta_2\cos\theta_2}\right) (2)$$
$$\left. \times \left(\frac{k^2(E_1+p_1)E_2(1-\beta_2\cos\theta_2)}{2E_1^2 m^2} - 1\right) \right]. \quad (8)$$

If we compare this to Eq. (3), we see that the spin effect is given by the additional last two terms. Table II shows the effect of these terms by giving the ratio of $d^2\sigma/d\Omega_1 dp_1$ for spin $\frac{1}{2}$ to $d^2\sigma/d\Omega_1 dp_1$ for spin 0, both produced coherently on beryllium at 0°. The calculation included the nuclear form factor as given in Eq. (4). The ratios given in the table are for 10–18-GeV incident photons and a 9-GeV/c momentum secondary particle. When the photon energy is greater by several GeV than the energy of the produced particle, the factor is 2.5 or

[11] S. D. Drell and C. L. Schwartz, Phys. Rev. 112, 568 (1958).
[12] Equation (8) is obtained by a reduction (setting one particle angle to zero) of the more general equation for the coherent production of a pair of spin-$\frac{1}{2}$ particles. See R. P. Feynman, *Quantum Electrodynamics* (W. A. Benjamin, Inc., New York, 1962), p. 113; W. Heitler, *Quantum Theory of Radiation* (Clarendon Press, Oxford, England, 1954).

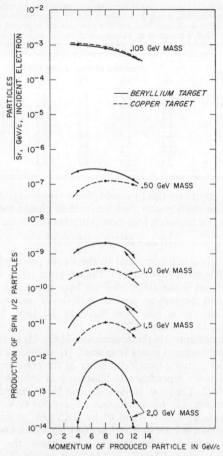

FIG. 3. Photoproduction of pairs of spin-$\frac{1}{2}$ particles by 18-GeV electrons incident on a 10-radiation-length target. The production angle is 0°. The mass number attached to each curve is the mass of the produced particles.

so. But as the photon energy approaches its threshold value, the factor increases. When the bremsstrahlung spectrum is taken into account, the spin-$\frac{1}{2}$ production is three or four times the spin-0 production at low masses and about two times at high masses.

To get the 0° yield of particles from a thick target, we must integrate Eq. (7) over the bremsstrahlung spectrum and the target thickness. The bremsstrahlung spectrum in a thick target has been discussed thoroughly by Tsai and Whitis.[13] For those photons radiated directly by the incident electron (first generation photons), they deduce the approximate expression

$$I_\gamma^{(1)}(t,k) = \frac{1}{k} \frac{(1-k/E_0)^{(4/3)t} \times e^{-(7/9)t}}{7/9 + \frac{4}{3}\ln(1-k/E_0)}, \quad (9)$$

where $I_\gamma^{(1)}(t,k)$ is the flux at depth t of first generation photons of energy k due to an electron incident with

[13] Y. S. Tsai and V. Whitis, Phys. Rev. 149, 1248 (1966).

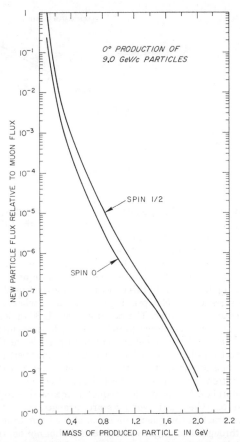

FIG. 4. Ratio of new particle flux to muon flux for various masses, taking the new particle as either spin 0 or spin ½. The flux is for 0° production and 9.0-GeV/c particle momentum from a 10-radiation-length-long beryllium target with 17.5-GeV incident electrons.

energy E_0 at $t=0$. This spectrum is shown in Fig. 2. We make the further approximation of neglecting particle production by the bremsstrahlung of electrons themselves produced in the target by the first generation photons (second generation photons), and all subsequent generations. Tsai and Whitis show that these secondary photons make an appreciable contribution to the spectrum only for large t and small k. Their contribution to the production of high-energy secondary particles can therefore be neglected, and Eq. (9) used directly.

We have calculated the yields for spin-0 particle production by the method described above, and we have used the results of Tsai and Whitis[1] (which are based on a similar method) for spin-½ particle production. In Fig. 3 are presented their results for the production at 0° of spin ½, pure Dirac particles from a 10-radiation-length target with an incident electron beam energy of 18 GeV. The calculations are presented for both beryllium and copper targets. There are three characteristics of the production which hold equally well for the production of spin-0 particles. First, as we expect, the production decreases rapidly as the mass increases. Second, for larger masses, beryllium is a better target than copper, because the incoherent production is relatively more important in beryllium. Third, the yield has a maximum at roughly half the incident electron momentum.

The most useful way to give the production calculation results is in terms of the ratio of a hypothetical particle flux to the muon flux at a fixed secondary beam momentum. The experimental data were taken at 5.05 and 8.99 GeV/c secondary beam momentum. Figure 4 gives the ratios for 9.0 GeV/c with a 17.5 GeV/c incident electron beam, a 10-radiation-length beryllium target, and 0° production. Results for both spin-½, pure Dirac particles and spin-0 particles are shown, with masses from 0.1 to 2.0 GeV. Figure 5 gives the ratio of the new particle flux to the muon flux for a secondary-beam momentum of 5.0 GeV/c, for masses below 0.6 GeV, under the same conditions.

V. EXPERIMENTAL METHOD

A. Outline of Method

The experiment is shown schematically in Fig. 6. A 17.5-GeV electron beam struck a thick target. The negative secondaries produced at 0° were formed into a momentum-analyzed beam which passed through two differential gas Čerenkov counters, H and J, set to count particles of a specific velocity, and hence mass. Two counters were used in coincidence in order to give better rejection of unwanted particles. At the end of the beam,

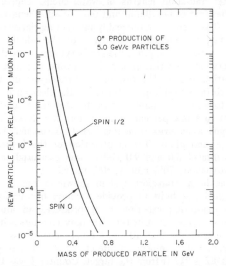

FIG. 5. Ratio of new particle flux to muon flux for various masses, taking the new particle as either spin 0 or spin ½. The flux is for 0° production and 5.0-GeV/c particle momentum from a 10-radiation-length beryllium target with 17.5-GeV incident electrons.

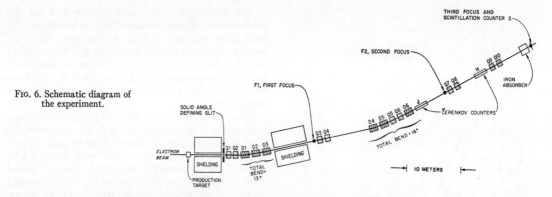

FIG. 6. Schematic diagram of the experiment.

a scintillation counter S was placed behind an iron absorber 5 ft thick. Weakly-interacting particles would have the signature HJS. The experiment consisted of fixing the beam momentum and varying the pressure of the gas in the Čerenkov counters in order to sweep through a range of masses, while recording HJ and HJS. The known particles provide indications of the operation of the system. In particular, the muons and pions provide a basic normalization of the experiment which does not depend upon the acceptance of the transport system or the efficiencies of the Čerenkov counters. Since the muon yield has been measured separately[14] and is understood theoretically, the muon normalization is particularly useful.

B. Apparatus

The target in which the secondaries were produced consisted of 3.6 radiation lengths of beryllium followed by ten radiation lengths of water-cooled copper, a further foot of beryllium, and ten radiation lengths of lead. The production of weakly-interacting particles in this target is adequately described by the calculations given in Sec. IV for production on beryllium, since there is very little particle production beyond the first 3.6 radiation length. The rest of the target was used to absorb the power (up to 20 kW) in the electron beam, and to reduce the number of electrons in the secondary beam to a few percent of the muon flux. Negatively charged secondaries from this target consist mainly of muons and pions. The composition of the beam at momenta of 5.0 and 9.0 GeV/c was measured to be approximately 70% muons, 30% pions.

The beam transport system shown in Fig. 6 was designed and built to provide a muon beam[15] for a muon-scattering experiment. It produces an almost dispersion-free beam in the Čerenkov counters with a diameter of less than 10 cm, a divergence of less than 4 mrad, and a momentum bite of ±1.5%. The second focus F2 is 212 ft from the target. Counter J was 19 ft upstream from F2; counter H was 33 ft downstream from F2. The scintillation counter S was at the third focus, 63 ft downstream from F2.

The differential Čerenkov counters were modeled closely on a counter described by Kycia and Jenkins.[16] The present counters are designed to operate at pressures up to 960 psi. In this experiment, CO_2 was used at pressures up to 600 psi. Figure 7(a) is a schematic diagram of a counter. The radiator region is 80 in. long and the counter is designed to be used with beams up to 12.5 cm in diam. Čerenkov light from particles of the correct velocity is focused onto an annular ring aperture. The aperture is split in two across a diameter and the light from each half is collected separately onto two phototubes. A coincidence is required for a particle to be counted. The quartz windows are arranged so that a stray track in the general direction of the beam cannot go through both. Light which falls near, but not on, the annular aperture is reflected from a spherical mirror in which the aperture is set and is collected onto a phototube put in anticoincidence. Without this, a particle of the wrong velocity at an angle to the beam could be counted, as illustrated in Fig. 7(b). The width of the annular aperture was chosen to give an angular acceptance of ±10 mrad about a mean Čerenkov angle of 75 mrad. This dominated the mass resolution of the counters, giving $\Delta m/m \sim 0.075(p^2/m^2) \times 10^{-2}$, where m is the mass and p is the momentum of the particle. This resolution was adequate to separate out the peaks of the known particles, but allowed a finite mass range to be covered at each pressure setting and sufficient tolerance so that we did not have difficulty in operating the two counters together. The pressure vessels of the two counters were connected together by a common feed pipe. We found that no special precautions were necessary to make the mass peaks coincide in the two counters, although the counters were located out of doors and the ambient temperature varied from 5°C at night to 27°C during the day.

Block diagrams of the electronic circuits are shown in Fig. 8. The three tubes on each counter were fed through

---

[14] A. Barna *et al.*, Phys. Rev. Letters **18**, 360 (1967).
[15] SLAC Users Handbook, Part D (unpublished); Stanford Linear Accelerator Center Laboratory Report No. SLAC-PUB 434 (unpublished); see also Ref. 13.
[16] T. F. Kycia and E. W. Jenkins, *Nuclear Electronics* (International Atomic Energy Agency, Vienna, 1963).

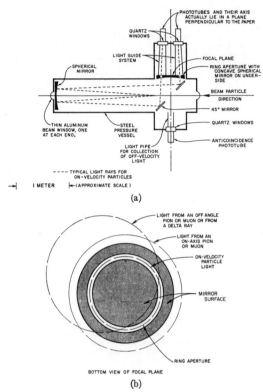

FIG. 7. (a) Schematic diagram of Čerenkov counter optics. (b) Illustration of focal-plane behavior of the Čerenkov light-ring images from different particles.

limiters and discriminators to a coincidence circuit. The discriminator thresholds were set high on the coincidence inputs and low on the veto, following a suggestion of Kycia. This resulted in some inefficiency, but gained more than an order of magnitude in rejection for particles of the wrong velocity. The coincidences HJ and HJS were each formed twice in different ways.

The over-all efficiency of the two counter systems at the π-μ peak was measured to be 80%.

C. Background and Rate Problems

In order to search with high sensitivity, it is necessary to operate at high intensity. A reasonable time to count at one pressure setting is $\frac{1}{2}$ h. This experiment used a pulse repetition rate of 180 pulses/sec, each about 1.2–1.4 μsec long. Thus to search through 10^7 muons in $\frac{1}{2}$ h meant operating with an instantaneous flux of 2.5×10^7 muons/sec. For this reason, we did not define the beam through the Čerenkov counters with scintillators. We used two Čerenkov counters in coincidence, since these are inherently low-rate devices and could be expected to give only those accidental coincidences which resulted from background effects, such as δ rays, off-angle particles, etc., in both counters.

We found a background associated with the kaon and antiproton peaks at the level of about 10^{-7} of the muon flux.

Independently of the S scintillation counter, the maximum usable intensity was limited by the singles rates in the veto phototubes. The light-collection system for the veto operates in such a way that properly aligned beam particles should give veto signals when the pressure is just above or just below the setting required for them to count in the coincidence channel, or when they interact or produce δ rays inclined at an angle to the beam. Otherwise, veto signals should result only from particles at an angle to the axis or with the wrong momentum. We estimated that the veto rate from δ rays should be less than 1% of the flux of pions and muons. However, with the discriminator levels set low on the veto channels, we found that the veto singles rate in each counter was approximately 10% of the beam flux, for pressures far removed from the settings required to count the known particles. We could not raise the discriminator levels without a serious effect on the rejection efficiency of the system. At an instantaneous rate of 2.5×10^7 beam particles/sec, and with veto pulses stretched to 70 nsec for maximum efficiency, the random veto off-time was therefore 33%, and no advantage would be gained by increasing the rate.

VI. EXPERIMENTAL PROCEDURE

The mass range 0.5–1.8 GeV was covered using a beam momentum of 9 GeV/c, roughly half the momentum of the incident electron beam, which the pro-

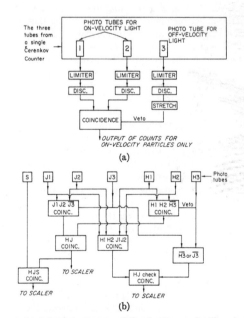

FIG. 8. (a) Schematic diagram of the circuitry used with a single Čerenkov counter; (b) simplified diagram of all the electronics.

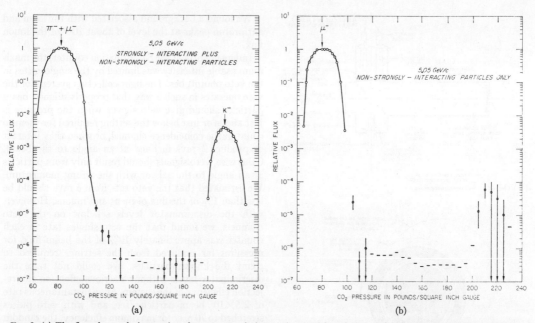

FIG. 9. (a) The flux of strongly-interacting plus non-strongly-interacting particles relative to the pion-plus-muon flux is plotted as a function of the pressure for the 5.05-GeV/c runs. Open circles represent counts and the vertical lines are the statistical errors, which are shown when they are larger than the circle diameter. Short horizontal bars are shown where *no* counts were found. The relative flux indicated by the bar is the flux *if one count* had been seen. At the pion and kaon peak, lines are used to connect the circles to guide the eye. The pressure is in psi gauge. (b) The flux of non-strongly-interacting particles only relative to the muon flux at 5.05 GeV/c. The notation is the same as in (a).

duction calculations showed would give the maximum flux of secondary particles. For the mass range below 0.5 GeV, it was of more importance to have good separation between pions and kaons, and this part of the search was carried out at a momentum of 5 GeV/c.

The process of taking data was relatively simple. The momentum of the secondary beam was fixed at either 9.0 or 5.0 GeV/c. The CO_2 pressure in both counters was set at the muon-pion peak (76–80-psi gauge, depending on the momentum). The timing of all the circuits was checked out and the efficiency of the system for strongly-interacting and non-strongly-interacting particles was determined. Then the pressure was varied from 70 to 100-psi gauge in 4-psi steps across the muon-pion peak. The shapes of the peaks in each counter and of the combined peaks were examined to see that the counters were operating properly. Then the pressure was raised in 5-psi steps. This ensured that at least three steps would be taken to cover the mass peak of any new particle. In each 8-h period of data taking, an upper mass peak (kaon or antiproton) was reached and swept through. This was done to make sure the system was operating properly both with respect to the position and shape of the mass peak. The number of particles per pulse passing through the apparatus was 30–50 during the main part of the data taking between the mass peaks. At the mass peaks, the rate was lower, particularly at the pion-muon peak, where about one particle per pulse was used to reduce dead-time and resolution-time corrections. The number of particles in the beam at these low rates was measured with large auxiliary scintillation counters placed in the beam.

To determine the flux to be used for the normalization of the HJ data at high rates, the singles counts from J1 and H1 were used indirectly. Once off the pion-muon peaks, the J1 or H1 counts were less than 1% of the total number of beam particles, and were observed to be proportional to the beam intensity measured at low rates in the scintillation counters. Therefore, J1 or H1 by themselves had negligible rate corrections. Using the auxiliary beam counters, J1 and H1 were calibrated at very low-beam rates. J1 and H1 were then used as monitors at the high-beam ratios. Of course, this normalization was pressure-dependent, and a calibration was performed regularly. For the HJS channel, dead-time corrections are avoided by normalizing directly to the singles count rate in S. Noise and nonbeam contributions to S singles were observed to be negligible at low rates. Accidental coincidences between HJ and S were monitored by recording coincidences between HJ and a delayed signal from S, and were subtracted.

During the early stages of the experiment, we found that a sensitivity of 10^{-7} relative to the muon flux could be attained, but it would be difficult to go much lower. The limitations were in part due to the maximum allowable muons per pulse being under 50, and in part

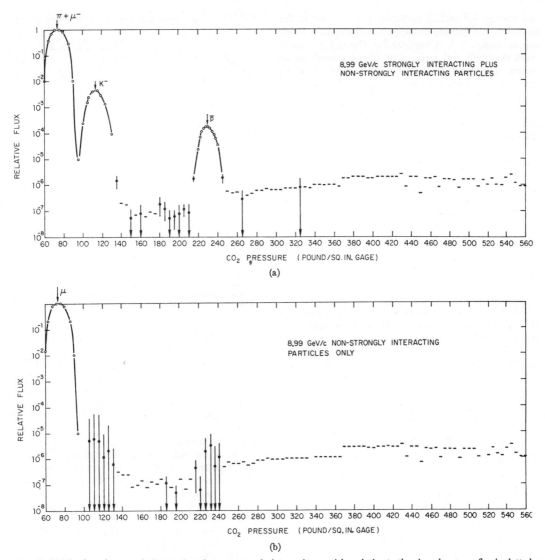

FIG. 10. (a) The flux of a strongly-interacting plus non-strongly-interacting particles relative to the pion-plus-muon flux is plotted as a function of the pressure for the 8.99-GeV/c runs. (b) The flux of non-strongly-interacting particles only relative to the muon flux at 8.99 GeV/c. The notation is the same as in Fig. 9.

due to a background to be discussed in Sec. VII. Now, as shown in Fig. 4, the relative flux of spin-0 particles to muons is 10^{-6} at 0.96-GeV particle mass and 10^{-7} at 1.32-GeV particle mass. We therefore decided to make a definitive search only up to the mass of the proton, i.e., a search in which the sensitivity is considerably better than the flux predicted from purely electromagnetic photoproduction of pairs. Above 1-GeV mass, we made a search with a sensitivity of only 10^{-6} to 3×10^{-6} relative to the muon flux. This part of the search would then depend upon some special mechanism to produce the particles. It was carried out up to a mass of 1.83 GeV.

The kinematical mass limit for coherent pair production on beryllium by 17.5 GeV/c photons is 5.4 GeV, and on free protons, 2.3 GeV.

VII. RESULTS AND ANALYSIS

The results presented in this paper are for negative particles. Figure 9(a) shows the combined data for all the 5-GeV/c runs. HJ counts, normalized to the HJ counts at the pion-muon peak, are plotted as a function of pressure. This is called the relative flux. The HJ counts represent the sum of strongly-interacting plus non-strongly-interacting particles. The notation in the figure has the following meaning. The circles show where counts were found. A vertical line through the circle gives the statistical error, if it is big enough to be shown. The short horizontal lines indicate that no counts were observed at that pressure. If *one* count had been found, the relative flux would be that given by the short horizontal line. The pion-muon and negative kaon peaks have a curve drawn between the circles to guide the eye.

Figure 9(b) shows the relative flux of non-strongly-interacting particles (HJS counts normalized to HJS at the muon peak) at 5 GeV/c. The corresponding results for 9-GeV/c particles are given in Fig. 10. The pion-muon, kaon and antiproton peaks are clearly seen in Fig. 10(a). The notation is the same as that of Fig. 9(a). We make the following observations from Figs. 9 and 10.

(a) The counts observed in Figs. 9(b) and 10(b) at pressures corresponding to the masses of the kaon and antiproton are due to accidental coincidences between the kaons or antiprotons in HJ, and muons in S. The large statistical errors are due to the subtraction of accidental counts from a purposely miss-timed parallel HJS channel. There was no practical way to reduce this accidental rate because at a lower beam intensity it would have taken too long to acquire data. Therefore, we have reduced sensitivity for non-strongly-interacting particles with masses close to the kaon or the antiproton.

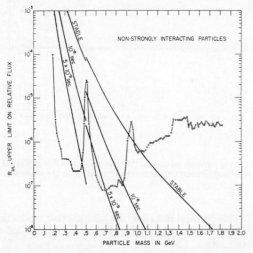

FIG. 11. The upper limit on the relative flux at the apparatus with 90% confidence is plotted against the particle mass in GeV for non-strongly-interacting particles only. The closed circles represent data taken from the 5.05-GeV/c runs and the open circles represent data from 8.99-GeV/c runs. The change comes at 0.5-GeV mass. The slanting lines are the predicted fluxes at the equipment for photoproduction of pairs of spin-0, unit charged, particles which do not *themselves* have form factors. As the notation on the lines indicates, the predicted fluxes are for stable particles, particles with 10^{-8} sec lifetime and particles with 5×10^{-9} sec lifetime. These lines break at 0.5 GeV because of the momentum change.

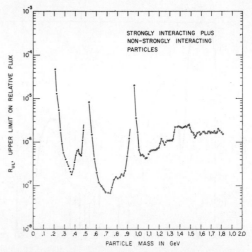

FIG. 12. The upper limit on the relative flux at the apparatus with 90% confidence is plotted against the particle mass in GeV for strongly-interacting plus non-strongly-interacting particles. The closed circles represent data taken from the 5.05-GeV/c runs and the open circles represent data from the 8.99-GeV/c runs.

(b) In Figs. 9(a) and 10(a) there are some counts to the low-mass side of both the kaon and the antiproton peaks, which are not present in Figs. 9(b) and 10(b). Thus, they are due to strongly-interacting particles. Their occurrence in both places makes one suspect a systematic experimental effect. One effect which could explain these counts would be the presence of some 10^{-4} of the kaons in the beam (5×10^{-4} of the antiprotons) with a momentum about 18% higher than the momentum of the beam. The presence of such a small tail of off-momentum particles can certainly not be excluded, although we can find no obvious reason why it should be present.

(c) The search in the region above 240-psi gauge in the 9-GeV/c data involved about 10^8 muons passing through the counters. There were no HJS counts at all in that region. Thus the rejection efficiency for muons is at least 10^8 at high pressures.

In order to compare the data quantitatively with the results of the calculations described in Sec. IV, we have calculated the upper limit which we can put to the relative flux with 90% confidence, using the experimentally observed mass resolution curves to take account of the fact that a particle would give counts at several neighboring pressures if it existed.

Figure 11 summarizes our results for non-strongly-interacting particles. It gives R_{ul}, the upper limit on the relative flux with 90% confidence, as a function of the mass of the particle sought. For 99.9% confidence, the upper limit should increase by a factor of about 5. The curve with solid dots up to about 0.5-GeV mass is from the 5-GeV/c runs. The curve with open circles above 0.5 GeV is from the 9-GeV/c runs. In using these curves.

one must remember that these are upper limits and that their shape depends upon the length of the runs and the background.

The slanted solid lines on the drawing are the predicted relative fluxes using the calculations described in Sec. III. These fluxes are for particles with the following properties: (a) zero spin, (b) unit charge, (c) non-strongly-interacting, and (d) no form factor of the particle itself.

If a particle has spin $\frac{1}{2}$ or higher, or has an anomalous magnetic moment, the production cross section and the relative flux will be greater. Therefore, these lines are the *minimum* relative fluxes for non-strongly-interacting particles with no form factor for the particle itself. The numbers on the slanted lines refer to the lifetime of the particle. The lines are broken at a mass of 0.5 GeV because of the change in momentum at that mass in the experiment.

With Fig. 11, we can make the following conclusions:

(1) There are no stable, unit-charge, non-strongly-interacting particles without form factors in the mass ranges 0.2–0.92 GeV and 0.97–1.03 GeV. By none, we mean that these particles do not exist in pairs capable of electromagnetic pair production.

(2) There are no unit-charge, non-strongly-interacting particles without form factors and with lifetimes greater than 10^{-8} sec in the mass range 0.2–0.86 GeV. There is a small hole in this range at 0.48–0.50 GeV. If the minimum lifetime is reduced to 5×10^{-9} sec, the mass range in which we can be confident that no new particle exists is about 0.2–0.46 and 0.55–0.70 GeV.

(3) Above 1.03 GeV, there could still be stable non-strongly-interacting particles but there is a limit on their production relative to muons. This limit, shown in detail in Fig. 11, ranges from 6×10^{-7} at near 1 GeV to 3×10^{-6} at 1.8 GeV relative to muon production.

(4) No evidence at all was found for the production of any new non-strongly-interacting particles.

Finally, we turn our attention to Fig. 12, which gives R_{u1} for strongly-interacting plus non-strongly-interacting particles. This includes all charged particles in the beam, produced by all sorts of mechanism, and with or without the particles themselves having form factors. It is not useful to put the predicted fluxes from pair photoproduction on this drawing, as was done in Fig. 11, because the strongly-interacting particles can have production cross sections much larger than the pair photoproduction cross section, or they can have smaller production cross sections because of their own form factors. We can only inquire if any new particles are observed.

(5) There is a slight possibility that a hitherto unknown, strongly-interacting particle of mass 0.42 GeV exists, but the evidence is weak and may be due to off-momentum kaons in our beam. If the particle exists, its relative flux was about 4×10^{-7} of the pion plus muon flux in our beam.

(6) There is a very slight possibility that a hitherto unknown, strongly-interacting particle with a mass of 0.8 GeV exists in our beam. However, it is likely that the apparent evidence for this particle is due to high-momentum antiprotons. The particle, if it exists, has a relative flux of 10^{-7} of the pion plus muon flux and is thus just at the edge of the sensitivity of the present experiment.

(7) There is no other evidence for new particles.

There are no earlier published searches for new non-strongly-interacting particles with which our results can be directly compared, because this is the first high-energy photoproduction search. The unpublished results that we have heard of were only qualitative statements that no new particles were seen, but no numerical statements of sensitivity were given.

From our results for strongly-interacting particles, we are able to say only that we did not see any new particles because we cannot predict the production cross section. But for non-strongly-interacting particles without form factors, we have been able to place definite limits of mass and lifetime on particles which could exist. These limits were contained in conclusions (1)–(3).

ACKNOWLEDGMENTS

We are extremely grateful to Dr. Y. S. Tsai for numerous discussions of how to understand and calculate the photoproduction of pairs and for providing us with detailed calculations for the spin-$\frac{1}{2}$ production. We are also very grateful to Dr. T. F. Kycia for providing us with a full set of drawings and advising us in detail of the considerations involved in building and operating the Čerenkov counters. L. Cooper and A. Newton were invaluable in the design and construction of the Čerenkov counters. We appreciate the work of Dr. R. Neal, Dr. E. Seppi, J. Harris and E. Keyser and the accelerator staff in the operation of the accelerator and the experimental areas. We are grateful to R. Vetterlein for the engineering design of the beam.

COMPARISON OF MUON-PROTON AND ELECTRON-PROTON DEEP INELASTIC SCATTERING *

W. T. TONER, T. J. BRAUNSTEIN, W. L. LAKIN, F. MARTIN, M. L. PERL ** and T. F. ZIPF
Stanford Linear Accelerator Center, Stanford University, Stanford, California 94305, USA

and

H. C. BRYANT and B. D. DIETERLE
Physics Department, University of New Mexico, Albuquerque, New Mexico 87106, USA

Received 5 July 1971

As a test of muon-electron universality we have compared muon-proton and electron-proton inelastic scattering cross sections for $|q^2|$ (square of the four-momentum transferred from the lepton) values up to 4.0 $(GeV/c)^2$ and for lepton energy losses up to 9 GeV. There is no experimentally significant deviation from muon-electron universality. If the muon is assigned the form factor $(1.0 + |q^2|/\Lambda_d^2)^{-1}$ relative to the electron, then with 97.7% confidence $\Lambda_d > 4.1$ GeV/c.

In this letter we report our recent measurements of 12 GeV/c muon-proton inelastic scattering and we compare them with measurements [1] of electron-proton inelastic scattering. Our purpose is to study the relationship between the muon and the electron usually called muon-electron universality. The muon and electron, neither of which are hadrons, have the same spin, same electric charge and same weak interaction coupling constant; they differ in their mass and in their lepton number. These relationships lead the physicist to speculate about possible connections between the muon and electron. Are they manifestations of a single particle split into two mass levels by unknown forces? Or are the electron and muon the lowest mass members of a larger family of charged leptons? With no theoretical guidance as to how to answer these questions, the experimentalist seeks clues to the answer by measuring known properties of the muon with increasing precision or by studying hitherto unexplored properties of the muon and comparing the results with the corresponding measurements on the electron. The inelastic scattering of leptons on protons is such an unexplored interaction.

The study of muon-electron universality through inelastic scattering has three novel features. (1) In elastic scattering, $\nu = |q^2|/(2M)$, where q^2 is the square of four-momentum transferred from the lepton, ν is the energy loss of the lepton in the laboratory frame and M is the proton mass. But in inelastic scattering where $\nu > |q^2|/(2M)$, ν and q^2 may be varied independently; thus allowing the exploration of a much larger kinematic region. (2) Measurements of inelastic lepton scattering in which only the scattered lepton is detected, place no restrictions upon the nature of the final hadronic state. It is conceivable that a violation of muon-electron universality involving hadrons would more easily be seen in inelastic scattering than in elastic scattering. (3) It is possible that one or both of the charged leptons, like the proton, have vertex form factors which are decreasing functions of $|q^2|$. One of the more unexpected results of μ-p and e-p inelastic scattering was the large cross section, compared to elastic scattering, at high $|q^2|$. Hence inelastic scattering can provide a greater sensitivity to lepton form factors through the large range of q^2 which can be covered easily in a single experiment.

The experiment was carried out at the Stanford Linear Accelerator Center using a 12 GeV/c, positive muon beam [2]. The apparatus [2-4] consisted of a liquid-hydrogen target, a large analyzing magnet, optical spark chambers and scintillation counters. The small momentum width ($\pm$ 1.5%) and small phase space (3×10^{-3}

---

* Work supported by the U. S. Atomic Energy Commission.
** John Simon Guggenheim Memorial Foundation Fellow.

cm² sr) of the muon beam allowed inelastic events to be defined by measuring just the scattering angle and final momentum of the muon. The spark chambers which provided this information were triggered whenever three planes of scintillation counters indicated a muon scattering angle greater than 30 mr. The beam at the hydrogen target contained less than 3×10^{-6} pions per muon. An additional pion rejection factor of 50 was obtained through the requirement that the scattered muon pass through a series of iron plates and spark chambers without nuclear interaction.

The data presented here result from 2.4×10^{10} muons incident upon the full hydrogen target. Empty target background subtraction runs were taken with 0.5×10^{10} incident muons. 10 950 inelastic events with target full (and 89 with target empty) were found in the kinematic region reported in this paper. The data were corrected by 2.5% for scanning, measuring and spark chamber inefficiencies and by about 2% for electronic dead time. We have also allowed for a systematic error due to an uncertainty of $\frac{1}{2}$% in the beam momentum. When we combine this uncertainty with estimated errors due to all other corrections and the uncertainty in the normalization procedure, we estimate a total systematic normalization uncertainty of ± 4%. However we find we must increase this estimate to ± 6% when we examine the internal consistency of our data and when we compare the 12 GeV/c measurements reported in this letter with the smaller sample of 10 GeV/c measurements previously reported [4].

The inelastic scattering of changed leptons on protons occurs through the emission of a virtual photon by the lepton [3] *; this photon interacts with the nucleon leading to the production of hadrons. For a point-like lepton the virtual photon emission is completely specified by quantum electrodynamics [3]. Muon-electron universality may therefore be tested by comparing the properties of the virtual photon-proton interaction derived from muon-proton inelastic scattering with those properties derived from electron-proton inelastic scattering. If muon-electron universality is valid, those properties should be the same in both cases. In making such a comparison it is necessary to establish that known effects would not produce a difference. Therefore radiative corrections have already been made in the analysis of both the muon and the electron data, and the contributions to the uncertainty in the results are included in the estimates of the errors. Finally, the contribution of two photon exchange to the inelastic interaction is at most of the order of a few percent *.

The inelastic differential cross section [3] $d^2\sigma/dq^2 d\nu$ is the product of, somewhat arbitrary, kinematic factors and two independent functions of q^2 and ν; these two functions must be experimentally determined. Two such functions are, $\sigma_T(q^2, K)$ and $\sigma_S(q^2, K)$, which may be thought of as the total cross sections for the interaction of transverse and scalar photons respectively with protons [6]. Here $K = \nu - |q^2|/(2M)$. K is the energy that a real photon must have to give the same total energy in the photon-proton center-of-mass system. $\sigma_T(q^2, K)$ and $\sigma_S(q^2, K)$ are defined by

$$d^2\sigma_l/dq^2 d\nu = d^2\sigma_l/dq^2 dK =$$
$$= \Gamma_T(q^2,K,p_l,m_l)\sigma_T(q^2,K) + \Gamma_S(q^2,K,p_l,m_l)\sigma_S(q^2,K)$$
$$= \Gamma_T(q^2,K,p_l,m_l)[\sigma_T(q^2,K) + \epsilon(q^2,K,p_l,m_l)\sigma_S(q^2,K)].$$

Γ_T and Γ_S are the virtual photon fluxes for transverse and scalar photons, respectively [3,4]. Γ_T, Γ_S and $\epsilon = \Gamma_S/\Gamma_T$ are known functions [3,4] of q^2, K, p_l and m_l. m_l is the lepton mass, p_l is the laboratory momentum of the incident lepton, and l stands for μ (muon) or e (electron). As q^2 goes to zero, $\sigma_S(q^2, K)$ goes to zero and $\sigma_T(q^2, K)$ goes to $\sigma_{\gamma p}(K)$ - the total cross section for the interaction of a physical photon of energy K with a proton. In our muon experiment we cannot seperate σ_T from σ_S. Therefore we report and use for the comparison only the combination

$$\sigma_{\exp,l}(q^2,K,p_l) =$$
$$= \sigma_T(q^2,K) + \epsilon(q^2,K,p_l,m_l)\sigma_S(q^2,K)$$
$$= \sigma_T(q^2,K)[1 + \epsilon(q^2,K,p_l,m_l)R(q^2,K)],$$

where
$$R(q^2, K) = \sigma_S(q^2, K)/\sigma_T(q^2, K).$$

---

* There are no comprehensive experimental studies of the validity of the one photon exchange assumption in inelastic scattering. The assertion that two photon exchange is less than a few percent effect in inelastic scattering is primarily based on theoretical considerations and on the lack of any detectable two photon contribution in elastic scattering. Some indication that two photon exchange in inelastic scattering is at most a few percent effect (but not necessarily an undetectable effect) has been given by experiments searching for T-violation in inelastic electron scattering.

Table 1
12 GeV/c muon-proton inelastic scattering cross sections. $d^2\sigma/dq^2dK$ is the measured differential cross section. $\sigma_{exp,\mu}$ is the "virtual photon-proton total cross section, defined in the text". Δ_{RAD} is the percentage subtracted from the raw data for radiative corrections.

K GeV	$\|q^2\|$ (GeV/c)^2	$d^2\sigma/dq^2 dK$ nb/(GeV^3/c^2)	$\sigma_{exp,\mu}$ μb	Δ RAD %	K GeV	$\|q^2\|$ (GeV/c)^2	$d^2\sigma/dq^2 dK$ nb/(GeV^3/c^2)	$\sigma_{exp,\mu}$ μb	Δ RAD %
6 to 1.5	.3 - .4	483. ± 27.	140.1 ± 7.7	2.7	3.5 to 5.0	.25 - .4	121. ± 7.5	98.4 ± 6.1	6.0
	.4 - .6	204. ± 6.4	100.3 ± 3.2	2.3		.4 - .6	37.7 ± 2.2	55.9 ± 3.3	6.5
	.6 - .8	93.6 ± 4.0	82.1 ± 3.5	1.4		.6 - .8	22.4 ± 2.0	50.6 ± 4.4	5.4
	.8 - 1.2	39.6 ± 2.1	64.0 ± 3.4	.5		.8 - 1.2	11.3 ± 1.2	38.8 ± 4.0	4.8
	1.2 - 1.6	13.6 ± 1.6	42.0 ± 4.8	- .3		1.2 - 1.6	5.3 ± 1.1	28.5 ± 6.1	4.2
	1.6 - 2.0	5.8 ± 1.1	29.3 ± 5.5	- .7		1.6 - 2.0	3.6 ± .6	28.7 ± 5.1	3.6
	2.0 - 3.0	2.2 ± .38	19.1 ± 3.2	- 1.5		2.0 - 2.6	1.8 ± .4	20.1 ± 4.6	3.4
	3.0 - 4.0	.17 ± .13	4.0 ± 2.9	- .7		2.6 - 3.4	1.0 ± .3	17.2 ± 5.0	3.0
1.5 to 2.5	.3 - .4	192. ± 14.2	84.1 ± 6.2	4.3	5.0 to 7.0	.1 - .2	125. ± 11.9	76.8 ± 7.3	15.6
	.4 - .6	109. ± 4.4	76.6 ± 3.1	3.3		.2 - .4	48.6 ± 2.2	61.0 ± 2.7	12.1
	.6 - .8	53.4 ± 3.2	60.9 ± 3.6	2.5		.4 - .6	23.2 ± 1.5	52.5 ± 3.3	9.4
	.8 - 1.2	24.1 ± 1.9	46.2 ± 3.6	1.8		.6 - .8	14.6 ± 1.4	49.5 ± 4.7	7.9
	1.2 - 1.6	13.0 ± 1.3	43.1 ± 4.2	.9		.8 - 1.2	7.7 ± .8	39.1 ± 4.2	7.3
	1.6 - 2.0	4.0 ± .9	21.0 ± 4.7	.8		1.2 - 1.6	3.1 ± .8	23.3 ± 6.4	7.3
	2.0 - 3.0	1.4 ± .32	13.2 ± 3.1	.4		1.6 - 2.0	2.0 ± .5	21.6 ± 5.1	6.5
	3.0 - 4.0	1.1 ± .31	20.1 ± 5.7	- .5	7.0 to 8.3	.1 - .2	89. ± 9.4	81.3 ± 8.5	21.5
2.5 to 3.5	.3 - .4	98.7 ± 10.3	66.0 ± 6.9	6.1		.2 - .4	29.9 ± 2.4	54.7 ± 4.3	18.5
	.4 - .6	66.9 ± 3.6	69.2 ± 3.7	4.6		.4 - .6	14.6 ± 1.8	46.4 ± 5.8	14.4
	.6 - .8	33.8 ± 2.6	52.2 ± 4.0	3.9		.6 - .8	10.6 ± 1.6	49.3 ± 7.4	11.4
	.8 - 1.2	18.7 ± 1.4	46.2 ± 3.4	3.0		.8 - 1.0	5.6 ± 1.9	33.8 ± 11.6	12.1
	1.2 - 1.6	7.8 ± 1.1	31.2 ± 4.5	2.5					
	1.6 - 2.0	5.9 ± .9	35.4 ± 5.6	1.8					
	2.0 - 3.0	2.0 ± .4	21.0 ± 4.1	1.6					
	3.0 - 4.0	.036 ± .78	.37 ± 8.0	- 6.9					

In our data $\sigma_{exp,\mu}(q^2, K, p_\mu)$ are only weakly dependent on p_μ because ϵ is always close to 1.

In table 1 we list our values of $d^2\sigma_\mu/dq^2 dK$ and $\sigma_{exp,\mu}$. The quoted errors are statistical and must be combined with the overall normalization uncertainty of ± 6%. These cross sections have been corrected for radiative effects [4].

In comparing $\sigma_{exp,\mu}$ to $\sigma_{exp,e}$ we must note three factors. First, $\sigma_{exp,l}$ depends on p_l and, very weakly, on m_l. Second, the electron and muon data were obtained at different incident lepton energies. Third, the muon data were acquired over a continuous q^2, K kinematic region while the electron data were acquired at almost discrete points. To allow for the first two factors we have modified the electron data through the equation,

$$\sigma_{exp,e}(q^2, K, p_\mu) = \left[\frac{1+\epsilon(q^2, K, p_\mu, m_\mu)R(q^2,K)}{1+\epsilon(q^2, K, p_e, m_e)R(q^2,K)}\right]\sigma_{exp,e}(q^2, K, p_e).$$

This procedure is subject to error due to uncertainties in R. At $q^2 = 0$, R must equal zero, but measurements of R have only been made at a few values of q^2, K in the region of this experi-

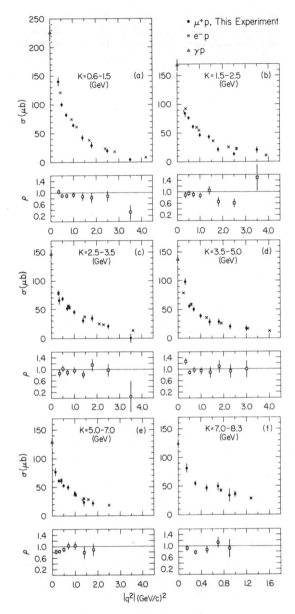

Fig. 1. For each K interval the upper plot gives the experimental values of $\sigma_{\exp,\mu}(q^2, K, p_\mu)$ denoted by a solid circle, $\sigma_{\exp,e}(q^2, K, p_\mu)$ denoted by an × and $\sigma_{\gamma p}(K)$ denoted by a triangle; p_μ = 12 GeV/c. These quantities are defined in the text. $\sigma_{\exp,e}(q^2, K, p_\mu)$ is taken from ref. [1] as described in the text. For each K interval the lower plot gives the values of $\rho(q^2, K) = \sigma_{\exp,\mu}(q^2, K, p_\mu)/\sigma_{\exp,e}(q^2, K, p_\mu)$. The error bars represent only statistical errors. In most cases the errors in $\sigma_{\exp,e}$ are too small to be displayed.

ment. These measurements are consistent [1] with R = 0.18 or with $R = |q^2|/16$ in the region of interest. Fortunately, for the data used in this comparison $\sigma_{\exp,e}(q^2, K, p_\mu)$ is rather insensitive to R; even if $R = 1 \pm 1$, the uncertainty in $\sigma_{\exp,e}$ is for the most part less than 1%. We have made the comparison assuming $R = 0.18$ and also with $R = 0$, 1 and $|q^2|/16$. The changes in the fits and the confidence levels, which we present later, are negligible. To take account of the third factor listed above, we interpolated and averaged the electron data to obtain $\sigma_{\exp,e}(q^2, K, p_e)$ for K bins corresponding to those used for the muon data.

In fig. 1, $\sigma_{\exp,\mu}(q^2, K, p_\mu)$ and $\sigma_{\exp,e}(q^2, K, p_\mu)$ for p_μ = 12 GeV/c are shown as functions of q^2 for various K intervals. It is obvious that any possible muon-electron differences are small. To quantify those differences we define the ratio

$$\rho(q^2, K) = \sigma_{\exp,\mu}(q^2, K, p_\mu)/\sigma_{\exp,e}(q^2, K, p_\mu),$$

p_μ = 12 GeV/c .

To compute this ratio we have made a fit to the electron data, as represented by $\sigma_{\exp,e}(q^2, K, p_\mu)$, and to the $\sigma_{\gamma p}(K)$ values. It was necessary to use $\sigma_{\gamma p}(K)$ [7] because our muon data extends to lower $|q^2|$ values than the electron data used in this comparison [1]. This ratio is plotted in fig. 1, the errors are the combined statistical errors only. We see that ρ is usually close to 1.0; but ρ is less than 1.0 more frequently than it is greater than 1.0.

To combine the data to search for less obvious differences, we need a model of how the two sets of measurements might differ. A common model assumes the leptons have a form factor $F_l(q^2) = (1.0 + |q^2|/\Lambda_l^2)^{-1}$. Then

$$\rho(q^2, K) = \frac{\sigma_{\exp,\mu}(q^2, K, p_\mu)}{\sigma_{\exp,e}(q^2, K, p_\mu)} = \frac{(1.0 + |q^2|/\Lambda_\mu^2)^{-2}}{(1.0 + |q^2|/\Lambda_e^2)^{-2}} \approx$$

$$\approx 1/(1.0 + |q^2|/\Lambda_d^2)^2 \quad (1)$$

where

$$\Lambda_d^{-2} = \Lambda_\mu^{-2} - \Lambda_e^{-2}.$$

Because this comparison uses data from two very different experiments, one might also allow for a normalization difference N^2 in the cross sections, generalizing eq. (1) to

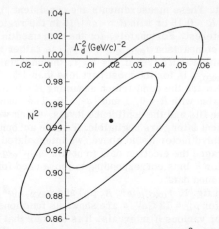

Fig. 2. Contours plots for the parameters N^2 and $\Lambda_{\bar{d}}^{-2}$ obtained by fitting the experimental values of the ratio $\rho(q^2, K)$ to the equation $\rho(q^2, K) = N^2/(1.0 + |q^2|/\Lambda_{\bar{d}}^2)^2$. The inner ellipse represents one standard deviation and the outer ellipse represents two standard deviations in the fit.

$$\rho(q^2, K) = N^2/(1.0 + |q^2|/\Lambda_{\bar{d}}^2)^2 . \qquad (2)$$

The overall normalization uncertainty in the muon data is ± 6%, excluding the statistical uncertainty in the number of events. The overall normalization uncertainty in the electron data [1] is about 4%. Thus the combined overall normalization uncertainty (excluding statistical errors) in the comparison is ± 7% if the two uncertainties are combined in quadrature.

We have made a fit of $\rho(q^2, K)$ to eq. (2), using all K bins at once. Since N^2 and $\Lambda_{\bar{d}}^{-2}$ are correlated parameters, we display the fit through the contour plot of fig. 2 based on statistical errors only. The ± 7% relative normalization uncertainty is not included. The effect of this normalization uncertainty is to allow the N^2 scale to be shifted up or down by an amount as large as 0.07. The best fit to eq. (2) is $\Lambda_{\bar{d}}^{-2} = 0.021 \pm 0.021$ (GeV/c)$^{-2}$ and $N^2 = 0.946 \pm 0.042$ with $\chi^2 = 41.1$ for 42 degrees of freedom. These numbers, if one ignores the errors, mean that the overall muon-proton inelastic cross section is less than the electron-proton inelastic cross section; and that the muon cross section falls off very slightly faster with $|q^2|$ than the electron cross section. However, considering the normalization uncertainty and the extent of the one and two standard deviation ellipses, it is quite possible that $N^2 = 1$ and $\Lambda_{\bar{d}}^{-2} = 0$. If we constrain $\Lambda_{\bar{d}}^{-2}$ to be zero, then $N^2 = 0.917 \pm 0.024$ with a $\chi^2 = 42.1$ for 43 degrees of freedom. Finally it is conventional to quote a 2 standard deviation lower limit in Λ_d. Allowing N^2 to take *any* value, $\Lambda_d > 4.1$ (GeV/c) with 97.7% confidence. We are able to set this high lower limit on Λ_d because the muon data has such a "long lever arm" in $|q^2|$. Thus we have found no experimentally significant deviation from muon-electron unversality. On the other hand the agreement with muon-electron universality is not that one might hope for. An exhaustive analysis of our data has not shown any additional sources of error beyond those which we have already taken into account.

Various other experiments have searched for muon-electron differences, but the only experiments which measure quantities similar to those measured in our experiment are the muon-proton elastic scattering experiments of Camilleri et al. [8] and Ellsworth et al. [9]. Both of these experiments found that the μ-p elastic cross sections were smaller than the e-p elastic cross sections up to a maximum $|q^2|$ of about 1 (GeV/c)^2. Camelleri et al., found $\Lambda_d > 2.4$ GeV/c, Ellsworth found $\Lambda_d > 2.0$, both with 95% confidence. In addition Camilleri gives a fit with $\Lambda_{\bar{d}}^{-2} = 0$ of $N^2 = 0.92$. Our experiment cannot be compared directly with the precision measurements [10, 11] of the gyromagnetic ratio of the muon (g_μ). However, we note that if the muon is assigned an electromagnetic form factor $(1.0 + |q^2|/\Lambda^2)^{-1}$, then the g_μ experiment requires [11] with 95% confidence the $\Lambda_\mu > 7$ GeV/c.

We conclude with a speculative observation. We have analyzed our results using eq. (2) which is just a simple function representing the belief that possible behavioral differences between the muon and the electron can be enhanced by going to larger values of $|q^2|$. Now if we consider our experiment and the two elastic experiments, we see that none of these experiments demand a muon-electron difference which increases steadily as $|q^2|$ increases. Therefore we should not rule out the possibility that any muon-electron differences which may exist will appear at relatively low $|q^2|$ values and will not increase steadily with $|q^2|$. Thus we might replace the form factor used in eqs. (1) and (2) by

$$F_\mu(q^2) = (1-b) + b/(1 + |q^2|/\Lambda^2) = \qquad (3)$$
$$= 1 - (b|q^2|)/(\Lambda^2 + |q^2|) \quad 0 \leq b \leq 1$$

If b were small, say 0.04, then in these scattering experiments all that we could see, with present statistics, is an apparent normalization difference when $|q^2|$ approaches Λ^2. But ρ would never fall below $(1-b)^2$. Such a form factor might result from a model in which most of the muon mass was, like the electron, concentrated into a point particle; but where some of the mass was distributed in a halo. Of course the parameters of such a model must not contradict the results of the g_μ experiment.

An alternative way to obtain eq. (3) is to postulate that the muon has a special interaction which connects the muon to the hadrons [e.g. 12]; an interaction nor possessed by the electron. The interference of this special interaction with the electromagnetic interaction can then lead to the second form of eq. (3) and an apparent muon form factor. Since the postulated special interaction is between the muon and hadrons, the g_μ experiment, with its present precision, may not substantially limit the parameters which can be used in this model. These speculations suggest that experimenter might search for muon-electron differences in elastic and inelastic scattering by making high precision measurements at moderate q^2 values, rather than going to high q^2 values, as was done in the present experiment. In such a high precision, moderate q^2 experiment, the limits on the systematic errors would have to be substantially reduced below the limits which now hold for present muon and electron scattering experiments.

We wish to acknowledge the kindness and help of the Stanford Linear Accelerator Center and the Massachusetts Institute of Technology electron scattering groups in providing us with their data. Obviously without their extensive and precise data we could not have made this comparison.

References

[1] E. E. Bloom et al., Phys. Rev. Letters 23 (1969) 930;
R. E. Taylor, in: Proc. 4th Intern. Symp. on Electron and photon interactions at high energy, Daresbury Nuclear Physics Laboratory, Daresbury, England (1969);
E. D. Bloom, Private communication.
For the comparison we have used the so-called 6° and 10° data reported in these papers.
[2] J Cox et al., Nucl. Instr. Methods 69 (1969) 77.
[3] W. T. Toner, in: Proc. 4th Intern. Symp. on Electron and photon interactions at high energy, Daresbury Nuclear Physics Laboratory, Daresbury, England (1969).
[4] B. D. Dieterle et al., Phys. Rev. Letters 23 (1968) 1187.
[5] S. Rock et al., Phys. Rev. Letters 24 (1970) 748.
[6] L. N. Hand, Phys. Rev. 129 (1963) 1834.
[7] D. O. Caldwell et al., Phys. Rev. Letters 25 (1970) 609;
H. Meyer et al., Phys. Letters 33B (1970) 189;
J. Ballam et al., Phys. Letters 23 (1969) 498;
J. Ballam et al., Phys. Rev. Letters 21 (1968) 1544.
[8] L. Camilleri et al., Phys. Rev. Letters 23 (1969) 153.
[9] R. W. Ellsworth et al., Phys. Rev. 165 (1968) 1449.
[10] E. Picasso, in: High-energy physics and nuclear structure (Plenum Press, New York, 1970).
[11] F. J. M. Farley, unpublished talk entitled "The status of quantum electrodynamics" (Royal Military College of Science, Shrivenham, Swindin, Wiltshire, England 1969).
[12] D. Kiang and S. H. Ng, Phys. Rev. D2 (1970) 1964.

* * * * *

SLAC-PUB-1496
October 1974
(T/E)

THE SEARCH FOR HEAVY LEPTONS AND MUON-ELECTRON DIFFERENCES*#

Martin L. Perl and Petros Rapidis
Stanford Linear Accelerator Center
Stanford University, Stanford, California 94305

Revised version of review paper originally presented at the Muon Physics
Conference, Colorado State University, September 6-10, 1971

---

*To be published in the Proceedings of the Muon Physics Conference.
#Work supported by the U.S. Atomic Energy Commission
This paper replaces and supersedes SLAC-PUB-982 and SLAC-PUB-1062.

		Page
I.	Introduction	1
II.	Some Possible Types of Heavy Leptons	4
	A. Heavy Sequential Leptons: μ', μ''...	4
	B. Heavy Excited Leptons: e^*, μ^*	5
	C. Special Pairs of Heavy Leptons	6
	D. Stable and Very Long-Lived Heavy Leptons	9
	E. Other Possibilities	9
	F. A Word on Notation	9
III.	The Decay Properties of the Heavy Leptons	10
	A. Charged Heavy Sequential Leptons	10
	B. Neutral Heavy Sequential Leptons	10
	C. Special Pairs of Leptons	20
	D. Charged Heavy Excited Leptons	22
IV.	Searches for Low Mass or Stable Heavy Leptons	22
	A. Searches in the Decay Modes of the Pion and Kaon	23
	1. The Method	23
	2. Past Searches	23
	3. Future Searches	24
	B. Searches in Particle Beams for Short-Lived Charged Heavy Leptons	24
	1. The Method	24
	2. Past Searches	24
	3. Future Searches	25
	C. Searches for Stable Charged Heavy Leptons in Particle Beams	25
	1. The Method	25
	2. Past Searches	26
	3. Future Searches	26

V.	Searches for Unstable Heavy Leptons with Masses Greater Than About 0.5 GeV/c^2	26
	A. Detection Method	26
	B. Electron-Positron Colliding Beams Production of Heavy Leptons	28
	1. The Method	28
	2. Past Searches	31
	3. Present and Future Searches	33
	C. Photoproduction of Heavy Leptons	36
	1. The Method	36
	2. Past Searches	39
	3. Future Searches	39
	D. Heavy Lepton Production in Proton-Proton Collisions	42
	E. Searches for Heavy Excited Leptons Using Charged Lepton-Proton Elastic Scattering	44
	1. The Method	44
	2. Past Searches	44
	3. Future Searches	46
	F. Searches for Heavy Excited Leptons in Lepton Bremsstrahlung	46
	1. The Method	46
	2. Past Searches	46
	3. Future Searches	46
	G. Searches for Heavy Leptons Using Charged Lepton-Proton Inelastic Scattering	47
	H. Searches for Special Pair Heavy Leptons Using Neutrino-Nucleon Inelastic Scattering	47
VI.	Comparison of Some Static Properties of the Muon and the Electron	49
	A. Electric Charge	49
	B. Gyromagnetic Ratio	
VII.	Mu-Mesic Atoms	53

VIII.	High Energy Reactions of Muons and Electrons	55
IX.	Muon-Proton Elastic Scattering and Form Factors	58
	A. Theoretical Background	58
	B. Experimental Results	64
X.	Muon-Proton Inelastic Scattering	66
	A. Theoretical Background	66
	B. Experimental Results	67
XI.	Charged Lepton Form Factors in Colliding Beam Experiments	70
	A. Elastic Electron-Electron Scattering: $e^- + e^- \to e^- + e^-$	70
	B. Bhabha Scattering: $e^- + e^+ \to e^- + e^+$	71
	C. Muon Pair Production: $e^- + e^+ \to \mu^- + \mu^+$	73
XII.	Speculations	74
	A. Searches for Heavy Leptons	74
	B. Charged-Lepton Form Factors	74
	C. Anomalous Lepton-Hadron Interactions	76
	References	81

I. INTRODUCTION

This paper is a review of some recent experimental studies on the fundamental nature of the muon and the electron, and on the relationship between these particles. The paper begins with a summary of our present knowledge as to the existence of higher mass charged leptons or of non-zero mass neutral leptons. This is followed by a brief discussion of some of the static and atomic properties of the muon and electron. The final portion of the paper is then concerned with the high energy behavior of the charged leptons in electromagnetic and strong interaction processes. Only a few references will be made to the behavior of the charged leptons in weak interaction processes.

For this audience there is no need to present an extended description of what we know about the fundamental nature of the muon and the electron. Therefore, we simply present in Table I a summary of our present knowledge -- or better a portrait -- of the muon and the electron. The set of properties (listed in Table I) that the muon and electron possess in common are collectively described by the phrase "muon-electron universality."

This portrait of the charged leptons leads to numerous questions. Are there heavier charged leptons? If there are no heavier charged leptons, why are there two charged leptons? Are the charged leptons really point particles, or do they have a structure which has not yet been detected? Are the electron and muon related in any profound way, or are they simply unrelated particles, both of which just happen to obey the Dirac equations? Are there differences between the muon and the electron other than those listed in Table I?

Table I

Property	Comparison Between Muon and Electron	If Property is Different Muon	Electron
Intrinsic spin	both 1/2		
Statistics	both Fermi-Dirac		
Fundamental equation	both Dirac equation		
Structure	both point particles (within present experimental precision as discussed in this article)		
Interact through the strong interactions	both no (within present experimental precision as discussed in this article)		
Interact through the electromagnetic interaction	both yes		
Magnitude of electric charge	same for both		
Sign of electric charge	both + or -, neither 0		
Gyromagnetic ratio	both given by quantum electrodynamics and particle's mass		
Interact through the weak interactions	both yes		
Magnitude of weak interaction coupling constant	same for both		
Associated neutrino	yes but different neutrinos	ν_μ	ν_e
Mass (MeV/c^2)		106	0.51

We must admit that at present we do not possess a fundamental theory which can provide answers to these questions. We must also admit that we do not even possess a theory which can guide us as to how we might try

- 2 -

to answer these questions experimentally. Therefore the experimenter is on his own in searching for answers to these questions. These searches, which in their very nature must be speculative, have taken two directions. One direction consists of attempts to find heavier members of the electron-muon family. The other direction consists of comparative measurements of the properties of the muon and of the electron in the hope that hitherto unknown differences between the two particles will be discovered. Of course, for this second direction to be fruitful, one must measure known properties with greater precision or one must measure properties which have not been previously measured. The recent high precision measurements of the gyromagnetic ratio of the muon are an illustration of the first type of measurement.[1] The deep inelastic scattering experiment, which we will describe later, is an illustration of the second type of measurement.

A number of comprehensive reviews of the properties of the muon and the electron have appeared in the last ten years.[2,3,4] We shall not repeat the material contained in those reviews, but only summarize their conclusions. Thus our emphasis will be on very recent experimental results. These new results have not altered the portrait, presented in Table I, in an experimentally significant way. But these new results do indicate what could be the most fruitful directions for future investigation. This forms the subject of the last section of this paper -- the section entitled "Speculations."

- 3 -

II. SOME POSSIBLE TYPES OF HEAVY LEPTONS

To simplify the discussion of past and future searches for heavy leptons, we shall consider four experimental classes of these hypothetical particles. The first three classes are defined according to the leptonic properties of the leptons.

A. Heavy Sequential Leptons: μ', μ'' ...

Suppose heavy leptons exist in the mass sequence

$$e, \mu, \mu', \mu'' \ldots , \tag{1}$$

with associated neutrinos

$$\nu_e, \nu_\mu, \nu_{\mu'}, \nu_{\mu''} \ldots . \tag{2}$$

Further suppose that each charged lepton and its associated neutrino possess a unique lepton number property which is different from the lepton number possessed by every other charged lepton-neutrino pair, and that these lepton numbers ($n_{\mu'}$) are <u>separately conserved</u> in strong, electromagnetic and weak interactions. In particular there are <u>no</u> electromagnetic vertexes of the form shown in Fig. 1;

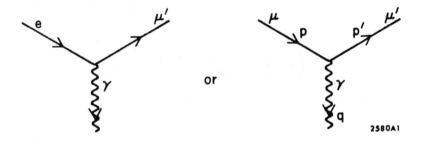

Fig. 1

and the charged heavy sequential leptons cannot decay electromagnetically. We refer to the charged leptons or their associated neutrinos as <u>heavy sequential</u> leptons and use the respective symbols μ' and $\nu_{\mu'}$. The use of the term sequential emphasizes that the properties of these heavy leptons follow in the main sequence of the e and the μ. In particular, unless we state otherwise, we assume that the mass of the associated neutrino is zero. And following the e and μ conventions, the lepton number $n_{\mu'} = +1$ is associated with the μ'^- and $\nu_{\mu'}$. In writing reactions we shall generally use the μ'^- as an example and for brevity omit the corresponding μ'^+ reaction.

B. <u>Heavy Excited Leptons: e*, μ*</u>

Suppose there exists a heavy charged lepton which possesses the same lepton number property as the electron, or the muon, of the same sign of charge. Then the electromagnetic couplings just pictured are allowed, and the electromagnetic decays

$$e^{*\pm} \to e^{\pm} + \gamma, \qquad \mu^{*\pm} \to \mu^{\pm} + \gamma \qquad (3)$$

occur. As was pointed out by Low[5], and discussed in more detail by Barut et al.[6], the simplest form for the electromagnetic coupling of the e to the e* in the Lagrangian (or the μ to the μ*) which obeys current conservation is[7]

$$e\left(\frac{\lambda}{M*}\right) \left[\sigma_{\mu\nu} F^{\mu\nu} + h.c.\right]. \qquad (4)$$

- 5 -

λ is an unknown dimensionless constant which measures the relative strength of this special electromagnetic interaction compared to the conventional electromagnetic interaction. The mass of the heavy excited lepton, M^*, is inserted simply to make λ dimensionless. Equation 4 means that an $e^- - e^{*-}$-photon vertex in a momentum space Feynman diagram has the form

$$V^\mu = -e \left(\frac{\lambda}{M^*} \right) \bar{u} \sigma^{\mu\nu} q_\nu u \qquad (5)$$

with q the momentum of the photon (as shown in Fig. 1b), rather than the form

$$V^\mu = -i e \bar{u} \gamma^\mu u \qquad (6)$$

of a conventional $e^- - e^-$-photon vertex

It is convenient and perhaps stimulating to the imagination to think of the e^* and μ^* as the respective excited states of the e and μ. Hence we designate the e^* and μ^* by the term heavy excited lepton. We may also consider the existence of a neutral heavy excited lepton, namely the heavy neutrino, ν^*, with the same lepton number as is possessed by the ν_e or the ν_μ. However, we shall not require that the e^* and ν^*, or the μ^* and ν^*, occur in pairs. In discussing the heavy excited charged leptons, we shall generally use the e^* as the example.

C. Special Pairs of Heavy Leptons

Lipmanov[8] and others[9-13] have suggested that a lepton, say the e', might have the same lepton number as the e of the opposite electric charge. The lepton number scheme would be

$$e^-, \nu_e, e'^+, \bar{\nu}_{e'} \quad \text{have} \quad n=+1 \quad ,$$

$$e^+, \bar{\nu}_e, e'^-, \nu_{e'} \quad \text{have} \quad n=-1 \quad . \qquad (7)$$

- 6 -

A special case of this hypothesis is the assumption that the μ and the e form such a pair,[11-15] namely,

$$e^-, \nu_e, \mu^+, \bar{\nu}_\mu \quad \text{have} \quad n=+1 \quad ,$$
$$e^+, \bar{\nu}_e, \mu^-, \nu_\mu \quad \text{have} \quad n=-1 \quad .$$

In the last few years there has been a strong revival of interest in some special pairs in connection with unified gauge theories of electromagnetic and weak interactions.[16] In these theories the four particle point interaction of the classic Fermi formulation of weak interactions, Fig. 2a, is replaced by the interaction in Fig. 2b:

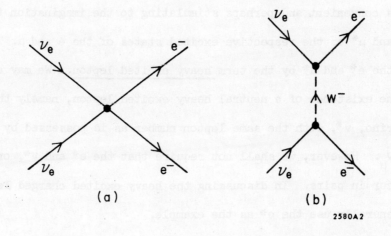

Fig. 2

The $W^\pm$ is at present a hypothetical spin 1 boson with only weak and electromagnetic interactions. Its electron lepton number and muon lepton number are both zero. However, to prevent the non-physical energy

- 7 -

dependence at very high energies of diagrams such as the one in Fig. 2b, and to make the theory renormalizable, additional particles must be hypothesized. Thus a neutral spin 1 boson analogous in properties to the $W^{\pm}$ usually called the Z^O may be assumed to exist. Either in addition to the Z^O or just by themselves, pairs of heavy leptons may also be assumed to exist. The following pairs of heavy leptons are usually assumed.[17]

E^+, E^O with lepton numbers of e^- ,

M^+, M^O with lepton numbers of μ^- ,

along with their antiparticles. To illustrate the need for such leptons and/or the Z^O, as discussed in Refs. 17, these heavy leptons are used to cancel the unphysical high energy behavior of the diagram in Fig. 3a by means of the diagrams in Fig. 3b:

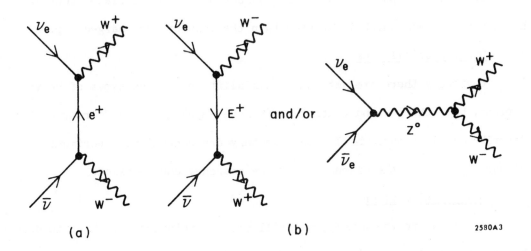

Fig. 3

For the experimenter, the most immediate significance of theories requiring E^+, E^o and M^+, M^o heavy leptons is that there is an additional incentive for searching for heavy leptons. Only if some aspects of these theories are verified--the existence of the W being proven and its mass measured, for example--do these theories provide useful guides as to what might be the masses and properties of these leptons.[18,19]

D. <u>Stable and Very Long Lived Heavy Leptons</u>

One may always assume that there is a special conservation rule or a special set of circumstances which give a heavy lepton a very long life or allows it to be stable.[20-22] Such special conditions are necessary to prohibit the decay processes which lead to the short lifetimes discussed in the next section. A simple way to obtain a stable heavy charged lepton is to assume that the lepton has a unique lepton number as in case of heavy sequential leptons, but to also assume that the associated neutrino has a nonzero mass which is greater than the mass of the charged lepton.

E. <u>Other Possibilities</u>

Obviously there are yet other possibilities for new types of heavy leptons, both charged and neutral. We have emphasized the four types described in II.A through II.D because they are convenient experimental classes upon which the discussion of the searches can be based.

F. <u>A Word on Notation</u>

As an aid to the memory, we shall use the notation μ', μ''... to denote heavy sequential leptons; we shall use e* and μ* to denote heavy excited leptons; and we shall use the script ℓ to denote any type of heavy lepton.

III. THE DECAY PROPERTIES OF THE HEAVY LEPTONS

A. <u>Charged Heavy Sequential Leptons</u>

The charged heavy sequential lepton will decay, through the weak interactions, in the leptonic modes

$$\mu'^- \to \nu_{\mu'} + e^- + \bar{\nu}_e \quad , \tag{8}$$

$$\mu'^- \to \nu_{\mu'} + \mu^- + \bar{\nu}_\mu \quad ; \tag{9}$$

and, depending on the μ' mass, in the hadronic modes

$$\mu'^- \to \nu_{\mu'} + \pi^- \quad , \tag{10}$$

$$\mu'^- \to \nu_{\mu'} + K^- \quad , \tag{11}$$

and

$$\mu'^- \to \nu_{\mu'} + 2 \text{ or more hadrons} \tag{12}$$

In all of those reactions we assume that the $\nu_{\mu'}$ mass is sufficiently small to allow the decay to occur. The decay rates (Γ) for the leptonic modes, Eqs. 8 and 9, have been discussed by many authors.[17,23-29] The calculation is straightforward if conventional, first order, weak interaction theory is used. The Feynman diagram for the decay is given in Fig. 4a

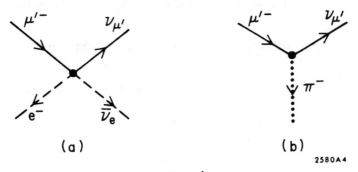

Fig. 4

- 10 -

where the solid and dashed lines distinguish the $n_{\mu'} = +1$ lines from the $n_e = \pm 1$ lines.

We find for

$$M_{\mu'} \gg M_\mu, \quad M_{\nu_{\mu'}} = 0$$

$$\Gamma(\mu'^- \to \nu_{\mu'} + \mu^- + \bar{\nu}_\mu) \approx \Gamma(\mu'^- \to \nu_{\mu'} + e^- + \bar{\nu}_e) = \frac{G^2 M_{\mu'}^5}{192\pi^3} \quad (13)$$

where

$$G = 1.02 \times 10^{-5}/M_p^2 \quad (14)$$

and M_p is the mass of the proton. That is, the decay rate for the leptonic decay modes is given by the <u>same</u> formula as is used for the $\mu^- \to e^- + \nu_\mu + \bar{\nu}_e$ decay; the μ' mass simply being substituted for the μ mass.

Two variations of Eq. 13 may occur:

(a) The μ' may have a different weak interaction coupling constant from that of the μ or e. Or, more generally, there may be two different coupling constants corresponding to the vector and axial vector interactions of the μ'. Therefore, the effective G^2 in Eq. 13 may not have the value given in Eq. 14. But we do not expect the effective G^2 to be different by more than a factor of 2 or 3.

(b) If the mass of the μ' neutrino ($M_{\nu_{\mu'}}$) is not zero, the decay rate will be smaller.[17] Setting $z = M_{\nu_{\mu'}}/M_{\mu'}$, the right hand side of Eq. 13 will be multiplied by a function[17] of z which contains one or more powers of $(1-z^2)$.

- 11 -

The single hadron decay modes, Eq. 10 and 11, can also be understood using conventional weak interaction theory.[23] Thus the decay rate for $\mu' \to \nu_{\mu'} + \pi^-$ is calculated using the diagram in Fig. 4b; the strength of the weak interaction coupling of the π being obtained from $\pi^- \to \mu^- + \bar{\nu}_\mu$.

When the mass of the charged sequential lepton exceeds 1 GeV/c^2 or so, the multi-hadron decay modes, Eq. 12, become important--perhaps most important. Yet it is just these modes which we <u>know</u> least how to calculate. We don't know how to calculate the cross hatched region in Fig. 5a from basic principles--not even in conventional weak interaction theory.

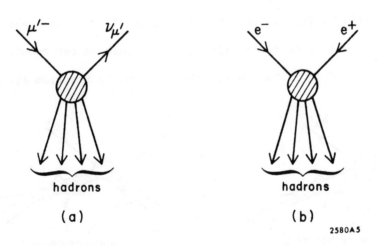

Fig. 5

However, if we accept the conserved vector current hypothesis--a concept which connects some aspects of weak interactions with electromagnetic interactions--then there is a relation between

$$\mu'^{-} \to \gamma_{\mu'} + \text{hadrons} \qquad (15a)$$

and

$$e^{-} + e^{-} \to \text{hadrons} \qquad (15b)$$

The reaction in Eq. 15b, the production of hadrons by positron-electron annihilation, is shown diagramically Fig. 5b.   The conserved vector current hypothesis states that there is a single hadronic transition current with vector spatial properties whose charge changing component is the I (isospin) = 1, strangeness conserving, part of the weak hadronic current and whose charge conserving part is the I = 1 part of the electromagnetic hadronic current. Redrawing Fig. 5 to emphasize the hadronic transition current; Fig. 6 shows that we can connect the weak decay process of Eq. 15a to the electromagnetic reaction Eq. 15b.

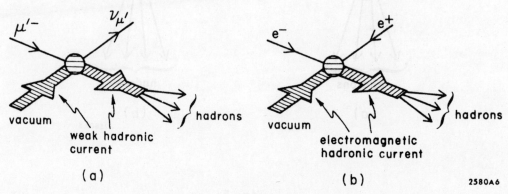

Fig. 6

- 13 -

The quantitative relation in Eq. 16 has been given by Tsai,[23] using some additional assumptions about the contributions of $I = 0$, axial vector, and strangeness changing currents.

$$\Gamma(\mu'^- \to \nu_{\mu'} + \text{hadrons}) \approx \frac{3G^2}{2^8 \pi^4 \alpha^2 M^3} \int_{s_0}^{M^2} ds \left[s(M^2-s)^2 (M^2+2s) \right.$$
$$\left. \times \sigma_{e^+e^- \to \text{had}}(s) \right] \quad (16)$$

Here $\sigma_{e^+e^- \to \text{had}}(s)$ is the total cross section for the reaction $e^+e^- \to$ hadrons when $\sqrt{s}$ is the total energy in the center of mass. M is the mass of the μ'. $\sqrt{s_0}$, the smallest invariant mass of hadrons which contribute to the decay $\mu' \to \nu_{\mu'} +$ hadrons, is taken to be about 1 GeV/c^2. α is the fine structural constant (1/137).

In thinking about $\sigma_{e^+e^- \to \text{had}}(s)$, it has become conventional to relate it to the cross section $\sigma_{e^+e^- \to \mu^+\mu^-}(s)$ for the reaction $e^+e^- \to \mu^+\mu^-$ by defining

$$R(s) = \frac{\sigma_{e^+e^- \to \text{had}}(s)}{\sigma_{e^+e^- \to \mu^+\mu^-}(s)} \quad (17a)$$

using

$$\sigma_{e^+e^- \to \mu^+\mu^-}(s) = \frac{4\pi\alpha^2}{3s}, \quad s \gg M_\mu^2 \quad (17b)$$

and Eq. 13, Eq. 16 take the simple form

$$\frac{\Gamma(\mu'^- \to \nu_{\mu'} + \text{hadrons})}{\Gamma(\mu'^- \to \nu_{\mu'} + e^- + \bar{\nu}_e)} \approx 3 \int_{1/M^2}^{1} dx \left[(1-x^2)(1+2x) R(s) \right] \quad (18)$$

$$x = s/M^2$$

Finally defining

$$\bar{R} = \int_{1/M^2}^{1} dx \left[(1-x)^2 (1+2x) R(s)\right] \bigg/ \int_{1/M^2}^{1} dx \left[(1-x)^2 (1+2x)\right] \quad (19)$$

we obtain

$$\frac{\Gamma(\mu'^{-} \to \nu_{\mu'} + \text{hadrons})}{\Gamma(\mu'^{-} \to \nu_{\mu'} + e + \bar{\nu}_e)} \approx \frac{3}{2} \bar{R} \quad (20)$$

$\sigma_{e^+e^- \to \text{had}}(s)$ has now been measured[30,31] for $\sqrt{s} \leq 5$ GeV. A rough fit to Eq. 17 yields (for s in units of GeV^2)

$$R(s) \approx \frac{5}{4}\sqrt{s} - \frac{3}{4}, \quad 1 \text{ GeV} \leq \sqrt{s} \leq 5 \text{ GeV} \quad (21a)$$

Putting all this together we obtain the fractional decay rates in Fig. 7, and the lifetime in Fig. 8. Here we have assumed

$$R(s) = 5.5, \sqrt{s} > 5 \text{ GeV} \quad (21b)$$

In Fig. 7 we note the dominance of the multi-hadron decay mode.

To remind us how speculative this is, we also show the fractional decay rates, Fig. 9, if we use a quark-parton model[23] in which

$$R(s) = \text{constant} = 2/3 \quad (22)$$

This model, gives the wrong value of R as measured in $e^+ + e^- \to$ hadrons.

R will turn out to be a useful concept when discussing methods of searching for heavy leptons. However we don't know if the preceeding theory connecting the multi-hadron decay mode of a heavy lepton with the reaction $e^+ + e^- \to$ hadrons is correct. Therefore in analogy to Eq. 20 we define for any hypothetical charged sequential lepton

$$\frac{\Gamma(\mu'^{-} \to \nu_{\mu'} + \text{hadrons})}{\Gamma(\mu'^{-} \to \nu_{\mu'} + e + \bar{\nu}_e)} = \frac{3}{2} R_{\text{decay}} \quad (23)$$

If the preceeding theory is correct $R_{\text{decay}} = \bar{R}$, otherwise we are in lieu of other knowledge free to choose any value for R_{decay}.

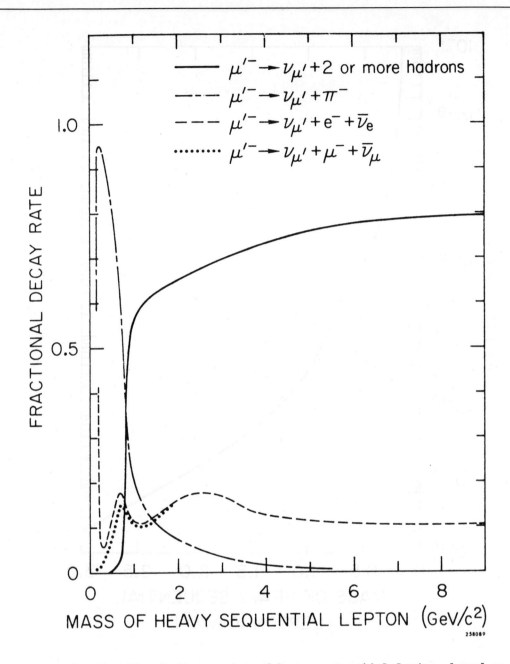

Fig. 7. Fractional decay rates of heavy sequential leptons based on calculations discussed in the text with R given by Eq. 21. The $\nu_{\mu'}$ mass is assumed to be zero. The low mass behavior of the 2 or more hadrons decay mode has been smoothed over the individual decay channel thresholds.

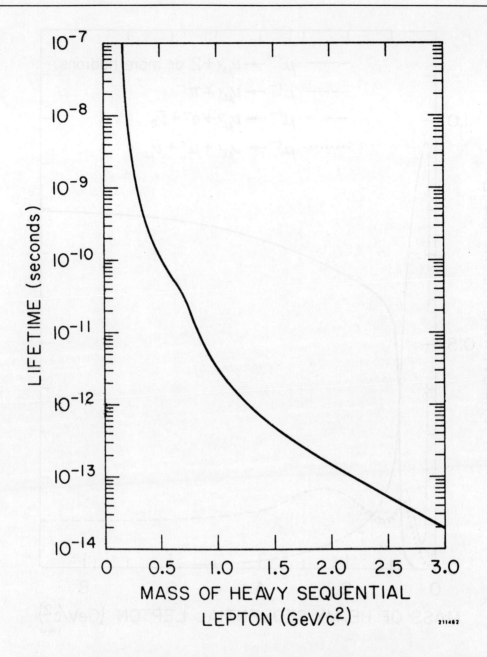

Fig. 8. Lifetime versus mass of heavy sequential leptons based on calculations discussed in the text using Eq. 21 and zero mass for the ν_μ.

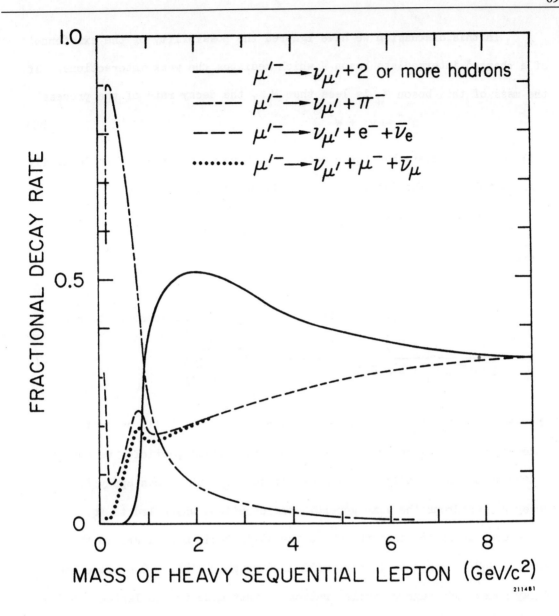

Fig. 9. Fractional decay rates of heavy sequential leptons based on calculations discussed in the text using R = 2/3 as in Eq. 22. The mass of the $\nu_{\mu'}$ is assumed to be zero. The low mass behavior of the 2 or more hadrons decay modes has been smoothed over the individual decay channel thresholds.

- 18 -

In this discussion we have ignored the possibility of the existence of a charged intermediate boson W which mediates the weak interactions. If the mass of this boson M_W is less than $M_{\mu'}$, the decay rate of the process

$$\mu'^- \to \nu_{\mu'} + W^- \tag{24}$$

will be much greater than the processes we have considered. This is easily seen as follows. The decay rate for this process is[23]

$$\Gamma(\mu' \to \nu_{\mu'} + W^-) = GM_{\mu'}^3 \left[\frac{(1-r^2)^2 (1+2r)}{8\pi\sqrt{2}} \right] \tag{25}$$

where

$$r = \left(M_W^2 / M_{\mu'}^2 \right) < 1 \quad . \tag{26}$$

Comparison of Eq. 25 with Eq. 14 yields the approximate ratio

$$\frac{\Gamma(\mu'^- \to \nu_{\mu'} + W^-)}{\Gamma(\mu'^- \to \nu_{\mu'} + e^- + \bar{\nu}_e)} \sim \frac{1}{GM_{\mu'}^2} \sim 10^5 \left(\frac{M_p}{M_{\mu'}} \right)^2 \tag{27}$$

Therefore, if $M_W < M_{\mu'}$, the observed decay modes of the μ' will be just the decay modes of the W. Since the W is still a hypothetical particle, we can say little about the details of its hadronic modes. But we can apply to the decay $W^- \to$ hadrons the same general considerations which led to Eq. 18. Therefore, we expect that the μ' and the W, if their masses are in the several GeV/c^2 range or larger, will have roughly similar 2 or more hadron decay modes and roughly similar ratios of that mode to the leptonic modes. Thus the search experiment that first finds the μ' will probably *not* show if the intermediate boson forms a step in the decay chain through the process of Eq. 24.

The lifetime of the μ' will of course depend upon whether or not there is an intermediate boson W with $M_W < M_{\mu'}$. In Fig. 8 we show the lifetime of the μ' assuming that there is <u>no</u> W with $M_W < M_{\mu'}$ and using Eqs. 13 through 21. If such a W exists, the μ' has the very short lifetime

$$\tau \approx 2.5 \times 10^{-23} \left(\frac{M_p}{M_{\mu'}}\right)^3 \left[\frac{1}{(1-r^2)^2 (1+2r)}\right] \text{ sec} \quad . \tag{28}$$

B. <u>Neutral Heavy Sequential Leptons</u>

If the mass of the neutral heavy sequential lepton ($M_{\nu_{\mu'}}$), that is of the neutrino, is less than that of the charged lepton ($M_{\mu'^\pm}$), as was assumed in the previous section, then like the conventional neutrinos the $\nu_{\mu'}$ is stable! A most peculiar object if $M_{\nu_{\mu'}}$ is large.

If
$$M_{\nu_{\mu'}} > M_{\mu'^\pm} \tag{29}$$

then the $\nu_{\mu'}$ has decay modes completely analogous to that in the previous section:

$$\nu_{\mu'} \to \mu'^{-} + e^{+} + \nu_e \tag{30}$$

$$\nu_{\mu'} \to \mu'^{-} + \pi^{+} \tag{31}$$

$$\nu_{\mu'} \to \mu'^{-} + 2 \text{ or more hadrons} \tag{32}$$

and so forth. The decay ratio will be the same as those for the analogous μ'^{-} decay modes given in the previous section. But now, we remind the reader, the $\mu'^{\pm}$ will be stable! An even more peculiar object!

C. <u>Special Pairs of Leptons</u>

The decay mode calculations for special pairs of leptons[17] are, in many ways, similar to those for heavy sequential leptons. To illustrate the discussion, we use the E^{+}, E^{0} pair which has the lepton numbers of the e^{-}.

If

$$M_{E^{+}} > M_{E^{0}} \tag{33}$$

analogous to Eqs. 8 and 12, we have

$$E^{+} \to E^{0} + e^{+} + \nu_e \tag{34}$$

$$E^{+} \to E^{0} + 2 \text{ or more hadrons} \tag{35}$$

Unlike the sequential heavy lepton even if $M_{E^{+}} < M_{E^{0}}$, the E^{+} is unstable because there are the decay modes

$$E^{+} \to \nu_e + e^{+} + \nu_e \tag{36}$$

$$E^{+} \to \nu_e + 2 \text{ or more hadrons} \tag{37}$$

The E^{0} is also always unstable, because of the decay modes

$$E^{0} \to e^{-} + (\text{leptons or hadrons}) \tag{38}$$

as well as the modes (when $M_{E^{0}} > M_{E^{+}}$)

- 21 -

$$E^o \rightarrow E^+ + \text{(leptons or hadrons)} \qquad (39)$$

and since neutral weak interaction currents exist we can also expect

$$E^o \rightarrow \nu_e + \text{(leptons or hadrons)} \qquad (40)$$

The decay rates for the E^+, E^o have been discussed by Bjorken and Llewellyn Smith[17]. The calculational methods and the predictions are similar to those given for the sequential heavy leptons. For crude estimates, Figs. 7 through 9 can be used.

D. <u>Charged Heavy Excited Leptons</u>

The very dominant decay mode for the charged heavy excited leptons is

$$e^* \rightarrow e + \gamma ; \qquad (41)$$

all other decay modes have ratios which are smaller by at least the factor α^2. The lifetime for $M_{e^*} \gg M_e$ is[32]

$$\tau = \frac{1}{\alpha M_{e^*} \lambda^2} = \frac{.9 \times 10^{-22}}{M_{e^*} \lambda^2} \text{ sec}, \qquad (42)$$

where M_{e^*} is in GeV/c^2.

IV. SEARCHES FOR LOW MASS OR STABLE HEAVY LEPTONS

When the mass of the heavy lepton is sufficiently small, there are two convenient search methods. First, if the charged or neutral heavy lepton has a mass M_ℓ less than the pion mass M_π or kaon mass M_K, the heavy lepton ℓ can appear in the decay of the pion or kaon respectively. Second, for heavy sequential leptons, a small mass leads to a lifetime sufficiently long to permit detection if the lepton is in a particle beam of conventional length--namely tens of meters.

A. **Searches in the Decay Modes of the Pion and Kaon**

1. The Method

Most past searches of this type have involved direct observation in a bubble chamber, but counters have been used.[33] Some recent tests of the Ramm effect[34] discussed below use wire spark chambers. There is no uniform method and the reader should consult the references which are cited in this section.

2. Past Searches

With one exception, which will be discussed below, <u>no</u> evidence for heavy leptons, either charged or neutral has been found in the study of the decay modes of the pion or kaon. Rothe and Wolsky[29] summarized the situation in 1968. An earlier summary is provided by Beier,[33] who carried out a search for heavy sequential leptons with masses just below the kaon mass.

The exception to these null results is the work of Ramm[34] who reports evidence for a neutral heavy meson in the decay $K_L^o \to \mu^{\pm} + \pi^{+} + \nu_{\mu}$. The heavy lepton appears as a narrow resonance in the $\mu^{o*} \to \mu^{\pm}\pi^{+}$ mass spectrum with mass $.422 < M_{\mu\pi} < .437$ GeV/c^2. Ramm[34] also reports evidence in neutrino interactions and in muon bremsstrahlung[35,36] for a charged counterpart with the decay mode $\mu^* \to \mu^- + \gamma$. Unfortunately, this work has not been verified by other studies of the K_L^o decay spectrum. While most of these studies with null results have not been published, Clark <u>et al</u>.[37] have published a very high statistics study, also reporting a null result. The weight of the evidence appears at present to be <u>against</u> the existence of the Ramm neutral heavy lepton.

- 23 -

3. Future Searches

We know of no plans for further searches for heavy leptons in pion or kaon decay modes.

B. Searches in Particle Beams for Short-Lived Charged Heavy Leptons

1. The Method

For a charged heavy lepton to be directly detected in a particle beam, the lepton's decay length in the laboratory frame must be of the order of magnitude of, or greater than, say 10 meters. This requires a laboratory decay time of at least 3×10^{-8} seconds. If the particle has a laboratory frame energy E_ℓ, the time dilation factor (E_ℓ/M_ℓ) may permit detection of particles with particle rest frame lifetimes (τ_ℓ) as short as $(M_\ell/E_\ell)(3 \times 10^{-8})$ seconds.

Searches at electron accelerators are most useful because, as is discussed in Sec. V.C, the heavy leptons are pair produced by photons with a known cross section, if they have unit charge. Searches at proton accelerators are less definitive, because the production cross section of the heavy leptons is not known; this production uncertainty is discussed in Sec. V.D.

2. Past Searches

Examples of searches at electron accelerators all with null results are the experiment of Coward et al.[38] who searched in the mass range of .5 to 90 MeV/c^2, and the experiment of Barna et al.[21] who studied the mass range[39] of .2 to 1.0 GeV/c^2. Because of their uncertain sensitivity, searches at proton accelerators are usually not reported, although we may be sure that any unknown charged particle which was found in a beam would have been reported. We can set a crude upper limit to the mass range of these searches by noting that at the older proton accelerators $(M_\ell/E_\ell) \gtrsim .01$. Therefore, $\tau_\ell \gtrsim 3 \times 10^{-10}$ seconds and, from Fig. 8, $M_\ell \lesssim .35$ GeV/c^2.

In closing this section we note an experiment of Ansorge et al.[40] in which a search was made for charged and neutral particles with masses less than .1 GeV/c^2. The search method involved the study of electron pair production and electron-like bremsstrahlung in a hydrogen bubble chamber. No new particles were found.[40]

3. Future Searches

We know of no special plans for future direct searches for low mass, short-lived, heavy leptons. The problem is that the new higher energy accelerators do not enlarge substantially the mass range over which charged heavy sequential leptons can be <u>directly</u> found in particle beams. For example, increasing E_ℓ by a factor of 10 decreases the lower limit on τ_ℓ by a factor of 10. But since τ_ℓ decreases at least as fast as M_ℓ^{-5}, this only extends the upper limit on M_ℓ to $10^{1/5}$ $(.35) = .56$ GeV/c^2.

C. <u>Searches for Stable Charged Heavy Leptons in Particle Beams</u>

1. The Method

The only question with respect to this type of search is whether its sensitivity is sufficient. Assuming unit charge for the leptons, this sensitivity can be determined quite well for particle beams produced by electrons or photons, Sec. V.C.; but for beams produced by protons, the very uncertain concepts in Sec. V.D. result in an indeterminate sensitivity.

- 25 -

2. Past Searches

No stable heavy leptons have been found. At electron accelerators the search has been conducted[21,22] with sufficient sensitivity up to a heavy lepton mass of 1.0 GeV. References 41 are examples of older searches at proton accelerators. Recent searches[42-44] include that of B. Alper et al.[42] at the CERN 30 GeV proton-proton colliding beams facility; and that of J. W. Cronin et al.[43] using a 300 GeV primary proton beam on a copper target at NAL. In the latter search, no charged heavy lepton was found in an equivalent yield of 10^9 pions.

3. Future Searches

We assume that everyone who is studying the spectra of particles produced at any of the new accelerators will continue to search for stable heavy leptons, but with steadily decreasing enthusiasm. We also note that it is quite easy to look for stable charged heavy leptons at electron-positron colliding beam facilities.

V. SEARCHES FOR UNSTABLE HEAVY LEPTONS WITH MASSES GREATER THAN ABOUT .5 GeV/c^2

A. Detection Method

In the last section we showed that the search for heavy leptons with masses less than .4 or .5 GeV/c^2 was just about complete. And we showed that no heavy leptons with masses less than .4 or .5 GeV/c^2 have been proven to exist. Therefore the future of heavy lepton searches belongs to the mass range above .5 GeV/c^2. As discussed in Sec. III, heavy sequential leptons, special pairs, and of course heavy excited leptons, above this mass will have lifetimes less than 10^{-9} or 10^{-10} seconds. These heavy leptons must, therefore, be detected through their decay products. And this detection must be carried out in the midst of a relatively large background of particles which have nothing to do with the heavy lepton production reaction. The same initial states, such as $e^+ + e^-$ or $p + p$, which can lead to heavy lepton production also lead to much more copious hadron production.

In searches for heavy sequential leptons or for special lepton pairs, most detection methods make use of special properties of one or more of the decay modes.

For example, the production of heavy sequential leptons or of special pairs will be relatively easy to observe *if* the leptonic decay modes can be used. The clearest signature is a μe pair with no other charged particles present; the μ comes from the decay of one heavy lepton, the e comes from the decay of the other heavy lepton. Although there are some background problems, μμ, πμ, or πe pairs with no other charged particles present also provide good signatures; the π coming from the $\ell \to \nu + \pi$ decay mode. We will mention later other detection methods which depend upon the e or μ momentum spectrum. Of course, the sensitivity depends upon the fractional decay rates to these single charged particle states. As R_{decay}, Eq. 23, increases, the sensitivity of this method decreases.

Indeed if R_{decay} is as large as 10 or so, the detection of the leptonic or single π decay modes becomes very difficult. One might hope then to make use of the (ν_e + 2 or more hadrons) decay mode. But here nature is unusually

- 27 -

cruel--there is no simple way to separate the multi-hadron decay modes from the hadronic background. This is the problem that plagues present day heavy lepton searches; we will come upon it repeatedly in the remainder of this section.

Turning now to the search for excited leptons, the e* or μ*, the experimenter has the advantage that he seeks only the (e +γ) or (μ +γ) decay modes. He has the disadvantage that:

(a) There are no other decay modes to confirm the existence of the e* or μ*.

(b) There are backgrounds which can simulate the desired signal. For example the process

$$e^+ + e^- \rightarrow e^+ + e^{-*}, \; e^{-*} \rightarrow e^- + \gamma \qquad (43)$$

can be confused with radiative e^+e^- scattering

$$e^+ + e^- \rightarrow e^+ + e^- + \gamma \qquad (44)$$

B. Electron-Positron Colliding Beams Production of Heavy Leptons

1. The Method

The production of charged heavy leptons pairs ℓ^-, ℓ^+ in electron-positron colliding beams takes place through the reaction

$$e^- + e^+ \rightarrow \ell^- + \ell^+ \qquad (45)$$

The dominant Feynman diagram for this process is given in Fig. 10.

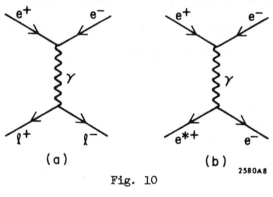

Fig. 10

Assuming the heavy leptons are Dirac point particles with unit charge, the total production cross section is

$$\sigma(e^-e^+ \to \ell^-\ell^+) = \frac{\pi\alpha^2}{2E^2} \beta[1-\beta^2/3] \quad , \tag{46}$$

where E is the energy of either the electron or positron beam, and β is the velocity of the ℓ. We shall also use $s = 4E^2$.

Given that the electron-positron colliding beams have sufficient energy for lepton production, $E > M_\ell$, this method provides in our view the best general method of searching for charged heavy leptons. It has two advantageous properties:

(a) The production cross section, Eq. 44, is relatively large. Once E is somewhat larger than M_ℓ, β approaches 1, and

$$\sigma(e^-e^+ \to \ell^-\ell^+) \approx \sigma(e^-e^+ \to \mu^-\mu^+) \approx \frac{2 \times 10^{-32}}{E^2} \text{ cm}^2 \quad , \tag{47}$$

where E is in GeV. Thus $\ell^-\ell^+$ pairs are produced almost as copiously as $\mu^-\mu^+$ pairs. Now we recall from Eqs. 17 and 21 that

$$\sigma(e^-e^+ \to \text{hadrons}) = R(s)\, \sigma(e^-e^+ \to \mu^-\mu^+) \tag{48}$$

where

$$1 \lesssim R \lesssim 5$$

Therefore, the cross section for heavy lepton production is within a factor of 10 of the dominant hadron production cross section (Fig. 11). This is a much better situation than occurs in the production of heavy leptons by hadrons or photons (Secs. V.C and V.D.).

- 29 -

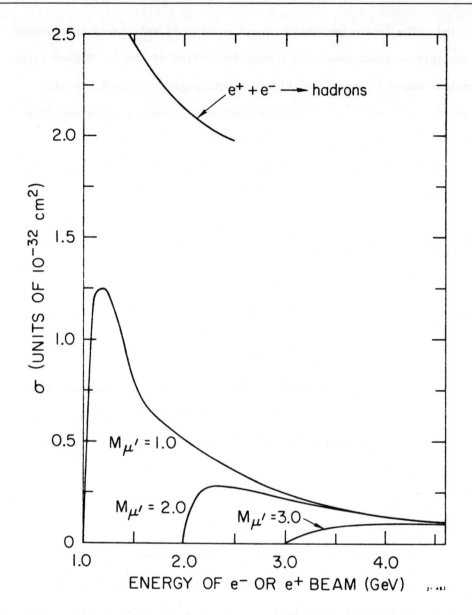

Fig. 11. Total cross section for the production of heavy leptons of mass M, at an electron-positron colliding beam facility, through the reaction $e^+ + e^- \rightarrow \mu'^+ + \mu'^-$. For comparison the uppermost solid line shows the total cross section for the production of hadrons in $e^+ + e^- \rightarrow$ hadrons; based on Refs. 30 and 31.

(b) The lepton production cross section is relatively independent of the spin or electromagnetic moment properties of the ℓ. Higher spins, anomalous magnetic moments or higher electromagnetic moments usually increase $e^-e^+ \to \ell^-\ell^+$. The cross section will only be substantially smaller than that given in Eq. 46, if the electric charge of the ℓ is much less than that of the electron, or if the ℓ has a form factor which is much less than unity for $q^2 = 4E^2$.

We should also remind you that for the excited electron, e*, the limit on the heavy lepton mass is <u>not</u> $M_{e*} < E$, but is

$$M_{e*} < 2E \qquad (49a)$$

for
$$e^+e^- \to e^{\pm *} + e^{\mp} \qquad (49b)$$
$$e^{\pm *} \to e^{\pm} + \gamma$$

This can occur because the vertex in Eq. 4 allows the e*e production diagram in Fig. 10b. In this case the production cross section of Eq. 47 will be reduced by roughly λ^2, and λ is of course unknown. A similar result holds for the μ*.

2. Past Searches

A series of searches for charged sequential leptons has been carried out by V. Alles-Borelli et al.[45] at ADONE; the e^+e^- colliding beams facility at Frascati. These experimenters looked for $e^{\pm}\mu^{\mp}$ pairs coming from the

- 31 -

sequence

$$e^- + e^+ \to \mu'^- + \mu'^+$$
$$\mu'^- \to \nu_{\mu'} + \mu^- + \bar{\nu}_\mu \qquad (50)$$
$$\mu'^+ \to \bar{\nu}_{\mu'} + e^+ + \nu_e \quad,$$

or the alternative set of leptonic decay modes. No heavy leptons were found in the mass range of 0.2 to 1.0 GeV/c^2. If the heavy leptons are assumed to couple only to leptons, the range extends to 1.4 GeV/c^2.

J. Feller et al.[46] conducted a search at the CEA e^+e^- colliding beams facility for charged sequential leptons with up to 2 GeV/c^2 mass. They were not able to rule out the existence of one heavy lepton in this mass range. But assuming $R_{decay} = 2$, they could show that the entire $e^+e^- \to$ hadrons cross section could not be due to the production of a series of heavy leptons.

Two searches[47,48] for excited heavy leptons have been carried out using e^+e^- colliding beams. In the first search,[47] conducted at Frascati, the reaction of Eq. 49b

$$e^+ + e^- \to e^+ + e^- + \gamma$$

was analyzed and the limit of

$$2\Lambda_{e*} = \frac{M^*}{\lambda} \geq 7.8 \text{ GeV}$$

was obtained with a 95% confidence level.

The second search,[48] at CEA, focussed on deviations from the QED prediction for the reaction

$$e^+ + e^- \to \gamma + \gamma \quad.$$

Two Feynman diagrams contribute, in the lowest order, to this process:

- 32 -

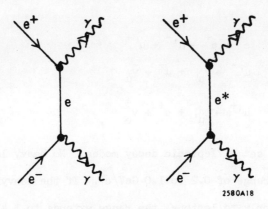

Fig. 12

An upper limit for λ^2 of 8×10^{-5} to 9×10^{-4} was set for a heavy electron with mass M* in the range of .6 to 2.4 GeV, or equivalently,

$$2\Lambda_{e^*} = \frac{M^*}{\lambda} > 30 \text{ to } 70 \text{ GeV}$$

3. Present and Future Searches

The authors and their colleagues are now conducting a search[49] for charged sequential and special pair leptons at SPEAR, the SLAC e^+e^- colliding beams facility. SPEAR has operated with a maximum energy of 2.6 GeV in each beam, and within a year the maximum energy will be increased to 4.8 GeV in each beam.

To give a feeling for the sensitivity of such a search, we note that a good luminosity,[50] L, is about 0.5 to 1.0×10^{31} events per cm^2 per sec. Using Eq. 47 we see that even for leptons with masses in the 4-5 GeV/c^2 we expect about 10 heavy lepton pairs per hour. Of course, the problem is how to detect them. We are not as sanguine as we once were about using the multi-hadron decay mode, Eq. 12. It is true that this mode contributes to the

- 33 -

total hadronic cross section. But the large values of R, Eq. 21 and Fig. 11, make it difficult to detect the additional contribution to hadron production when the beam energy is above the lepton production threshold.

Therefore, the search makes use of the leptonic and single π decay modes. When both members of the lepton pair decay into these modes, the final state shows only two charged particles and no π^o's. Even if there is misidentification of the e, μ, or π, events with this signature are rare in the high multiplicity hadronic background.

Another way to use the leptonic or single π decay modes is being pursued by J. Kirkby.[51] In this method one uses the momentum imbalance in the event due to the undetected momentum carried off by the neutrinos.

Similar searches will no doubt be carried out at the DESY e^+e^- storage ring, DORIS, which will have a maximum energy of 5 GeV in each beam.

The major limitations on the electron-proton colliding beam search method is that, except for the special case of Eq. 45, $M_\ell < E$ and E is less than 5 GeV for existing facilities. In contrast, as discussed in Sec. V.C., a 200 GeV photon beam at a 300 GeV proton accelerator may produce pairs of heavy leptons with M_ℓ as large as 10 GeV. But, as is also discussed in that section, photoproduction of pairs is not nearly as clean a search method as is the production by e^-e^+ colliding beams. For this reason, and for many other reasons, much higher energy electron-positron colliding beam facilities are now being discussed. For example, physicists at the Lawrence Berkeley Laboratory and at SLAC are considering

- 34 -

the design of a 15 GeV electron-positron colliding beam facility called
PEP; and Rutherford Laboratory physicists are considering a similar
facility called EPIC.

To conclude this section, we note that assuming the existence of
an intermediate neutral boson, the Z^o, neutral heavy leptons can be
produced[52] through the diagrams in Fig. 13. Here the E^o is used as
an example.

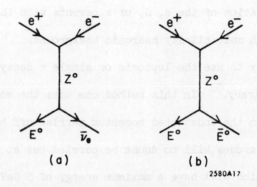

Fig. 13

Unfortunately, the experimenter finds such searches to be impractical.
Comparing the diagrams in Fig. 13 with the virtual photon production of
charged leptons diagram of Fig. 10, it is easy to estimate the change
in Eq. 47 --the lepton production cross section. The electromagnetic
coupling and photon projection factor e^2/s, where $s = 4E^2$, is replaced by

$$\frac{g^2}{s + M_{Z^o}^2} \sim G, \text{ for } M_{Z^o}^2 \gg s \tag{51}$$

Thus the right hand side of Eq. 47 is multiplied by 10

$$(G^2/\alpha^2)E^4 \sim 10^{-6} E^4$$

Hence

- 35 -

$$\sigma(e^+e^- \to E^0 + \bar{E}^0 \text{ or } E^0 + \tilde{\nu}_e) \sim 10^{-38} \, E^2 \, cm^2 \qquad (52)$$

where E is in GeV. Even E's in the 10-20 GeV range do not yield sufficient cross section.

C. Photoproduction of Heavy Leptons

1. The Method

For heavy leptons which are Dirac point particles with unit charge, the production cross section for the photoproduction of pairs of these particles can be calculated quite well.[53,54] One of the most recent complete calculation on these processes is that of Kim and Tsai.[53] They give the results of a typical calculation for a beryllium target for $.1 \leq M_\ell \leq 6$ GeV and k, the photon energy, up to 200 GeV. Figure 14 is a plot of the total cross section for production.

The differential cross section[53] for the production of one lepton at an angle θ and momentum p, summed over all allowed momenta and angles of the other lepton and all hadronic final status, is

$$\frac{d\sigma}{d\Omega dp} = \frac{2\alpha^3}{\pi k} \left(\frac{E^2}{M_\ell^4}\right) \left[\frac{2x^2 - 2x+1}{(1+r)^2} + \frac{4x(1-x)r}{(1+r)^4}\right] X \qquad (53)$$

Here k is the photon energy, M_ℓ is the lepton mass, E is lepton's total energy, $x = E/k$, $\gamma = E/M_\ell$, and

$$r = \gamma^2 \theta^2 = E^2 \theta^2 / M_\ell^2 \qquad (54)$$

X, defined exactly in Ref. 53, is a function of the minimum four-momentum transfer to the hadronic vertex and the form factors at that vertex. From Eq. (53) we observe the following well-known property of the photoproduction of pairs. When θ is small so that $r \ll 1$,

$$\frac{d\sigma}{d\Omega dp} \sim \frac{1}{M_\ell^4} \, . \qquad (55)$$

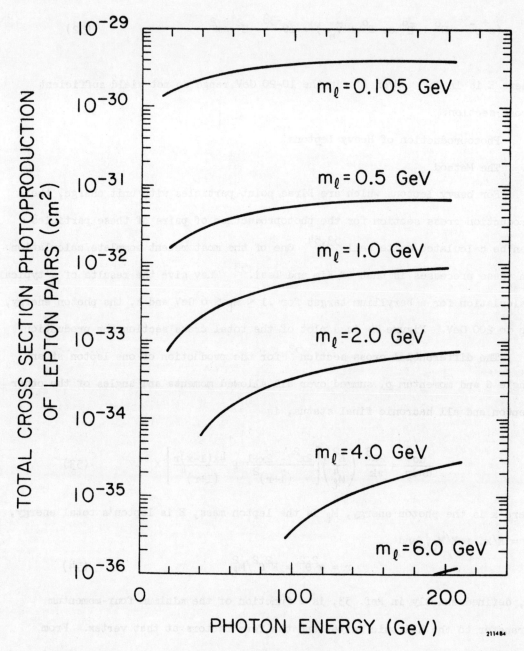

Fig. 14

Thus at small angles the production of high mass heavy leptons is much suppressed compared to electron or muon production. On the other hand, when $\theta E \gtrsim M_\ell$, then $r \gtrsim 1$ and $d\sigma/(d\Omega dp)$ is, except for the X term, independent of M_ℓ. Hence the crude rule that at large angles, all leptons will be produced in roughly equal numbers.

Separated photon beams need not be used to produce lepton pairs. For example, at electron accelerators, electrons hitting a target of 5 or 10 radiation lengths lead to a double process in the same target,[21,55] namely

$$e^- + \text{Nucleus} \to e^- + \text{Nucleus} + \gamma$$
$$\gamma + \text{Nucleus} \to \ell^+ + \ell^- + \text{hadrons} \quad . \tag{56}$$

At proton accelerators the following sequence of processes can occur in a thick target[56]

$$p + \text{Nucleus} \to \pi^\circ + \text{hadrons},$$
$$\pi^\circ \to \gamma + \gamma,$$
$$\gamma + \text{Nucleus} \to \ell^+ + \ell^- + \text{hadrons} \quad . \tag{57}$$

But the processes of Eq. 57 cannot always be distinguished from other hypothetical processes for making lepton pairs in proton-proton or proton-nucleus collisions. Therefore we postpone further discussions of all proton-proton or proton-nucleus searches to the next section.

Since the relatively high mass heavy leptons under discussion here have very short lifetimes, they must, of course, be detected through their decay modes. Their detection through their hadronic decay modes, if they are heavy sequential leptons or special pair leptons, is very difficult because of the copius direct production of hadrons in photoproduction. The total cross section for the production of hadrons by photons at high energy is about 10^{-28} cm^2, which is 10^4 to 10^8 times larger than the heavy lepton production cross section in Fig. 14. Therefore past and future searches for heavy sequential leptons almost always rely on the detection of the electrons, the muons, or the neutrinos produced in the leptonic decay modes. For the heavy excited lepton the electron, muon or γ must be detected. We now turn to these searches.

2. Past Searches

We know of only one search for short-lived heavy leptons carried out at an electron accelerator. This is a search[55,57] carried out at SLAC in which an 18 GeV, high intensity electron beam was used in the production process of Eq. 56. All charged particles were stopped very quickly by a thick wall of matter. However a heavy lepton with a lifetime shorter than 10^{-10} seconds would decay before stopping. The decay mode

$$\mu'^- \to \nu_{\mu'} + \pi^- \tag{58}$$

would yield high energy, $\nu_{\mu'}$, neutrinos which would penetrate the wall.[57] Some of these $\nu_{\mu'}$ neutrinos might then interact in optical spark chambers, placed downstream of the wall; and these interactions would lead to hadron productions through the production and immediate decay of the μ'. No events which require this explanation were found, and in general no other clear evidence was found for the production of the μ'. But the experiment did not have sufficient sensitivity.[55] And the production, by the processes of Eq. 56, of heavy sequential leptons with $M_{\mu'} > .5$ GeV was not excluded.

3. Future Searches

High energy photon beams have been produced at Serpukov and will be produced at FNAL and at the new CERN accelerator. A general summary of proposed FNAL heavy lepton searches by photoproduction, and by other methods, has been given by Heusch and Sandweiss.[56] As an example, Lee et al.[58] plan to use a photon beam with a

- 39 -

maximum energy of 400 GeV. The beam hits a 0.1 radiation length beryllium target. Spark chamber hodoscopes, an analyzing magnet, and scintillation and shower counters are used to select and identify e's, μ's and π's of greater than 30 GeV/c momentum produced at between 20 to 50 mrad to the photon beam direction. The trick is to search for those events in which <u>both</u> heavy leptons have decayed in the decay modes

$$
\begin{align}
\mu' &\to \nu_{\mu'} + \mu + \nu_\mu , \\
\mu' &\to \nu_{\mu'} + e + \nu_e , \\
\mu' &\to \nu_{\mu'} + \pi ;
\end{align}
\tag{59}
$$

these being the decay modes in which just <u>one</u> charged particle is produced. In particular the experimenters hope to make use of the sharp upper cutoff at $p_\perp = M_{\mu'}/2$ in the transverse momentum ($p_\perp$) spectrum of the e, μ, or π.

To give the reader a rough feeling for the sensitivity of such experiments we have shown in Fig. 15, the expected photon spectrum for a photon beam of maximum energy 400 GeV, produced at a proton accelerator. For a crude calculation of the total rate of production of a heavy lepton of 4.0 GeV/c^2 mass we take an average production cross section of 3×10^{-35} cm^2 from Fig. 14; and we assume that the accelerator will on the average provide 10^{13} protons per pulse and 1000 pulses per hour to the individual experiment. Then in a heavy lepton production target consisting of 1.0 radiation length of beryllium, one would obtain

$$\frac{\text{total number of mass 4.0 GeV/c}^2 \text{ leptons}}{\text{hour}} \sim 100 .$$

Experiments using small solid angle detectors, such as the ones being described in this section, would appear to have a fractional acceptance for each decay particle from the heavy lepton of about 10^{-1} to 10^{-3}. This includes the angular

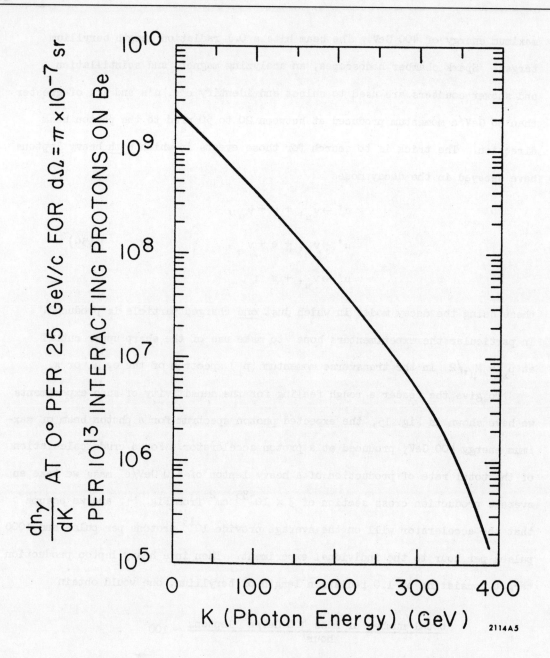

Fig. 15

and momentum acceptance; and the fractional decay into the desired decay mode. Therefore

$$\frac{\text{detected number of mass 4.0 GeV/c}^2 \text{ leptons}}{\text{hour}} \sim 1 \text{ to } 10^{-4}$$

Of course higher mass leptons would have even lower detection rates; and to compensate larger solid angle detectors would have to be used.

D. <u>Heavy Lepton Production in Proton-Proton Collisions</u>

The interactions of the full intensity primary proton beam of a very high energy proton accelerator with a thick target is the most copious source of secondary particles. And among those particles one might hope to find heavy leptons. Also proton-proton colliding beams facilities, although not as potentially a copious a source of heavy leptons, have the highest upper limit on the mass of the lepton which is kinematically allowed. Therefore proton-proton and proton-nucleus reactions are certainly a good place to look for heavy leptons. However the experimenter is plagued by two severe uncertainties.

(a) The production cross section is <u>not</u> known. We might expect heavy lepton pair production through a timelike virtual photon (γ_v) as shown in Fig. 16. But the

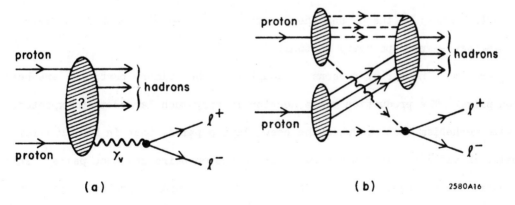

Fig. 16

cross section calculation depends on as yet untested models. An example is provided by the parton model calculations[59-61] using the diagram in Fig. 16b and the Drell-Yan theory.[59] In Fig. 16b the dashed lines are partons or antipartons, and it is the parton-antiparton annihilation which produces the virtual photon. But at present we don't know if parton model calculations are meaningful in proton-proton collisions.

(b) It is not clear if a recently discovered flux of elections and muons from proton-proton and proton-nucleus collisions is <u>related</u> to heavy lepton production or is another processs which will obscure heavy lepton searches. Briefly, experimenters[62-64] at FNAL and at the ISR have found that in the aforementioned collisions there are about 10^{-4} electrons per pion and about 10^{-4} muons per pion at transverse momenta of greater than 1 GeV/c At present we know very little about the production of these leptons. Are they produced in association with their charged antiparticle or in association with their neutrino? If they belong to charged pairs, are these pairs the same phenomena seen earlier at lower energies by Christenson <u>et al.</u>?[65] leptons produced by the mechanism in Fig. 16a, or do they come from the decay of a more primary particle? Are these primary particles something ordinary like vector mesons — the ρ^o has the decay $\rho^o \to \mu^+\mu^-$ — or are these primary something new — perhaps heavy leptons?

There are a number of reasons to doubt that the primary particles are heavy lepton pairs. The production cross section is very much larger than expected. The same mechanism that produces the heavy lepton pairs, that in Fig. 16a for example, should directly produce even more electron pairs and muon pairs. This is because the propagator of γ_v in Fig. 16a is $1/q^2$, and to produce a pair of mass M_ℓ leptons we require $q^2 > 4M_\ell^2$.

- 43 -

However until more is known about these electrons and muons we must reserve judgement as to their origin. We simply don't know if they will be a blessing or a curse to the heavy lepton hunter.

E. Searches for Heavy Excited Leptons Using Charged Lepton-Proton Elastic Scattering

1. The Method

If the vertex of Eq. 4 is assumed then the reactions

$$e + p \rightarrow e^* + p, \quad e^* \rightarrow e + \gamma \quad (60a)$$

or

$$\mu + p \rightarrow \mu^* + p, \quad \mu^* \rightarrow \mu + \gamma \quad (60b)$$

can occur.[66] The cross section depends of course on λ^2. The reaction can be identified by looking for a sharp peak other than the elastic peak in the momentum spectrum of the recoil proton. Therefore this search method does not require any knowledge of the nature of the e^* or μ^*, it only requires the assumption of an electromagnetic coupling of the e or μ to the heavy lepton.

2. Past Searches

Experimenter at Orsay,[67] DESY,[68] and CEA[69,70] have used this method to search for e^* heavy excited leptons. But <u>none</u> have been found. Figure 17a summarizes the limits on λ set by these searches. Also shown in Fig. 17b are the λ limits found in the colliding beam searches[47,48] for the e^* discussed in Sec. V.B2.

H. Gittleson et al.,[71] at the Brookhaven AGS used the reaction in Eq. 60b to look for the μ^*. No μ^* was found in a mass search range of 0 to 2 GeV/c^2; the limits on λ are given in Fig. 17 and to quote these authors[71] "This (experimental result) would appear to rule out the Ramm particle (discussed in Sec. IV.A2.)" The g-2 limits on λ in Fig. 17 come from the measurement discussed in Sec. VI.B.

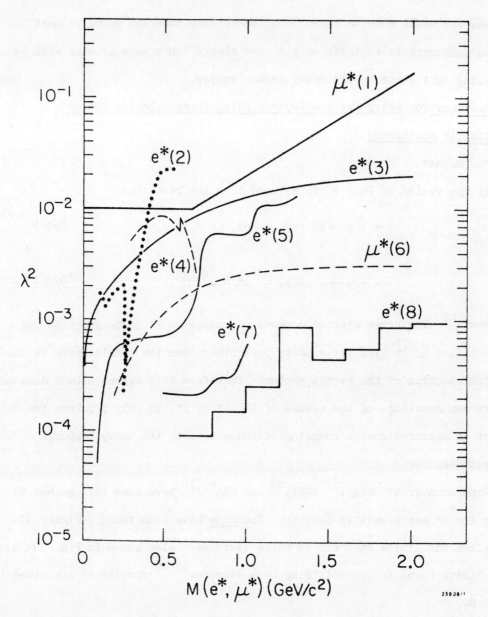

Fig. 17. Comparison of upper limits on the value of λ^2, Eq. 4, for the e* and μ*. References are: 1. H. Gittleson et al., Phys. Rev D., to be published, (μ-p scattering); 2. C. Betourne et al., Phys. Letters 17, 70(1965), (e-p scattering); 3. A. De Rújula and B. Lautrup, Letter Nuovo Cimento 3, 49(1972), (g-2 of electron); 4. R. Budnitz et al., Phys. Rev. 141, 1313(1966), (e-p scattering); 5. C.D. Boley et al., Phys. Rev. 167, 1275(1968), (e-p scattering); 6. A. De Rújula and B. Lautrup, Lettere Nuovo Cimento 3, 49 (1972), (g-2 of muon); 7. H.J. Behrend et al., Phys. Rev. Letters 15, 900 (1965), (e-p scattering); 8. C. Bacci et al., Phys. Letters 44B, 530(1973), (e + e → e + e + γ).

- 45 -

3. Future Searches

Higher energy lepton beams are now available at the new higher energy accelerators. For a primary lepton beam of energy E this search method extends to excited lepton masses $M_{\ell*} \approx \sqrt{2EM_p}$. However, one should remember that the square of the minimum four-momentum-transfer $|q^2|_{min}$ goes as

$$q^2_{min} = \frac{M_{\ell*}^4}{4E^2} \quad . \tag{61}$$

The form factors at the proton vertex will lead to a rapid loss of sensitivity as $|q^2|_{min}$ increases.

F. Searches for Heavy Excited Leptons in Lepton Bremsstrahlung

1. The Method

This search method, which is closely related to the method discussed in the last section, consists simply of the study of the invariant mass $M_{e\gamma}$ in the bremsstrahlung process

$$e + nucleus \rightarrow e + \gamma + nucleus \quad . \tag{62}$$

An exactly similar method can be used to search for the μ* in muon bremsstrahlung.

2. Past Searches

No evidence for the e* has been found in studies of electron bremsstrahlung.[32,72,73] The most extensive study covers the mass range of .1 to 1.2 GeV/c^2. But as pointed out by Lichtenstein,[32] these bremsstrahlung experiments are quite a bit less sensitive for heavy lepton searches compared to the lepton-proton scattering experiments previously descirbed.

3. Future Searches

A very high energy search for the e* in electron bremsstrahlung has been proposed by J.F. Crawford et al.[74] This experiment, to be carried out at FNAL, is designed for electron beams of several hundred GeV energy.

G. **Searches for Heavy Leptons Using Charged Lepton-Proton Inelastic Scattering**

Charged lepton-proton inelastic scattering can be used to search for heavy leptons. However we postpone the discussion of this search method to Sec. X where we discuss lepton-proton inelastic scattering in detail.

H. **Searches for Special Pair Leptons Using Neutrino-Nucleon Inelastic Scattering**

The availability of high energy, high intensity neutrino beams at the new proton accelerators, and the interest in theories uniting weak and electromagnetic which may require heavy leptons (Sec. II.C) have stimulated searches for the E^+, E^0 and M^+, M^0 members of special lepton pairs. Recalling from Sec. II.C that the M^+, M^0 are assigned the lepton numbers of the μ^-, ν_μ, we see that the M^+ or M^0 can be produced through

$$\nu_\mu + \text{nucleon} \to M^+ + \text{hadrons} \qquad (63a)$$

$$\nu_\mu + \text{nucleon} \to M^0 + \text{hadrons} \qquad (63b)$$

The M^+ or M^0 are to be detected through their decay products. For example, the decay $M^+ \to \nu_\mu + \mu^+ + \nu_\mu$ allows the two-step process

$$\nu_\mu + \text{nucleon} \to M^+ + \text{hadrons} \to \mu^+ + 2\nu_\mu + \text{hadrons} \qquad (64a)$$

Here a μ^+ is produced in contrast to the usual one-step inelastic process

$$\nu_\mu + \text{nucleon} \to \mu^- + \text{hadrons} \qquad (64b)$$

where a μ^- is produced. A thorough discussion of the various search methods including those for the E^0 and M^0 has been given by Cline.[75]

There are three advantages of this method, assuming that the weak interaction coupling constants of the M and E are not to different from that of the e. First, the production cross section can be estimated. Second the desired signal is not hopelessly buried in background as it may be in proton-proton production of heavy leptons (Sec. V.D). Third, the probably dominant multi-hadron decay

- 47 -

modes, such as Eq. 37, can be used if the two-step process

$$\nu_\mu + \text{nucleon} \to M^+ + \text{hadrons} \to \nu_\mu + \text{hadrons} \tag{65}$$

can be distinguished from the one-step neutral current interaction

$$\nu_\mu + \text{nucleon} \to \nu_\mu + \text{hadrons} \tag{66}$$

The disadvantage of the method is that it is restricted to looking for special pair leptons which have the lepton numbers of the e or μ.

No such heavy leptons have been found. For example, Barish et al.[76] searched for the M^+ up to masses of 10 GeV/c^2 using ν_μ neutrinos at FNAL. The sensitivity, as we expect, depends upon what one assumes for the M^+ coupling constant and R_{decay} (Eq. 23).

VI. COMPARISON OF SOME STATIC PROPERTIES OF THE MUON AND THE ELECTRON

In this section we compare some of the static properties of the muon and the electron, properties which are particularly relevent to later discussions. We adopt the point of view that the property of the electron is the standard and that the corresponding property of the muon requires comment.

A. Electric Charge

Four properties of the muon — the electric charge e_μ, the mass m_μ, the magnetic moment μ_μ and the gyromagnetic ratio g_μ — are connected by the relation

$$\mu_\mu = \left(\frac{g_\mu}{2}\right)\left(\frac{e_\mu h}{2m_\mu c}\right) \qquad (67)$$

g_μ and μ_μ have been determined with great precision;[1,4,77] .3 parts per million and 12 parts per million respectively. Therefore e_μ can be determined if m_μ is known from an independent measurement. Such a measurement is provided by the study of the mu-mesic atom, an atom in which a negative muon is captured in an atomic orbit.[78] Ignoring relativistic corrections, fine structure and hyperfine structure, the nth energy level of such an atom is given by the Bohr formula

$$E_n = \frac{-m_\mu e_\mu^2 (Ze_p)^2}{2n^2 \hbar^2} \qquad (68)$$

We have distinguished the muon charge e_μ from the charge on the nucleus Ze_p. By measuring the energy difference between levels, m_μ or more precisely the combination $m_\mu e_\mu^2$ can be determined. This measured value of $m_\mu e_\mu^2$ combined with Eq. 67 and the known values of μ_μ, g_μ, $\hbar$, c and e_e (the charge on the electron) yields[4]

$$e_\mu/e_e = 1 \pm (4 \times 10^{-5})$$

- 49 -

But a much lower limit can be obtained[79] by observing that <u>if charge is conserved</u> in the muon decay process

$$\mu \to e + \nu_\mu + \bar{\nu}_e$$

then one or both neutrinos will have a nonzero charge if $e_\mu \neq e_e$. However neutrinos could then be pair produced by low energy photons, leading to an additional mechanism for energy loss in stars! Astrophysical considerations then set an upper limit on the charge that could be possessed by a neutrino. This limit leads to the conclusion that

$$e_\mu/e_e = 1 \pm 1 \times 10^{-13}$$

B. <u>Gyromagnetic Ratio</u>

The gyromagnetic ratio, g_μ, can be calculated exactly from quantum electrodynamics, once the muon mass is known, if strong interactions are ignored. (Fortunately the influence of the strong interactions on g_μ is small; we will give the estimated size of the effect below.) The Dirac relativistic theory of the electron or the muon yields $g = 2$. The Feynman diagram for the interaction of a muon with an external magnetic field (which yields $g = 2$) is given in Fig. 18

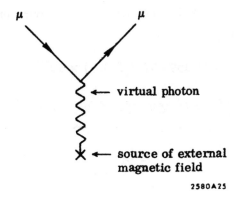

Fig. 18

But quantum electrodynamics shows that there is an anomalous magnetic moment so that g is not exactly 2. It is conventional to set

$$(g_\mu - 2)/2 = a_\mu = \tfrac{1}{2}\left(\frac{\alpha}{\pi}\right) + A_2\left(\frac{\alpha}{\pi}\right)^2 + A_3\left(\frac{\alpha}{\pi}\right)^3 + \ldots$$

The coefficients A_i are all of the order of magnitude of 10 or less so that α_μ is very small. Nevertheless it has been measured to great accuracy. The measurement of a_μ is a measurement of the combined effect of terms like those in Fig. 19.

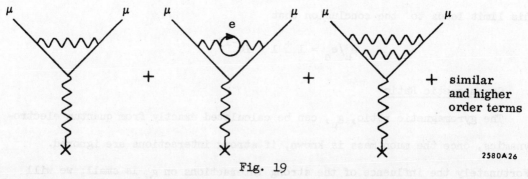

Fig. 19

The most recent results of Farley, Picasso and their colleagues[1] at CERN yield

$$a_\mu^{exp} = (116\ 616 \pm 31) \times 10^{-8}$$

and quantum electrodynamics yields[2,77]

$$a_\mu^{theory} = (116\ 588 \pm 2) \times 10^{-8}$$

Thus experiment and theory are in agreement. Even more precise agreement is found for the electron.[77]

$$a_e^{exp} = (1\ 159\ 657.7 \pm 3.5) \times 10^{-9}$$

$$a_e^{theory} = (1\ 159\ 655.3 \pm 2.5) \times 10^{-9}$$

- 51 -

Therefore with respect to the measurement of g-2, once the mass of the muon is taken into account, there is not observable difference between the muon and the electron.

Incidently, if the M^o special pair lepton exists, the diagram in Fig. 20a will contribute to g_μ, where W is the hypothetical charged

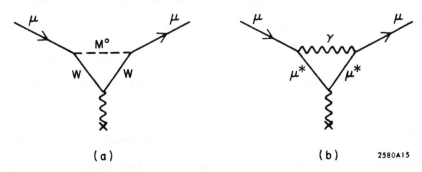

Fig. 20

intermediate boson.[16] Therefore the comparison of g_μ^{exp} with g_μ^{theory} limits the contribution of this diagram.[19] Unfortunately the sensitivity as a function of M^o mass also depends upon assumptions about the W mass and the relevent weak interaction coupling constants.[19] No useful limits on the M^o's existence can be set as yet by this method. A similar remark applies to searching for the E^o by using g_e comparisons.

Turning next to Fig. 18b we see that the existence of a μ^* can contribute to g_μ-2. Similarly there is an e^* contribution to g_e-2. The limits on the corresponding λ's of Eq. 4 are given in Fig. 17. Obviously there is no evidence for an e^* or μ^* from gyromagnetic ratio measurements.[80]

VII. MU-MESIC ATOMS

If the muon is a point Dirac particle the energy levels of the mu-mesic atom[78] will be given by Eq. 68 corrected for fine structure, hyperfine structure, relativistic effects, quantum electrodynamic effects, nuclear size and nuclear charge distribution. All but the nuclear corrections can be calculated from known and accepted theory. The corrections for the size and charge distribution of the nucleus must be determined by experiment. In fact the major purpose of mu-mesic atom experiments is to measure those properties of the nucleus.

By measuring many mu-mesic X-ray lines from a mu-mesic atom, a large amount of interrelated information on the spacing of the energy levels is obtained. Some of this information, particularly that coming from lower energy levels, can be used to derive the relevant nuclear properties. These derived nuclear properties can then be used to calculate the nuclear corrections in the higher energy levels, where those corrections are relatively small. In this way one can attempt a self-consistent calculation of all the mu-mesic X-ray lines. If such a self-consistent calculation cannot be made, the usual hypothesis is that the theory of atomic energy levels contains an error or that the theory of how to derive and correct for the nuclear properties contains an error. If neither of these errors could be found, then one would have to assume that the problem lay with the muon. The muon might not be a point particle or the muon might have an anomalous interaction with the nucleus. Thus high precision measurements of the X-ray lines from mu-mesic atoms provide a test of muon-electron universality.

The search for anomalous effects will be most sensitive if the distance between the nucleus and the muon is relatively small. For if the muon is not a point particle, this will be most evident for small distances. Also any anomalous muon-nucleus interaction is likely to fall off rapidly with distance if, as discussed in Secs. IX, X, and XI, the interaction is strongest at large four-momentum transfers. Thus the search for anomalous effects in mu-mesic atoms is best carried out with high-Z atoms such as lead.

Three high precision measurements of X-rays from high-Z mu-mesic atoms have been made.[81] Two of these experiments, Dixit et al., and Walter et al., find small discrepancies from theoretical predictions[82] — the measured energy difference between two atomic levels being smaller than the theoretical prediction. The discrepancies are very small; for example, about 70 ev for the 438,000 ev energy difference between the $5g_{7/2}$ and $4f_{5/2}$ levels in lead. And the discrepancies are only about 3 standard deviations. A thorough review of the relation of these discrepancies to anomalous muon-hadron interactions has been given by Okun and Zakharov.[83] We don't know how to interpret these possible discrepancies and shall not comment further on them, but this does not mean that they should be ignored.

VIII. HIGH ENERGY REACTIONS OF MUONS AND ELECTRONS

Although the static and atomic properties of the charged leptons show no unexplained differences, one might hope that differences will appear when the dynamic properties of the charged leptons are measured at high energy. For high energies were required to reveal the richness and complexities of the strong interactions. Might not high energies also reveal unsuspected complexities in muon and electron physics? The high energy reactions of the charged leptons may be divided into three classes.

1. One class consists of those reactions in which a neutrino is absorbed or produced. Those reactions as presently measured show no violation of muon-electron universality.[3,84] But the high energy experiments[84] in this class only have precisions of the order of 10 or 20 percent. Therefore we have not yet had stringent tests of muon-electron universality in this class of reactions.

2. Another class consists of purely electromagnetic reactions in which no hadron participates or in which the hadron has only an auxiliary role acting as an almost static source of electric charge. Examples are the colliding beam reactions

$$e^+ + e^- \to e^+ + e^- \quad , \tag{69a}$$

$$e^+ + e^- \to \mu^+ + \mu^- \quad ; \tag{69b}$$

and muon bremsstrahlung

$$\mu + p \to \mu + p + \gamma \tag{70}$$

with very small momentum transfer to the proton. Many of these experiments have been reviewed[2] with respect to tests of quantum electrodynamics. Some have also been reviewed at this conference by M. J. Tannenbaum[85] with respect to the search for muon-electron differences. In particular he discussed a recent experiment[86] confirming that muons, like electrons, obey Fermi-Dirac statistics. Except for some early experiments, all experiments in this class confirm that the charged

- 55 -

leptons are point Dirac particles obeying quantum electrodynamics. Thus all these experiments confirm electron-muon universality in purely electromagnetic reactions. We will review in Sec. IX some of the electron-positron colliding beam experiments which are relevant to the major concerns of this paper.

3. The third class of reactions, those which we shall emphasize in the remainder of this article, consist of reactions in which hadrons play an intimate role. Our interest in this class of reactions has two origins. First, as we shall discuss later, these ractions provide a way to search for spatial structure in the charged leptons; a way to test if the charged leptons are truly point Dirac particles. (Some Class 2 reactions also test for spatial structure.) Second, a speculation which particularly intrigues us is that the leptons may in some very reduced manner take part directly in the strong interactions. After all, the mass difference between the muon and the electron is almost a pion mass and thus could be caused by the strong interactions. To see if the charged leptons in any way directly take part in the strong interactions, it is desirable to have hadrons present — hadrons act as a source for the strong interactions.

In the interaction of muons (or electrons) with protons we can consider two kinds of processes; <u>elastic scattering</u> where

$$\mu + p \to \mu + p$$

and <u>inelastic scattering</u> where

$$\mu + p \to \mu + \text{(any set of 2 or more hadrons)}$$

Examples of inelastic scattering are:

$$\mu + p \to \mu + p + \pi^o$$
$$\mu + p \to \mu + n + \pi^+ + \pi^o$$
$$\mu + p \to \mu + \Sigma^o + K^+$$

In these elastic or inelastic scattering reactions, the charged lepton is not altered in the reaction. This distinguishes these processes from neutrino induced reactions of Class 1.

IX. MUON-PROTON ELASTIC SCATTERING AND FORM FACTORS

A. Theoretical Background

We will consider first elastic scattering, and to set the stage we will discuss electron-proton elastic scattering. To a precision of a few percent all data on electron-proton elastic scattering is explained by the Feynman diagram in Fig. 21.

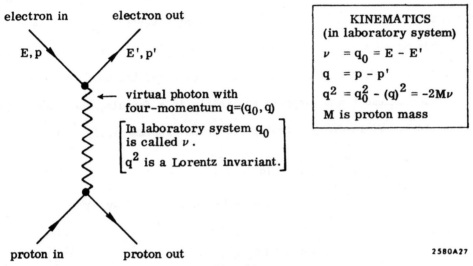

Fig. 21

in which only one photon is exchanged. All experiments agree[87] that the differential cross section for this elastic scattering process is described by the equation

$$\left(\frac{d\sigma}{dq^2}\right)_{ep,elas} = \left(\frac{d\sigma}{dq^2}\right)_{NS} \left[\frac{G_E^2(q^2) + \tau G_M^2(q^2)}{1 + \tau} + 2\tau G_M^2(q^2)\tan^2\frac{\theta}{2}\right] \quad (71)$$

This equation, the Rosenbluth formula, assumes that the electron is a point Dirac particle with only electromagnetic and weak interactions. The equation is written for scattering in the laboratory system, θ is the electron scattering angle and

- 58 -

$\tau = |q^2|/4M^2 \cdot (d\sigma/dq^2)_{NS}$ is the differential cross section for the scattering of an electron by a spin-zero point proton; NS denotes no spin. $(d\sigma/dq^2)_{NS}$ is a function only of the total energy of the system and θ; it is completely specified by quantum electrodynamics. $G_E(q^2)$ and $G_M(q^2)$ are the proton form factors. They take into account that the proton has nonzero spatial extent, and that the proton has strong interactions. If the proton were a point Dirac lepton G_E and G_M would both equal unity for all values of q^2. The crucial variable is q^2, the square of the four-momentum transferred from the lepton vertex. q^2 is always <u>spacelike</u> in this process and in <u>our metric is negative</u>. In this article energy units will always be GeV, momentum units will be GeV/c and the units of q^2 will be $(\text{GeV}/c)^2$. Also unless $\hbar$ and c appear explicitly in a formula, they have both been set equal to 1. We remind you that it is found experimentally[87] that

$$G_E(q^2) \approx 1/[1 + |q^2|/.71]^2 \qquad (72)$$

units are $(\text{GeV}/c)^2$

and

$$G_M(q^2) \approx 2.79\, G_E(q^2) \qquad (73)$$

G_E and G_M are functions of q^2 which is a Lorentz scalar. Thus they express in a relativistically correct way the effects of the hadronic and non-pointlike nature of the proton. When $|q^2|$ is small compared to M^2, we can treat the proton nonrelativistically and provide a simple physical picture of the meaning of these G's.[88,89] For $|q^2| \ll M^2$, $|q^2| \approx |q|^2$ where q is the three-momentum transferred to the proton. Then $G_E(q^2) \approx G_E(|q|^2) = G_E(\underline{q})$ where $G_E(\underline{q})$ is the three-dimensional Fourier transform of the electric distribution. Explicitly

$$G_E(q^2) \approx G_E(\underline{q}) = \int \rho_E(\underline{r}) e^{i\underline{q} \cdot \underline{r}}\, d^3r \qquad (74a)$$

$\rho_E(r)$ is the charge density distribution and is normalized by

$$\int \rho_E(\underline{r})\, d^3r = 1$$

$G_M(q^2)$ can be similarly interpreted. If the proton is contained within a sphere of R, then for $|q| R \ll 1$

$$G_E(q) = 1 - (\tfrac{1}{6})|q|^2 \langle r^2 \rangle_E + (\tfrac{1}{120})|q|^4 \langle r^4 \rangle_E + \ldots \qquad (74b)$$

Here $\langle r^2 \rangle_E$ and $\langle r^4 \rangle_E$ are the average values of r^2 and r^4 respectively over the charge distribution of the proton. It is also possible to expand the function $G_E(q^2)$ in the relativistically invariant form

$$G_E(q^2) = 1 + a_1 q^2 + a_2 (q^2)^2 + \ldots \qquad (74c)$$

But the coefficients in Eq. 74c can be rigorously assigned[89] their corresponding meanings in Eq. 74b only if $|q^2| \ll M^2$.

Now if the muon is a pure Dirac point particle we can use Eq. 71 for muon-proton elastic scattering. There are small effects due to the muon mass, which we have not exhibited explicitly; but these are known. Then

$$\left(\frac{d\sigma}{dq^2}\right)_{\mu p, elas} = \left(\frac{d\sigma}{dq^2}\right)_{ep, elas} \qquad (75)$$

But suppose the muon is not a point particle; suppose the muon, like the proton, has a form factor $G_\mu(q^2)$. Then Eq. 75 becomes

$$\left(\frac{d\sigma}{dq^2}\right)_{\mu p, elas} = \left(\frac{d\sigma}{dq^2}\right)_{ep, elas} G_\mu^2(q^2) \qquad (76)$$

Of course the most general modification[90] of Eq. 75 would require the introduction of two form factors corresponding to G_E and G_M. But our very primitive knowledge of the structure of the muon does not warrant such a refinement. We have no theoretical guidance to what $G_\mu(q^2)$ might be. But the data reviewed in Sec. VI show that with great precision the static properties of the muon are those of a point Dirac particle. Therefore at $q^2 = 0$ we must have $G_\mu(0) = 1$. Conventionally we take a form analogous to the proton form factor and write

$$G_\mu(q^2) = 1/\left[1-q^2/\Lambda_\mu^2\right] . \tag{77a}$$

When q^2 is spacelike, and hence negative in our metric, we write Eq. (77a) in the form

$$G_\mu(q^2) = 1/\left[1+|q^2|/\Lambda_\mu^2\right] . \tag{77b}$$

Note however that unlike Eq. 72, only the first power $\left[1+|q^2|/\Lambda_\mu^2\right]$ appears in the denominator. Λ_μ is a sort of inverse measure of the deviation of the muon from a point particle. The smaller Λ_μ, the greater the deviation. The form of Eq. 77 is actually not as restrictive as it might appear to be. As we shall see later in this paper, all experiments have led to values of Λ_μ^2 which are much larger than the $|q^2|$ values occurring in the experiment. Therefore Eq. (77b) is well approximated by

$$G_\mu(q^2) \approx 1-|q^2|/\Lambda_\mu^2 \tag{77c}$$

Therefore we are actually allowing $G_\mu(q^2)$ to differ from 1 by a term linear in $|q^2|$; this is certainly a simple enough assumption.

Comparing Eq. 77c with Eq. 74b we are tempted to make the identification $\langle r \rangle_\mu/6 = 1/\Lambda_\mu^2$

or

$$\sqrt{\langle r^2 \rangle_\mu} = \sqrt{6/\Lambda_\mu^2} = (.48/\Lambda_\mu) \times 10^{-13} \text{ cm} \tag{78}$$

- 61 -

where Λ_μ is in GeV/c. But this identification is only rigorous[89] if $|q^2| \ll (\text{mass}_\mu)^2$. We shall see that in the high energy experiments $|q^2| \gg (\text{muon mass})^2$. Therefore for these experiments we shall have to be cautious in our interpretation of $1/\Lambda_\mu$ as an indication of the size of the muon.

But a muon might differ in other ways from an electron. There might be a special particle, the X particle, that couples to muons and hadrons but not to electrons. Then muon-proton elastic scattering would be the result of two amplitudes whose diagrams are given in Fig. 22

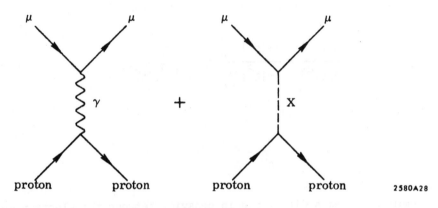

Fig. 22

This would produce some deviation from Eq. 75, but the nature of the deviation cannot be determined because we do not know what X is. Therefore we continue to use $G_\Lambda(q^2)$ in Eq. 76 to express the deviation of muon-proton elastic scattering from electron-proton elastic scattering. In doing so we are making an assumption to which we shall return at the end of the article. We are assuming that the deviation between muon-proton and electron-proton elastic scattering will increase as $|q^2|$ increases.

Nonrelativistic quantum mechanics provides some insight into the relation between the anomalous interaction concept and the form factor concept.[89]

Nonrelativistically the form factor of Eq. 77b is the three-dimensional Fourier transform of

$$\rho_\mu(r) = \frac{1}{4\pi r} e^{-r\Lambda_\mu} \qquad (77d)$$

Thus the discovery of a form factor for the muon of the type of Eq. 77b could also be interpreted as the discovery of an anomalous interaction of the Yukawa form with range $1/\Lambda_\mu$.

In all of this we have assumed that the electron is a pure Dirac point particle. There is <u>no</u> need for this assumption. We can ascribe a form factor $G_e(q^2) = 1/(1+|q^2|/\Lambda_e^2)$ to the electron. Then to order $|q^2|$

$$\frac{G_\mu(q^2)}{G_e(q^2)} = \frac{1+|q^2|/\Lambda_e^2}{1+|q^2|/\Lambda_\mu^2} \approx \frac{1}{1+|q^2|/\Lambda_d^2} \qquad (79)$$

where

$$\frac{1}{\Lambda_d^2} = \frac{1}{\Lambda_\mu^2} - \frac{1}{\Lambda_e^2}$$

Then Λ_d simply measures a difference in behavior between the electron and the muon. From now on we shall use Λ_d. Defining

$$\rho_{elastic}(q^2) = G_\mu^2(q^2)/G_e^2(q^2) = 1/{1+|q^2|/\Lambda_d^2}^2 \qquad (80a)$$

Eq. 76 becomes

$$\rho_{elastic}(q^2) = (d\sigma/q^2)_{\mu p, elas}/(d\sigma/dq^2)_{ep, elas} \qquad (80b)$$

Eq. 80b is not exactly true, there is a slight correction due to the muon mass which is not explictly exhibited.

- 63 -

B. Experimental Results

In the experimental determination of $\rho_{elastic}$, the absolute normalizations of the two cross sections present a special problem because of the very different techniques used to measure the cross sections. The electron cross section measurements use[87] a high intensity beam, short hydrogen target, and small angle spectrometer. The muon experiments use[91-94] a low intensity beam, long hydrogen target and large solid angle detector. To allow for relative normalization uncertainties, Eq. 30a is changed to the form

$$\rho_{elastic}(q^2) = \frac{N^2}{\left[1+|q^2|/\Lambda_d^2\right]^2} \qquad (80c)$$

Table I presents the fits to Eq. (80c) of the three comparisons[91-93] which has been done

TABLE I

Least-squares fit parameters to world data of μ-p elastic scattering. The normalization has been constrained to N = 1.0 ± 0.1. Taken from Ref. 93.

$\langle P_{incident}\rangle$ (GeV/c)	q^2 (GeV/c)^2		N	$1/\Lambda^2$ (GeV/c)$^{-2}$
5.8	Ref. 93	1.02	1.07 ± 0.09	+0.042 ± 0.046
7.3	Ref. 93	1.52	0.97 ± 0.10	+0.034 ± 0.042
5.8	Ref. 93 recoil protons	0.95	1.00 ± 0.10	+0.037 ± 0.066
2.1	Ref. 91	0.62	0.97 ± 0.08	+0.107 ± 0.075
6, 11, 17	Ref. 92	0.26	0.96 ± 0.03	+0.054 ± 0.051

Muon-electron universality requires

$$\rho_{elastic}(q^2) = 1, \text{ for all } q^2 \qquad (81)$$

- 64 -

Hence it requires

$$1/\Lambda^2 = 0 \qquad (82)$$

regardless of normalization problems.

Table I shows that no individual experimental determination yields a statistically significant deviation from Eq. 82. However the combined data[93] from these experiments, Eq. 83, shows some weak evidence for a deviation.

$$1/\Lambda^2 = +0.051 \pm 0.024 \ (\text{GeV}/c)^{-2} \qquad (83)$$

This is not a strong effect — only two standard deviations — certainly not strong enough to seriously challenge muon-electron universality. Still, we should not ignore Eq. 83. We return to it in Sec. XII.

X. COMPARISON OF ELECTRON-PROTON AND MUON-PROTON INELASTIC SCATTERING

A. Theoretical Background

Another very general way to search for anomalous behavior of the electron or the muon is to compare electron-proton and muon-proton inelastic scattering. The kinematics of lepton-proton inelastic scattering are a bit complicated and we shall digress for a moment to discuss the experimental method and kinematics. The relevant kinematics, for the muon case, are shown in Fig. 23 for one-photon-exchange.

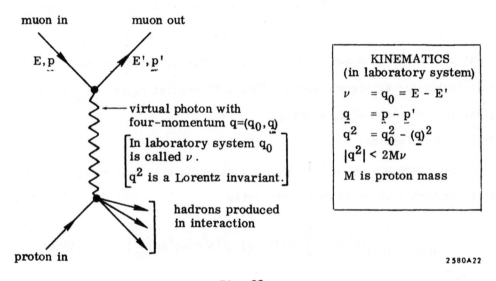

Fig. 23

In the experiments used in the comparison only the inelastically scattered charged lepton is detected. <u>No</u> attempt is made to detect any of the hadrons produced. This inelastic scattering experiment then sums experimentally over the different hadronic states which can be produced. As may be deduced from Fig. 23 the reaction is then completely described by three independent kinematic quantities. For using p, p', and P to represent the four-momentum of the incident lepton, final lepton and incident proton respectively, we can define the four-momentum of the virtual photon

$$q = p-p' \quad (84a)$$

and the three independent Lorentz scalars

$$s = (p+P)^2, \quad q^2 = (p-p')^2, \quad P \cdot q = P \cdot (p-p') \quad (84b)$$

However it turns out to be more convenient to use another set of independent quantities

$$E, \quad q^2, \quad \nu = E-E' \quad (84c)$$

E is the laboratory energy of the incident lepton and ν, the laboratory energy of the virtual photon, represents a Lorentz scalar since $\nu = P \cdot q/M$. The experiment consists of the measurement of the double differential cross section of the inelastically scattered lepton. This differential cross section, $d^2\sigma/dq^2 d\nu$, is a function of E, ν and q^2.

B. Experimental Results

To compare muon-proton (μp) inelastic scattering with electron-proton (ep) inelastic scattering we again define the ratio

$$\rho_{inelastic,p}(q^2,\nu) = \left[d^2\sigma/dq^2 d\nu\right)_{\mu p}/(d^2\sigma/dq^\sigma d\nu)_{ep}\right] \quad (85a)$$

The effect of the muon-electron mass difference, which must be taken into account, is not shown explicitly.

Two medium energy muon-proton inelastic scattering experiments, one at SLAC by Braunstein et al.[95,96] and one at BNL by Entenberg et al.[94,97] have been carried out. No statistically significant deviations from $\rho_{inelastic}(q^2,\nu) = 1$ have been found in any portion of the q^2-ν kinematic region. Furthermore no statistically significant deviation have been found if the data is summed over one of the variables, q^2 or ν, a search being made for a deviation from

$\rho_{inelastic} = 1$ in the other variable.

The same experimenters who measured muon-proton inelastic scattering at BNL,[97] have also measured muon-deuteron inelastic scattering and compared it[98] with electron-deuteron inelastic scattering. They find no statistically significant dependence of $\rho_{inelastic,d}$ on q^2 or on ν; but they do find that the average value of $\rho_{inelastic,\,d}$ is less than 1 by about 2 standard deviations.

Finally, the recent very high energy (53 and 150 GeV) muon-proton inelastic scattering experiment of Fox et al.[99] at FNAL can be used to test muon-electron universality if the inelastic structure functions νW_2 and W_1 are functions only of $W = 2M\nu/|q^2|$ — that is if one assumes Bjorken scaling.[89] Scaling must be assumed because there is no electron-proton inelastic scattering data at that energy. This experiment is a potentially important test of muon-electron universality because $|q^2|$ values up to $50(\text{GeV}/c)^2$ are obtained, whereas the experiments discussed above only reach $|q^2|$ values of 3 or $4(\text{GeV}/c)^2$. However uncertainties about scaling at the 10 or 20% level prevent precise use of this data; and once again one finds no statistically significant violation of muon-electron universality.

To quantify the observations we have just made and to connect the inelastic scattering comparison with the elastic comparison, we define

$$\rho_{inelastic}(q^2,\nu) = N'^2 / \left[1 + |q^2|/\Lambda'^2\right]^2 \qquad (85b)$$

just as in Eq. (80c). But here we average over ν to obtain N' and Λ'. The parameters are primed to indicate that they apply to inelastic scattering. Again $\Lambda'^{-2} = \Lambda'^{-2}_e - \Lambda'^{-2}_e$. Table II gives the values of N' and Λ'_d found in the inelastic experiments.

TABLE II. Fits to Eq. 85b for comparing muon-hadron and electron-hadron inelastic scattering. The 150 GeV/c data assumes Bjorken scaling of the inelastic structure functions.

$\langle P_{incident}\rangle$ GeV/c	target	Ref.	$\lvert q^2\rvert$ range $(\text{GeV/c})^2$	N'	$1/\Lambda'^2$ $(\text{GeV/c})^{-2}$
5.8 and 7.3	p	97	0.4 – 3.6	0.997 ± 0.043	$+0.006 \pm 0.016$
12	p	96		0.946 ± 0.042	$+0.021 \pm 0.021$
150	p	99	6 – 44	$1.10 ^{+0.12}_{-0.06}$	$+0.004 ^{+0.003}_{-0.002}$
7.3	d	98	0.4 – 3.4	0.925 ± 0.038	-0.019 ± 0.016

Hence we find <u>no</u> statistically significant deviation from muon-electron universality in charged lepton-hadron inelastic scattering.

XI. CHARGED LEPTON FORM FACTORS IN COLLIDING BEAM EXPERIMENTS

A. <u>Elastic Electron-Electron Scattering: $e^- + e^- \to e^- + e^-$</u>

This reaction occurs through the diagram in Fig. 24

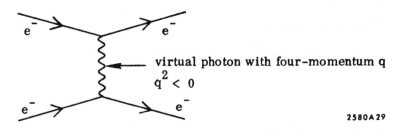

Fig. 24

To obtain larger values of $|q^2|$ it is necessary to use an electron-electron colliding beam apparatus, and such an experiment has been carried out by Barber <u>et al</u>.[100]

If we ascribe an electron form factor

$$g_{es}(q^2) = 1/\left[1-q^2/\lambda_{es}^2\right] \quad , \tag{86a}$$

as in Eq. 77b, the differential cross section predicted by pure quantum electrodynamics will be multiplied by the factor $1/\left[1-q^2/\lambda_{es}^2\right]^4$. The λ_{es} parameter in this form factor is lower case to distinguish it from the Λ_e parameter which occurs in Eq. 79. The latter applies to electron-proton elastic scattering. An anomalous electron-proton interaction could lead to an electron form factor differeing from unity in electron-proton elastic scattering, but would not effect electron-electron scattering. Therefore λ_{es} and Λ_e need not be the same. The s subscript indicates that q^2 is negative and hence spacelike in the diagram in Fig. 24. As we see next, an electron-virtual photon vertex with q^2 positive and hence timelike can also occur; and we allow a different form factor in that case.

Barber et al.[100] find with 95% confidence the lower limit

$$\lambda_{es} > 6.1 \text{ GeV/c} \tag{87}$$

Thus this experiment agrees with the assumption that the electron is a point particle.

B. **Bhabha Scattering:** $e^- + e^+ \to e^- + e^+$

This process takes place through the two diagrams in Fig. 25.

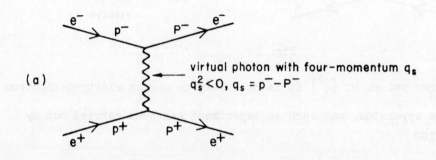

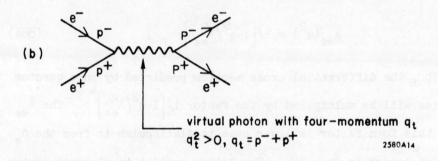

Fig. 25

We could ascribe the space-like form factor g_{es}, Eq. 86a, to each vertex in the diagram in Fig. 25a; and the time-like form factor

$$g_{et}(q^2) = 1/\left[1 - q^2/\lambda_{et}^2\right] \tag{86b}$$

to each vertex in Fig.25b. But Eq. 86b is a bit peculiar because now $q^2 > 0$; hence, we seem to be looking for an enhancement in the cross section, rather than the reduction due to a conventional form factor. It seems best to allow either possibility by defining

$$g^{\mp}_{es} = 1/\left[1 \mp q^2/\lambda^{\mp 2}_{es}\right] \quad (86c)$$

$$g^{\mp}_{et} = 1/\left[1 \mp q^2/\lambda^{\mp 2}_{et}\right] \quad (86d)$$

As we shall see in a moment, we find $\lambda \gg |q^2|$ in all experiments. Thus we are really making a linear fit

$$g \approx 1+cq^2 \quad (86e)$$

with c allowed to be positive or negative.

Many electron-positron colliding beam experiments on Bhabha scattering have been carried out.[101-103] They all agree with the conventional view of the electron as a **point** particle in both the space-like and time-like regions. We give in Table III the lower limits on the λ's obtained in recent experiments at SLAC.[102,103] These being the highest energy and largest statistics experiments yet performed, are the most sensitive.

TABLE III
Lower limits with 95% confidence on $\lambda^{\pm}_{es}$ and $\lambda^{\pm}_{et}$ in Eqs. (86c) and (86d). λ in unites of GeV/c

	Lower limit on					
λ^{+}_{es}	λ^{-}_{es}	λ^{+}_{et}	λ^{-}_{et}	if $\lambda^{+}_{es} = \lambda^{+}_{et}$	if $\lambda^{-}_{es} = \lambda^{-}_{et}$	Ref.
14	14	10	17	16	13	103
				23	14	102

C. <u>Muon Pair Production: $e^- + e^+ \rightarrow \mu^- + \mu^+$</u>.

This process takes place only through the diagram in Fig. 26 and thus involves only time-like form factors. Several colliding beam studies of this reaction have been made.[102-104] They all agree with the assumption that the muon is pointlike in the time-like region.

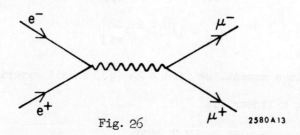

Fig. 26

Using Eq. 86d, only $g^{\pm}_{\mu t}$ occurs in Fig. 26, and allowing different λ's for the muon and electron; experiments[102,103] show

$$\lambda^{\pm}_{\mu s} \gtrsim 15 \text{ GeV/c} \qquad (88)$$

with 95% confidence. Thus the muon also acts like a point particle in the reaction $e^+ + e^- \rightarrow \mu^+ + \mu^-$; and muon-electron universality is maintained in the lepton pair production reaction $e^+ + e^- \rightarrow \ell^+ + \ell^-$.

XII. SPECULATIONS

We have come to the end of this long review of experiments without coming upon any new facts or even new clues as to the nature of the muon and electron or as to their relation to each other. Table I in the Introduction still encompasses all that is known of the charged leptons. Nevertheless physics is an experimental science and experiments seeking new facts or new clues about heavy leptons will go on. Therefore in our view it seems useful to speculate as to what might be the most fruitful experimental directions or at least to give our prejudices as to fruitful directions.

A. Searches for Heavy Leptons

Those who have searched for heavy leptons, the authors included, have to some extent ignored the probable dominance of the ν + hadrons decay modes. We have concentrated on the leptonic and single pion decay modes because we understood how to search for these modes. But as the searches extend to higher lepton masses, the search sensitivities may decrease drastically unless we learn how to detect the multi-hadron decay modes. We need some new experimental ideas in this direction.

B. Charged-Lepton Form Factors

In the original version of this paper written about three years ago we speculated that the muon electromagnetic form factor might deviate from 1. This speculation was based (1) on the earlier muon-proton elastic and inelastic scattering measurements, Refs. 91, 92 and 95, all having N or N' less than 1; and (2) was allowed by the precision of then existing measurements of $\lambda^{\pm}_{es}$, $\lambda^{\pm}_{et}$, and $\lambda^{\pm}_{\mu s}$ in e^+e^- colliding beam experiments. However the recent measurements of these quantities, Sec. XI, yield 95% confidence lower limits of about 15 GeV/c. This is to be compared with the possible values of Λ or Λ' in the elastic or inelastic scattering comparisons (Sec. IX and X) which are of the order of 5 GeV/c.

Therefore it is unreasonable to attempt to attribute any possible anomalies in the elastic or inelastic scattering comparisons to a form factor effect at the virtual photon-lepton vertex.

The new e^+e^- colliding beam measurements of $\lambda^{\pm}_{es}$, $\lambda^{\pm}_{et}$ and $\lambda^{\pm}_{\mu s}$ also surpass in precision the limit on the muon form factor set by the measurement of the muon's gyromagnetic ratio, g_μ (Sec. VI.B). This limit is set through the diagram in Fig. 27

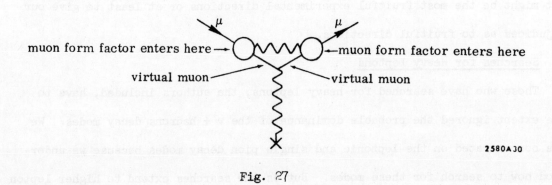

Fig. 27

The q^2 which enters the muon form factor in this diagram comes from the virtual muons, and the important q^2 values are those whose magnitude is smaller than m_μ^2; m_μ is the muon mass. Hence the measurement of g_μ both space-like and time-like regions must be selected. It is conventional to select the form factor given by Eq. 86a. The g_μ measurement requires[1,105] that in Eq. 86a.

$$\lambda_{\mu g} > 7.0 \text{ GeV/c} \qquad (89)$$

with 95% confidence. The $\lambda_{\mu g}$ lower limit in Eq. 89 is less than half the $\lambda^{\pm}_{\mu s}$ given in Eq. 88.

Of course from a very general viewpoint we should not compare a very high precision, very small $|q^2|$ search for a non-unity form factor with a relatively lower precision, very high $|q^2|$ search. A deviation in the form factor from unity

- 75 -

could occur in one q^2 range and not in the other q^2 range. And the formula $N/(1 \pm |q^2|/\lambda^2)$ used to test for deviations has no physical significance until a deviation is found. Therefore it seems best to regard Eqs. 88 and 89 as complementary measurements.

To summarize: all electrodynamic measurements confirm unity form factors for the electron and muon. There is not even a hint of a deviation from the simple concept that the electron and muon and point Dirac particles. Nor is there even a hint of deviation from muon-electron universality in electromagnetic reaction.

C. Anomalous Lepton-Hadron Interactions

Indeed the reactions which show even a hint of a deviation from muon-electron universality are those which take place in the presence of a hadron. There are two such reactions which have been studied:

1. In charged lepton-proton elastic scattering, Sec. IX, there is a two standard deviation effect -- the muon-proton differential cross section decreases more rapidly than the comparable electron-proton cross section as $|q^2|$ increases.

2. In the charged lepton-hadron inelastic scattering comparison, Sec. X, N' averages less than 1 for the three medium energy experiments. The use of the 150 GeV experiment to test muon-electron universality is provisional until possible deviations from scaling are better understood.

Thus we are left with the vague possibility that there is an anomalous electron-hadron or an anomalous muon-hadron interaction or both -- these anomalous interactions resulting in muon-hadron cross sections being slightly less than comparable electron-hadron cross sections. By anomalous we mean that the interaction is neither that described by standard quantum electrodynamics, nor is it a weak interaction.

As a speculative example for this vague possibility suppose that the muon has a special interaction with the hadrons, an interaction *not* possessed by the *electron*. Muon-proton elastic or inelastic scattering would take place through the sum of the two diagrams in Fig. 28

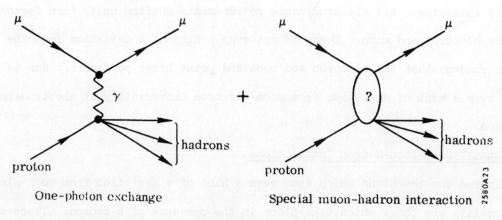

Fig. 28

The second diagram would result in a difference between muon-proton and electron-proton inelastic cross sections, because only the first diagram would enter in electron-proton inelastic scattering.

The problem is how to find this special interaction -- how to enhance the effects of this special interaction. The usual prescriptions are to measure to greater precision or to carry out reactions at greater $|q^2|$ values. But at present we don't know how to improve the precision of the elastic and inelastic comparisons. Systematic errors at the 5% level plague the experiments.

And there is no guarantee that larger $|q^2|$ values will enhance the anomalous effect. For example, assume that the muon interacts with the hadrons through the exchange of particle X with spin 1 and mass M_X as in Fig. 29.

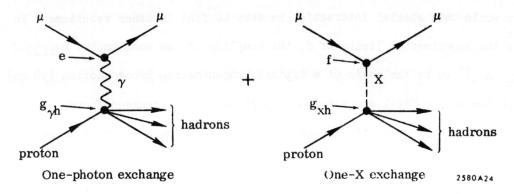

Fig. 29

A conventional speculation [106,107] is that the X particle is some undiscovered heavy photon, but we prefer the speculation that the X particle is itself a <u>hadron</u>. More generally the X particle might be taken to represent the summation of the interaction of different kinds of hadrons with the muon. The coupling constants are indicated in the diagrams; thus e is the electric charge. Those at the lower vertices are to be regarded only as very crude measures of the strength of the coupling of the virtual photon or the X particle to hadrons. Then[106]

$$\rho_{elastic} \text{ or } \rho_{inelastic} \approx \left[1+a \left(\frac{|q^2|}{|q^2| + M_x^2} \right) \right] \quad (90a)$$

$$a = \left(\frac{f}{e}\right)\left(\frac{g_{xn}}{g_{\gamma n}}\right) \quad (90b)$$

Suppose that b ~ -0.05 and M_x ~ 5 GeV/c². Then we can get the deviations possibly seen in the elastic and inelastic comparisons; yet even for $|q^2| \gg M_x^2$ we only get a 10% effect.

- 78 -

Nor would this special interaction be easy to find in other reactions. To estimate the experimental limits of f, the coupling of the muon to the hadron X, take $(g_{xh}/g_{\gamma h})^2$ to be the ratio of a typical hadron-hadron cross section (30 mb) to the photon-proton total cross section (0.12 mb). As discussed above, exisiting muon-proton scattering measurements easily allow b to be as large as .05. Then

$$|f/e| \lesssim 0.05/\sqrt{250} \lesssim 1/300$$

Thus in this "X=hadron" model, the coupling of the muon to the hadrons is much weaker than the electromagnetic coupling. The contribution of this special interaction to electromagnetic processes such as $e^+ + e^- \to \mu^+ + \mu^-$ would be 10^{-7} of the one-photon exchange process, Fig. 26.

Returning to the general example of Fig. 28, or to the converse example of a special electron-hadron interaction, how then is one to isolate this interaction? What is needed is a pair of reactions in which the one-photon exchange process is suppressed. Perhaps such a reaction pair is

$$p + p \to \mu^+ + \mu^- + \text{hadrons} \quad , \qquad (91a)$$
$$p + p \to e^+ + e^- + \text{hadrons} \quad , \qquad (91b)$$

the reactions discussed in Sec. V.D as a background source. Very preliminary data[62-64] indicates that the muon-electron ratio is roughly one. Still for the immediate future the reactions in Eq. 91 are perhaps the best comparative experimental way to search for anomalous interactions of charged leptons with hadrons.

We emphasized the phase comparative experimental in the last sentence because there has been a good deal of speculation[108] recently as to whether the large $e^+ + e^- \to$ hadrons cross section -- large compared to simple parton model predictions (Secs. III.A and V.B) -- indicated an anomalous electron-hadron interaction. This speculation is that in addition to the one-photon exchange diagram of Fig. 30a there is a special electron-hadron interaction which leads to the diagram in Fig. 30b.

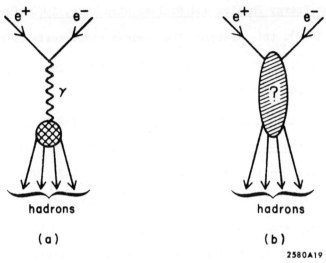

Fig. 30

However this is not a <u>comparative experimental</u> search for a special electron-hadron interaction because we have no sure theory for calculating the cross hatched photon-hadron vertex in Fig. 30a. It is only the use of a simple parton model for calculating that vertex which leads to the invoking of the diagram in Fig. 30b. It may be that the high energy reactions

$$e^+ + e^- \rightarrow \text{hadrons}$$

will turn out to be the best way to study anomalous charged lepton-hadron interactions; but first we have to convince ourselves that we cannot explain the cross sections for these reactions using conventional theory.

REFERENCES

1. E. Picasso in *High-Energy Physics and Nuclear Structure, 1970* (Plenum Press, New York, 1970), this article also reviews other tests of muon-electron universality; J. Bailey et al., Nuovo Cimento, $\underline{9A}$, 369(1972) J. Bailey et al., Phys. Letters $\underline{28B}$, 287(1968).

2. S.J. Brodsky in *Proceedings of the 4th International Symposium on Electron and Photon Interactions at High Energy*, Daresbury, England (1969); S.J. Brodsky and S.D. Drell, Ann. Rev. Nucl. Sci. $\underline{20}$, 147(1970). These articles review quantum electrodynamic tests of muon-electron universality.

3. T.D. Lee and C.S. Wu, Ann. Rev. Nucl. Sci. $\underline{15}$, 381(1965). This article reviews weak interaction tests of muon-electron universality.

4. G. Feinberg and L.M. Lederman, Ann. Rev. Nucl. Sci. $\underline{13}$, 431(1963).

5. F.E. Low, Phys. Rev. Letters $\underline{14}$, 238(1965).

6. A.O. Barut et al., Phys. Rev. $\underline{182}$, 1844(1969).

7. The notation used here is that given in J.D. Bjorken and S. Drell, *Relativistic Quantum Mechanics* (McGraw-Hill, New York, 1964) or in M.L. Perl, *High Energy Hadron Physics* (Wiley, New York, 1974).

8. E. Lipmanov, Zh. Eksp. Teor. Fiz. $\underline{43}$, 893(1962) (Sov. Phys. JETP $\underline{16}$, 634 (1963)).

9. Ya. B. Zel'Dovich, Usp. Fiz. Nauk. $\underline{78}$, 549(1962) (Sov. Phys. Usp. $\underline{5}$, 931(1963)).

10. S.S. Gerstein and V.N. Folomeshkin, Yad. Fiz. $\underline{8}$, 768(1968) (Sov. J. Nucl. Phys. $\underline{8}$, 447(1969)).

11. B. Pontecorvo, Zh. Eksp. Teor. Fiz. $\underline{53}$, 1717(1967) (Sov. Phys. JETP $\underline{26}$, 984(1968)).

12. R.E. Marshak, Riazuddin, and C.P. Ryan, *Theory of Weak Interactions in Particle Physics* (Interscience, New York, 1969).

13. E.J. Konopinski and H.M. Mahmoud, Phys. Rev. $\underline{92}$, 1045(1953).

14. L.S. Kisslinger, Phys. Rev. Letters $\underline{26}$, 998(1971).

15. D.A. Bryman et al., Phys. Rev. Letters $\underline{28}$, 1469(1972).

16. It would be pointless to attempt to list here even the most important papers in the vast literature on these theories. Recent review papers with extensive lists of references are:

    E.S. Abers and B.W. Lee, Phys. Reports $\underline{9C}$, 1(1973)

    M.A.B. Beg and S. Sirlin, Gauge Theories of Weak Interactions, Rockefeller Univ. Report No. COO-2232B-47 (1974), to be published in the Annual Review of Nuclear Science, Vol. 24.

17. Two papers of particular value in summarizing the position of heavy leptons in these theories are the J.D. Bjorken and C.H. Llewellyn Smith, Phys. Rev. $\underline{D7}$, 88(9773) and C.H. Llewellyn Smith in Proceedings of the Fifth Hawaii Topical Conference in Particle Physics (1973) Vol. 1 (Univ. of Hawaii Press, Honolulu, 1974).

18. C.H. Albright, C. Jarlskg, M.O. Tjia, CERN preprint TH-1887 (1974), as well as the first paper of Ref. 17 provide an extensive list of theoretical models and their associated heavy leptons. We chose to discuss only the more conventional models that include only the pairs mentioned in the text.

19. J.R. Primack and H.R. Quinn, Phys. Rev. $\underline{D6}$, 3171(1972).

20. S.S. Gershtein et al., Institute of High Energy Physics Report No. IHEP 72-115.

21. A. Barna et al., Phys. Rev. $\underline{173}$, 1391(1968); M.L. Perl, Proceedings of the 1967 International Symposium on Electron and Photon Interactions at High Energies (Stanford Linear Accelerator Center, Stanford, 1967).

22. See for example: B. Pontecorvo, Zh. ETF. Pis. Red. $\underline{13}$, 218(1971) (Sov. Phys. JETP Letters $\underline{13}$, 199(1971)).

23. Y.S. Tsai, Phys. Rev. $\underline{D4}$, 2821(1971).

24. E.M. Lipmanov, Zh. Eksp. Teor. Fiz. 46, 1917(1964) (Sov. Phys. JETP 14, 859(1962)).

25. E.W. Beier, Lettere Nuovo Cimento 1, 1118 (1971).

26. A.K. Mann, Lettere Nuovo Cimento 1, 486(1971).

27. J.J. Sakurai, Lettere Nuovo Cimento 1, 624(1971).

28. H.B. Thacker and J.J. Sakurai, Phys. Letters 36B, 103 (1971).

29. K.W. Rothe and A.M. Wolsky, Nucl. Phys. 10B, 241(1969).

30. G. Tarnopolsky et al., Phys. Rev. Letters 32, 432(1974).

31. B. Richter Proceedings of the 17th International Conference on High Energy Physics, 1974 (to be published).

32. C.A. Lichtenstein, Thesis, Harvard University (1970), unpublished.

33. E.W. Beier, Thesis, University of Illinois (1966), unpublished.

34. C.A. Ramm, Nature 227, 1323(1970); Nature Physical Science 230, 145(1971); Addendums 2 through 6 to CERN Report No. NPA 69-6 (1971-1972).

35. A.D. Liberman et al., Phys. Rev. Letters 22, 663(1969).

36. A.D. Liberman, Thesis, Harvard University (1969), unpublished. In this thesis Liberman notes the evidence for the $\mu^{*\pm}$ in $\mu^{\pm}\gamma$ mass spectrum, but comments that its statistical significance is only 2 standard deviations.

37. A.R. Clark, Nature 237, 388(1972).

38. D.H. Coward et al., Phys. Rev. 131 1782(1963).

39. This search was only definitive up to a mass of about .4 GeV/c^2. For higher masses the lifetime of the heavy sequential lepton would be too short.

40. R.E. Ansorge et al., Phys. Rev. D7, 26(1973).

41. D.E. Dorfan et al., Phys. Rev. Letters 14, 999(1965), P. Franzini et al., Phys. Rev. Letters 14, 196(1965).

42. B. Alper et al., Experiments on High Energy Particle Collisions, (American Institute of Physics, New York, 1973).

43. J.W. Cronin et al., Enrico Fermi Institute Preprint EFI-74-29 (1974).

44. S. Frankel et al., Phys. Rev. $\underline{D9}$, 1932(1974).

45. V. Alles-Borelli et al., Lettre Nuovo Cimento $\underline{4}$, 1156(1970); M. Bernardini et al., Nuovo Cimento $\underline{17}$, 383(1973).

46. J. Feller et al., paper presented to the 17th International Conference on High Energy Physics, London, 1974).

47. C. Bacci et al., Phys. Letters $\underline{44B}$, 530(1973).

48. G. Hanson et al., Lettere Nuovo Cimento $\underline{7}$, 587(1973).

49. M.L. Perl, Bull. Amer. Phys. Soc. $\underline{19}$, 542(1973).

50. The luminosity of a colliding beam facility is the quantity which when multiplied by a reaction cross section gives the number of events produced by that reaction per unit time. The maximum depends on the beam energy -- we are using a convenient average value here.

51. J. Kirkby et al., SLAC Proposal SP-18 (1974), unpublished.

52. See for example the papers presented by A. Ali and by A. Soni to the 17th International Conference on High Energy Physics -- London, 1974.

53. K.J. Kim and Y.S. Tsai, Phys. Letters $\underline{40B}$, 665(1972).

54. Some classic references to photoproduction of particle pairs are W. Pauli and V.F. Weisskopf, Helv. Phys. Acta $\underline{7}$, 709(1934); and W. Heitler, Quantum Theory of Radiation (Clarendon Press, Oxford, England, 1954).

55. A. Rothenberg, Thesis, Stanford University (1962), unpublished.

56. C.A. Heusch and J. Sandweiss, 1969 Summer Study for National Accelerator Laboratory, Vol. 4, edited by A. Roberts (national Accelerator Laboratory, Batavia, Illinois, 1970), p. 111.

57. M. Schwartz, Reports on Progress in Physics $\underline{27}$, 61(1965).

58. W. Lee et al., Proposal No. 87A to the National Accelerator Laboratory (1970), unpublished.

59. S.D. Drell and T.M. Yan, Phys. Rev. Letters $\underline{25}$, 316(1970) and references contained therein; Ann. Phys. N.Y. $\underline{66}$, 578(1971).
60. S.M. Berman et al., Phys. Rev. $\underline{D4}$, 3388(1971).
61. G. Ghiu and J.F. Gunion, LBL Report No. LBL-3020 (1974), submitted to Phys. Rev.
62. J.A. Appel et al., Phys. Rev. Letters $\underline{33}$, 722(1974).
63. J.P. Boymond et al., Phys. Rev. Letters $\underline{33}$, 112(1974).
64. F.W. Busser et al., Proceedings of the 17th International Conference on High Energy Physics, 1974 (to be published).
65. J.H. Christenson et al., Phys. Rev. Letters $\underline{25}$, 1523(1970).
66. For a specific calculation of this process and an analysis of the search results, see F. Gutbrod and D. Schildsnecht, Zeit. Phys. $\underline{192}$, 271(1966).
67. C. Betourne et al., Phys. Letters $\underline{17}$. 70(1965).
68. H.J. Behrend, Phys. Rev. Letters $\underline{15}$, 900(1965).
69. R. Budnitz et al., Phys. Rev. $\underline{141}$, 1313(1966).
70. C.D. Boley et al., Phys. Rev. $\underline{167}$, 1275(1968).
71. H. Gittleson et al., Phys. Rev. D (to be published).
72. C. Bernadini et al., Lettere Nuovo Cimento $\underline{1}$, 15(1969).
73. C.A. Lichtenstein et al., Phys. Rev. D1, 825(1970).
74. J.F. Crawford et al., Proposal No. 164 to the National Accelerator Laboratory (1971), unpublished.
75. D.B. Cline, Proceedings of the IVth International Conference on Neutrino Physics and Astrophysics, Philadelphia, 1974, to be published.
76. B.C. Barish et al., Phys. Rev. Letters 32, 1387(1974).
77. A. Rich and J.C. Wesley, Rev. Mod. Phys. $\underline{44}$, 250(1972).
78. C.S. Wu and L. Wilets, Ann. Rev. Sci. $\underline{19}$, 527(1969).
79. J. Bernstein, M. Ruderman and G. Feinberg, Phys. Rev. $\underline{132}$, 1227(1963).

80. A. DeRujula and B. Lautrup, Nuovo Cimento 3, 49(1972).

81. G. Backenstoss et al., Phys. Letters 31B, 233(1970); M.S. Dixit et al., Phys. Rev. Letters 27, 878(1971); H.K. Walter et al., Phys. Letters 40B, 197(1972).

82. Recent references and G.A. Rinker, Jr. and L. Wilets, Phys. Rev. Letters 31, 1559(1973); J. Arafune, Phys. Rev. Letters 32, 560(1974); L.S. Brown et al., Phys. Rev. Letters 32, 562(1974).

83. L.B. Okun and V.I. Zakharov, Nuclear Phys. B57, 252(1973).

84. For recent summaries see Proceedings of the 17th International Conference on High Energy Physics, 1974 (to be published).

85. M.J. Tannenbaum, paper presented to the Muon Physics Conference, Colorado State University, 1971.

86. J.J. Russell et al., Phys. Rev. Letters 26, 46(1971).

87. J.G. Rutherglen in Proceedings of the 4th International Symposium on Electron and Photon Interactions at High Energy. (Daresbury Nuclear Physics Laboratory, Daresbury, England, 1969).

88. R. Hofstadter, Ann. Rev. Nucl. Sci. 7, (1957).

89. M.L. Perl, High Energy Hadron Physics (Wiley, New York, 1974), Chapters 19 and 20.

90. K.J. Barnes, Nuovo Cimento 27, 228(1963).

91. R.W. Ellsworth et al., Phys. Rev. 165, 1449(1968).

92. L. Camilleri et al., Phys. Rev. Letters 23, 153(1969); L. Camilleri, Columbia University Nevis Laboratories Report No. 176.

93. I. Kostoulas et al., Phys. Rev. Letters 32, 489(1974).

94. A.B. Entenberg, Ph.D. Thesis, Univ. of Rochester Report No. UR-476 (1974), unpublished.

95. T.J. Braunstein et al., Phys. Rev. D6, 106(1972).

96. M.L. Perl, Physics Today, July, 34(1971).

97. A. Entenberg et al., Phys. Rev. Letters 32, 486(1974).

98. I.J. Kim et al., Phys. Rev. Letters 33, 551(1974).

99. D.J. Fox et al., Phys. Rev. Letters (to be published).

100. W.C. Barber et al., Phys. Rev. D3, 2796(1971).

101. H. Newman et al., Phys. Rev. Letters 32, 483(1974); V. Silvestrini, Proceedings of the 16th International Conference on High Energy Physics (National Accelerator Laboratory, Batavia, 1973).

102. B.L. Beron et al., Phys. Rev. Letters 33, 663(1974).

103. A.M. Boyarski, Bull. Am. Phys. Soc. 19, 542(1973); B. Richter, Proceedings of the 17th International Conference on High Energy Physics (London, 1974).

104. B. Borgia et al., Lettere Nuovo Cimento 3, 115(1972); V. Alles-Borelli et al., Nuovo Cimento 7A, 330(1972).

105. F.J.M. Farley, unpublished paper entitled, The Status of Quantum Electrodynamics (Royal Military College of Science, Shrivenham, Swindon, England, 1969).

106. For the application of this equation to muon-proton elastic scattering see D. Kiang and S.H. Ng, Phys. Rev. D2, 1964(1970) and the references contained in that paper.

107. L. Yu. Kobzarev and L.B. Okun', Soviet Physics JETP 14, 859(1962); J. Exptl. Theoret. Phys. (U.S.S.R.) 41, 1205(1961).

108. B. Richter, Invited talk at the Irvine Conference, December 1973 (unpublished); I.I.Yi Bigi and and J.D. Bjorken, Phys. Rev. (to be published); M.A.B. Beg and G. Feinberg, Phys. Rev. Letters 33, 606(1974); J.C. Pate and A. Salan, Phys. Rev. Letters 32, 1083(1974).

LECTURES ON ELECTRON-POSITRON ANNIHILATION -- PART II

ANOMALOUS LEPTON PRODUCTION

Martin L. Perl

Stanford Linear Accelerator Center
Stanford University
Stanford, Calif.

This is the second part of a series of tutorial lectures delivered at the Institute of Particle Physics Summer School, McGill University, June 16-21, 1975.

The two parts have separately numbered equations and references. Referrals to Part I are prefixed by a I.

Table of Contents -- Part II

1. Introduction — 2
   A. Heavy Leptons
   B. Heavy Mesons
   C. Intermediate Boson
   D. Other Elementary Bosons
   E. Other Interpretations

2. Experimental Method — 6

3. Search Method and Event Selection — 8
   A. The 4.8 GeV Sample
   B. Event Selection

4. Backgrounds — 11
   A. External Determination
   B. Internal Determination

5. Properites of eμ Events — 15

6. Cross Sections of eμ Events — 17

7. Hypothesis Tests and Remarks — 18
   A. Momenta Spectra
   B. θ_{coll} Distribution
   C. Cross Sections and Decay Ratios

8. Compatibility of ee and μe Events — 27

9. Conclusions — 29

ANOMALOUS LEPTON PRODUCTION

1. INTRODUCTION

In this second part I shall continue the informal style of the earlier lectures. The data analyzed here all comes from the LBL-SLAC Magnetic Detector Collaboration (Ref. I1). Unlike the data presented in Part I, this data is not yet published. However, because of the interest in this work and its possible significance I believe it is worthwhile to present the data and my analysis of that data even though that analysis is still in progress. Any errors of fact or analysis in this presentation are my responsibility.

Most of this lecture concerns evidence for events of the form.

$$e^+ + e^- \to e^{\pm} + \mu^{\mp} + \text{missing momentum} \quad (1.1)$$

in which no other particles are detected. Other anomalous lepton production processes such as

$$e^+ + e^- \to e^{\pm} + \text{hadrons} \quad (1.2)$$

or

$$e^+ + e^- \to \mu^{\pm} + \text{hadrons} \quad (1.3)$$

are discussed briefly in Sec. 7.

Anomalous lepton production in $e^+ - e^-$ annihilation might occur if various types of hypothetical particles exist. I consider some examples.

1.A Heavy Leptons

Suppose the electron ($e^{\pm}$) and muon ($\mu^{\pm}$) are the lowest mass members of a sequence of leptons,[1-3] each lepton ($\ell^{\pm}$) having a unique quantum number n_ℓ and a unique associated neutrino (ν_ℓ). Such sequential[1] heavy leptons have the purely leptonic decay modes:

$$\ell^- \to \nu_\ell + e^- + \bar{\nu}_e , \qquad (1.4a)$$

$$\ell^- \to \nu_\ell + \mu^- + \bar{\nu}_\mu ; \qquad (1.4b)$$

assuming the quantum number n_ℓ must be conserved as are n_μ and n_e. (The ℓ^+ has corresponding decay modes.) If the ℓ has a sufficiently large mass it will also have semileptonic decay modes.

$$\ell^- \to \nu_\ell + \pi^- \qquad (1.5a)$$

$$\ell^- \to \nu_\ell + K^- \qquad (1.5b)$$

$$\ell^- \to \nu_\ell + \rho^- \qquad (1.5c)$$

$$\ell^- \to \nu_\ell + 2 \text{ or more hadrons} \qquad (1.5d)$$

Another example is provided by charged heavy leptons associated with unified guage theories of electomagnetic and weak interactions[2,4] These leptons have purely leptonic decay modes such as

$$E^+ \to e^+ + 2\nu_e \qquad (1.6)$$

The reaction in Eq. 1.1 could come from lepton pair production processes such as

$$e^+ + e^- \to \ell^+ + \ell^- \text{ or } e^+ + e^- \to E^+ + E^- \qquad (1.7)$$

For convenience I shall always take the heavy lepton to have spin 1/2 although higher half integral spins are possible.

1.B Heavy Mesons

If new charged mesons, $M^\pm$, exist which have relatively large leptonic decay modes (due to the inhibition of purely hadronic decay modes) then the purely leptonic decay modes.

$$M^- \to e^- + \bar{\nu}_e \qquad (1.8a)$$

$$M^- \to \mu^- + \bar{\nu}_\mu \qquad (1.8b)$$

can lead to the reaction in Eq. 1.1, thru

$$e^+ + e^- \to M^+ + M^- \qquad (1.9)$$

Such charged mesons are predicted by theories which introduce the charmed quark.[5,6] Of course, in an experimental search we need not restrict the interpretation of Eqs. 1.8 and 1.9 to a particular theory -- indeed we shall not a priori restrict the mass or spin of M in this discussion.

1.C Intermediate Boson

Although the mass[7] of the intermediate boson, which is supposed to mediate the weak interaction ($W^\pm$), if it exists, is probably too high to allow pair production

$$e^+ + e^- \to W^+ + W^- \qquad (1.10)$$

at the energies discussed in this paper; the decay mode

$$W^- \to e^- + \bar{\nu}_e \qquad (1.11a)$$

$$W^- \to \mu^- + \bar{\nu}_\mu \qquad (1.11b)$$

can lead to the reaction of Eq. 1.1.

1.D Other Elementary Bosons

We may also consider other types of elementary bosons -- not necessarily the intermediate boson W. The difference between an elementary boson and a heavy meson is that we suppose the former to be a point particle with a form factor always equal to unity. As we shall discuss briefly in Sec. 7, the heavy meson does have a form factor. We use B to denote all elementary boson including the W. The B is assumed to have the purely leptonic decay modes

$$B^- \to e^- + \bar{\nu}_e \tag{1.12a}$$

$$B^- \to \mu^- + \bar{\nu}_\mu \tag{1.12b}$$

It may also have hadronic decay modes

$$B \to \text{hadrons} \tag{1.13}$$

1.E Other Interpretations

The signal

$$e^+ + e^- \to e^\pm + \mu^\mp + \text{missing momentum}$$

need not come from the production of a pair of particles purely leptonic decay modes. One can consider a resonance (R) with the weak decay mode

$$R \to e^+ + \nu_e + \mu^- + \bar{\nu}_\mu \tag{1.14}$$

Or one can think about the higher order weak interaction process

$$e^+ + e^- \to e^+ + \nu_e + \bar{\mu} + \bar{\nu}_\mu \tag{1.13}$$

However the observed cross sections -- of the order of .01 to .02 nb (Sec. 6) -- appears to be much too large for this conjecture.

- 6 -

2. EXPERIMENTAL METHOD

The magnetic detector (Fig. 1) used in the search has cylindrical symmetry about the beam axis. A 4 kg magnetic field is produced by a coil of radius 1.65 m and length 3.6 m. Most of the space inside the coil, that is the magnetic field region, is occupied by cylindrical magnetostrictive spark chambers. The azimuthal angle, θ, subtended by these chambers extends from $50°$ to $130°$ relative to the $e^{\pm}$ beam direction. The full cylindrical angle of 2π is covered. Just inside the coil are 48, 2.6 m long, scintillation counters, and just outside the coil are 24, 3.1 m long, lead plastic-scintillator shower counters. The scintillation and shower counters cover the full 2π cylindrical angle. Outside the shower counters is the iron magnetic flux return which is 20 cm thick. As shown in Fig. 1 the flux return consists of eight iron plates forming an octagon. Finally on the outside of each of these plates are two single-plane magnetostrictive spark chambers referred to, as the muon detection chambers. Most hadrons are absorbed by the iron of the flux return and do not reach the muon chambers.

Electrons are identified by requiring a large pulse in the shower counters. Quantitatively, an $e^{\pm}$ is required to have a pulse height greater than 50. on a scale for which a 1.5 GeV/c $e^{\pm}$ produces an average pulse height of 131. 97% of all 1.5 GeV/c $e^{\pm}$ have pulse heights above 50. Muons are identified by two requirements. First, the $\mu^{\pm}$ must produce a spark in at least one of the two muon chambers. Since the $\mu^{\pm}$ multiple scatters in the iron, some allowance must be made for a deviation of the sparks position from the extrapolated muon track. Up to 4 standard deviations is allowed. Actually 96% of all $\mu^{\pm}$ give sparks in both muon chambers, as measured by events from the reaction $e^+ + e^- \rightarrow \mu^+ + \mu^-$. The second requirement for $\mu^\pm$ identification is that the shower counter pulse height be less than 50.. Indeed, as measured by $e^+ + e^- \rightarrow \mu^+ + \mu^-$ events, the average $\mu^\pm$ pulse height is 13.. The crucial question of the probability of a hadron being misidentified as an $e^\pm$ or $\mu^\pm$ will be taken up later.

The shower counters also detects photons (γ). For γ energies above 200 MeV, their photon detection efficiency is about 95%. Between 150 and 50 MeV, the detection efficiency decreases and becomes dependent on the longitudinal position of the γ in the counters.

3. SEARCH METHOD AND EVENT SELECTION

3.A The 4.8 GeV Sample

To illustrate the method of searching for the reaction $e^+ + e^- \to e^{\pm} + \mu^{\mp} +$ missing momentum, and to provide specific information on the events selection criteria, I shall consider our largest statistical sample -- data taken at a total energy (W) of 4.8 GeV. W is given by

$$W = 2E_{beam} \qquad (3.1)$$

where E_{beam} is the energy of the e^+ or e^- beam. About 80% of the 4.8 GeV data presented here was taken at the full magnetic field of 4.0 kg. The remaining 20% was taken at 2.0 or 2.4 kg. The "full" field and "half" field data are consistent.

To give you a feeling for the 4.8 GeV sample, it contains

$$22{,}600 \text{ collinear } e^+ e^- \text{ pairs} \qquad (3.2a)$$

from the reaction

$$e^+ + e^- \to e^+ + e^- \qquad (3.2b)$$

and

$$1{,}700 \text{ collinear } \mu^+ \mu^- \text{ pairs} \qquad (3.3a)$$

from the reaction

$$e^+ + e^- \to \mu^+ + \mu^- \qquad (3.3b)$$

There are

$$9{,}550 \text{ 3-or-more-prong hadronic events} \qquad (3.4)$$

which are the primary source for our studies of hadron production in $e^+ - e^-$ annihilation. (A prong is a charged track in the detector which comes from a vertex.)

To study 2-prong hadronic events

$$e^+ + e^- \rightarrow 1 + 2 \text{ (no other charged tracks but } \gamma\text{'s allowed)}, \quad (3.5)$$

or other 2-prong events, we define a coplanarity angle (θ_{copl})

$$\cos \theta_{copl} = -(\underline{n}_1 \times \underline{n}_{e^+}) \cdot (\underline{n}_2 \times \underline{n}_{e^+}) \quad (3.6)$$

$\underline{n}_1$, $\underline{n}_e$, $\underline{n}_\ell$ are unit vectors along the direction of particles 1, 2, and e^+. For a coplanar event $\cos \theta = 1$, $\theta = 0$. The contamination of events from the reactions in Eqs. 3.2b and 3.3b is greatly reduced if we require

$$20^\circ < \theta_{copl} \quad (3.7)$$

This provides a class of 2493 events at 4.8 GeV. And it is in this class that we search for the μ-e signature.

3.B **Event Selection**

To penetrate the iron plates, Fig. 1, a particle must have a momentum greater than about 0.55 GeV/c. Therefore muons can only be identified at higher momenta. Also electrons of momentum below 0.5 GeV/c will be misidentified as pions more than half the time. Therefore to select e-μ events we require that the momenta of particle 1 (p_1) and of particle 2 (p_2) each be greater than 0.65 GeV/c. This reduces the 2030 events to the 513 shown in Table I. Thus the selection criteria for Table I are

(1) 2-prongs
(2) $\theta_{copl} > 20^\circ$
(3) $p_1 > 0.65$ GeV/c and $p_2 > 0.65$ GeV/c.

In Table I the events are classified according to

(1) Total charge (Q) : 0, ± 2

(2) Number of photons associated with event : 0, 1, or > 1

(3) The charged particle nature e, μ, or h (for hadron). Any particle not an e or a μ is called an h.

We make the following observations

(1) There are very few $Q = \pm 2$ events and we focus our attention on the $Q = 0$ events.

(2) If there were no particle misidentification, no decays in flight, and no anomalous events we should see only

   (a) $e^+ e^-$ events from $e^+ + e^- \rightarrow e^+ + e^- + \gamma$, $e^+ + e^- \rightarrow e^+ + e^- + 2\gamma$, or from^8 $e^+ + e^- + \mu^+ + \mu^-$

   (b) $\mu^+ \mu^-$ events from similar reactions

   (c) hh events

(3) The 24 eμ events in column 1 catch our attention immediately. We shall refer to them as the <u>signature</u> eμ events. If they cannot be explained by particle misidentification or decays in flight they constitute the anomalous leptonc signal of the reaction in Eq. 1.1. Incidently they cannot come from the two-virtual-photon process,^{8}

$$e^+ + e^- \rightarrow e^+ + e^- + \mu^+ \mu^-, \tag{3.8}$$

since we should see equal numbers of $e^+ \mu^+$ or $e^- \mu^-$; and we see none (column 4 of Table I). Our task is to calculate the background for eμ events to see if we can explain away the 24 eμ events.

4. BACKGROUNDS

Continuing with our study of the 4.8 GeV sample we calculate the backgrounds in two ways.

4.A External Determination

For an external determination of the backgrounds I turn to the 9,550 3-or-more prong hadronic events, (Eq. 3.). I overestimate the background by assuming that every particle in these events which was called an e or a µ by the detector was either (1) a misidentified hadron, or (2) came from the decay of a hadron. Examples of (1) are a K^+ penetrating the iron and being identified as a μ^+ or a π^- producing a large shower counter pulse height. Examples of (2) are a μ^+ from π^+ decay or an e^- from K_{e3} decay. Thus the possibility of anomalous lepton production in 3-or-more prong events is ignored, any such production being included in this background calculation. Using $P_{h \to b}$ to designate the misidentification probability, $P_{h \to e}$ and $P_{h \to \mu}$ based on the 3-or-more prong data are given in Table II. I also give $P_{h \to h}$ the probability of not misidentifying a hadron. The P's are momentum dependent. To obtain the proper average value, I use all the eµ, µh and hh events in column 1 of Table I to calculate a "hadron" spectrum; and weight the P's accordingly. These average values are also given in Table II.

We also need to know $P_{e \to a}$ and $P_{\mu \to a}$. These are determined by studying collinear ee and µµ pairs at various incident beam energies. We find

$$\begin{aligned} P_{e \to h} &= .056 \pm .02 \\ P_{e \to \mu} &= .011 \pm .01 \\ P_{\mu \to h} &= .08 \pm .02 \\ P_{\mu \to e} &< .01 \end{aligned} \quad (4.1)$$

- 12 -

As shown in Table III, $P_{e \to \mu}$ or $P_{\mu \to e}$ are negligible sources of $e\mu$ background. The major effect of e misidentification is to send ee events into the eh category; and the corrected number of eh events is 13.3 ± 4.3

I now come to the major question -- can the $e\mu$ events be due to $P_{h \to \mu}$ or $P_{h \to e}$. First a rough calculation. Let us suppose that all $e\mu$, eh, μh, and hh events (after correction for $P_{e \to h}$ and $P_{\mu \to h}$) are actually hh events. Then

$$N_{hh,true,approximate} = 61.4 \qquad (4.2)$$

And the predicted $e\mu$ background is

$$N_{e\mu,background} = 2\, P_{h \to \mu}\, P_{h \to e}\, N_{hh,true,approximate} = 4.4 \qquad (4.3)$$

Thus only 4 or 5 of the 24 events can be explained in this way! A more exact calculation which makes no assumption about the $e\mu$ events uses

$$N_{hh,true} = \frac{N_{eh} + N_{\mu h} + N_{hh}}{P_{h \to h}(P_{h \to h} + 2 P_{h \to e} + 2 P_{h \to \mu})}$$

$$= 44.9 \pm 8.0 \qquad (4.4)$$

Then

$$N_{e\mu,background} = 3.3 \pm 0.6 \qquad (4.5)$$

Putting everything together we calculate the total $e\mu$ background to be

$$N_{e\mu,background,total} = 4.3 \pm 1.2 \qquad (4.6)$$

- 13 -

The statistical probability of such a number yielding the 24 data event

$$N_{e\mu,\text{data}} = 24 \qquad (4.7)$$

is very small. The crucial question is : have we calculated the background correctly? Is it possible that the 3-or-more prong events are not representative of the 2-prong events in Table I? To try to answer these questions, I turn to a study of the 2-prong matrix, Table I -- this is the internal background determination.

4.B **Internal Determination**

Looking at Table I we make a number of observations

(1) The $e\mu$ background calculations fit within statistics to the number of $e\mu$ events in columns 2 or 3 of Table I.

(2) Our $P_{e \to \mu}$ values, Eq. 4.1, cannot be too low. If $P_{e \to \mu}$ were large $N_{e\mu}$ would not decrease while N_{ee} increases as we go from column 1 to column 2.

(3) We can even estimate $N_{e\mu,\text{background}}$ using just column 1. Assuming all eh and μh events (after correcting for $P_{e \to h}$ and $P_{\mu \to \mu}$ as in Table III) are misidentified hh events, we can calculate $P_{h \to e}$ and $P'_{h \to \mu}$ from formulas like

$$N_{eh} = \frac{2\, P'_{h \to e}\, P'_{h \to h}}{(P'_{h \to h})^2} N_{hh} \qquad (4.8)$$

Indeed the convenient equation

$$N_{e\mu} = \frac{N_{eh}\, N_{\mu h}}{2 N_{hh}} = \frac{(13.3)(12.2)}{2(12.9)} \qquad (4.9)$$

- 14 -

leads to the background estimate

$$N_{e\mu,background} = 6.3 \pm 3.1 \qquad (4.10)$$

just using column 1 of Table I. This calculation argues against the possibility that the hadronic events in column 1 of Table I are vastly different in character from those in the other columns of Table I or from those in the 3-or-more prong events.

(4) The charge distributions of the eµ events are randomly distributed as shown in Table IV.

(5) The isolation of the 24 eµ events depends upon the use of the number of detected photons. It might be argued that the number of photons associated with an events is randomly distributed; and that the 24 eµ events with 0 photons is just a fluctuation. Table V presents an argument against this. If the number of photons is randomly distributed we expect

$$\frac{(46)(40)}{189} = 9.7 \pm 2.2 \qquad (4.11)$$

eµ events with 0 photons, not 24!

5. PROPERTIES OF eµ EVENTS

We turn next to the properties of the 24 signature eµ events in the 4.8 GeV sample; remembering that there are some background events in this sample. Letting p_e, p_μ, p_i be respectively the four-momentum of the e, the µ and of the entire initial state; we define the invariant mass squared

$$M_i^2 = (p_e + p_\mu)^2 \quad ; \tag{5.1}$$

and the missing mass squared

$$M_m^2 = (p_i - (p_e + p_\mu))^2 \quad . \tag{5.2}$$

The distributions in M_i^2 and M_μ^2 are shown in Fig. 2. As shown by the figures in Sec. 8, the σ of M_m^2 is roughly 0.6 GeV^2. The distribution in Fig. 2 means that in the reaction

$$e^+ + e^- \rightarrow e^\pm + \mu^\mp + \text{missing momentum} \tag{5.3}$$

at least <u>two</u> particles are not detected.

Figure 3 shows the p_e and p_μ distributions. For use in the next section I note that p_e and p_μ both extend up to 1.8 GeV/c or so; that neither momentum distribution piles up against the cut; and that

$$\langle p_e \rangle = 1.19 \text{ GeV/c}, \langle p_\mu \rangle = 1.29 \text{ GeV/c}; \tag{5.4}$$

each about 1/4 the total energy.

Next I present the relative angular distributions. Figure 4 shows the θ_{copl} distribution; θ_{copl} is defined in Eq. 3.6. A dynamically more significant angle is θ_{coll} defined by

$$\cos\theta_{coll} = -\underline{p}_e \cdot \underline{p}_\mu/(|\underline{p}_e||\underline{p}_\mu|) \qquad (5.5)$$

When the e and μ are moving in exactly opposite directions $\theta_{coll} = 0$. The θ_{coll} distribution is shown in Fig. 5.

The $\theta_{copl} > 20°$ cut appears explictly in Fig. 4. The small angle behavior of the θ_{coll} distribution is also due to the θ_{copl} cut. All $\theta_{coll} < 20°$ are eliminated and larger θ_{coll} are partially lost.

The absence of large θ_{coll} events in Fig. 5 has dynamic significance and is discussed at the end of the next section. The absence of large θ_{copl} events in Fig. 4 is caused by the absence of large θ_{coll} events for the following reason. In our apparatus in which the angle between an observed particle and the e^+ beam direction is restricted to the range of 50° - 130°, there is a maximum value of θ_{copl} ($\theta_{copl,max}$) for a fixed θ_{coll}. Roughly

$$0 < \theta_{copl,max} \lesssim \theta_{coll} \qquad (5.6)$$

This relation is exact if $\underline{p}_e$ or $\underline{p}_\mu$ is perpendicular to the e^+ beam direction. Therefore the absence of large θ_{coll} events results in an absence of large θ_{copl} events.

6. CROSS SECTIONS FOR eμ EVENTS

Until this point I have been discussing the 4.8 GeV sample. Similar analysis have been performed at 3.0, 3.8, 4.1, and 4.45 GeV. All this data was acquired using the experimental configuration described in Sec. 2. I call this Configuration 1.

More recently some data has been acquired under two other conditions:
<u>Configuration 2</u>: The two side muon chambers were removed to build a new, high precision muon detector. The remaining muon chamber coverage was 0.70 of that in Configuration 1.
<u>Configuration 3</u>: The three lower muon chambers were temporarily inoperative due to an electrical fire in a spark chamber pulser. The remaining muon chamber coverage was 0.35 of that in Configuration 1.

Table VI lists the energy range, the number of data eμ events (corrected for background as in Sec. 4.A), and the equivalent luminosity (the luminosity multiplied by the relative muon chamber coverage of the configuration). The observed cross sections are shown in **Fig. 6**.

7. HYPOTHESES TESTS

In one sense the proposal of explanations for the $e\mu$ events is premature. The analysis is still in progress, data is still being acquired, and we are still seeking a conventional explanation. However, in another sense the analysis is aided by hypotheses. The testing of an hypothesis leads to further examination of the data and the backgrounds.

The most natural hypothesis is that discussed in Secs. 1.A through 1.D; pair production of new particles

$$e^+ + e^- \to X^+ + X^- \qquad (7.1)$$

X may be a heavy sequential lepton (L), a heavy meson (M), or an elementary boson (B). I will discuss these hypotheses with respect to the momentum spectra, the angular distributions, and the observed cross sections.

7.A Momenta Spectra

(1) Heavy Meson M or Elementary Boson B

The two-body decays

$$X^- \to e^- + \bar{\nu}_e$$
$$X^- \to \mu^- + \bar{\nu}_e \qquad (7.2)$$

yield a square momentum spectrum if one ignores the e and μ mass and if the X is unpolarized. The spectrum extends from p_{min} to p_{max} where

$$p_{max} = E_X (1 + \beta)$$
$$p_{min} = E_X (1 - \beta) \qquad (7.3)$$

Here $E_X = E_{beam}$ is the energy of either incident beam;

and $\beta = (1 - (M_X/E_{beam})^2)^{1/2}$ is the velocity of X. The spectrum for M_X = 2 GeV and E_{beam} = 2.4 GeV is shown in Fig. 7. As we shall discuss in Sec. 7.B it is possible that the X has spin 1 and that there is an $\epsilon_+ \cdot \epsilon_-$ correlation between the X^+ and X^-. This spectrum is also shown in Fig. 7. Note that p_{max} is unchanged.

(2) Heavy Lepton L

In the rest frame of the heavy lepton, L, the three-body decays

$$L^- \to \nu_\ell + e^- + \bar{\nu}_e$$

$$L^- \to \nu_\ell + \mu^- + \bar{\nu}_\mu \qquad (7.4)$$

have the momentum distribution

$$P_{e \text{ or } \mu, \text{rest frame of } L}(y) = 2y^2 (3 - 2y), \qquad (7.5)$$

$$y = 2P/M_L ;$$

assuming L has spin 1/2, using conventional first order weak interaction theory, and neglecting the e or μ mass. The laboratory frame spectrum for M_L = 2.0 GeV, E_L = 2.4 GeV is shown in Fig. 8.

(3) Comparison With Data

Figure 9 shows the comparison of the heavy meson (M) elementary boson (B) or heavy lepton (L) hypothesis with the combined p_e and p_μ momentum spectra. For M or B the best fitting mass is 2.0 or 2.1 GeV, for L the best fitting mass is 1.9 GeV. But the statistics are low, and masses

- 20 -

$$1.6 \lesssim M_X \leq 2.3 \text{ GeV} \tag{7.6}$$

are acceptable. Also any attempt to distinguish between a boson (M or B) and a lepton (L) on the basis of Fig. 9 is premature.

7.B θ_{coll} Distribution

I find the most disquieting aspect of the data to be the absence of 4.8 GeV events with $\theta_{coll} > 80°$, Fig. 5. For bosons (M or B) or heavy leptons present explanations ultimately require some eμ events to occur with $\theta_{coll} > 80°$

(1) Heavy Meson or Elementary Boson

Although I have not yet mentioned it, the reader may have already realized that if the M or B has spin 0, the $e^- + \bar{\nu}_e$ decay mode will be strongly suppressed compared to the $\mu^- + \bar{\nu}_\mu$ decay mode. This is an helicity effect and is seen in the $\pi^{\pm}$ and $K^{\pm}$. Since all our discussion is predicated on roughly equal $e^- + \bar{\nu}_e$ and $\mu^- + \bar{\nu}_\mu$ decay modes, the M or B must have spin 1 or greater. Therefore spin-spin correlations between the bosons may occur.[9] To see why these correlations are necessary to explain Fig. 5; consider a pair of bosons X, each with mass M_X, produced at threshold

$$W_{threshold} = 2M_X \tag{7.7}$$

The rest frame of each X then coincides with the laboratory frame; and the e and μ from the decay are uncorrelated in their directions of flight. Hence θ_{coll} would be evenly uniformly distributed from 0 to 180°. (The $\theta_{copl} > 20°$ cut would of course eliminate or reduce the number of small θ_{coll} events.) As W increases substantially above $W_{threshold}$ the X's will become very relativistic; and small θ_{coll} angles will be favored. However at W = 4.8 GeV for $M_X \approx 2$ GeV, this purely kinematic effect is not enough to explain Fig. 5.

However if we assume the X's are spin 1 bosons and assume some spin-spin correlations we obtain the θ_{coll} distribution of Fig. 10a.

In Fig. 10a the angular and momentum cuts of the detector and event selection are also included so that it can be directly compared with Fig. 5. The fit is acceptable although a few $\theta_{coll} > 80°$ events are predicted.

(2) Heavy Lepton

The 0.65 GeV lower limit on the e or μ momentum strongly affects the θ_{coll} distribution for the heavy lepton hypothesis. The e or μ can only exceed this limit when their production angles are quite forward along the direction of motion of their parent heavy lepton. Incidently the .65 lower limit causes a large loss of events if the X is a heavy lepton. The acceptance (A) including all angular, momentum, and detector cuts is only

$$A_L(W = 4.8 \text{ GeV}, M = 2.0 \text{ GeV}) = 0.131 \qquad (7.8a)$$

For a spin 1 boson with the spin-spin correlation discussed above

$$A_{spin\ 1}(W = 4.8 \text{ GeV}, M = 2.0 \text{ GeV}) = 0.279 \qquad (7.8b)$$

- 22 -

Returning to the heavy lepton the predicted θ_{coll} distribution, Fig. 10b, is also to be compared with **Fig. 5.** In making this calculation I have not taken into account the spin-spin alignment of the heavy leptons[3] which must occur in

$$e^+ + e^- \to L^+ + L^- \qquad (7.9)$$

7.C <u>Cross Sections and Decay Ratios</u>

(1) ee and µµ Events

If the process

$$e^+ + e^- \to X^{\pm} + X^{\mp} \qquad (7.10a)$$

$$X^{\pm} \to e^{\pm} + \text{1-or-more neutrinos} \qquad (7.10b)$$

$$X^{\mp} \to \mu^{\mp} + \text{1-or-more neutrinos} \qquad (7.10c)$$

is being observed; then we should also see events of the form

$$e^+ + e^- \to e^+ + e^- + \text{missing momentum} \qquad (7.11a)$$

and

$$e^+ + e^- \to \mu^+ + \mu^- + \text{missing momentum} \qquad (7.11b)$$

The evidence for such events is considered in Sec. 8.

Assuming that these events also exist and that the e and and µ decay mode rates are equal we have for these events

$$\sigma_{ee,\text{"observed"}} = \sigma_{\mu\mu,\text{"observed"}} = (1/2)\sigma_{e\mu,\text{observed}} \qquad (7.12)$$

Defining

$$\sigma_{\text{leptonic decay,"observed"}} = \sigma_{ee,\text{"observed"}} + \sigma_{\mu\mu,\text{"observed"}} + \sigma_{e\mu,\text{observed}}; \qquad (7.13)$$

for the 4.8 GeV sample

$$\sigma_{\text{leptonic decay, observed}} = 0.042 \pm 0.011 \text{ nb} \qquad (7.14)$$

(2) Heavy Lepton

Using the acceptance factor of Eq. 7.8a, we have at 4.8 GeV

$$\sigma_{\text{leptonic decay}} = 0.32 \pm 0.08 \text{ nb} \qquad (7.15)$$

The production cross section for a heavy lepton is given by

$$\sigma_{e^+e^- \to L^+L^-} = \frac{2\pi\alpha^2}{s} \beta \left[1 - \beta^2/3\right] \qquad (7.16)$$

At 4.8 GeV, for $M_L = 2.0$ GeV

$$\sigma_{e^+e^- \to L^+L^-} = 2.80 \text{ nb} \qquad (7.17)$$

Hence the ratio of the decay rate

$$\frac{\Gamma(L^- \to \nu_L + e^- + \bar{\nu}_e)}{\Gamma(L^- \to \text{all modes})} = \frac{\Gamma(L^- \to \nu_L + \mu^- + \bar{\nu}_\mu)}{\Gamma(L^- \to \text{all modes})} = 0.17 \pm 0.02 \qquad (7.18)$$

Such ratios are compatable with conventional theories of heavy lepton decay as shown in Figs. 11 and 12 take from Ref. 1.

Accepting Eqs. 7.15, 7.17, and 7.18 we should observe in our 4.8 GeV sample corrected cross sections of 0.63 nb for

$$e^+ + e^- \to e + \text{hadrons,}$$

or (7.19)

$$e^+ + e^- \to \mu + \text{hadrons;}$$

when one L decays purely leptonically and the other semi-leptonically. With our relatively large $P_{h \to e}$ and $P_{h \to \mu}$ coefficients in the 3-or-more prong events (Sec. 4) and other uncertainties in our data such a prediction can be easily encompassed in our 4.8 GeV sample. We do not know if we can make a significant test of this prediction.

Incidently, if a heavy lepton with a mass in the vicinity of 2 GeV exists this means that

$$R = \frac{\sigma(e^+ + e^- \to \text{hadrons})}{\sigma(e^+ + e^- \to \mu^+ + \mu^-)} \tag{7.20}$$

is not 5 at $W \gtrsim 5$ GeV, but is actually 4. (See Part 1 of these lectures.)

(3) Heavy Meson

Following the reasoning which led to Eq. 7.14, but using Eq. 7.8b we obtain

$$\sigma_{\text{leptonic decay}} = 0.15 \pm 0.04 \text{ nb} \tag{7.21}$$

The calculation of $\sigma_{e^+e^- \to M^+M^-}$, unlike that for $\sigma_{e^+e^- \to L^+L^-}$, is very model dependent. It depends upon what we take to be the form factor at the γMM vertex. Indeed if we take the M to be just like the π or K we expect $\sigma_{e^+e^- \to M^+M^-}$ to be much less than .01 nb. The only way in which we can obtain a $\sigma_{e^+e^- \to M^+M^-}$ large enough to accomodate Eq. 7.21 is to use a general idea taken from charm-quark theories. As illustrated in Fig. 13, we assume (1) that there is a heavy quark-parton (q_c) which can only lead to hadron production when

$$W_t \sim 2(\text{Mass of } q_c); \tag{7.22}$$

- 25 -

and (2) that as W rises above W_t the dominant channel available for $q_c - \bar{q}_c$ annihilation is M^+M^- pair production

To see how large $\sigma_{e^+e^- \to M^+M^-}$ is allowed to be at 4.8 GeV we note that about half of

$$\sigma_{e^+e^- \to \text{had}} \approx 20 \text{ nb} \tag{7.23}$$

might be due to new particle production (See Part 1 of these lectures.) Then

$$\sigma_{e^+e^- \to M^+M^-} \lesssim 10 \text{ nb} \tag{7.24}$$

As in the heavy lepton case, such a prediction is compatable with our present knowledge.

(4) Elementary Boson

The discussion of the elementary boson follows that of the heavy meson, except there is no form factor problem! $\sigma_{e^+e^- \to B^+B^-}$ depends upon the coupling assumptions; and this lecture is already too long to allow a discussion of these questions. We only note that a $\sigma_{e^+e^- \to B^+B^-}$ of 1 to 10 nb in in magnitude is compatable with the 4.8 GeV data.

(5) Energy Dependence of Cross Sections

In Fig. 6 I have drawn a

$$\sigma_{e^+e^- \to X^+X^-} = \frac{\text{constant}}{s} \tag{7.25}$$

curve thru the 4.8 GeV point. A 1/s dependence is consistent with the heavy lepton or elementary boson concept. It would be some what surprising if the form factor of the heavy meson did not lead to a more rapid decrease of $\sigma_{e^+e^- \to \mu^+\mu^-}$ with s.

8. COMPATIBILITY OF ee AND $\mu\mu$ EVENTS

As discussed in Sec. 7.C, the pair production hypothesis requires Eq. 7.12. Once again we consider the 4.8 GeV sample. In terms of the raw data, column 1 of Table I, the ee and $\mu\mu$ categories should each have buried in them approximately 10 events having the same properties as the eμ events. To test this possibility we must first remove events from the reactions

$$e^+ + e^- \to e^+ + e^- + \gamma \qquad (8.1)$$
$$e^+ + e^- \to \mu^+ + \mu^- + \gamma$$

Such events can appear in column 1 of Table I if the γ escapes through the ends of the detector. For these events, the missing mass squared should obey

$$M_m^2 = 0 \qquad (8.2)$$

within the mass resolution. Figures 13 and 14 show a peak at $M_m^2 = 0$ for ee and $\mu\mu$ events in both the number photons = 0 and number photons = 1 category. (Those ee and $\mu\mu$ events in the $M_m^2 = 0$ peak with number = 0 have been examined and indeed the photon should escape detection.) To exclude the reactions in Eq. 8.1 we require $M_m^2 > 2.0$ GeV. Then restricting our attention to number photon = 0 events we find the following numbers.

	ee	$\mu\mu$
$M_m^2 < 2.0$ GeV^2	27	6
$M_m^2 > 2.0$ GeV^2	13	10
TOTAL	40	16

The 13 ee events and 10 μμ events might correspond to the 24 signature eμ events. Actually only 20 of the 24 eμ events have $M_m^2 > 2.0$ GeV^2.

However we are not yet prepared to say that these 13 ee and 10 μμ events have the same origin as the signature eμ events. Unlike the eμ events there are other conventional sources for ee and μμ events with large M_m^2. These sources are the two-real-photon production processes

$$e^+ + e^- \to e^+ + e^- + \gamma + \gamma,$$
$$e^+ + e^- \to \mu^+ + \mu^- + \gamma + \gamma; \quad (8.3)$$

and the two-virtual photon processes[8]

$$e^+ + e^- \to e^+ + e^- + e^+ + e^-$$
$$e^+ + e^- \to e^+ + e^- + \mu^+ + \mu^- \quad (8.4)$$

We are now in the process of calculating the cross section for these processes with the angular, momentum, and M_m^2 cuts we use. The theory of such calculations is straightforward; but the cuts, particularly the noncoplanarity cut, make the calculation tedious.

9. CONCLUSIONS

1) No conventional explanation for the signature eμ events has been found.

2) The hypothesis that the signature eμ events come from the production of a pair of new particles -- each of mass about 2 GeV -- fits almost all the data. Only the θ_{coll} distribution is somewhat puzzling.

3) The assumption that we are also detecting ee and μμ events coming from these new particles is still being tested.

References to Part II

1. A review as of October, 1974 on heavy lepton theories and searches is M.L. Perl and P. Rapidis, SLAC-PUB-1496 (1974) (unpublished).

2. J.D. Bjorken and C.H. Llewellyn Smith, Phys. Rev. $\underline{D7}$, 88 (1973).

3. Y.S. Tsai, Phys. Rev. $\underline{D4}$, 2821 (1971).

4. M.A. Beg and A. Sirlin, Ann. Rev. Nucl. Phys. $\underline{24}$, 379 (1974).

5. M.K. Gaillard, B.W. Lee, and J.L. Rosner, Rev. Mod. Phys. $\underline{47}$, 277 (1975).

6. M.B. Einhorn and C. Quigg, FERMILAB-PUB-75/21-THY, to be published in Phys. Rev.

7. B.C. Barish et al., Phys. Rev. Letters $\underline{31}$, 180 (1973).

8. For reviews of this subject see V.M. Budner et al., Phys. Rept. $\underline{15C}$, 182 (1975); H. Terazawa, Rev. Mod. Phys. $\underline{46}$, 615 (1973). We are indebted to S. Brodsky for very useful discussions on this reaction.

9. Y.S. Tsai and A.C. Hearn, Phys. Rev. $\underline{140}$, B721 (1965). I am greatly indebted to Y.S. Tsai for pointing out to me the importance of spin effects in the θ_{coll} distribution.

TABLE I

Distribution of 513, 4.8 GeV, 2-prong, events which meet the criteria: $p_e > 0.65$ GeV/c, $p_\mu > 0.65$ GeV/c, $\theta_{copl} > 20°$.

	Total Charge = 0			Total Charge = ± 2		
Number Photons =	0	1	> 1	0	1	> 1
ee	40	111	55	0	1	0
eμ	24	8	8	0	0	3
μμ	16	15	6	0	0	0
eh	18	23	32	2	3	3
μh	15	16	31	4	0	5
hh	13	11	30	10	4	6
Sum	126	184	162	16	8	17

TABLE II

Misidentification Probabilities for 4.8 GeV Sample

Momentum range (GeV/c)	$P_{h \to e}$	$P_{h \to \mu}$	$P_{h \to h}$
0.6 - 0.9	.130 ± .005	.161 ± .006	.709 ± .012
0.9 - 1.2	.160 ± .009	.213 ± .011	.627 ± .020
1.2 - 1.6	.206 ± .016	.216 ± .017	.578 ± .029
1.6 - 2.4	.269 ± .031	.211 ± .027	.520 ± .043
weighted average using hh, µh, and eµ events	1.83 ± .007	.198 ± .007	.619 ± .012

Table III

Backgrounds in 4.8 GeV, total charge = 0, number photons = 0, 2-prong event sample. A background less than 0.1 event is called 0.

type	data	background from misidentified ee	background from misidentified μμ	data corrected for ee and μμ background	background from misidentified hh
ee	40		0		1.5 ± 0.3
eμ	24	1.0 ± 1.0	< 0.3	23.0 ± 5.0	3.3 ± 0.6
μμ	16	0			1.8 ± .3
eh	18	4.7 ± 0.8	0	13.3 ± 4.3	10.2 ± 1.8
μh	15	0	2.8 ± .7	12.2 ± 3.9	11.0 ± 2.0
hh	13	0.1 ± .02	0	12.9 ± 3.6	

- 34 -

Table IV

Charge distributions of 24 eμ events in column 1 of Table I with reference to direction of incident e^+ beam.

	μ^+	μ^-	e^+	e^-
Forward hemisphere	6	4	7	6
Backward hemisphere	5	9	6	5
Sum	11	13	13	11

Table V

Classification of 4.8 GeV total charge = 0, 2-prong events from Table I to examine effect of number of photon classification.

Number of photons	0	≥ 1	0 or ≥ 1
eµ	24	16	40
eh + µh + hh	46	143	189

- 36 -

Table VI

The numbers (N) of signature eµ events, the equivalent luminosity ($L_{equiv.}$), and the observed cross section ($\sigma_{e\mu,obs}$) for those events. N and $\sigma_{e\mu,obs}$ are corrected for background. The error given for N includes the statistical error of the raw signature eµ events.

Total Energy (GeV)	N	$L_{equiv.}$ (nb^{-1})	$\sigma_{e\mu,obs}$ (10^{-36} cm^2)	Configuration
3.0	0.0	151	0.0	1
3.8	2.2 ± 2.0	421	5.2 ± 5.0	1
4.1	$0.0 ^{+1.0}_{-0.0}$	114	$0.0 ^{+9.0}_{-0.0}$	1
4.45	$0.8 ^{+1.0}_{-0.8}$	91	$9.0 ^{+11.0}_{-9.0}$	1
4.8	19.7 ± 5.0	937	21.0 ± 5.4	1
5.4 to 6.8	3.6 ± 1.8	300	12.0 ± 6.0	2
6.2 to 7.8	7.2 ± 3.0	550	13.0 ± 5.4	3

- 37 -

Figure Captions

1. Cross section of magnetic detector.

2. Invariant mass squared (M_i^2) and missing mass squared (M_m^2) distributions for signature eμ events at 4.8 GeV.

3. Distribution of the momenta of the $e(p_e)$ and the $\mu(p_\mu)$ for the 4.8 GeV signature eμ events.

4. Distribution of θ_{copl} for the 4.8 GeV signature eμ events.

5. Distribution of θ_{coll} for the 4.8 GeV signature eμ events.

6. The observed cross section $\sigma_{e\mu,obs}$ for the signature eμ events. The two high energy measurements (dashed lines) are preliminary.

7. Momentum spectrum of the charged lepton produced in the decay $X^- \to e^- + \bar{\nu}_e$ or $X^- \to \mu^- + \bar{\nu}_\mu$ of an elementary boson or heavy meson. The solid curve is the theoretical spectrum; the dashed curve shows the effect of the angle and momentum cuts used to select the eμ events.

8. Momentum spectrum of the charged leptons produced in the decay $L^- \to \nu_L + e^- + \bar{\nu}_e$ or $L^- \to \nu_L + \mu^- + \bar{\nu}_\mu$. The solid curve is the theoretical spectrum; the dashed curve shows the effect of the angle and momentum cuts used to select the eμ events.

9. Comparison of heavy lepton momentum spectrum (solid curve) and elementary boson or heavy meson spectrum (dashed curve) with 4.8 GeV data.

10. Two possible θ_{coll} distributions including effects of angle and momentum cuts.

11. Fractional decay rates of sequential heavy leptons using an asymptotic value of R = 2/3 in Eq. 7.20. Taken from Ref. 1.

12. Fractional decay rates of sequential heavy leptons using an asymptotic value of R = 5 in Eq. 7.20. Taken from Ref. 1.

- 38 -

13. A model for X pair production just above the $q_c \bar{q}_c$ threshold.

14. Distribution of missing mass squared (M_m^2) for e^+e^- events in 4.8 GeV sample.

15. Distribution of missing mass squared (M_m^2) for $\mu^+\mu^-$ events in 4.8 GeV sample.

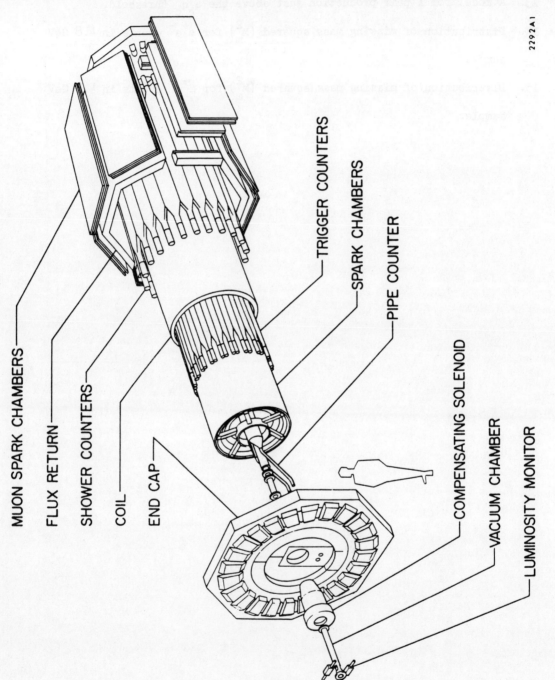

Fig. 1

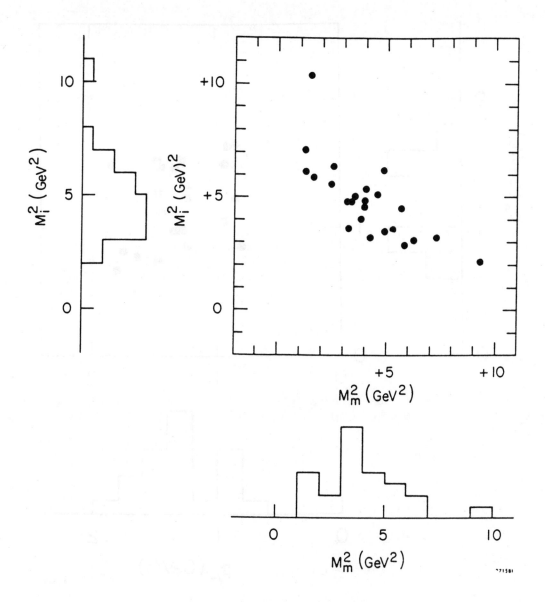

Fig. 2

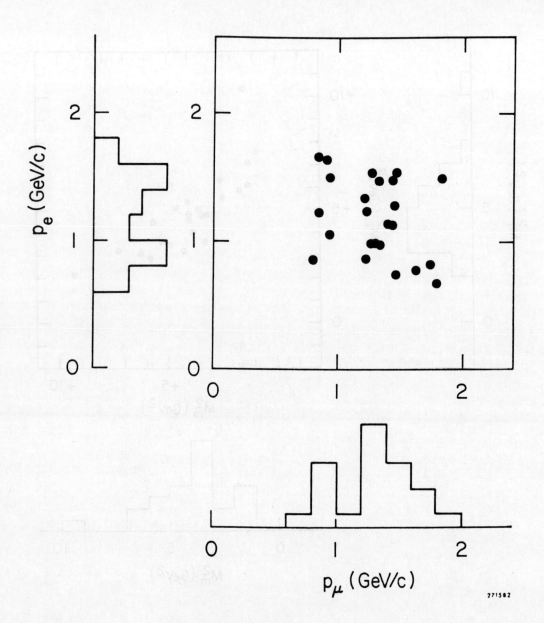

Fig. 3

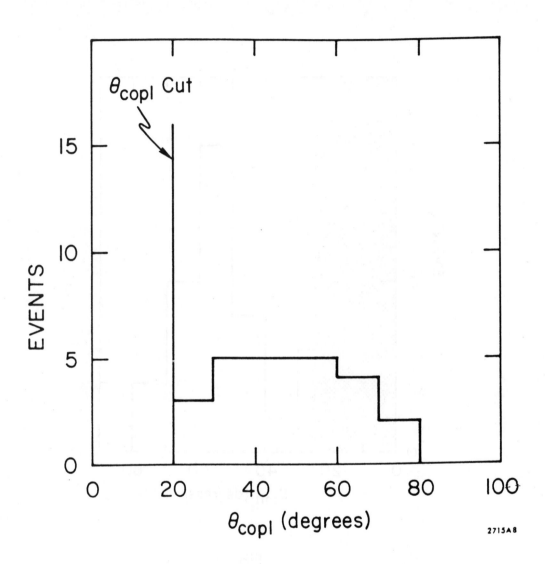

Fig. 4

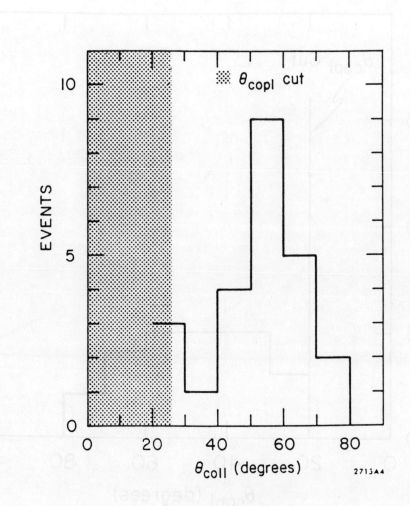

Fig. 5

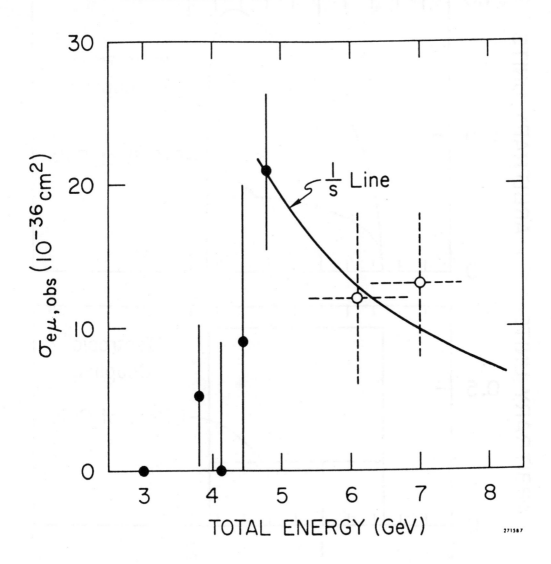

Fig. 6

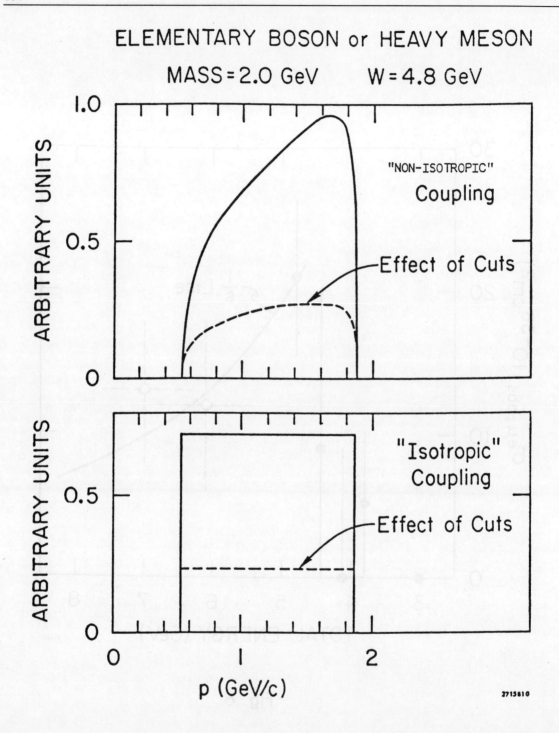

Fig. 7

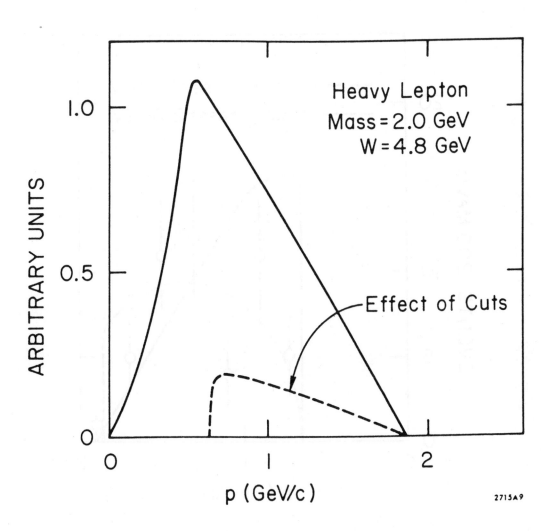

Fig. 8

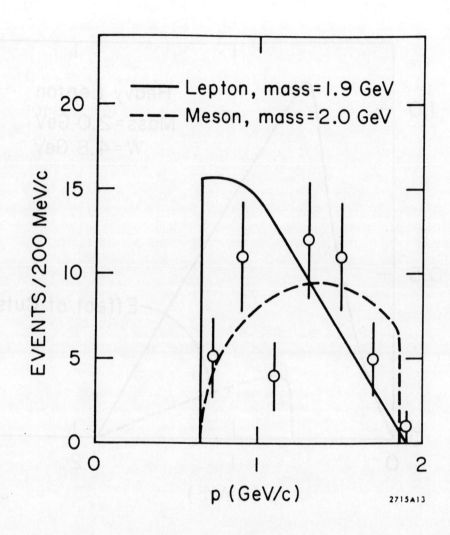

Fig. 9

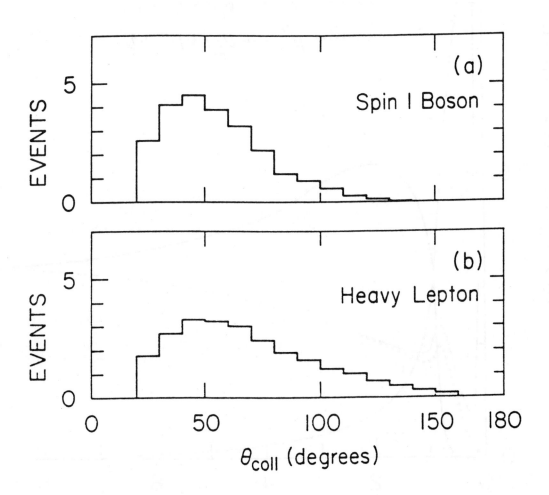

Fig. 10

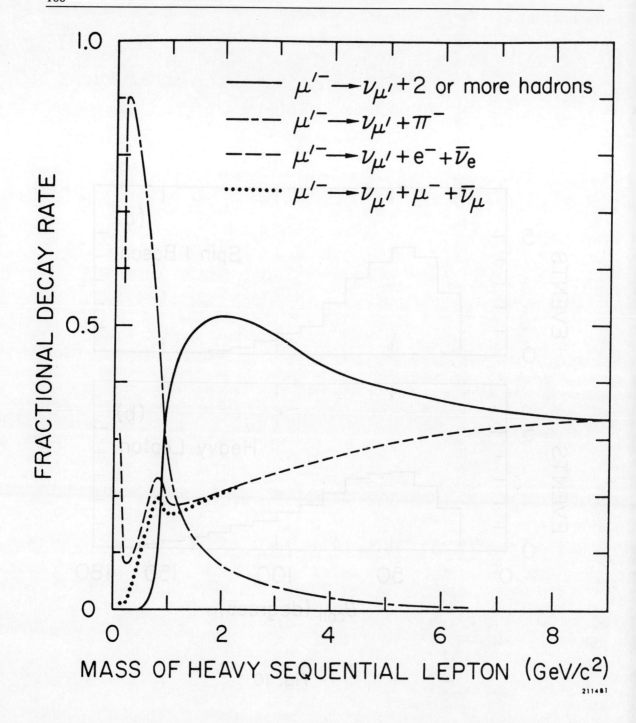

Fig. 11

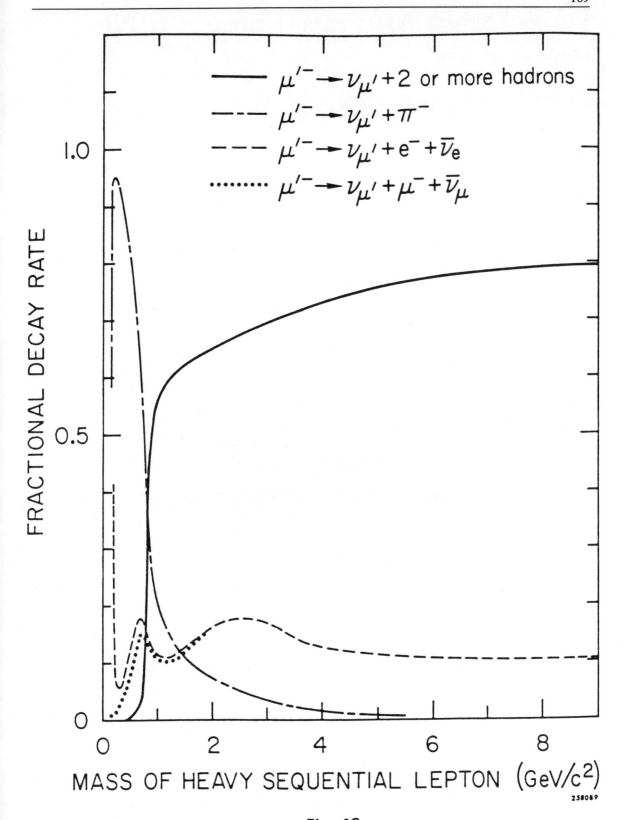

Fig. 12

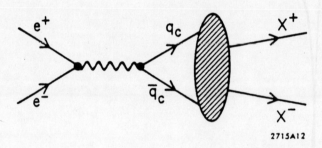

Fig. 13

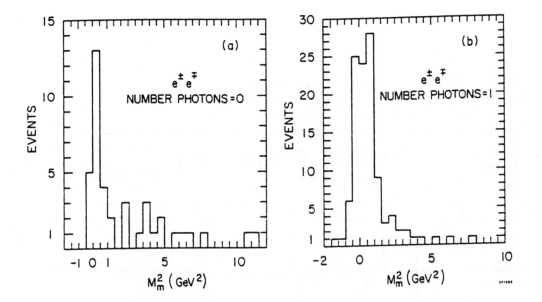

Fig. 14

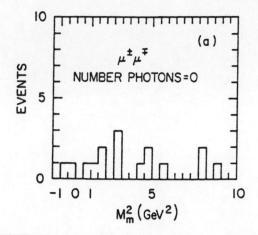

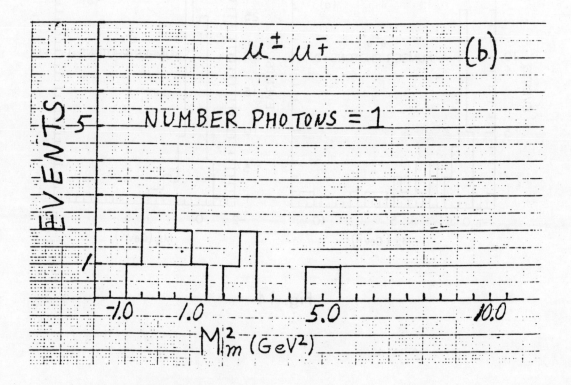

Fig. 15

Evidence for Anomalous Lepton Production in e^+-e^- Annihilation*

M. L. Perl, G. S. Abrams, A. M. Boyarski, M. Breidenbach, D. D. Briggs, F. Bulos, W. Chinowsky,
J. T. Dakin,† G. J. Feldman, C. E. Friedberg, D. Fryberger, G. Goldhaber, G. Hanson,
F. B. Heile, B. Jean-Marie, J. A. Kadyk, R. R. Larsen, A. M. Litke, D. Lüke,‡
B. A. Lulu, V. Lüth, D. Lyon, C. C. Morehouse, J. M. Paterson,
F. M. Pierre,§ T. P. Pun, P. A. Rapidis, B. Richter,
B. Sadoulet, R. F. Schwitters, W. Tanenbaum,
G. H. Trilling, F. Vannucci,‖ J. S. Whitaker,
F. C. Winkelmann, and J. E. Wiss

Lawrence Berkeley Laboratory and Department of Physics, University of California, Berkeley, California 94720, and Stanford Linear Accelerator Center, Stanford University, Stanford, California 94305

(Received 18 August 1975)

> We have found events of the form $e^+ + e^- \rightarrow e^{\pm} + \mu^{\mp}$ + missing energy, in which no other charged particles or photons are detected. Most of these events are detected at or above a center-of-mass energy of 4 GeV. The missing-energy and missing-momentum spectra require that at least two additional particles be produced in each event. We have no conventional explanation for these events.

We have found 64 events of the form

$$e^+ + e^- \rightarrow e^{\pm} + \mu^{\mp} + \geq 2 \text{ undetected particles} \quad (1)$$

for which we have no conventional explanation. The undetected particles are charged particles or photons which escape the 2.6π sr solid angle of the detector, or particles very difficult to detect such as neutrons, K_L^0 mesons, or neutrinos. Most of these events are observed at center-of-mass energies at, or above, 4 GeV. These events were found using the Stanford Linear Accelerator Center–Lawrence Berkeley Laboratory (SLAC-

LBL) magnetic detector at the SLAC colliding-beams facility SPEAR.

Events corresponding to (1) are the signature for new types of particles or interactions. For example, pair production of heavy charged leptons[1-4] having the decay modes $l^- \to \nu_l + e^- + \bar{\nu}_e$, $l^+ \to \bar{\nu}_l + e^+ + \nu_e$, $l^- \to \nu_l + \mu^- + \bar{\nu}_\mu$, and $l^+ \to \bar{\nu}_l + \mu^+ + \nu_\mu$ would appear as such events. Another possibility is the pair production of charged bosons with decays $B^- \to e^- + \bar{\nu}_e$, $B^+ \to e^+ + \nu_e$, $B^- \to \mu^- + \bar{\nu}_\mu$, and $B^+ \to \mu^+ + \nu_\mu$. Charmed-quark theories[5,6] predict such bosons. Intermediate vector bosons which mediate the weak interactions would have similar decay modes, but the mass of such particles (if they exist at all) is probably too large[7] for the energies of this experiment.

The momentum-analysis and particle-identifier systems of the SLAC-LBL magnetic detector[8] cover the polar angles $50° \leq \theta \leq 130°$ and the full 2π azimuthal angle. Electrons, muons, and hadrons are identified using a cylindrical array of 24 lead-scintillator shower counters, the 20-cm-thick iron flux return of the magnet, and an array of magnetostrictive wire spark chambers situated outside the iron. Electrons are identified solely by requiring that the shower-counter pulse height be greater than that of a 0.5-GeV e. Incidently, the e's in the e-μ events thus selected give no signal in the muon chambers; and their shower-counter pulse-height distribution is that expected of electrons. Also the positions of the e's in the shower counters as determined from the relative pulse heights in the photomultiplier tubes at each end of the counters agree within measurement errors with the positions of the e tracks. Hence the e's in the e-μ events are not misidentified combinations of $\mu + \gamma$ or $\pi + \gamma$ in a single shower counter, except possibly for a few events already contained in the background estimates. Muons are identified by two requirements. The μ must be detected in one of the muon chambers after passing through the iron flux return and other material totaling 1.67 absorption lengths for pions. And the shower-counter pulse height of the μ must be small. All other charged particles are called hadrons. The shower counters also detect photons (γ). For γ energies above 200 MeV, the γ detection efficiency is about 95%.

To illustrate the method of searching for events corresponding to Reaction (1), we consider our data taken at a total energy ($\sqrt{s}$) of 4.8 GeV. This sample contains 9550 three-or-more-prong events and 25 300 two-prong events which include $e^+ + e^- \to e^+ + e^-$ events, $e^+ + e^- \to \mu^+ + \mu^-$ events, two-prong hadronic events, and the e-μ events described here. To study two-prong events we define a coplanarity angle

$$\cos\theta_{\text{copl}} = -(\vec{n}_1 \times \vec{n}_{e^+}) \cdot (\vec{n}_2 \times \vec{n}_{e^+})/ |\vec{n}_1 \times \vec{n}_{e^+}||\vec{n}_2 \times \vec{n}_{e^+}|, \quad (2)$$

where $\vec{n}_1$, $\vec{n}_2$, and $\vec{n}_{e^+}$ are unit vectors along the directions of particles 1, 2, and the e^+ beam. The contamination of events from the reactions $e^+ + e^- \to e^+ + e^-$ and $e^+ + e^- \to \mu^+ + \mu^-$ is greatly reduced if we require $\theta_{\text{copl}} > 20°$. Making this cut leaves 2493 two-prong events in the 4.8-GeV sample.

To obtain the most reliable e and μ identification[9] we require that each particle have a momentum greater than 0.65 GeV/c. This reduces the 2493 events to the 513 in Table I. The 24 e-μ events with no associated photons, called the signature events, are candidates for Reaction (1). The e-μ events can come conventionally from the two-virtual-photon process[10] $e^+ + e^- \to e^+ + e^- + \mu^+ + \mu^-$. Calculations indicate that this source is negligible, and the absence of e-μ events with charge 2 proves this point since the number of charge-2 e-μ events should equal the number of charge-0 e-μ events from this source.

We determine the background from hadron misidentification or decay by using the 9550 three-or-more-prong events and assuming that every particle called an e or a μ by the detector either was a misidentified hadron or came from the decay of a hadron. We use $P_{h \to l}$ to designate the sum of the probabilities for misidentification or decay causing a hadron h to be called a lepton l. Since the P's are momentum dependent[9] we use all the

TABLE I. Distribution of 513 two-prong events, obtained at $E_{\text{c.m.}} = 4.8$ GeV, which meet the criteria $|\vec{p}_1| > 0.65$ GeV/c, $|\vec{p}_2| > 0.65$ GeV/c, and $\theta_{\text{copl}} > 20°$. Events are classified according to the number N_γ of photons detected, the total charge, and the nature of the particles. All particles not identified as e or μ are called h for hadron.

Particles	N_γ 0	1	>1	0	1	>1
	Total charge = 0			Total charge = ±2		
e-e	40	111	55	0	1	0
e-μ	24	8	8	0	0	3
μ-μ	16	15	6	0	0	0
e-h	20	21	32	2	3	3
μ-h	17	14	31	4	0	5
h-h	14	10	30	10	4	6

e-h, μ-h, and h-h events in column 1 of Table I to determine a "hadron" momentum spectrum, and weight the P's accordingly. We obtain the momentum-averaged probabilities $P_{h \to e} = 0.183 \pm 0.007$ and $P_{h \to \mu} = 0.198 \pm 0.007$. Collinear e-e and μ-μ events are used to determine $P_{e \to h} = 0.056 \pm 0.02$, $P_{e \to \mu} = 0.011 \pm 0.01$, $P_{\mu \to h} = 0.08 \pm 0.02$, and $P_{\mu \to e} < 0.01$.

Using these probabilities and assuming that all e-h and μ-h events in Table I result from particle misidentifications or particle decays, we calculate for column 1 the contamination of the e-μ sample to be 1.0 ± 1.0 event from misidentified e-e,[11] < 0.3 event from misidentified μ-μ,[11] and 3.7 ± 0.6 events from h-h in which the hadrons were misidentified or decayed. The total e-μ background is then 4.7 ± 1.2 events.[12,13] The statistical probability of such a number yielding the 24 signature e-μ events is very small. The same analysis applied to columns 2 and 3 of Table I yields 5.6 ± 1.5 e-μ background events for column 2 and 8.6 ± 2.0 e-μ background events for column 3, both consistent with the observed number of e-μ events.

Figure 1(a) shows the momentum of the μ versus the momentum of the e for signature events.[14] Both p_μ and p_e extend up to 1.8 GeV/c, their average values being 1.2 and 1.3 GeV/c, respectively. Figure 1(b) shows the square of the invariant e-μ mass (M_i^2) versus the square of the missing mass (M_m^2) recoiling against the e-μ system. To explain Fig. 1(b) at least two particles must escape detection. Figure 1(c) shows the distribution in collinearity angle between the e and μ ($\cos\theta_{\text{coll}} = -\vec{p}_e \cdot \vec{p}_\mu / |\vec{p}_e||\vec{p}_\mu|$). The dip near $\cos\theta_{\text{coll}} = 1$ is a consequence of the coplanarity cut; however, the absence of events with large θ_{coll} has dynamical significance.

Figure 2 shows the *observed* cross section in the range of detector acceptance for signature e-μ events versus center-of-mass energy with the background subtracted at each energy as described above.[9] There are a total of 86 e-μ events summed over all energies, with a calculated background of 22 events.[12] The corrections to obtain the true cross section for the angle and momentum cuts used here depend on the hypothesis as to the origin of these e-μ events, and the corrected cross section can be many times larger than the observed cross section. While Fig. 2 shows an apparent threshold at around 4 GeV, the statistics are small and the correction fac-

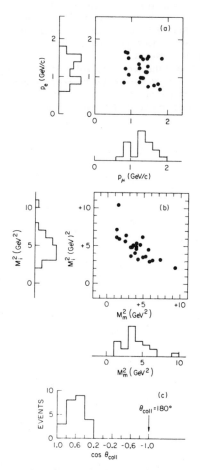

FIG. 1. Distribution for the 4.8-GeV e-μ signature events of (a) momenta of the e (p_e) and μ (p_μ); (b) square of the invariant mass (M_i^2) and square of the missing mass (M_m^2); and (c) $\cos\theta_{\text{coll}}$.

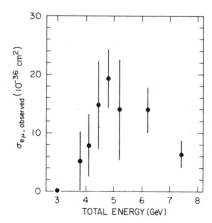

FIG. 2. The *observed* cross section for the signature e-μ events.

tors are largest for low $\sqrt{s}$. Thus, the apparent threshold may not be real.

We conclude that the signature e-μ events cannot be explained either by the production and decay of any presently known particles or as coming from any of the well-understood interactions which can conventionally lead to an e and a μ in the final state. A possible explanation for these events is the production and decay of a pair of new particles, each having a mass in the range of 1.6 to 2.0 GeV/c^2.

---

*Work supported by the U. S. Energy Research and Development Administration.

†Present address: Department of Physics and Astronomy, University of Massachusetts, Amherst, Mass. 01002.

‡Fellow of Deutsche Forschungsgemeinschaft.

§Centre d'Etudes Nucléaires de Saclay, Saclay, France.

‖Institut de Physique Nucléaire, Orsay, France.

[1]M. L. Perl and P. A. Rapidis, SLAC Report No. SLAC-PUB-1496, 1974 (unpublished).

[2]J. D. Bjorken and C. H. Llewellyn Smith, Phys. Rev. D 7, 887 (1973).

[3]Y. S. Tsai, Phys. Rev. D 4, 2821 (1971).

[4]M. A. B. Beg and A. Sirlin, Annu. Rev. Nucl. Sci. 24, 379 (1974).

[5]M. K. Gaillard, B. W. Lee, and J. L. Rosner, Rev. Mod. Phys. 47, 277 (1975).

[6]M. B. Einhorn and C. Quigg, Phys. Rev. D (to be published).

[7]B. C. Barish et al., Phys. Rev. Lett. 31, 180 (1973).

[8]J.-E. Augustin et al., Phys. Rev. Lett. 34, 233 (1975); G. J. Feldman and M. L. Perl, Phys. Rep. 19C, 233 (1975).

[9]See M. L. Perl, in Proceedings of the Canadian Institute of Particle Physics Summer School, Montreal, Quebec, Canada, 16–21 June 1975 (to be published).

[10]V. M. Budnev et al., Phys. Rep. 15C, 182 (1975); H. Terazawa, Rev. Mod. Phys. 45, 615 (1973).

[11]These contamination calculations do not depend upon the source of the e or μ; anomalous sources lead to overestimates of the contamination.

[12]Using only events in column 1 of Table I we find at 4.8 GeV $P_{h\to e}=0.27\pm 0.10$, $P_{h\to\mu}=0.23\pm 0.09$, and a total e-μ background of 7.9 ± 3.2 events. The same method yields a total e-μ background of 30 ± 6 events summed over all energies. This method of background calculation (Ref. 9) allows the hadron background in the two-prong, zero-photon events to be different from that in other types of events.

[13]Our studies of the two-prong and multiprong events show that there is no correlation between the misidentification or decay probabilities; hence the background is calculated using independent probabilities.

[14]Of the 24 events, thirteen are $e^+ +\mu^-$ and eleven are $e^- +\mu^+$.

REVIEW OF HEAVY LEPTON PRODUCTION IN e^+e^- ANNIHILATION*

Martin L. Perl
Stanford Linear Accelerator Center
Stanford University, Stanford, California 94305

ABSTRACT

The existing data on $e^{\pm}\mu^{\mp}$, $e^{\pm}x^{\mp}$, $\mu^{\pm}x^{\mp}$, and related events produced in e^+e^- annihilation are reviewed. All data are consistent with the existence of a new charged lepton, $\tau^{\pm}$, of mass $1.9 \pm .1$ GeV/c^2.

---

*Work supported by the Department of Energy.

I. INTRODUCTION

Since the discovery[1,2] of anomalous $e^{\pm}\mu^{\mp}$ events at SPEAR two and one half years ago, there has been a steady increase in the data on these events and the related two-charged prong $e^{\pm}x^{\mp}$ and $\mu^{\pm}x^{\mp}$ events. All such data which has been published, or has been presented at this or previous conferences, agree on the following points.

a. <u>Anomalous</u> two-charged prong leptonic events ($e^{\pm}\mu^{\mp}$, $e^{\pm}x^{\mp}$, $\mu^{\pm}x^{\mp}$, e^+e^-, $\mu^+\mu^-$) are produced in e^+e^- annihilation.
b. Most of these events do <u>not</u> come from the decays of charmed particles.
c. The behavior of these events is consistent with the hypothesis that a new charged lepton, τ, exists with a mass of 1.9 ± 0.1 GeV/c^2.

Points a and b have been thoroughly discussed by the individual speakers using their own data; and so with respect to these points I will only summarize their data and conclusions. In this paper I will put more emphasis on point c, the consistency of the data with the τ hypothesis; and on using the τ hypothesis to deduce a variety of properties of the τ.

I will try to give a complete set of experimental references in this paper. I will give very few theoretical references because I have given complete lists of older theoretical references in two review articles;[3,4] and T. F. Walsh[5] will provide an up-to-date theoretical summary.

An excellent and recent experimental review[6] of the heavy lepton in e^+e^- annihilation was given by G. Flugge at the 1977 Experimental Meson Spectroscopy Conference; and I gave an earlier review[7] at the XII Rencontre de Moriond.

II. SUMMARY OF THEORY

A. Sequential Lepton Model

In discussing the evidence for the τ I shall distinguish several possible types of leptons. First there is the <u>sequential</u> type:

Charged lepton	Associated neutrinos
$e^{\pm}$	$\nu_e, \bar{\nu}_e$
$\mu^{\pm}$	$\nu_{\mu}, \bar{\nu}_{\mu}$
$\tau^{\pm}$	$\nu_{\tau}, \bar{\nu}_{\tau}$
.	.
.	.

(1)

in which the τ^- and its associated neutrino, ν_{τ}, have a unique lepton number which is conserved in all interactions. This is a simple way to <u>prevent</u> the electromagnetic decays $\tau^- \to e^-\nu_{\tau}, \mu^-\nu_{\tau}$. The purely leptonic decay modes are

$$\tau^- \to \nu_{\tau} + e^- + \bar{\nu}_e$$
$$\tau^- \to \nu_{\tau} + \mu^- + \bar{\nu}_{\mu}$$

(2)

Depending on the τ mass, m_τ, and the nature of the coupling there will also be semileptonic decay modes containing hadrons[8,9] such as:

$$\tau^- \to \nu_\tau + \pi^- \qquad (3)$$

$$\to \nu_\tau + \rho^- \qquad (4)$$

$$\to \nu_\tau + \pi^- + \pi^+ + \pi^- \qquad (5)$$

B. Paralepton Model

Another simple way to suppress the electromagnetic decay of the τ is to assume it is a paralepton[10] where the τ has the lepton number of the oppositely charged e or μ.[11] Specifically:

E^- has the same lepton number as e^+

M^- has the same lepton number as μ^+

C. Ortholepton Model

In principle the τ could have the same lepton number as the same sign e or μ. We then call it an ortholepton.[10] Specifically:

e^{*-} has the same lepton number as e^-

μ^{*-} has the same lepton number as μ^- $\qquad (7)$

Then the $e\gamma e^*$ or $\mu\gamma\mu^*$ coupling must be strongly suppressed to make the electromagnetic decay rate small compared to the weak decay rate, as is required by the data (see 10c). I shall not discuss other models.[3,5,11,12]

III. SIGNATURES FOR NEW CHARGED LEPTONS PRODUCED IN e^+e^- ANNIHILATION

A. $e^\pm \mu^\mp$ Events

The cleanest signature for new charged lepton production is

$$e^+ + e^- \to \tau^+ \quad + \quad \tau^- \qquad (8)$$
$$\qquad\qquad\qquad \downarrow \qquad\qquad \downarrow$$
$$\qquad\qquad\bar\nu_\tau e + \nu_e \quad \nu_\tau \mu^- \nu_\mu$$

Such events must have:
  i. an $e^+\mu^-$ or $e^-\mu^+$
  ii. no other charged particles
  iii. no photons
  iv. missing energy
  v. a "hard" heavy lepton momentum spectra for the e and μ as shown in Fig. 1.

- 148 -

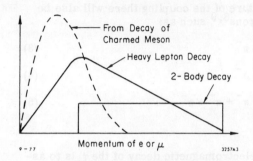

Fig. 1 Schematic comparison of the momentum spectrum for a lepton from a heavy lepton decay compared to the lepton spectrum from a charmed particle semileptonic decay or from a two-body decay.

B. $e^{\pm}x^{\mp}$, $\mu^{\pm}x^{\mp}$ Events

The decay of a τ with a mass of 1.9 GeV/c^2 is expected to yield only one charged particle (an e, μ, or hadron) a large fraction of the time; perhaps as much as 85% of the time. This leads to a two-charged prong event with or without photons:

$$e^+ + e^- \to \tau^+ + \tau^- \qquad (9)$$
$$\downarrow \downarrow$$
$$\bar{\nu}_\tau e^+ \nu_e \quad \nu_\tau x^- + \geq 0\,\gamma\text{'s}$$

where x is an e, μ, or charged hadron.

IV. $e^{\pm}\mu^{\mp}$ DATA

Table I lists the eμ data reported previously or at this conference; and Figs. 2-4 show the lepton momentum spectra. All the sets of eμ events in Table I have the following properties:

a. Their production cross section and properties are consistent with their sole source being the pair production of a mass 1.9 ± 0.1 GeV/c^2 charged lepton.

b. No other explanation for these events has been put forth which fits their production cross section and properties.

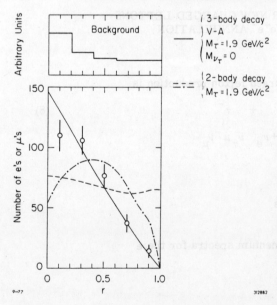

Fig. 2 The momentum spectrum for eμ events, with $3.8 \leq E_{c.m.} \leq 7.8$ GeV, from the SLAC-LBL Magnetic Detector Collaboration[7,15] corrected for background. Here $r = (p-0.65)/(p_{max}-0.65)$ where p is the e or μ momenta in GeV/c. The solid theoretical curve is for the 3-body leptonic decay of a mass 1.9 GeV/c^2 τ; the dashed theoretical curve is for the 2-body decay of an unpolarized boson; and the dash-dotted theoretical curve is for the 2-body decay of a boson produced only in the helicity = 0 state.

TABLE I

Data on $e\mu$ events. In addition to the lower limits on p_e and p_μ all these sets of p_μ events have acoplanarity requirements such as 10° or 20°. The references should be consulted for details on the event selection criterion.

Experimental group or detector	$E_{c.m.}$ range (GeV)	Lower limits on p_e p_μ (GeV/c)	Total number of $e\mu$ events	Number of background events	Comment	Ref.
M. Bernardini et al.	1.2 to 3.0				Early search at ADONE, lepton mass ≥ 1.0 GeV/c^2	13
S. Orioto et al.	2.6 to 3.0				Early search at ADONE, lepton mass ≥ 1.15 GeV/c^2	14
SLAC-LBL magnetic detector	3.8 to 7.8	0.65 0.65	190	46	First evidence. Used to determine m_τ, m_{ν_τ}, τ-ν_τ coupling.	1,2, 15, 16
PLUTO Group	3.6 to 5.0	0.3 1.0	23	1.9	Very clean. Strong argument against charm.	6,17, 18, 19
LBL-SLAC lead glass wall	3.7 to 7.4	0.4 0.65	22	0.4	Very clean. Low p_e cutoff.	20, 21
DASP Group	4.0 to 5.2	0.15	11	0.7	Good γ detection. Good hadron identification.	22

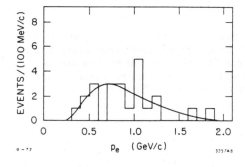

Fig. 3 The electron momentum spectrum for $e\mu$ events with $4.0 \leq E_{c.m.} \leq 5.0$ GeV from the PLUTO Group[6,17,19]; compared with the theoretical curve for the 3-body leptonic decay of a mass 1.9 GeV/c^2 τ.

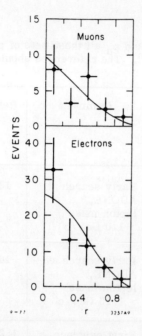

Fig. 4 The muon and electron spectra for $e\mu$ events with $4.0 \leq E_{c.m.} \leq 7.4$ GeV from the LBL-SLAC Lead Glass Wall Experiment[20,21]; compared with the theoretical curve for the 3-body leptonic decay of a mass 1.9 GeV/c² τ. Here $r = (p-p_{cut})/(p_{max}-p_{cut})$ where $p_{cut} = 0.65$ GeV/c for the muons and 0.40 GeV/c for the electrons.

In Figs. 2 and 4

$$r = (p - p_{cut})/(p_{max} - p_{cut}) \tag{10}$$

is a variable used to consolidate the lepton momentum spectra from different $E_{c.m.}$ energies. Here p is the momentum of the e or μ in GeV/c; p_{max} is its maximum value which depends on $E_{c.m.}$ and m_τ; and p_{cut} is the low momentum cutoff used in the selection of the $e\mu$ events.

V. $\mu^{\pm} x^{\mp}$ DATA

These two-charged prong events have the form

$$e^+ + e^- \to \mu^{\pm} + x^{\mp} + \geq 0 \text{ photons: } x = e \text{ or hadron} \tag{11}$$

Note that unlike the $e\mu$ events, photons are allowed in these events to allow contributions from decay modes like $\tau^- \to \rho + \nu_\tau \to \pi^- + \gamma + \gamma + \nu_\tau$. In these events $\mu^{\pm}\mu^{\mp}$ pairs we excluded either by direct identification of the x as a μ or by μ pair background subtraction. The $\mu^{\pm} x^{\mp}$ data reported previously and at this conference is summarized in Table II.

Figures 5 and 6 show the SLAC-LBL magnetic detector data.[25] Note in Fig. 5 that the 2-prong events have a considerably larger production cross section than any other single multiplicity. This is also true for other μx and ex data and is one of the basic reasons why the 2-prong μx and ex events require a lepton source explanation. Figure 7 shows the beaufitul data of the PLUTO Group.[26]

TABLE II

Data on $\mu^{\pm}x^{\mp}$ events (Eq. (11)) as described in the references. These sets of events have acoplanarity cuts.

Experimental group or detector	$E_{c.m.}$ range (GeV)	Lower limits on p_μ p_x (GeV/c)	Number μx events above background	Comments	Ref.
Maryland-Princeton-Pavia	4.8	1.0 ~0.1		First evidence. Small statistics.	23, 24
SLAC-LBL magnetic detector	4.0 to 7.8	1.0 0.2	103 ± 18 above $E_{c.m.}$ = 5.8 GeV	Strong signal above 5.8 GeV. Clearly different from μ+ > 2 charged particle events.	25
PLUTO Group	4.0 to 5.0	0.7 ~0.1	~230	Strong signal in 3 $E_{c.m.}$ ranges in 4-5 GeV regions.	6, 17, 19, 26
DASP Group	4.0 to 5.2	0.7 ~0.1	≈12	Can be directly compared to ex events.	22
Maryland-Princeton-Pavia	7	1.15 ~0.1	8^{+4}_{-3}	Good charged prong detection.	27, 28

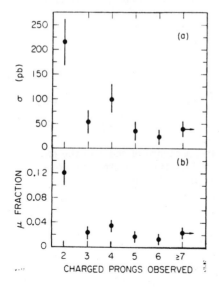

Fig. 5 (a) Anomalous muon production cross section and (b) ratio of anomalous muons to candidates versus the number of observed charged prongs in the $E_{c.m.}$ range 5.8 to 7.8 GeV from the SLAC-LBL Magnetic Detector Collaboration.[25]

- 152 -

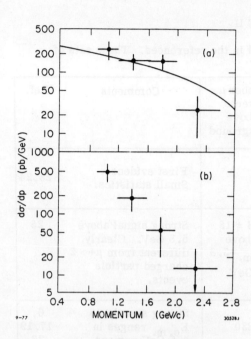

Fig. 6 Differential cross section for anomalous muon production versus momentum for (a) two-prong events and (b) multiprong events in the $E_{c.m.}$ range 5.8 to 7.8 GeV from the SLAC-LBL Magnetic Detector Collaboration.[25] The solid curve represents the expected cross section from the decays of a mass 1.9 GeV/c^2 τ.

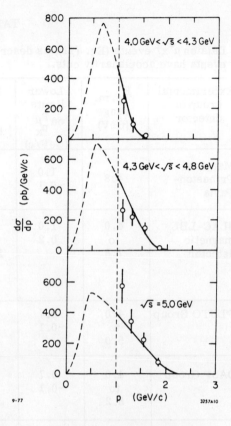

Fig. 7 The muon spectrum for μx events from the PLUTO Group[6,17,19] compared to the theoretical curve for the 3-body leptonic decay of a mass 1.9 GeV/c^2.

VI. $e^{\pm} x^{\mp}$ DATA

These events are of the form

$$e^+ + e^- \to e^{\pm} + x^{\mp} + \geq 0 \text{ photons} \quad x = \mu \text{ or hadron} \tag{12}$$

and are listed in Table III.

Figures 8 and 9 show the preliminary momentum spectrum of the e in the ex event from the DASP[21] and DELCO[30] group respectively. Both spectra are consistent with that expected from a 1.9 ± 0.1 GeV/c^2 charged lepton.

TABLE III

Data on $e^{\pm}x^{\mp}$ events. See references for the acoplanarity cut.

Experimental group or detector	$E_{c.m.}$ range (GeV)	Lower limits on p_e p_x (GeV/c)	Number of ex events above background	Comments	Ref.
LBL-SLAC lead glass wall	3.7 to 7.4	0.4 0.65	70	See hadronic decay modes of τ.	20, 21
DASP Group	4.0 to 5.2	.2 .2	60	See hadronic decay modes of τ. K/π ratio = 0.07 ± 0.06 compared to 0.24 ± 0.05 for ≥ 3 charged prong e events	22, 29
DELCO	3.7 to 7.4	.1 .3	230	Very, very clean e selection with large solid angle	30

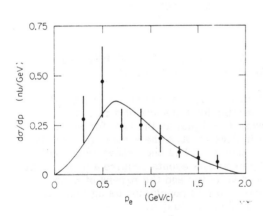

Fig. 8 The electron spectrum for ex events from the DASP Group (Refs. 22, 29) compared to the theoretical curve for the 3-body leptonic decay of a mass 1.9 GeV/c² τ.

VII. e^+e^- AND $\mu^+\mu^-$ DATA

If the τ hypothesis is correct one should observe noncoplanar events of the form

$$e^+ + e^- \to e^+ + e^- + \text{missing energy}$$
$$e^+ + e^- \to \mu^+ + \mu^- + \text{missing energy} \quad (13)$$

which are not from QED processes. It is difficult to isolate such anomalous events because of contamination from QED processes such as

$$e^+ + e^- \to \mu^+ + \mu^- + \gamma$$
$$e^+ + e^- \to \mu^+ + \mu^- + \gamma + \gamma \quad (14)$$
$$e^+ + e^- \to \mu^+ + \mu^- + e^+ + e^-$$

in which only the $\mu^+\mu^-$ pair is detected. Two results have been reported.

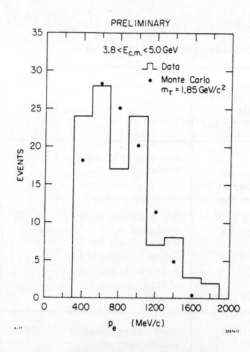

Fig. 9 Preliminary data from the DELCO Group[30] on the electron spectrum for ex events, compared to a theoretical Monte Carlo calculation for the 3-body leptonic decay of a mass 1.85 GeV/c² τ.

e^+e^- and $\mu^+\mu^-$ pairs in SLAC-LBL magnetic detector data[31]—e^+e^- and $\mu^+\mu^-$ pairs were selected requiring $p_e > 0.65$ GeV/c, $p_\mu > 0.65$ GeV/c and $\theta_{copl} > 20°$. After corrections (which are large) for QED processes and for hadronic backgrounds, the following ratios of number of events is found.

$$\frac{\text{Number } ee}{\text{Number } e\mu} = 0.52 \pm .10 \pm \begin{matrix}.16\\.19\end{matrix}$$

$$\frac{\text{Number } \mu\mu}{\text{Number } e\mu} = 0.63 \pm .10 \pm .19$$

(15)

Here the first error is one standard deviation in the statistical error; and the second error is the limits on the systematic errors added in quadrature.

$\mu^+\mu^-$ pairs in Colorado-Pennsylvania-Wisconsin "Iron Ball" experiment at SPEAR[32,28]—Using $p_\mu > 1.2$ GeV/c and $\theta_{copl} > 10°$ this experiment finds 25 $\mu^+\mu^-$ events. The expected background from QED processes and hadronic contamination is 14 events, leaving 11 anomalous $\mu^+\mu^-$ events. The authors report[32] that this number is consistent with that expected from the τ.

VIII. WHY THESE ANOMALOUS TWO-CHARGED PRONG EVENTS ARE NOT FROM CHARMED PARTICLE DECAYS

There are two reasons why there is a natural tendency to try to explain these anomalous two-prong events as due to the semileptonic decays of a pair of charmed particles. First the $e\mu$ events were found just as the hunt for singly charmed mesons began. Second as shown in Section IX. A the τ mass lies within 100 MeV/c² of the D meson masses. Nevertheless it has been shown repeatedly that almost all of $e\mu$ events and most of the ex and μx events require a non-charm explanation. The best way to see why this is so is to read the papers of each experimental group to see why they each came to this conclusion using their own data. Here I will summarize the reasons for this conclusion.

A. Summary of Why Anomalous Two-Prong Events are not From Charm

i. Very few or no ≥ three-charged prong $e^\pm\mu^\mp$ events have been found compared to the number of two-charged prong, 0 photons $e^\pm\mu^\mp$ events.[15,17,19] Since charm will produce more ≥ three-charged prong $e\mu$ events than two-charged prong $e\mu$ events, particularly at high $E_{c.m.}$ energy, the two-charged prong, 0 photon, $e\mu$ events cannot come from charm.

ii. The momentum spectra of the e or μ in eμ, ex and μx events is too hard for charm (the charm e or μ spectra is now known experimentally[19,21,22,30]).

iii. The ratio of eK to eπ events is too small for charm.[22,33]

iv. The production cross sections for eμ, ex, and μx events are all compatible with the point particle production of a mass 1.9 ± 0.1 lepton.[6,7,19,21,22,30] These production cross sections do not follow the ups and downs of the charm production cross section. A recent example is the production cross section for ex events presented by the DELCO Group[30] at this conference. There is a sharp dip in the ≥ 3 prong e events (the signature for charm events) at about 4.28 GeV/c. But there is no dip in the 2-prong e events (the signature for τ events) at this point. At the $\psi(3772)$ there is a peak in the raw number of 2-prong e events, but according to Kirkby[30] about half of these events are from charm because the e and x momentum go down to the hundred MeV/c range. Once this correction is made, there is no peak in the 2-prong e event at $\psi(3772)$.

In the next section I will show in more detail the production cross section for eμ events including new data at $\psi(3772)$.

B. New Data on eμ Production Cross Sections

The LBL-SLAC lead glass wall experiment[20,21] found 8 $e^{\mp}\mu^{\pm}$ events ($p_e > 0.65$ GeV/c, $p_\mu > 0.65$ GeV/c, $\theta_{copl} > 20^\circ$, no other charged tracks, 0 photons) at $E_{c.m.} = 3.772$ GeV, which is "the peak of the ψ". One of these events had its e in the lead glass wall. There are three types of backgrounds:

a. background from hadronic events = 2.3 events
b. background from joint semileptonic decays of a $D\bar{D}$ pair ≤ 0.2 events
c. background from the semileptonic decay of a D and the misidentification of a hadron ≤ 1.9 events.

There may be some double counting between a and c because a study of ≥ 3-prong $e^{\pm}\mu^{\mp}$ events at ψ'' finds 65 events with a calculated background of 64; and this background of 64 events is calculated using the same method as the 2.3 events in a. The small statistics and the presence of the ψ'' make it impossible to prove we have heavy lepton eμ events at the ψ''. The observed eμ production cross section is $5.4 ^{+2.8}_{-4.7}$ where the lower error takes into account the uncertainty as to how to do the backgrounds.

To display this result in comparison with earlier eμ data we define

$$R_{e\mu,\text{observed}} = \sigma_{e\mu,\text{observed}} / \sigma_{ee \to \mu\mu} \quad (16)$$

Note that $R_{e\mu,\text{observed}}$ is corrected for background contamination but is not corrected for acceptance or triggering efficiency. Figure 10 shows $R_{e\mu,\text{observed}}$. The points at the $\psi(3772)$ is consistent with a monotonic rise in $R_{e\mu,\text{observed}}$ and shows no effect of the peak in R (Fig. 11) at that resonance. Comparing Fig. 10a with Fig. 11, we also see no peak in $R_{e\mu,\text{observed}}$ at the 4.1 or 4.4 peaks in R. Thus $R_{e\mu,\text{observed}}$ does not follow charm production as reflected in the variations in R.

In Fig. 12 we define

$$R_\tau = R_{e\mu,\text{observed}} / (2 B_e B_\mu A_{e\mu}) \quad (17)$$

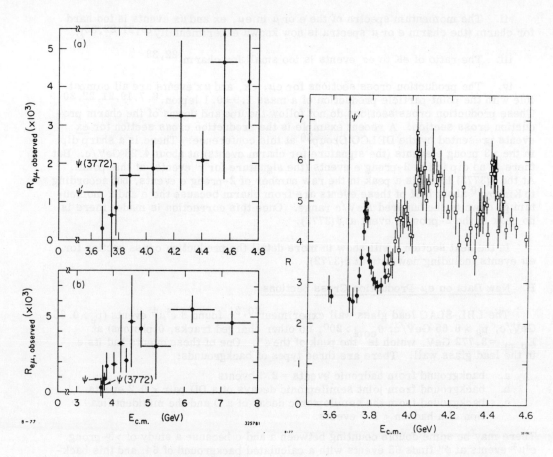

Fig. 10 $R_{e\mu, observed}$ for (a) $3.6 \leq E_{c.m.} \leq 4.8$ GeV and (b) $3.6 \leq E_{c.m.} \leq 7.8$ GeV.

Fig. 11 R for $3.6 \leq E_{c.m.} \leq 4.6$ GeV.

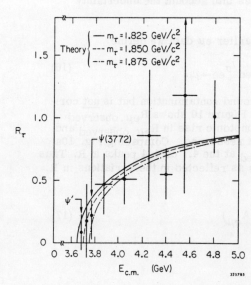

Fig. 12 R_τ compared to theoretical R_τ curves for various τ masses.

where the branching ratios to e and μ are taken to be[16] $B_e = B_\mu = .186$ and $A_{e\mu}$ is the product of the acceptance, the trigger efficiency, and various particle loss corrections. From Fig. 12 we see that if we take R_τ at the $\psi(3772)$ as being its nonzero value, the τ mass lies in the range of 1800 to 1875 MeV/c^2. In any case we see that R_τ is a monotonic function of $E_{c.m.}$ as it must be for the heavy lepton.

IV. PROPERTIES OF THE τ

A. τ Mass

Table IV gives those m_τ values which have been reported. I have not included information where data is said to be consistent with a certain m_τ but no error on that m_τ is given.

TABLE IV

Measurements of m_τ assuming V-A coupling and $m_{\nu_\tau} = 0.0$.

Experiment	Data Used	Method	τ Mass (GeV/c^2)	Comment	Ref.
SLAC-LBL magnetic detector	$e\mu$	p	$1.91 \pm .05$	Statistical error	7, 16
		$\cos\theta_{coll}$	$1.85 \pm .10$	Statistical error	
		r	$1.88 \pm .06$	Statistical error	
		composite	$1.90 \pm .10$	Statistical and systematic error	
PLUTO Group	μx	$\sigma_{\mu x}$	$1.93 \pm .05$		19
LBL-SLAC lead glass wall	$e\mu$'s at 3.772	p_e, p_μ	1.800 to 1.875	If $e\mu$'s at 3.772 are from τ	this paper
LBL-SLAC lead glass wall	ex at 3.772	σ_{ex}	1.800 to 1.875	If ex's at 3.772 are from τ	21

B. ν_τ Mass

Two upper limits have been set on m_{ν_τ}. Using $e\mu$ events[7,16]: $m_{\nu_\tau} \leq 0.6$ GeV/c^2 with 95% CL. Using μx events[19]: $m_{\nu_\tau} \leq 0.54$ GeV/c^2 with 95% CL.

C. τ-ν_τ Coupling

Using Fig. 13 we find[7,16] that V+A coupling has a χ^2 probability of less than 0.1% to fit the r distribution (Eq. (10)) of the $e\mu$ events. V-A coupling has a 60% χ^2 probability. If we ignore the r=.1 point the χ^2 probability for V+A is 5%. An

- 158 -

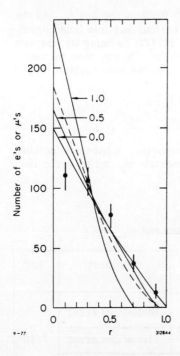

Fig. 13 r for all eμ events from the SLAC-LBL Magnetic Detector Collaboration. r is defined in the caption of Fig. 2.

additional argument against V+A coupling is that one <u>cannot</u> obtain a consistent m_τ value as shown in Table V.

TABLE V

m_τ for V+A coupling and $m_{\nu_\tau} = 0.0$.

Method	$p_\perp$	$\cos\theta_{coll}$	r
Mass (GeV/c²)	2.12 ± .05	1.95 ± .10	Upper limit is 1.76 with 95% CL.

Using μx events, the PLUTO Group also finds[6,19] the V-A coupling is favored over V+A coupling. Neither experiment is able to say anything about coupling intermediate between V+A and V-A such as pure V or pure A.

X. DECAY MODES OF τ

A. Purely Leptonic Decay Modes

Table VI gives the existing data on the purely leptonic decay rates: B_e for $\tau^- \to e^- \nu_\tau \bar{\nu}_e$ and B_μ for $\tau^- \to \mu^- \nu_\tau \bar{\nu}_\mu$.

We note that these purely leptonic branching ratios are in agreement within the errors. This is a very pleasing result considering the wide variety of methods and the difficulty of working with these small signals. These measurements are also in agreement with the theoretical calculations for a $m_\tau = 1.9$ GeV/c, $m_{\nu_\tau} = 0.0$, V-A coupling, sequential charged lepton, and Table VII.

B. Semileptonic Decay Modes

Table VIII gives the existing information on semileptonic decay modes of the τ. Comparing this table with Table VII we see that several of the predicted decay semileptonic modes of the τ have been seen, and within errors they have the expectant branching ratios. The $\tau^- \to \pi^- + \nu_\tau$ has not been seen; using $B_e = .2$, the DASP Group finds[22] a preliminary result $B_\pi = .02 \pm .025$. If further experiments confirm this relatively low value of B_π then the present theory of the nature of the τ lepton will have to be revised. For example: the τ might not have V-A coupling to the conventional weak currents.

Since this is the first presentation of the "A_1" + ν_τ decay mode by the SLAC-LBL Collaborators;[35] I will show some preliminary graphs. (Incidently the notation "A_1" is used because the expected spin (1^+) of the A_1 has not been tested and

TABLE VI

The measured fractional decay rates B_e and B_μ. V-A coupling, $m_\tau = 1.9$ GeV/c and $m_{\nu_\tau} = 0.0$ was used to calculate acceptances.

Experimental group or detector	Data Used	B_e or B_μ	Comment	Ref.
SLAC-LBL magnetic detector	eμ	$0.186 \pm .010 \pm .028$	Assume $B_e = B_\mu$. First error is statistical, second is systematic.	7,16
SLAC-LBL magnetic detector	μx	$0.175 \pm .027 \pm .030$	Assume $B_x = 0.85$. First error is statistical, second is systematic.	7,16
PLUTO Group	μx	$B_\mu = 0.14 \pm .034$		19
PLUTO Group	μx, eμ	$B_e = 0.16 \pm .06$		19
LBL-SLAC lead glass wall	eμ	$0.224 \pm .032 \pm .044$	Assume $B_e = B_\mu$. First error is statistical, second is systematic.	20,21
DASP Group	eμ	$0.20 \pm .03$	Assume $B_e = B_\mu$.	22
DELCO Group	ex	0.15	No error given.	30
Iron Ball	μμ	$0.22 ^{+.07}_{-.08}$		32
Maryland-Princeton-Pavia	μx	$0.20 \pm .10$		27

TABLE VII

Predicted branching ratios for a τ^- sequential charged heavy lepton with a mass 1.9 GeV/c^2, an associated neutrino mass of 0.0, and V-A coupling. The predictions are based on Refs. 8 and 9 as discussed in Ref. 34. The hadron continuum branching ratio assumes a threshold at 1.2 GeV for production of $\bar{u}d$ quark pairs whose final state interaction leads to the hadron continuum. From the third column it is predicted that 85% of the decays of the τ will contain only one charged particle.

Decay mode	Branching ratio	Number of charged particles in final states
$\nu_\tau \bar{e}^- \bar{\nu}_e$	.20	1
$\nu_\tau \nu^- \bar{\nu}_\mu$	.20	1
$\nu_\tau \pi^-$	.11	1
$\nu_\tau K^-$	.01	1
$\nu_\tau \rho^-$	.22	1
$\nu_\tau K^{*-}$	.01	1
$\nu_\tau A_1^-$	.07	1,3
ν_τ (hadron continuum)$^-$	.18	1,3,5

TABLE VIII

Observed semileptonic decay modes of the τ. V-A coupling, $m_\tau = 1.9$ GeV/c^2 and $m_{\nu_\tau} = 0.0$ was used to calculate acceptances. Here h means hadron.

Experimental group or detector	Decay mode (for τ^-)	Branching ratio	Ref.
LBL-SLAC lead glass wall	$h^- + \nu_\tau + \geq 0\ \gamma$'s	0.45 ± 0.19	20,21
DASP Group	$\rho^- + \nu_\tau$	$0.24 \pm .09$	22
DASP Group	$\pi^- + \nu_\tau$	$B_e B_\pi = 0.004 \pm .005$	22
PLUTO Group	"A_1"$^- + \nu_\tau$	$0.11 \pm .04 \pm .03$ for "A_1" $\to$ all	19
LBL-SLAC lead glass wall and SLAC-LBL magnetic detector	"A_1"$^- + \nu_\tau$		35

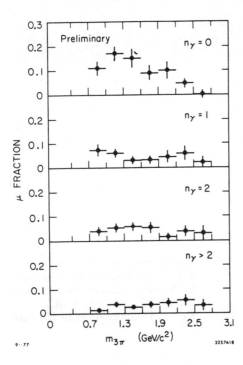 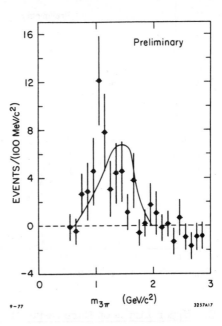

Fig. 14 μ/μ candidates versus invariant mass of remaining three prongs in 4-prong events.

Fig. 15 3-particle invariant mass distribution opposite muons corrected for hadron misidentification. The curve gives the mass distribution expected from nonresonant production ($\tau \to \nu\pi\rho$), corrected for acceptance effects, and normalized to the data in the range $.7 < m_{3\pi} < 1.8$.

and because the evidence from hadronic experiments on the A_1 is confusing. The SLAC-LBL analysis which was carried out by J. Jaros selects events using the following criteria

    i. $E_{c.m.} > 6$ GeV
    ii. 4-charged prongs with total charge 0
    iii. one of the prongs must be identified as a muon by the muon tower or mini-muon tower of the magnetic detector
    iv. $p_\mu > 0.9$ GeV/c.

Figure 14 shows the ratio μ/μ candidates versus the mass of the 3π system. (The non-μ particles are assumed to be pions.) Only the 0 photon data shows a ratio greater than the ~0.05 expected from π decay, K decay, and punchthru. Figure 15 shows the 3π mass spectra of the 0 photon events corrected for background. The peak in the 1.-1.2 GeV/c^2 region is too narrow to come from the nonresonant $\tau \to \nu_\tau + \pi + \rho$ decay mode. Figure 16 shows that the μ in these events have "hard" spectrum required for the τ.

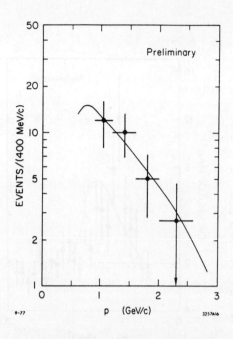

Fig. 16 Muon spectrum opposite tri-pions with $1.0 < m_{3\pi} < 1.3$. The curve is the Monte Carlo prediction for $m_\tau = 1.85$, $m_{\nu_\tau} = 0$, V-A, $E_B = 3.5$, normalized to the data.

C. Upper Limits on Rare Decay Modes

TABLE IX

Upper limits on rare decay modes of the τ using V-A coupling, $m_\tau = 1.9$ GeV/c^2, $m_{\nu_\tau} = 0.0$ for acceptance calculations.

Experimental group or detector	Mode	Upper limit on branching ratio	C.L.	Ref.
PLUTO Group	$\tau^- \to$ (3 charged particles)$^-$	0.01	95%	6
PLUTO Group	$\tau^- \to$ (3 charged leptons)$^-$	0.01	95%	6
SLAC-LBL magnetic detector	$\tau^- \to$ (3 charged leptons)$^-$	0.006	90%	36
SLAC-LBL magnetic detector	$\tau^- \to \rho^- + \pi^0$	0.024	90%	37
PLUTO Group	$\tau^- \to e^- + \gamma$ $\tau^- \to \mu^- + \gamma$	0.12	90%	6
LBL-SLAC lead glass wall	$\tau^- \to e^- + \gamma$	0.026	90%	38
LBL-SLAC lead glass wall	$\tau^- \to \mu^- + \gamma$	0.013	90%	38

XI. CONCLUSIONS

a. All data on anomalous $e\mu$, ex, μx, ee and $\mu\mu$ events produced in e^+e^- annihilation is <u>consistent</u> with the existence of a mass 1.9 ± 0.1 GeV/c^2 charged lepton, the τ.

b. This data <u>cannot</u> be explained as coming from charmed particle decays.

c. Many of the expected decay modes of the τ have been seen. A very important problem is the existence of the $\tau^- \to \nu_\tau \pi^-$ decay mode.

d. There has not been the space to discuss it here, but ν_μ experiments[6,7] say that the τ <u>cannot</u> be a muon-related ortholepton or paralepton with conventional coupling strengths. The results in Eq. (15) say that the τ is not an electron-related paralepton[7,31] using the theoretical work of Ali and Yang.[40] The τ may be a sequential lepton or an electron-related ortholepton.

REFERENCES

(1) M. L. Perl in Proceedings of the Summer Institute on Particle Physics (SLAC, Stanford, California, 1975).
(2) M. L. Perl et al., Phys. Rev. Lett. 35 (1975) 1489.
(3) G. J. Feldman and M. L. Perl, Phys. Reports (to be published); also issued as a SLAC preprint.
(4) M. L. Perl and P. Rapidis, Stanford Linear Accelerator Center preprint SLAC-PUB-1496(1974)(unpublished).
(5) T. F. Walsh, this conference.
(6) G. Flügge in Proceedings of the 1977 Experimental Meson Spectroscopy Conference (Northeastern University, Boston, 1977); also issued as DESY 77/35.
(7) M. L. Perl in Proceedings of the XII Rencontre de Moriond, Flaine, 1977, edited by Tran Thanh Van (R. M. I. E. M., Orsay).
(8) H. B. Thacher and J. J. Sakurai, Phys. Lett. 36B (1971) 103.
(9) Y. S. Tsai, Phys. Rev. D 4 (1971) 2821.
(10) C. H. Llewellyn Smith, Oxford University preprint 33/76 (1976), submitted to Proc. Roy. Soc.
(11) J. D. Bjorken and C. H. Llewellyn Smith, Phys. Rev. D 7 (1973) 88.
(12) S. Weinberg, this conference.
(13) M. Bernardini et al., Nuovo Cimento 17 (1973) 383.
(14) S. Orioto et al., Phys. Lett. 48B (1974) 165.
(15) M. L. Perl et al., Phys. Lett. 63B (1976) 466.
(16) M. L. Perl et al., Stanford Linear Accelerator Center preprint SLAC-PUB-1997 (1977), submitted to Phys. Lett.
(17) H. Meyer in Proceedings of the Orbis Scientiae, Coral Gables, 1977.
(18) J. Burmester et al., DESY 77/25.
(19) G. Knies, this conference.
(20) A. Barbaro-Galtieri et al., Lawrence Berkeley Laboratory preprint 6458.
(21) A. Barbaro-Galtieri, this conference.
(22) S. Yamada, this conference.
(23) M. Cavalli-Sforza et al., Phys. Rev. Lett. 36 (1976) 588.
(24) G. Snow, Phys. Rev. Lett. 36 (1976) 766.

(25) G. J. Feldman et al., Phys. Rev. Lett. 38 (1976) 177.
(26) J. Burmester et al., DESY 77/24.
(27) D. H. Badtke et al., this conference.
(28) H. Sadrozinski, this conference.
(29) R. Brandelis et al., DESY 77/36.
(30) J. Kirkby, this conference.
(31) F. B. Heile et al., to be published.
(32) U. Camerini et al., this conference.
(33) R. Brandeli et al., DESY 77/36.
(34) G. J. Feldman in Proceedings of the 1976 Summer Institute on Particle Physics (SLAC, Stanford, California, 1976); also issued as SLAC-PUB-1852 (1976).
(35) J. Jaros et al., to be published.
(36) G. J. Feldman, private communication.
(37) H. K. Nguyen, private communication.
(38) J. Jaros, private communication.
(39) M. Murtagh, this conference.
(40) A. Ali and T. C. Yang, Phys. Rev. D 14 (1976) 3052.

Part B

The Physics of the Tau Lepton and Tau Neutrino

THE PHYSICS OF THE TAU LEPTON AND TAU NEUTRINO IN 1995

CONTENTS

1. Introduction — 220
2. τ Production and Related τ Properties — 220
   2.1. $e^+ + e^- \to \tau^+ + \tau^-$ — 221
      2.1.1. Threshold Region and m_τ — 221
      2.1.2. Above Threshold to About 10 GeV — 223
      2.1.3. Above 10 GeV to Below Z^0 Resonance — 223
      2.1.4. Z^0 Resonance — 225
      2.1.5. Above the Z^0 Resonance — 226
      2.1.6. Some τ Studies Related to $e^+ + e^- \to \tau^+ + \tau^-$ — 226
   2.2. Photoproduction: $\gamma + N \to \tau^+ + \tau^- + N'$ — 227
   2.3. Particle Decays to τ and ν_τ — 229
      2.3.1 $W^+ \to \tau^+ + \nu_\tau$ — 229
      2.3.2 D Decays to τ and ν_τ — 229
      2.3.3 B Decays to τ and ν_τ — 230
   2.4. τ Production in Hadron Collisions — 231
      2.4.1. τ Production in $p + N$ Collisions — 231
      2.4.2. $\tau^+\tau^-$ Production in Heavy Ion Collisions — 231
3. General Discussion of τ Decays — 232
   3.1. Overview of τ Decay — 232
   3.2. Overview of Branching Fractions — 232
   3.3. Topological Branching Fractions — 233
   3.4. Unconventional Decays — 234
      3.4.1. No-ν_τ — 234
      3.4.2. With X^0 — 235
      3.4.3. Non-W Exchange — 236
4. Leptonic Decays — 237
   4.1. Overview of Decay Widths and Branching Fractions — 237
   4.2. Crude Calculation of B_e, B_μ, and B_{had} — 237
   4.3. Precise Calculation of Γ_e and Γ_μ — 238
   4.4. Aside on Radiative Decays — 239
   4.5. Comparison of B_e, B_μ, and T_τ Measurements — 241
   4.6. Momentum Spectrum in Leptonic Decays — 242
5. Hadronic Decays — 243
   5.1. $\tau^- \to \pi^- + \nu_\tau$, $\tau^- \to K^- + \nu_\tau$ — 244
   5.2. Application of Quantum Number Conservation in Non-Strange Hadronic Decays — 246
   5.3. $\tau^- \to \rho^- + \nu_\tau$ and Other Vector Decay Modes — 248
   5.4. Measurements of Modes Containing K Mesons — 250

5.5.	Comparison of B_1 and B_3 with $\sum_i B_i$	250
5.6.	Measurements of Strong Interaction Constant α_s	252
6.	Tau Spin Phenomena	254
6.1.	τ Spin Alignment and Decay Correlation	254
6.2.	Polarization at the Z^0	255
6.3.	Search for CP Violation in τ Production	256
6.4.	Search for CP Violation in τ Decay	256
6.5.	Tau Magnetic Moment	256
7.	The τ Neutrino ν_τ	257
7.1.	ν_τ Mass Limits	257
7.2.	ν_τ as Dark Matter	258
7.3.	ν_τ Lifetime Limits	258
7.4.	ν_τ Weak Interactions	259
7.5.	The ν_τ and Neutrino Mixing	259
7.6.	ν_τ-Nucleon Interactions	260

1. Introduction

> ··· *they are ill discoverers that think there is no land when they can see nothing but sea.*
>
> Francis Bacon

The quotation from Francis Bacon which heads this section describes the standard model of particle physics, a uniform and endless sea which seems to surround us. Perhaps the tau will provide the island, the new land, which will enable us to climb out of that sea.

This review is an updated revision of my 1993 paper on tau physics (Perl 1993). There is a great deal of new data on the τ from the experiments at CESR and LEP1, but very little new data on the ν_τ. Other recent reviews are Weinstein and Stroynowski (1993), the 1994 Review of Particle Properties (Montanet et al 1994), Patterson (1994), the last two proceedings of the Workshops on Tau Lepton Physics (Gan 1993, Rolandi 1995), the proceedings of the Third Workshop on the Tau-Charm Factory (Kirkby and Kirkby 1994), and the proceedings of the Workshop on the Tau-Charm Factory in the Era of B-Factories and CESR (Beers and Perl 1994).

2. τ Production and Related τ Properties

> *He had brought a large map representing the sea,*
> *Without the least vestige of land:*
> *And the crew were much pleased when they found it to be*
> *A map they could all understand.*
>
> Lewis Carroll, *The Hunting of the Snark*

The map is the standard model, we can all understand it, but it does not tell us where to look for the land outside the standard model. To use the τ as a possible guide to that land we must make τ's and experiment with them. This section describes the way we have made τ's, through $e^+ + e^- \to \tau^+ + \tau^-$ and through particle decays, it also describes possible future methods.

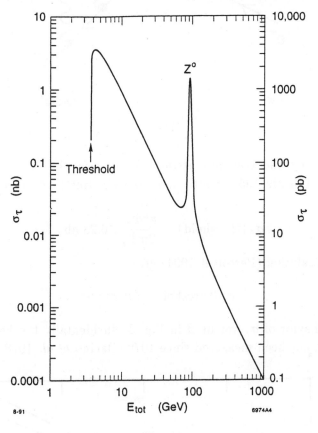

Fig. 1

2.1. $e^+ + e^- \to \tau^+ + \tau^-$

Figure 1 shows the six energy regions for τ pair production through

$$e^+ + e^- \to \tau^+ + \tau^-. \qquad (2.1)$$

2.1.1. Threshold Region and m_τ

At threshold the total pair production cross section is

$$\sigma_\tau = \frac{4\pi\alpha^2}{3s} \frac{\beta(3-\beta^2)}{2} F_c \qquad (2.2a)$$

$$F_c = \frac{\pi\alpha/\beta}{1-\exp(-\pi\alpha/\beta)} \qquad (2.2b)$$

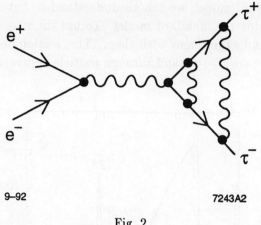

Fig. 2

where F_c is caused by the coulomb attraction between the τ^+ and τ^- as shown in Fig. 2 (Landau and Lifshitz 1958, Smith and Voloshin 1994). At threshold $s = 4m_\tau^2$ and $\beta = 0$

$$\sigma_\tau(\text{threshold}) = \frac{\pi^2 \alpha^3}{2m_\tau^2} = 0.23 \text{ nb}.$$

A more accurate calculation (Perrottet 1994) gives

$$\sigma_\tau(\text{threshold}) = 0.20 \text{ nb} \qquad (2.2c)$$

and leads to the behavior of σ_τ versus β in Fig. 3. Incidentally the behavior of σ_τ in the threshold region has not been measured since 1978 (Bacino *et al.* 1978).

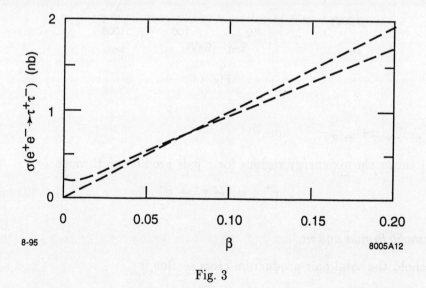

Fig. 3

The classic way to measure the τ mass, m_τ, is to find the threshold energy, $E_{\text{threshold}} = 2m_\tau$, using the first part of the cross section curve in Fig. 3. The newest measurement from the BEPC collider is (Qi 1995)

$$m_\tau = 1776.96\,{}^{+0.18}_{-0.19}\,{}^{+0.20}_{-0.16} \tag{2.3a}$$

There have been two other new measurements of m_τ. Albrecht *et al.* (1992a) used the spectrum of the invariant mass of the 3π's in

$$\tau^- \to \pi^- + \pi^+ + \pi^- + \nu_\tau$$

to find

$$m_\tau = 1776.3 \pm 2.4 \pm 1.4 \text{ MeV}/c^2. \tag{2.3b}$$

The CLEO experimenters, Balest *et al.* (1993), used $\pi^- + n\pi^0$ invariant mass spectra in

$$\tau^- \to \pi^- + n\pi^0 + \nu_\tau, \ 0 \leq n \leq 2$$

to find

$$m_\tau = 1777.8 \pm 0.7 \pm 1.7 \text{ MeV}/c^2. \tag{2.3c}$$

Stroynowski (1995) gives the world average value for m_τ

$$m_\tau = 1777.02 \pm 0.25 \text{ MeV}/c^2. \tag{2.3d}$$

Since the original papers of Kirkby (1987) and Jowett (1987) a great deal of thought has been given to an e^+-e^- circular collider which would have high luminosity in this threshold region, a tau-charm factory. The concept and potential physics have been summarized by Kirkby and Rubio (1992). Three proceedings on the tau-charm factory contain much information: Beers (1989), Kirkby and Kirkby (1994), and Beers and Perl (1994).

2.1.2. Above Threshold to About 10 GeV

In this energy range τ pair production is dominated by the γ exchange diagram in Fig. 4a and

$$\sigma_\tau = \frac{4\pi\alpha^2}{3s}\frac{\beta(3-\beta^2)}{2} = \frac{86.8}{s}\frac{\beta(3-\beta^2)}{2} \text{ nb} \tag{2.4}$$

where s is in GeV^2 in the rightmost formula. This is the energy region where the τ was discovered and where a great many studies of τ physics have been carried out at the SPEAR, DORIS, CESR and BEPC e^+e^- colliders. This will continue to be an important region for τ studies at CESR and BEPC, at the two B-factories, PEP II and KEKB, now under construction, and if it is built, at a tau-charm factory. As shown in Fig. 1 σ_τ has its maximum value in this energy region.

2.1.3. Above 10 GeV to Below Z^0 Resonance

In this energy region the Z^0 exchange amplitude, Fig. 4b, contributes through interference with the γ exchange amplitude. In the past this energy region provided a vast

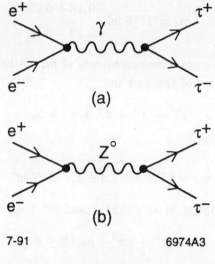

Fig. 4

amount of data on the τ experiments at the PETRA, PEP, and TRISTAN e^+e^- colliders. At present only the TRISTAN collider is still operating.

The data on the total and differential cross sections, σ_τ and $d\sigma_\tau/d\Omega$, for

$$e^+ + e^- \to \tau^+ + \tau^- \tag{2.5}$$

in this energy region was used extensively for searches for deviations from the conventional theory for the process in Eq. (2.5). As discussed in Perl (1992), two different models were used to parameterize deviations. One model, an old one (Feynman 1949, Drell 1958), allows for modifications of the photon propagator or $\tau - \gamma - \tau$ vertex in the diagram of Fig. 4a such as

$$\sigma_\tau(\text{modified}) = \sigma_\tau \, F_\pm^2(s) \tag{2.6a}$$

where

$$F_\pm^2(s) = 1 \mp \frac{s}{s - \Lambda_\pm^2} \tag{2.6b}$$

The other newer model (Eichten, Lane and Peskin 1983) assumes that the τ and e are composite particles and introduces an effective Lagrangian for a contact interaction between the constituent particles. Thus for a vector-vector interaction

$$L_{eff} = \pm \frac{g^2}{2\Lambda_\pm^{c2}} \, \bar{\psi}_2 \gamma^\mu \psi_2 \, \bar{\psi}_1 \nu_\mu \psi_1 \tag{2.7}$$

with $g^2/4\pi$ set equal to 1 to define Λ^c.

No deviations have been found, hence there are only lower limits on the parameters $\Lambda_\pm$ and $\Lambda_\pm^c$. Examples of 95% C.L. lower limits on $\Lambda_\pm$ and $\Lambda_\pm^c$ are given in Eq. (2.8). The $\Lambda_\pm$ limits are for the vector-vector interaction in Eq. (2.7).

Reference	Λ_+(GeV)	Λ_-(GeV)	Λ_+^c(TeV)	Λ_-^c(TeV)
Bartel *et al.* (1986)	285	210	4.1	5.7
Adeva *et al.* (1986)	235	205		
Behrend *et al.* (1989)	318	231		

(2.8)

However, these deviation models gives a false sense of the precision of such tests. Suppose there is a new particle χ^0 which contributes to $e^+ + e^- \to \tau^+ + \tau^-$ through the diagram in Fig. 5, and suppose the χ^0 mass is small or zero. This would not have been detected if it contributes less than about 5% to σ_τ or $d\sigma_\tau/d\Omega$. Since the contribution would be though interference with γ-exchange, the new process would not have been detected if

$$\frac{g_{ee\chi^0} \, g_{\tau\tau\chi^0}}{g_{ee\gamma} \, g_{\tau\tau\gamma}} \lesssim 5\% \tag{2.9}$$

Of course there are constraints on $g_{ee\chi^0}$ from other studies of the $ee\gamma$ vertex.

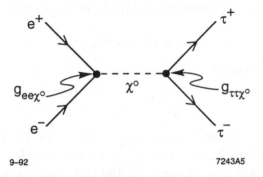

Fig. 5

2.1.4. Z^0 Resonance

At the Z^0 resonance, Fig. 1, the dominant process is Z^0-exchange in Fig. 4b. Ignoring γ-exchange and radiative corrections, the resonance is given by

$$\sigma_\tau \approx \frac{\Gamma_{z\tau\tau}}{\Gamma_z} \frac{\Gamma_z^2/m_z^2}{[s/m_z^2 - 1]^2 + \Gamma_z^2/m_z^2} \sigma_z(s = m_z^2) \tag{2.10}$$

using

$$\Gamma_{z\tau\tau}/\Gamma_z \approx 0.034, \quad \sigma_z(s = m_z^2) \approx 59 \text{ nb} \tag{2.11}$$

$$\sigma_\tau \text{ (no rad. corr., } s = m_z^2) \approx 2.0 \text{ nb} \tag{2.12}$$

with radiative correction

$$\sigma_\tau(s = m_z^2) \approx 1.4 \text{ nb} \tag{2.13}$$

The four experiments at the LEP e^+e^- collider have provided and will continue to provide a large amount of data on the τ. For example, their studies of

$$e^+ + e^- \to Z^0 \to \tau^+ + \tau^- \tag{2.14}$$

show that the $Z^0\tau^+\tau^-$ vertex obeys $e = \mu = \tau$ universality within experimental error (Montanet *et al* 1994)

$$B(Z^0 \to \tau^+\tau^-) = (3.360 \pm 0.015)\%$$
$$B(Z^0 \to e^+e^-) = (3.366 \pm 0.008)\% \tag{2.15}$$
$$b(Z^0 \to \mu^+\mu^-) = (3.367 \pm 0.013)\%$$

2.1.5. *Above the Z^0 Resonance*

Until this section, I have described energy regions which have been achieved and used for τ studies. In thinking about the energy region above the Z^0, Fig. 1, we must rely on the conventional theory of the processes in Fig. 4. If there are no higher mass resonances or other new physics in

$$e^+ + e^- \to \tau^+ + \tau^-$$

then far above the Z^0 resonance

$$\sigma_\tau = \frac{4\pi\alpha^2}{3s}[1 + 0.14] \sim \frac{0.1}{s} \text{ pb} \tag{2.16}$$

where s is in TeV^2. In the square bracket in Eq. (2.16), the 1 is from γ-exchange and the 0.14 is for Z^0-exchange. Thus far above the Z^0 resonance γ-exchange once again dominates.

The LEP II collider will reach to at least 180 GeV total energy and then linear e^+e^- collider will reach 0.5 to 1 TeV.

If the cross section, σ_τ, is as small as the conventional theory predicts in Eq. (2.16), then this energy range will not be useful for τ decay studies. For example a luminosity of $\mathcal{L} = 10^{33}$ cm^{-2} s^{-1} at 0.5 TeV yields only 4×10^3 τ pairs per year. This energy region will be useful to look for compositeness in the τ as in Eq. (2.7) or to look for other new physics in $e^+ + e^- \to \tau^+ + \tau^-$.

2.1.6. *Some τ Studies Related to $e^+ + e^- \to \tau^+ + \tau^-$*

In the course of studying $e^+ + e^- \to \tau^+ + \tau^-$ experimenters have looked in vain for non-conservation of the τ lepton number

$$e^+ + e^- \to \tau^\pm + e^\mp \tag{2.17}$$
$$e^+ + e^- \to \tau^\pm + \mu^\mp. \tag{2.18}$$

At $\sqrt{s} = 29$ GeV Gomez-Cadenas *et al.* (1991) found the 95% confidence level upper limits

$$\sigma(e^+e^- \to \tau^\pm e^\mp)/\sigma_\tau \leq 1.2 \times 10^{-3}$$
$$\sigma(e^+e^- \to \tau^\pm \mu^\mp)/\sigma_\tau \leq 4.1 \times 10^{-3}$$

and at the Z^0 (Vorobiev 1995) the 95% confidence level upper limits are

$$B(Z^0 \to \tau^\pm e^\mp)/B(Z^0 \to \ell^+\ell^-) \leq 2.6 \times 10^{-4} \quad (2.19)$$

$$B(Z^0 \to \tau^\pm \mu^\mp)/B(Z^0 \to \ell^+\ell^-) \leq 3.3 \times 10^{-4} \quad (2.20)$$

Here $\ell = e, \mu$, or τ.

Since the beginning of $e^+ + e^- \to \tau^+ + \tau^-$ studies there have been searches for the hypothetical excited τ, τ^*, defined by

$$\tau^{*\pm} \to \tau^\pm + \gamma \quad (2.21)$$

being the dominant decay. Searches at the Z^0 (Akrawy et al. 1990, Adeva et al. 1990, Decamp et al. 1990) provide the most stringent lower mass limits on m_{τ^*}. For the τ^* pair process

$$e^+ + e^- \to Z^0 \to \tau^{*+} + \tau^{*-} \to \tau^+ + \tau^- + \gamma + \gamma \quad (2.22)$$

the lower limit on m_{τ^*} is

$$m_{\tau^*} \gtrsim 45 \text{ GeV}/c^2 \quad (2.23)$$

The process

$$e^+ + e^- \to Z^0 \to \tau^{*\pm} + \tau^\mp \to \tau^+ + \tau^- + \gamma \quad (2.24)$$

depends not only upon the existence of the τ^*, but also upon the strength of the $Z^0 \tau^* \tau$ coupling. The searches using this process find

$$m_{\tau^*} \gtrsim 89 \text{ GeV}/c^2 \quad (2.25)$$

2.2. Photoproduction: $\gamma + N \to \tau^+ + \tau^- + N'$

Tsai (1979) has discussed the photoproduction of τ pairs, Fig. 6a,

$$\gamma + N \to \tau^+ + \tau^- + N' \quad (2.26)$$

where N is a target proton or nucleus and N' represents the final hadronic state. The behavior of the cross section, $\sigma_{\tau,\text{photo}}$, is sketched in Fig. 7. This method of producing τ's has not yet been used for experiments because it seems much more difficult to use than $e^+ + e^- \to \tau^+ + \tau^-$. However, it may have special uses, thus Tsai (1992a) has pointed out that it is a means of producing a ν_τ, $\bar{\nu}_\tau$ beam through decay of the τ's in Eq. (2.26).

Incidentally, electroproduction, Fig. 6b,

$$e^- + N \to e^- + \tau^+ + \tau^- + N' \quad (2.27)$$

might also be used.

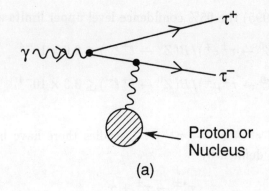

(a)

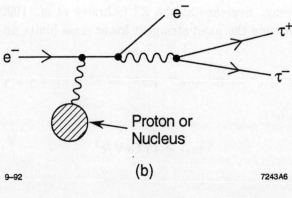

(b)

Fig. 6

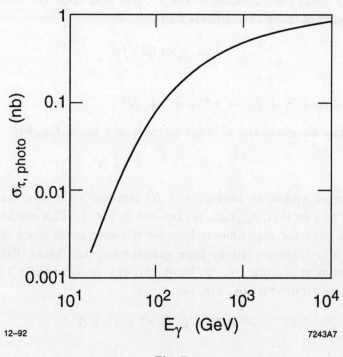

Fig. 7

2.3. Particle Decays to τ and ν_τ

2.3.1. $W^+ \to \tau^+ + \nu_\tau$

The decays

$$W^+ \to \tau^+ + \nu_\tau, \quad W^- \to \tau^- + \bar{\nu}_\tau \qquad (2.28)$$

have been used for two purposes. One purpose is to identify W's (Savoy-Navarro 1991).

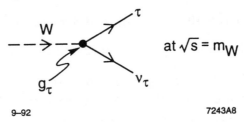

Fig. 8

The other purpose is the study of the $W\tau\nu_\tau$ vertex, Fig. 8, at $\sqrt{s} = m_W$. This is in contrast to the $W\tau\nu_\tau$ vertex in τ decays where $\sqrt{s} \leq m_\tau$. The basic question is whether the coupling constant g_τ at the $W\tau\nu_\tau$ vertex obeys e, μ, τ universality. Within the experimental errors, universality is obeyed as shown below.

$$\begin{array}{ll} g_\tau/g_e & \text{Reference} \\ 0.97 \pm 0.07 & \text{Abe } et\ al.\ 1992 \\ 1.02 \pm 0.6 & \text{Alitti } et\ al.\ 1992 \end{array} \qquad (2.29)$$

2.3.2. D Decays to τ and ν_τ

Of the pure leptonic decays of the $D^\pm$ and $D_s^\pm$

$$\left.\begin{array}{l} D^+ \to \ell^+ + \nu_\ell \\ D_s^+ \to \ell^+ + \nu_\ell \end{array}\right\} \quad \ell = e, \mu, \tau \qquad (2.30)$$

only $D_s^+ \to \mu^+ \nu_\mu$ has been observed and the branching ratio published. The decay width is

$$\Gamma(D^+, D_s^+ \to \ell^+ \nu_\ell) = \frac{G^2 + F}{8\pi} f_{D,D_s}^2\, m_{D,D_s}\, m_\ell^2$$

$$\times |V_{cd,cs}|^2 [1 - m_\ell^2/m_{D,D_s}^2]^2 \qquad (2.31)$$

Here f_{D,D_s} are the so-called weak decay constants of the D and D_s and take into account the strong interaction dynamics of cd and cs annihilation inside the meson. Theory estimates their size to be 150 to 350 MeV, but they must be measured through these decay processes. The m_ℓ^2 term in Eq. (2.31) leads to the τ mode having the largest Γ. Using

$$V_{cd} \approx 0.22, \qquad V_{cs} \approx 0.97 \qquad (2.32)$$

$$f_D \approx 200 \text{ MeV}, \qquad f_{D_s} \approx 200 \text{ MeV} \qquad (2.33)$$

I calculate the D, D_s branching fractions

$$B(D^+ \to \tau^+\nu_\tau) \approx 0.8 \times 10^{-3} \qquad (2.34a)$$

$$B(D_s^+ \to \tau^+\nu_\tau) \approx 3 \times 10^{-2} \qquad (2.34b)$$

The measurements of $B(D_s^+ \to \mu^+\nu_\mu)$ yield

$$B = (9.1 \pm 3.3) \times 10^{-3}: \qquad f_{D_s} = (344 \pm 37 \pm 52 \pm 42) \text{ MeV} \qquad (2.35a)$$

$$B = (4.0^{+2.6}_{-2.3}) \times 10^{-3}: \qquad f_{D_s} = (232 \pm 45 \pm 20 \pm 48) \text{ MeV} \qquad (2.35b)$$

from Acosta *et al* (1994) and Aoki *et al* (1993) respectively.

Tsai (1992b) has pointed out that the decay processes D^+, $D_s^+ \to \tau^+\nu_\tau$ provide polarized τ's, and given enough such events the τ decays can be used for special studies of the $\tau - W - \nu_\tau$ vertex.

The decays

$$\begin{aligned} D_s^+ &\to \tau^+ + \nu_\tau \\ D_s^- &\to \tau^- + \bar\nu_\tau \end{aligned} \qquad (2.36)$$

are crucial for fixed target production of ν_τ and $\bar\nu_\tau$ beams through the sequence

$$p + N \to D_s + \cdots \qquad (2.37)$$

$$D_s^+ \to \tau^+ + \nu_\tau, \qquad D_s^- \to \tau^- + \bar\nu_\tau$$
$$\downarrow \qquad\qquad\qquad\qquad \downarrow$$
$$\bar\nu_\tau \cdots \qquad\qquad\qquad \nu_\tau + \cdots \qquad (2.38)$$

where N is a nucleon or nucleus (Sec. 7).

Since

$$m_D - m_\tau = 92 \text{ MeV}/c^2 \qquad (2.39)$$

there are no semi-leptonic decays of the D to τ. But, the large mass of the D_s allows the not yet observed semi-leptonic decay

$$D_s^+ \to \tau^+ + \nu_\tau + \pi^0. \qquad (2.40)$$

2.3.3. *B Decays to τ and ν_τ*

The theory of leptonic decays of the B mesons

$$\begin{aligned} B^+ &\to \tau^+ + \nu_\tau \\ B_c^+ &\to \tau^+ + \nu_\tau \end{aligned} \qquad (2.41)$$

is analogous to that for D decays, but the smaller values of V_{ub} and V_{cb} reduce the decay widths. Alexander *et al.* (1994) have measured an upper limit of

$$B(B^+ \to \tau^+ \nu_\tau) < 2.2 \times 10^{-3}$$

compared to a theoretical prediction of about 5×10^{-5}.

The semileptonic decays of B to τ have substantial widths due to the large $m_B - m_\tau$ difference. References and some details are given in Sec. 4.3 of Perl (1992). The total semileptonic branching fraction is

$$B(B \to \tau + \nu_\tau + \cdots) = (4.08 \pm 0.76 \pm 0.62)\% \qquad (2.42)$$

as measured by Buskulic *et al* (1993). Theory predicts $(2.83 \pm 0.31)\%$ (Heiliger and Sehgal 1989).

2.4. τ Production in Hadron Collisions

2.4.1. τ Production in $p + N$ Collisions

As described in connection with Eqs. (2.37) and (2.38), τ's can be produced in $p + N$ collisions where N is a nucleon or nucleus. In addition to the route through D_S production and leptonic decay, there is the route through B production and semi-leptonic decay

$$B \to \tau^+ + \nu_\tau + \cdots \qquad (2.43)$$

The $p + N$ collisions can be from an external proton beam on a fixed target, from a circulating proton beam on a gas jet target, or from a proton-proton collider.

Excluding the production of $\nu_\tau, \bar{\nu}_\tau$ beams, I have not seen any arguments for studying τ physics this way rather than through $e^+ + e^- \to \tau^+ + \tau^-$ production; there are tremendous background problems when τ's are produced through hadron collisions. But there may be special uses.

2.4.2. $\tau^+\tau^-$ Production in Heavy Ion Collisions

Figure 9 shows how the virtual photons emitted in the collision of a pair of heavy ions can produce a $\tau^+\tau^-$ pair when the ions are at energies much greater than the τ mass.

$$\text{ion} + \text{ion} \to \text{ion} + \text{ion} + \gamma_{\text{virtual}} + \gamma_{\text{virtual}}$$
$$\gamma_{\text{virtual}} + \gamma_{\text{virtual}} \to \tau^+ + \tau^-. \qquad (2.44)$$

The ions would not be disrupted and the event would be clean. However, it could be difficult to detect the event because the decay products of the τ have transverse momentum less than m_τ. This process has been studied by Bottch and Strayer (1990), Amaglobeli *et al* (1991), Almeida *et al* (1991) and del Aguila *et al.* (1991).

As an example a Pb-Pb collisions at the LHC del Aguila *et al* (1991) has

$$\sigma_{Pb\,Pb\,\tau\tau} \approx 1 \text{ mb} \qquad (2.45)$$

and with $\mathcal{L} \approx 10^{28}$ cm^{-2}s^{-1} the number of τ pairs produced per month would be about 10^7.

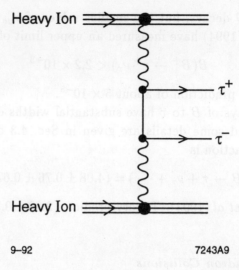

Fig. 9

3. General Discussion of τ Decays

*Dans les champs de l'observation le hasard ne
favorise que les esprits préparés.*

*Where observation is concerned, chance favours
only the prepared mind.*

Louis Pasteur

3.1. Overview of τ Decay

The conventional theory of τ decays is that they occur through the process, Fig. 10,

$$\tau^- \to \nu_\tau + W^-_{\text{virtual}}$$
$$W^-_{\text{virtual}} \to \text{final particles} \tag{3.1}$$

with lepton number separately conserved at each vertex. All experimental results in τ physics are compatible with this conventional theory.

3.2. Overview of Branching Fractions

Table 1 gives an overview of present knowledge of the major decay branching fractions of the τ. The symbol h means:

$$h^- = \pi^- \text{ or } K^-$$
$$h^+ = \pi^+ \text{ or } K^+ \tag{3.2}$$

Some remarks. The largest branching fraction modes are the leptonic modes, the modes with one π or three π's, and the ρ mode. The relatively small mass of the τ favors these

Leptonic

$\tau^- \to \nu_\tau + e^- + \bar{\nu}_e$

$\tau^- \to \nu_\tau + \mu^- + \bar{\nu}_\mu$

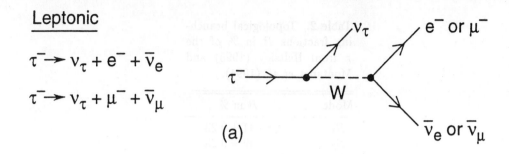

(a)

Semi leptonic or hadronic

$\tau^- \to \nu_\tau + \text{hadrons}$

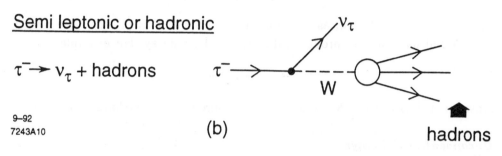

(b)

Fig. 10

Table 1. The branching fraction B in % of the major decay modes of the τ from Heltsley (1995), $\delta B/B$ is the fractional error.

Mode	B in %	$\delta B/B$
$e^- \bar{\nu}_e \nu_\tau$	17.79 ± 0.09	0.005
$\mu^- \bar{\nu}_e \nu_\tau$	17.33 ± 0.09	0.005
$h^- \nu_\tau$	11.77 ± 0.14	0.012
$h^- \pi^0 \nu_\tau$	25.36 ± 0.21	0.008
$h^- 2\pi^0 \nu_\tau$	9.18 ± 0.14	0.015
$h^- h^+ h^- \nu_\tau$	9.24 ± 0.21	0.023
$h^- 3\pi^0 \nu_\tau$	1.15 ± 0.15	0.13
$h^- h^+ h^- \pi^0 \nu_\tau$	4.45 ± 0.14	0.031

modes over modes with more pions or more massive resonances. The modes with a K or $K^*(890)$ are suppressed by the Cabibbo factor $\sin^2 \theta_c = 0.049$ relative to the corresponding π or ρ mode, Sec. 5.5.

3.3. *Topological Branching Fractions*

Although they have no precise physical significance the topological branching fractions in Table 2 are important in the methods for selecting and studying and studying τ events

Table 2. Topological branching fractions B in % of the τ from Heltsley (1995) and Montanet et al. (1994).

Mode	B in %
B_1	85.41 ± 0.23
B_3	14.49 ± 0.23
B_5	0.10 ± 0.01
B_7	< 0.0019 (90% CL)

produced in e^+e^- annihilation, as described in Sec. 5.2 of Perl (1992). The notation B_n means n charged particles are produced directly in the τ decay. For example

$$\tau^- \to \pi^- + K^0 + \nu_\tau \tag{3.3}$$

with subsequent decay of the $K^0 \to \pi^+ + \pi^-$ is counted as a one-charged particle decay.

3.4. Unconventional Decays

3.4.1. No-ν_τ

The class of unconventional τ decay modes usually discussed has no ν_τ in the mode, the particles occurring in the mode all being conventional. Examples of such hypothetical modes are

$$\begin{aligned} \tau^- &\to e^- + \gamma \\ \tau^- &\to \mu^- + \gamma \\ \tau^- &\to e^- + \pi^0 \\ \tau^- &\to e^+ + e^- + e^+ \\ \tau^- &\to \mu^+ + \mu^- + \mu^+ \end{aligned} \tag{3.4}$$

If such modes exist they violate τ lepton number conservation and either e or μ lepton number conservation. If $\bar{p}$ is substituted for an e or μ, the hypothetical modes

$$\begin{aligned} \tau^- &\to \bar{p} + \gamma \\ \tau^- &\to \bar{p} + \pi^0 \end{aligned} \tag{3.5}$$

violate τ lepton number conservation and baryon number conservation. If one wants to test just τ lepton number conservations then the non-conservation of total spin must be allowed, for example

$$\begin{aligned} \tau^- &\to \pi^- + \gamma \\ \tau^- &\to \pi^- + \pi^0 \end{aligned} \tag{3.6}$$

None of these no-ν_τ modes have been found (Eigen 1995), examples of the upper limits on the branching fractions are given in Table 3. Incidentally, it is easy to look for these modes

Table 3. Examples of 90% CL upper limits on the branching ratio B for no-ν_τ decay modes.

Mode	B	Reference
$e^-\gamma$	1.2×10^{-4}	Albecht et al. 1992c
$e^-\gamma$	1.4×10^{-4}	Stugu 1995
$\mu^-\gamma$	4.2×10^{-6}	Bean et al. 1993
$\ell^-\ell^+\ell^-$	3.3×10^{-6} to 4.3×10^{-6}	Bartelt et al. 1994
($\ell = e$ or μ)		
$e^-h^+h^-$	4.4×10^{-6} to 7.7×10^{-6}	Bartelt et al. 1994
$\mu^-h^+h^-$	7.4×10^{-6} to 20×10^{-6}	Bartelt et al. 1994
($h = \pi$ or K)		
ℓ^-h^0	4.2×10^{-6} to 11×10^{-6}	Bartelt et al. 1994
($\ell = e$ or μ)		
($h^0 = K^0$, K^{0*}, or ρ^0)		

since all the particles in the final state can be detected and their invariant mass must equal the τ mass

$$\left[\left(\sum_n E_n\right)^2 - \left(\sum_n \vec{p_n}\right)^2\right]^{\frac{1}{2}} = m_\tau \quad (3.7)$$

Here the sum is over all the particles in the final state.

The attainable lower limits on the branching fractions for these modes are set by the number of identified τ pairs in a data sample and by misidentification of normal τ decays. Misidentification of normal τ decays as no-ν_τ modes will occur if the neutrinos carry off so little energy that Eq. (3.7) is satisfied within experimental error. Thus the radiative decay

$$\tau^- \to e^- + \gamma + \nu_\tau + \bar{\nu}_e \quad (3.8a)$$

could be misidentified as

$$\tau^- \to e^- + \gamma \quad (3.8b)$$

3.4.2. With X^0

A class of unconventional decays which is much more difficult to study supposes that there is a small mass, weakly interacting boson X^0 which allows lepton number violation between τ and e or τ and μ. Then the unconventional modes

$$\tau^- \to e^- + X^0$$
$$\tau^- \to \mu^- + X^0 \quad (3.9)$$

and perhaps the modes

$$\tau^- \to e^- + \text{hadrons} + X^0$$
$$\tau^- \to \mu^- + \text{hadrons} + X^0 \quad (3.10)$$

could occur.

Such modes are very difficult to find because, unlike the modes in Eq. (3.4), the τ mass cannot be reconstructed. The problem of misidentification of normal modes is severe. For example, an event

$$\tau \to e^- + X^0 \tag{3.11a}$$

might actually be

$$\tau \to \pi^- + \nu_\tau \tag{3.11b}$$

with the π^- is misidentified as an e^-. Or, it might actually be

$$\tau^- \to e^- + \bar{\nu}_e + \nu_\tau \tag{3.11c}$$

where $\bar{\nu}_e + \nu_\tau$ taken as a single particle X^0.

In 1990 Albrecht *et al.* (1990) reported with a 95% C.L.

$$B(\tau^- \to e^- X^0) < 0.003, \quad m_{X^0} < 100 \text{ MeV} \tag{3.11d}$$

rising to 0.009 at $m_{X^0} = 500$ MeV. The limits on $B(\tau^- \to \mu^- X^0)$ are similar. There are no more recent results.

3.4.3. *Non-W Exchange*

A third class of unconventional decays involves non-W exchange. For example, in Fig. 11 an unknown particle, U, which couples only to leptons is involved in τ leptonic decays. As discussed by Tsai (1989a, 1989b) U might be a special kind of Higgs particle. The presence of this type of unconventional decay process cannot be detected by the presence of an unconventional decay mode, it can only be detected by a change in the properties of a conventional decay mode, for example, by a deviation in the kinematic distributions from those predicted by conventional theory.

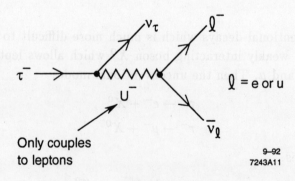

Fig. 11

4. Leptonic Decays

> *The aim of science is to seek the simplest explanation of complex facts. We are apt to fall into the error of thinking that the facts are simple because simplicity is the goal of our quest. The guiding motto in the life of every natural philosopher should be "Seek simplicity and distrust it".*
>
> Alfred North Whitehead

4.1. Overview of Decay Widths and Branching Fractions

I begin with some notations and definitions.

$$
\begin{array}{lcc}
\text{Mode} & \text{Decay Width} & \text{Branching Fraction} \\
\tau^- \to e^- + \bar{\nu}_e + \nu_\tau & \Gamma_e & B_e \\
\tau^- \to \mu^- + \bar{\nu}_\mu + \nu_\tau & \Gamma_\mu & B_\mu \\
\tau^- \to \text{hadrons} + \nu_\tau & \Gamma_{\text{had}} & B_{\text{had}}
\end{array}
\tag{4.1}
$$

The total width is

$$\Gamma = \Gamma_e + \Gamma_\mu + \Gamma_{\text{had}} \tag{4.2}$$

and

$$B_e = \Gamma_e/\Gamma, \quad B_\mu = \Gamma_\mu/\Gamma, \quad B_{\text{had}} = \Gamma_{\text{had}}/\Gamma. \tag{4.3}$$

At present we can precisely calculate Γ_e and Γ_μ (Sec. 4.3) but there is no way to precisely calculate Γ_{had}, hence at present we cannot calculate precisely any B_i. However, as described in Sec. 5, from theory and other data we can calculate precise decay widths for some hadronic modes

$$
\begin{array}{rcl}
\tau^- \to \pi^- + \nu_\tau & : & \Gamma_\pi \\
\tau^- \to K^- + \nu_\tau & : & \Gamma_K \\
\tau^- \to \rho^- + \nu_\tau & : & \Gamma_\rho
\end{array}
\tag{4.4}
$$

Since

$$B_i/B_j = \Gamma_i/\Gamma_j \tag{4.5}$$

we can predict precise value for ratios of branching fractions such as:

$$B_\mu/B_e, \quad B_\pi/B_e, \quad B_K/B_\pi, \quad B_\rho/B_e. \tag{4.6}$$

4.2. Crude Calculation of B_e, B_μ, and B_{had}

A crude calculation of B_e, B_μ, and B_{had} can be made using the diagram in Fig. 12, setting to 0 the masses of the e, μ and all quarks, taking all ν masses as 0, and ignoring the effects of the strong interaction on the conversion of quarks to hadrons. Then

$$B_e = B_\mu = \frac{1}{5} = 20\% \tag{4.7}$$

$$B_{\text{had}} = 1 - B_e - B_\mu = 60\% \tag{4.8}$$

I have ignored the Cabibbo-suppressed channel $\bar{u} + s$.

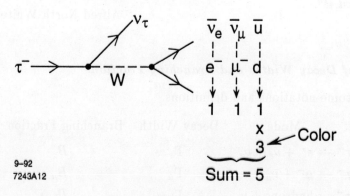

Fig. 12

It is surprising that these crude calculations give B's close to present average measured values, Table 1.

$$B_e = (17.79 \pm 0.09)\% \tag{4.9}$$

$$B_\mu = (17.33 \pm 0.09)\% \tag{4.10}$$

$$B_{\text{had}} = (64.88 \pm 0.13)\% . \tag{4.11}$$

Surprising, because this calculation uses quark counting in an energy region where half of Γ_{had} is due to two resonances, the π and the ρ.

It is instructive to carry out the same calculation for the decay of a real W. Then there are the additional decay channels.

$$\begin{aligned} \bar{\nu}_\tau, \tau &\to 1 \\ \bar{c}, s &\to 3 \end{aligned} \tag{4.12}$$

and

$$\begin{aligned} B_e = B_\mu = B_\mu &= \tfrac{1}{9} = 11\% \\ \Gamma_{\text{had}} = 1 - B_e - B_\mu - B_\tau &= 67\% \end{aligned} \tag{4.13}$$

4.3. *Precise Calculation of* Γ_e *and* Γ_μ

$$\Gamma_\ell = \frac{G_F^2 \, m_\tau^5}{192\pi^3} \, F_\ell(y) \, F_W \, F_{\text{rad}} \tag{4.14}$$

where

$$G_F = 1.166 \times 10^{-5} \text{ GeV}^{-2} \tag{4.15}$$

$$m_\tau = 1777.02 \pm 0.25. \tag{4.16}$$

The function

$$F_\ell(y) = 1 - 8y + 8y^2 - y^4 - 12y^2 \ln y \tag{4.17a}$$

is the correction for non-zero ℓ mass (Tsai 1971) and

$$y = m_\ell^2 / m_\tau^2. \tag{4.17b}$$

Specifically

$$F_e = 1.000, \quad F_\mu = 0.973. \tag{4.17c}$$

Furthermore in Eq. (4.14)

$$F_W = 1 + \frac{3}{5} \frac{m_\tau^2}{m_{W^2}} = 1.0003 \tag{4.18}$$

is the correction for m_W being finite, and

$$F_{\text{rad}} = 1 - \frac{\alpha_\tau}{2\pi}\left(\pi^2 - \frac{25}{4}\right) = 0.9957 \tag{4.19}$$

is the electromagnetic radiative correction (Marciano and Sirlin 1988).

The Γ_ℓ in Eq. (4.14) includes the basic decay

$$\tau^- \to \ell^- + \bar{\nu}_\ell + \nu_\tau, \tag{4.20a}$$

the radiative decay into γ's

$$\tau^- \to \ell^- + \bar{\nu}_\ell + \nu_\tau + n\gamma, \quad n \geq 1, \tag{4.20b}$$

and the radiative decay into $e^+ e^-$ pairs

$$\tau^- \to \ell^- + \bar{\nu}_\ell + \nu_\tau + e^+ + e^- \tag{4.20c}$$

(Sec. 4.4).

Using Eqs. (4.14)–(4.19), conventional theory predicts

$$\begin{aligned} F_e &= 4.030 \times 10^{-13} \text{ GeV} \\ F_\mu &= 3.922 \times 10^{-13} \text{ GeV}. \end{aligned} \tag{4.21}$$

The fraction error is $\pm 7 \times 10^{-4}$ due to the uncertainty in m_τ in Eq. (4.16).

4.4. *Aside on Radiative Decays*

Figure 13 shows the processes which lead to a radiative leptonic decay with one γ, the dominant process is radiation from the e or μ since these have the smallest masses

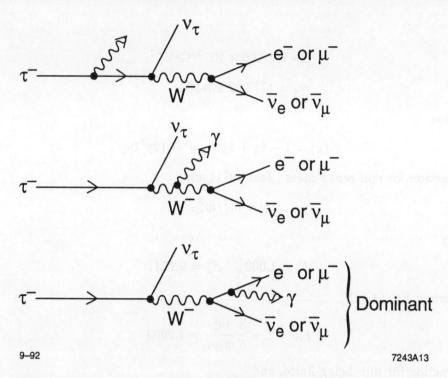

Fig. 13

(Wu 1990a, Marciano and Sirlin 1988). From the work of Kinoshita and Sirlin (1959) on radiative decay of the muon, for photons with energy

$$E_\gamma \lesssim m_\tau/2$$

$$\frac{d\Gamma(\tau^- \to \ell^- \bar{\nu}_\ell \nu_\tau \gamma)}{dy} \approx \Gamma(\tau^- \to \ell^- \bar{\nu}_\ell \nu_\tau)$$

$$\times \frac{1-y}{y}\left[\frac{\alpha}{\pi}\left(2\ell n \frac{m_\tau}{m_\ell} - \frac{17}{6}\right)\right] \quad (4.22)$$

where

$$y = 2E_\gamma/m_\tau. \quad (4.23)$$

The factor in the square bracket is 0.031 for $\ell = e$ and 0.0065 for $\ell = \mu$.

Returning to Eq. (4.20a) recall that Eq. (4.14) gives the total width for all these processes. If we make the Γ for the radiative decay in Eq. (4.20b) larger by going to smaller y in Eq. (4.22), then the Γ for the non-radiative decay in Eq. (4.20a) becomes smaller.

There are only two studies of τ radiative decays, Wu *et al.* (1990b) measured

$$\tau^- \to \mu^- + \bar{\nu}_\mu + \nu_\tau + \gamma; \quad (4.24)$$

and the CLEO experimenters (Mistry 1992) indirectly studied

$$\tau^- \to e^- + \bar{\nu}_e + \nu_\tau + \gamma. \quad (4.25)$$

A great deal of work remains to be done on τ radiative decays, not only to test conventional theory, but also to explore hadronic radiative decays such as

$$\tau^- \to \pi^- + \nu_\tau + \gamma \tag{4.26}$$

$$\tau^- \to \rho^- + \nu_\tau + \gamma. \tag{4.27}$$

As discussed by Decker and Finkemeir (1993) and the reference they give, we can learn about internal bremsstrahlung and structure-dependent radiation from distributions such as the γ energy spectra and the hadron-γ mass spectra in Eqs. (4.26) and (4.27). And there is always the possibility of finding "new physics" in radiative decays.

4.5. Comparison of B_e, B_μ, and T_τ Measurements

We expect

$$B_\mu / B_e = 0.973 \tag{4.28}$$

from Eq. (4.17c). Using the world average values of Table 1

$$\begin{aligned} B_e &= 17.79 \pm 0.09 \\ B_\mu &= 17.33 \pm 0.09 \end{aligned} \tag{4.29}$$

gives

$$B_\mu / B_e = 0.974 \pm 0.007 \tag{4.30}$$

which agrees with Eq. (4.28).

By definition

$$\begin{aligned} T_\tau &= \hbar B_e / \Gamma_e \\ T_\tau &= \hbar B_\mu / \Gamma_\mu \end{aligned} \tag{4.31}$$

where Γ_e and Γ_μ have been calculated as in Eq. (4.14) while B_e and B_μ must be measured. Using the values in Eq. (4.29)

$$\begin{aligned} T_\tau \text{ (from } B_e) &= 290.6 \pm 1.5 \text{ fs} \\ T_\tau \text{ (from } B_\mu) &= 290.8 \pm 1.5 \text{ fs} \end{aligned} \tag{4.32}$$

I remind you that 1 fs = 10^{-15} s.

Now I compare these values with directly measured values of T_τ. A recent compilation by Davier (1995) gives

$$T_\tau \text{ (measured)} = 291.6 \pm 1.6 \text{ fs}. \tag{4.33}$$

Thus the measured value of T_τ agrees with the values calculated from B_e and B_μ. This is a change from older comparisons in which the measured T_τ value was larger than the calculated value, by several standard deviations.

4.6. Momentum Spectrum in Leptonic Decays

There is a great deal that can be learned about τ leptonic decays from momentum spectra, angular distributions, and polarization information. I begin with a simple example, the momentum spectrum of the electron in

$$\tau^- \to e^- + \bar{\nu}_e + \nu_\tau . \tag{4.34}$$

If we suppose the $\tau W \nu_\tau$ vertex in Fig. 10a is not exactly V-A we can look for new physics in the matrix element

$$M = \frac{G}{\sqrt{2}}[\bar{u}_e \gamma^\mu (1-\gamma_5) v_{\bar{\nu}_e}] \times [\bar{u}_{\nu_e} \gamma_u (v_\tau + a_\tau \gamma_5) u_\tau] . \tag{4.35}$$

Then defining

$$x = 2E_e/m_\tau , \tag{4.36}$$

and setting

$$m_e = m_{\nu_e} = m_{\nu_\tau} = 0 , \tag{4.37}$$

the momentum spectrum in the τ rest frame is given by

$$\frac{d\Gamma_e}{\Gamma_e dx} = [12(x^2 - x^3)] + \left[\frac{8\rho_\tau}{3}(4x^3 - 3x^2)\right] \tag{4.38a}$$

$$\rho_\tau = \frac{3}{4} \frac{(v_\tau - a_\tau)^2}{(v_\tau - a_\tau)^2 + (v_\tau + a_\tau)^2} . \tag{4.38b}$$

In the standard model for τ, $v_\tau = +1$, $a_\tau = -1$ and

$$\rho_\tau \text{ (standard model)} = \frac{3}{4} . \tag{4.39}$$

To see how ρ_τ is measured, return to Eq. (4.38a) and call the second square bracket factor the ρ part of the momentum spectrum. Figure 14a shows $d\Gamma_e/\Gamma_e\, dx$ and the contribution of the ρ part when the τ is at rest in the laboratory frame, Fig. 14b shows the same quantities when the τ has high energy in the laboratory frame, $E_\tau \gg m_\tau$.

From measurements (Stroynowski 1995)

$$\rho = 0.751 \pm 0.039 \pm 0.027 . \tag{4.40}$$

Ignoring radiative effects, the more general leptonic spectrum is given by

$$\frac{d\Gamma}{dx\, d\cos\theta} \propto x^2 \left\{ 12(1-x) + \rho\left(\frac{32x}{3} - 8\right) \right.$$

$$\left. + \eta \frac{m_\ell}{m_\tau} \frac{24(1-x)}{x} - P_\tau \xi \cos\theta \left[4(1-x) + \delta\left(\frac{32x}{3} - 8\right)\right] \right\} . \tag{4.41}$$

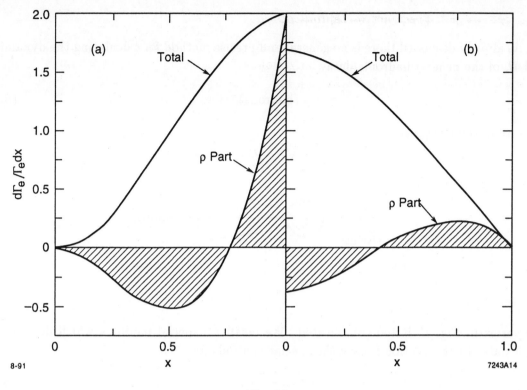

Fig. 14

Here $x = 2E_\ell/m_\tau$ and θ is the angle between the τ spin and ℓ momentum, both in the τ rest frame. P_τ is the τ polarization. The parameters ρ and γ can be determined from $d\Gamma/dx$. The parameters ξ and δ require either directly polarized τ's or the use of the τ spin correlation as discussed in Sec. 6. The average measured values are

$$y = -0.04 \pm 0.15 \pm 0.10 \tag{4.42a}$$

$$\delta = 0.74 \pm 0.19 \pm 0.10 \tag{4.42b}$$

$$\xi = 1.18 \pm 0.15 \pm 0.08, \tag{4.42c}$$

compared to standard values of 0, 1, and 3/4. The measurements in Eqs. (4.40) and (4.42a) are averaged over the e and μ decays, except η which can only be observed in the μ decay mode.

5. Hadronic Decays

> *False facts are highly injurious to the progress of science, for they often endure long; but false views, if supported by some evidence, do little harm, for everyone takes a salutary pleasure in proving their falseness; and when this is done, one path towards error is closed and the road to truth is often at the same time opened.*
>
> Charles Darwin

5.1. $\tau^- \to \pi^- + \nu_\tau$, $\tau^- \to K^- + \nu_\tau$

As already discussed there is no general and precise method for calculating the dynamics and B_i of the general hadronic decay

$$\tau^- \to (\text{hadrons})_i^- + \nu_\tau \tag{5.1}$$

because in the energy range

$$\sqrt{x} < m_\tau$$

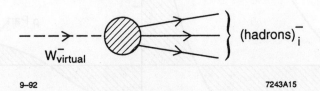

Fig. 15

the vertex in Fig. 15 is too complicated. We must use special methods which depend on other data. In this section I show the special methods for

$$\tau^- \to \pi^- + \nu_\tau \tag{5.2}$$

$$\tau^- \to K^- + \nu_\tau . \tag{5.3}$$

Figure 16a shows the diagram for the τ decay in Eq. (5.2). We cannot calculate the strength of the W_π vertex, but it is exactly the same vertex as in π decay, Fig. 16b,

$$\pi^- \to \mu^- + \bar{\nu}_\mu . \tag{5.4}$$

For Eq. (5.2)

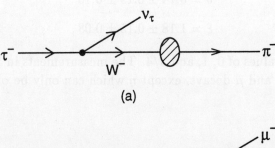

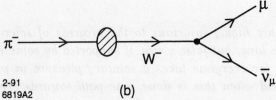

Fig. 16

$$\Gamma(\tau^- \to \pi^- \nu_\tau) = \frac{G_F^2 m_\tau^3 f_\pi^2 \cos^2 \theta_c}{16\pi} \left[1 - \frac{m_\pi^2}{m_\tau^2}\right]^2 \tag{5.5}$$

and for Eq. (5.4)

$$\Gamma(\pi^- \to \nu^- \bar{\nu}_\mu) = \frac{G_F^2 m_\pi m_\mu^2 f_\pi^2 \cos^2 \theta_c}{8\pi} \left[1 - \frac{m_\mu^2}{m_\pi^2}\right]^2. \tag{5.6}$$

In these equations f_π summarizes what we cannot calculate precisely about the W_π vertex. Radiative corrections which are of order α/π are ignored in these equations.

Then one of the branching fraction ratios of Eq. (4.6), B_π/B_e, is given by

$$\frac{B_\pi}{B_e} = \frac{12\pi^2 f_\pi^2 \cos^2 \theta_c}{m_\tau^2} \left[1 - \frac{m_\pi^2}{m_\tau^2}\right]^2 \tag{5.7}$$

using Eqs. (5.5) and (4.14), and again ignoring radiative corrections. Using Eq. (5.6) and the π lifetime

$$f_\pi = 132 \text{ MeV}, \tag{5.8}$$

the calculation in Eq. (5.7) gives

$$B_\pi/B_e \text{ (predicted)} = 0.61. \tag{5.9}$$

With radiative corrections

$$B_\pi/B_e \text{ (predicted)} = 0.6129 \pm 0.0018. \tag{5.10}$$

Using the world average values from Tables 1 and 4

$$B(\tau^- \to h^- \nu_\tau) = (11.77 \pm 0.14)\% \tag{5.11}$$

$$B(\tau^- \to K^- \nu_\tau) = (0.68 \pm 0.04)\% \tag{5.12}$$

we obtain

$$B(\tau^- \to \pi^- \nu_\tau) = (11.09 \pm 0.15). \tag{5.13}$$

Using Eq. (4.29)

$$B_\pi/B_e \text{ (measured)} = 0.623 \pm 0.009 \tag{5.14}$$

which is in good agreement with Eq. (5.10).

Next we consider

$$B_K/B_\pi = B(\tau^- \to K^- \nu_\pi)/B(\tau^- \to \pi \nu_\tau) \tag{5.15}$$

to test the effect of Cabibbo suppression. In analogy to Eq. (5.5)

$$\Gamma(\tau^- \to K^- \nu_\tau) = \frac{G_F^2 m_\tau^3 f_K^2 \sin^2 \theta_c}{16\pi} \left[1 - \frac{m_K^2}{m_\tau^2}\right]^2 \tag{5.16}$$

and from the lifetime for

$$K^- \to \mu^- + \bar{\nu}_\mu \tag{5.17}$$

$$f_K = 161 \text{ MeV} \tag{5.18}$$

Combining Eqs. (5.5) and (5.16), the prediction is

$$B_K/B_\pi = \tan^2 \theta_c \left(\frac{f_K}{f_\pi}\right)^2 \left[\frac{m_\pi^2 - m_K^2}{m_\tau^2 - m_\pi^2}\right]^2 = 0.071 \,.$$

With radiative corrections (Decker and Finkemeier 1993)

$$B_K/B_\pi = 0.066 \,. \tag{5.19}$$

The measured value of B_K/B_π from Eqs. (5.11) and (5.12) is

$$B_K/B_\pi \text{ (measured)} = 0.057 \pm 0.020 \tag{5.20}$$

which is in moderate agreement with Eq. (5.19).

5.2. *Application of Quantum Number Conservation in Non-Strange Hadronic Decays*

The rules from quantum number conservation which control non-strange hadronic decays of the τ have been frequently derived and discussed since the original work of Tsai (1971). I will not repeat the discussion here but simply quote the conclusions from Perl (1992).

The weak charged current in τ decay has the following properties:

$$\begin{aligned} \text{Isospin}: &\quad I = 1 &&\text{for vector and axial vector currents} \\ G-\text{parity}: &\quad G = +1 &&\text{for vector current} \\ &\quad G = -1 &&\text{for axial vector current} \\ \text{Spin}-\text{parity}: &\quad J^P = 1^- &&\text{for vector current} \\ &\quad J^P = 0^-, 1^+ &&\text{for axial vector current} \end{aligned} \tag{5.21}$$

The G-parity assignment <u>opposite</u> to that in Eq. (5.21) corresponds to a so-called second class current, the decay width is then suppressed by a factor of 10^{-4} to 10^{-6} as discussed below.

It is straightforward to apply the G and J^P requirements to the non-strange hadrons which are produced in τ decay:

$$\begin{aligned} \pi: &\quad G = -1, \quad J^P = 0^- \\ \eta: &\quad G = +1, \quad J^P = 0^- \\ \rho: &\quad G = +1, \quad J^P = 1^- \\ \omega: &\quad G = -1, \quad J^P = 1^- \end{aligned} \tag{5.22}$$

and so forth. For example in $\tau^- \to \nu_\tau \pi^-$ the π with $G = -1$, $J^P = 0^-$ is produced through the axial vector current decay. Conversely, the decay $\tau^- \to \nu_\tau \rho^-$ occurs through the vector current since $G = +1$. However the decay

$$\tau^- \to \nu_\tau + \pi^- + \eta \tag{5.23}$$

is forbidden since $G(\pi\eta) = -1$ requires an axial vector current with $J^P = 0^-$ or 1^+. But for $J = 0$, $P(\pi\eta) = +1$ and for $J = 1$, $P(\pi\eta) = -1$.

In a decay with n π's

$$\tau^- \to \nu_\tau + (n\pi)^- \tag{5.24}$$

$G = (-1)^n$. Hence the vector current produces states with an even number of π's.

Isopin conservation is also used to derive inequalities between different hadronic decay modes with the same I (Gilman and Rhie 1985). Consider for example the 3π modes

$$\begin{aligned} \tau^- &\to \nu_\tau + \pi^- + \pi^0 + \pi^0 \\ \tau^- &\to \nu_\tau + \pi^- + \pi^+ + \pi^- \end{aligned} \tag{5.25}$$

with $I = 1$. Gilman and Rhie (1985) show

$$\frac{\Gamma(\tau^- \to \nu_\tau \pi^- \pi^0 \pi^0)}{\Gamma(\tau^- \to \nu_\tau \pi^- \pi^0 \pi^0) + \Gamma(\tau^- \to \nu_\tau \pi^- \pi^+ \pi^-)} \leq \frac{1}{2}. \tag{5.26}$$

Hence

$$B(\tau^- \to \nu_\tau \pi^- \pi^0 \pi^0) \leq B(\tau^- \to \nu_\tau \pi^- \pi^+ \pi^-). \tag{5.27}$$

The G-parity rule in Eq. (5.21) depends upon ignoring the effect of the unequal masses of the u and d quarks, $m_u \neq m_d$, and ignoring the effect of electromagnetism. Once these effects are taken into account the τ decay can occur through the so-called second-class current. For second-class current decays,

$$\begin{aligned} \text{Vector}: \ &G = -1, \ J^P = 1^- \\ \text{Axial vector}: \ &G = +1, \ J^P = 0^-, \ 1^+. \end{aligned} \tag{5.28}$$

But the decay widths and hence the branching fractions are reduced by

$$\left(\frac{m_d - m^\mu}{m_\pi}\right)^2 \sim 10^{-4} \tag{5.29a}$$

or

$$\alpha^2 \sim 10^{-4} \tag{5.29b}$$

or even more (Leroy and Pestieau 1978, Pich 1987, Zachos and Meurice 1987). Using B_π of about 0.1 as a standard for first-class current decays, the second-class current branching fraction will be 10^{-5} or less. The present upper bound for the decay in Eq. (5.23) is (Artuso et al 1992),

$$B(\tau^- \to \nu_\tau \pi^- \eta) < 3.4 \times 10^{-4}, \quad 95\% \text{ CL}. \tag{5.30}$$

Quoting again from Perl (1992), there are two interests in observing and studying second-class current decays. First, what is the strength of a second-class current decay due to the electromagnetic correction, that is a decay within the standard model? Second, are there second-class current decays whose properties cannot be explained by the standard model? Interesting discussions are given by Berger and Lipkin (1987) and by Bramon et al. (1987).

5.3. $\tau^- \to \rho^- + \nu_\tau$ and Other Vector Decay Modes

The decay width of a τ vector decay mode can be calculated from the cross section from e^+e^- annihilation to a related final state (Tsai 1971, Gilman and Rhie 1985, Narison and Pich 1993), but the e^+e^- annihilation section must be measured. The calculation can be done because the unknown W-hadron vertex in the decay process is connected by the

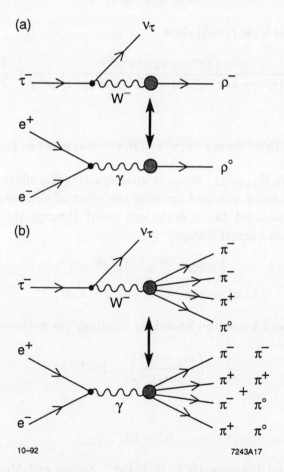

Fig. 17

conserved vector current hypothesis to the unknown γ-hadron vertex in the annihilation process, Fig. 17.

For example, the decay width, Γ, for

$$\tau^- \to \rho^- + \nu_\tau \tag{5.31}$$

is related to the cross section, σ, for

$$e^+ + e^- \to \rho^0 \tag{5.32}$$

through

$$\Gamma(\tau^- \to \rho^- \nu_\tau) = \frac{G_F^2 \cos^2\theta_c m_\tau^3}{384\pi^5 \alpha^2}$$

$$\times \int_0^{m_\tau^2} q^2\, dq^2 \left(1 - \frac{q^2}{m_\tau^2}\right)^2 \left(1 + 2\frac{q^2}{m_\tau^2}\right) \sigma_{I=1}\left(e^+ e^- \to \rho^0,\, q^2\right) \tag{5.33}$$

Thus Eidelman (1995) calculates

$$B(\tau^- \to \nu_\tau \pi^- \pi^0,\ \text{CVC prediction}) = (24.9 \pm 0.7)\% \tag{5.34a}$$

and Sobie (1995) calculates

$$B(\tau^- \to \nu_\tau \pi^- \pi^0,\ \text{CVC prediction}) = (24.3 \pm 1.1)\% \tag{5.34b}$$

These calculations agree well with

$$B(\tau^- \to \nu_\tau \pi^- \pi^0,\ \text{measured}) = (24.84 \pm 0.21)\%, \tag{5.34c}$$

this value is derived from $B(\tau^- \to \nu_\tau h^- \pi^0)$, Table 1, and $B(\tau^- \to \nu_\tau K^- \pi^0)$, Table 4. The $\tau^- \to \pi^- \pi^0 \nu_\tau$ decay mode is almost completely $\rho(770)$, there is a trace of $\rho(1450)$ (Weinstein 1995).

The 4π vector starts decay mode

$$\tau^- \to \pi^- + \pi^+ + \pi^- + \pi^0 + \nu_\tau \tag{5.35a}$$

can also be compared with the cross section for

$$e^+ + e^- \to 2\pi^- + 2\pi^+, \quad \pi^- + \pi^+ + 2\pi^0 \tag{5.35b}$$

using CVC (Eidelman 1995),

$$\sigma(\tau \to \nu_\tau 3\pi\pi^0,\ \text{CVC predicted}) = (4.20 \pm 0.29)\%. \tag{5.35c}$$

From Table 1 the measured cross section is

$$\sigma(\tau \to \nu_\tau 3h\pi^0,\ \text{measured}) = (4.45 \pm 0.14)\% \tag{5.35d}$$

which includes a few tenthes of a percent of K modes.

Similarly (Eidelman 1995)

$$\sigma(\tau \to \nu_\tau \pi 3\pi^0, \text{ CVC, predicted}) = (1.08 \pm 0.05)\% \qquad (5.36a)$$

and

$$\sigma(\tau \to \nu_\tau h 3\pi^0, \text{ measured}) = (1.15 \pm 0.15)\% \qquad (5.36b)$$

from Table 1.

5.4. *Measurements of Modes Containing K Mesons*

In the past several years there has been tremendous progress in measuring the branching fractions for modes containing K mesons; Table 4 from Heltsley (1995) summarizes these measurements. The sum of all these branching fractions is

$$\sum B(1 \text{ or more K's}) = (3.4 \pm 0.2)\% . \qquad (5.37)$$

The $2K$ modes contribute only 0.45% to this sum even though these modes are not Cabibbo suppressed.

Table 4. The branching fraction B in % of modes containing K mesons from Heltsley (1995), $\delta B/B$ is the fractional error.

Mode	B in %	$\delta B/B$
$K^-\nu_\tau$	0.68 ± 0.04	0.06
$K^{*-}\nu_\tau$	1.36 ± 0.08	0.06
$K^-\pi^+\pi^-\nu_\tau$	0.40 ± 0.09	0.2
$\bar{K}^0\pi^-\pi^0\nu_\tau$	0.41 ± 0.07	0.2
$K^-2\pi^0\nu_\tau$	0.09 ± 0.03	0.3
$K^+K^-\pi^-\nu_\tau$	0.20 ± 0.07	0.4
$K^0K^-\nu_\tau$	0.13 ± 0.04	0.3
$K^0K^-\pi^0\nu_\tau$	0.12 ± 0.04	0.3

5.5. *Comparison of B_1 and B_3 with $\sum_i B_i$*

Since the work of Gilman and Rhie (1985) and Truong (1984) the world of τ research has been faced with the question: Can we find and identify all the decay modes of the τ with branching fractions

$$B_i \gtrsim \text{few} \times 0.1\%$$

such that

$$\sum_i B_i = 100\% ? \qquad (5.38)$$

On the face of it Eq. (5.38) is an identity; the fundamental question is: Are there some unknown and unconventional τ decays such that

$$\sum_i B_i \text{ (known and measured}) < 100\% ? \qquad (5.39)$$

Historically the question was first asked about decay modes with 1-charged particle, B_{1i}, since these made up most τ decays. The topological 1 and 3-charged particle branching fractions, Table 2, are

$$\begin{aligned} B_1 &= (85.41 \pm 0.23)\% \\ B_3 &= (14.49 \pm 0.23)\% \,. \end{aligned} \qquad (5.40)$$

We usually break up the question in Eq. (5.39) into two questions. Does

$$\sum_i B_{1i} \text{ (known and measured)} = B_1 \qquad (5.41\text{a})$$

and does

$$\sum_i B_{3i} \text{ (known and measured)} = B_3 \,. \qquad (5.41\text{b})$$

Turning to data, Table 5 is a recent compilation by Heltsley (1995). The numbers in the tables are averages of measurements from several or even many experiments. To try to answer the questions in Eqs. (5.41) I give

$$\begin{aligned} \Delta_1 &= B_1 - \sum_i B_{1i} \text{ (known and measured)} = 0.75 \pm 0.44 \\ \Delta_3 &= B_3 - \sum_i B_{3i} \text{ (known and measured)} = 0.09 \pm 0.35 \,. \end{aligned} \qquad (5.42\text{a})$$

Remember, these compilations have many data sets in common, they are not statistically independent. Hayes in references (Hayes 1995) and (Montenat 1994) has given an important discussion of the problems in compiling such tables.

Thus within these errors Δ_1 and Δ_3 are consistent with zero. Thus the search begun ten years ago (Gilman and Rhie 1985, Truong 1984) for non-zero Δ's has ended for the present. Since the sum of the exclusive mode B_i's is within 1% of 100% there are no mysterious modes with $B \gtrsim 1\%$. But are there mysterious modes with $B \approx 10^{-3}$ or $B \approx 10^{-4}$. Can we begin to use our detectors as bubble chambers which were once used to pick out a few "new physics" events out of thousands of ordinary events? For example, is there a mysterious decay

$$\tau^- \to \nu_\tau + x^- + 3\gamma \qquad (5.43)$$

which does not come from

$$\tau^- \to \nu_\tau + h^- + \pi^0 + \text{fake } \gamma$$

or

$$\tau^- \to \nu_\tau + h^- + 2\pi^0 \, (\gamma \text{ lost})$$

or

$$\tau^- \to \nu_\tau + e^- + \bar{\nu}_e + 3\gamma \, ?$$

Table 5. Comparison of sums of individual branching fractions with the B_1, B_3 and B_5 topological branching fractions from Heltsley (1995).

Mode	B in %	$\delta B/B$
$e\bar{\nu}_e\nu_\tau$	17.79 ± 0.09	0.005
$\mu\bar{\nu}_\mu\nu_\tau$	17.33 ± 0.09	0.005
$h\nu_\tau$	11.77 ± 0.14	0.012
$h\pi^0\nu_\tau$	25.36 ± 0.21	0.008
$h2\pi^0\nu_\tau$	9.18 ± 0.14	0.015
$h3\pi^0\nu_\tau$	1.15 ± 0.15	0.13
$h4\pi^0\nu_\tau$	0.16 ± 0.07	0.4
$h\bar{K}^0\nu_\tau$	1.03 ± 0.09	0.09
$h\pi^0\bar{K}^0\nu_\tau$	0.53 ± 0.06	0.11
$h\bar{K}^0 K^0\nu_\tau$	0.08 ± 0.04	0.5
$h\pi^0\eta(\to\gamma\gamma)\nu_\tau$	0.07 ± 0.01	0.14
$h\omega(\to\pi^0\gamma)\nu_\tau$	0.18 ± 0.02	0.11
$h\omega(\to\pi^0\gamma)\pi^0\nu_\tau$	0.03 ± 0.01	0.3
$\sum_i B_{1i}$	84.66 ± 0.38	0.005
B_1	85.41 ± 0.23	0.003
Δ_1	0.75 ± 0.44	
$3h\nu_\tau$	9.24 ± 0.21	0.02
$3h\pi^0\nu_\tau$	4.45 ± 0.14	0.03
$3h2\pi^0\nu_\tau$	0.51 ± 0.05	0.10
$3h \geq 3\pi^0\nu_\tau$	0.20 ± 0.07	0.4
$\sum_i B_{3i}$	14.40 ± 0.27	0.02
B_3	14.49 ± 0.23	0.02
Δ_3	0.09 ± 0.35	
$5h\nu_\tau$	0.073 ± 0.007	0.1
$5h\pi^0\nu_\tau$	0.021 ± 0.006	0.3
$\sum_i B_{5i}$	0.09 ± 0.01	0.1
B_5	0.10 ± 0.01	0.1
Δ_5	0.01 ± 0.014	
$\sum_i B_i$	99.15 ± 0.46	0.005

5.6. Measurements of Strong Interaction Constant α_s

My research career in experimental particle physics began in strong interaction physics, but over the years my interest has shifted to weak and electromagnetic interactions mostly due to my discovery of the τ, but also because of the complexity of quantum chromodynamics (QCD). I have not studied that theory, so I limit my discussion of QCD in τ physics to reproducing what the experts say about the strong interaction constant, α_s, in τ physics.

As summarized by Narison (1994) the seminal work of Braaten (1988) defining

$$R_\tau = \frac{\Gamma(\tau \to \nu_\tau + \text{hadrons})}{\Gamma(\tau \to \nu_\tau e \bar{\nu}_e)} \quad (5.44)$$

leads to

$$R_\tau = 3(|V_{ud}|^2 + |V_{us}|^2) S_{EW} [1 + \delta_{EW} + \delta_{S,P} + \delta_{S,NP}]. \quad (5.45)$$

Here V_{ud} and V_{us} are the usual CKM matrix elements, $S_{EW} = 1.0194$ and $\delta_{EW} = 0.0010$ are electroweak contributions. The QCD terms are the perturbation term

$$\delta_{S,P} = \frac{\alpha_S}{\pi} + 5.20\left(\frac{\alpha_S}{\pi}\right)^2 + 26.4\left(\frac{\alpha_S}{\pi}\right)^3 \pm 50\left(\frac{\alpha_S}{\pi}\right)^4 \quad (5.46)$$

and the non-perturbation term

$$\delta_{S,NP} = 1(15 \pm 4) \times 10^{-3} \quad (5.47)$$

To illustrate using B_e and B_μ from Table 1

$$R_\tau = 3.65 \quad (5.48)$$

compared to the naive value $R_\tau = 3$, Eqs. (4.7) and (4.8). Ignoring δ_{SW} and $\delta_{S,NP}$ and setting $S_{EW} = 1$, the difference between 3.65 and 3 is attributed to

$$0.65 = 3\left[\frac{\alpha_S}{\pi} + 5.20\left(\frac{\alpha_S}{\pi}\right)^2 + 26.4\left(\frac{\alpha_S}{\pi}\right)^3\right] \quad (5.49)$$

which gives $\alpha_S/\pi \approx 0.11$, hence

$$\alpha_S \approx 0.35. \quad (5.50)$$

Precise considerations (Duflot 1995) give

$$\alpha_S = 0.355 \pm 0.021 \quad (5.51a)$$

and

$$\alpha_S = 0.309 \pm 0.024 \quad (5.51b)$$

from the ALEPH and CLEO experiments. These values of α_S are at the τ mass. To compare with α_S measured at the Z^0 mass one uses a formula whose approximate form is (Kane 1987)

$$\alpha_S(m_Z) = \frac{\alpha_S(m_\tau)}{1 + \frac{21\alpha_S(m_\tau)}{12\pi} \ln\left(\frac{m_Z^2}{m_\tau^2}\right)} \quad (5.52)$$

This results in (Duflot 1995)

$$\alpha_s(m_Z) = 0.121 \pm 0.002 \tag{5.53a}$$

$$\alpha_s(m_Z) = 0.114 \pm 0.003 \tag{5.53b}$$

corresponding to the $\alpha_s(m_\tau)$ values in Eqs. (5.51). Deep inelastic scattering measurements give (Montenat 1994)

$$\alpha_s(m_Z) = 0.112 \pm 0.004.$$

Recent discussions of the extraction of α_s from hadronic τ decays have been given by Narison (1995) and Altarelli (1995).

6. Tau Spin Phenomena

6.1. τ Spin Alignment and Decay Correlation

Tsai (1971) contains the seminal discussion on the alignment of τ spins and the correlation of τ decays below the Z^0 energy range. Close to threshold, the τ is nonrelativistic and the differential cross section is given by

$$\frac{d\sigma}{d\Omega} = \frac{\alpha^2 \beta}{8 E_{\text{tot}}^2} [1 + (\vec{s}^- \cdot \vec{n})(\vec{s}^+ \cdot \vec{n})]. \tag{6.1}$$

Here $\vec{s}^-$ and $\vec{s}^+$ are unit vectors in the spin directions of the τ^- and τ^+ in their rest frames, $\vec{n}$ is the unit vector in the direction of the e^- beam, and E_{tot} is the total energy. This formula is for unpolarized e^- and e^+ beams with the electron mass set to 0. We see $d\sigma/d\Omega$ is maximum when

$$\vec{s}^- \cdot \vec{n} = \vec{s}^+ \cdot \vec{n} = +1, \tag{6.2a}$$

both τ spins pointing along the e^- direction, or

$$\vec{s}^- \cdot \vec{n} = \vec{s}^+ \cdot \vec{n} = -1, \tag{6.2b}$$

both τ spins pointing along the e^+ direction. Thus there is τ spin alignment along the beam axis.

At very high energies, ignoring the Z,

$$\frac{d\sigma}{d\Omega} =_{\gamma \to \infty} \frac{\alpha^2 \beta}{16 E_{\text{tot}}^2} [(1 + \cos^2 \theta)(1 + S_z^- S_z^+) + \sin^2 \theta (S_x^- S_x^+ - S_y^- S_y^+)] \tag{6.3}$$

where the z axis is along the τ^- flight direction and the y axis is defined by $\vec{p}^- \times \vec{n}$. Here the cross section is maximum when

$$S_z^- = S_z^+ = +1, \quad S_x^- = S_x^+ = S_y^- = S_y^+ = 0, \tag{6.4a}$$

both τ spins point along the τ^+ flight direction,

$$S_z^- = S_z^+ = -1, \quad S_x^- = S_x^+ = S_y^- = S_y^+ = 0, \tag{6.4b}$$

both τ spins point along the τ^- flight direction. Thus at high energies there is τ spin alignment along the $\tau^- - \tau^+$ axis.

As has been discussed recently by Tsai (1994) polarized e^- or e^+ beams will produce polarized τ spins rather than just aligned τ spins, but this has not been done at low energies. A polarized e^- beam is of course used in the SLD-SLC experiment at the Z^0 (Abe et al 1994).

The spin direction of a tau can be determined statistically by the energy and angle spectra of τ decay products (Tsai 1971, Weinstein and Stroynowski 1993). For

$$\tau^- \to \nu_\tau + e^- + \bar{\nu}_e \qquad (6.5a)$$

in the τ rest frame, we have a special case of Eq. (4.41)

$$\frac{1}{\Gamma} \frac{d^2\Gamma}{d\cos\theta \, dx} = x^2(3 - 2x) - P_\tau \cos\theta(x^2 - 2x^3) \qquad (6.5b)$$

where as before $x = 2E_e/m_\tau$, θ is the angle between the direction of motion of the τ^- and the direction of motion of the e^-, and P_τ is the τ polarization in the direction of the τ^-. In the rest frame for

$$\tau^- \to \nu_\tau + \pi^- \qquad (6.6a)$$

$$\frac{1}{\Gamma} \frac{d\Gamma}{d\cos\theta} = 1 + P_\tau \cos\theta \qquad (6.6b)$$

where θ is the angle between the τ^- and the π^- and for the decay

$$\tau^- \to \nu_\tau + \rho^- \qquad (6.7a)$$

$$\frac{1}{\Gamma} \frac{d\Gamma}{d\cos\theta} = 1 + 0.46 P_\tau \cos\theta. \qquad (6.7b)$$

In all these equations we have assumed the τ and ν_τ are conventional leptons.

6.2. Polarization at the Z^0

Parity violation in

$$e^+ + e^- \to Z^0 \to \tau^+ + \tau^- \qquad (6.8)$$

results in the τ's being polarized along their direction of motion. Conventional weak interaction theory predicts the τ polarization will be

$$P_\tau = -\frac{2g_v^\tau g_a^\tau}{g_v^{\tau 2} + g_a^{\tau 2}} \qquad (6.9a)$$

where for the τ^-

$$\begin{aligned} g_v^\tau &= -\tfrac{1}{2} \\ g_a^\tau &= -\tfrac{1}{2} + 2\sin^2\theta_W \end{aligned} \qquad (6.9b)$$

Using (Montenat 1994) $\sin^2 \theta_W = 0.232$.

$$P_\tau(\text{predicted}) = -0.143. \tag{6.9c}$$

Recent measurements have been reviewed by Harton (1995), Bella (1995) and Matorvas (1995). As summarized by Kounine (1995)

$$P_\tau(\text{measured}) = -0.145 \pm 0.009 \tag{6.9d}$$

in good agreement with Eq. (6.9c). Of course present average values of $\sin^2 \theta_W$ include P_τ measurements.

6.3. Search for CP Violation in τ Production

The correlation between τ decay products have been used to search for CP violation in τ pair production (Bernreuther 1993 and ref. contained therein)

$$e^+ + e^- \to \tau^+ + \tau^-. \tag{6.10}$$

The method involves finding the average value of a CP-odd but CPT-even quantity such as

$$(\vec{P}_+ - \vec{P}_-)_i (\vec{P}_+ \times \vec{P}_-)_j + (i \leftrightarrow j). \tag{6.11}$$

Here $\vec{P}_+$ and $\vec{P}_-$ are the momenta of an e, μ, or π from the decay of the τ^+ and τ^-. A P or T transformation changes $\vec{P}$ to $-\vec{P}$ and C interchanges $\vec{P}_+$ and $\vec{P}_-$. If CP is conserved in Eq. (6.10) then the average value of the quantity is 0.

All measurements are consistent with 0 and present practice is to give the upper limit on CP violation in terms of a τ dipole moment. At the Z^0 the upper limits on a weak dipole moment d_τ^Z from LEP 1 experiments are (Stahl 1995)

$$\begin{aligned}|\text{Re}(d_\tau^Z)| &< 6.4 \times 10^{-18}\, \text{e cm}, \quad 95\% \text{ CL}\\ |\text{Im}(d_\tau^Z)| &< 4.5 \times 10^{-17}\, \text{e cm}, \quad 95\%.\end{aligned} \tag{6.12}$$

Below the Z^0 the lack of observed CP violation gives a limit on the electric dipole moment d_τ^γ (Montanet 1994)

$$|d_\tau^\gamma| < 5 \times 10^{-17}\, \text{e cm}, \quad 95\% \text{ CL}. \tag{6.13}$$

For further discussion see Weinstein and Stroynowski (1993).

6.4. Search for CP Violation in τ Decay

The τ provides almost the only way to search for CP violation in the decay of leptons (Lee 1994, Tsai 1994, Nelson 1995). Tsai (1994) has discussed using a longitudinally polarized e^- or e^+ beam at low energies: a tau-charm factory, a B-factory, or CESR. Nelson (1995) has discussed using τ spin correlations. At present there are no measurements on CP violation in τ decay.

6.5. Tau Magnetic Moment

If the τ is a conventional Dirac charged particle, its magnetic moment is given (Samuel et al. 1991) by

$$\mu_\tau = g_\tau \frac{eh}{2m_\tau c} \qquad (6.14a)$$

$$\frac{g_\tau - 2}{2} = \frac{\alpha}{2\pi} + O(\alpha^2) = a_\tau, \qquad (6.14b)$$

where

$$\frac{\alpha}{2\pi} = 1.16 \times 10^{-3}$$

is the Schwinger term. In Eq.(6.14b) Samuel *et al* (1991) give

$$a_\tau = 1.177 \times 10^{-3}.$$

Escribano and Masso (1993) have set 2σ limits on α_τ from LEP I experimental data:

$$-8 \times 10^{-3} \leq \alpha_\tau \leq 1 \times 10^2. \qquad (6.15)$$

These measured limits are ten times larger than the expected valued. Laursen *et al.* (1984) have suggested a method using leptonic radiative decay.

7. The τ Neutrino ν_τ

> *This is my letter to the world,*
> *That never wrote to me, —*
> *The simple news that Nature told,*
> *With simple majesty.*
>
> Emily Dickinson

Is the tau neutrino a simple, massless, stable, Dirac particle which obeys perfectly the conventional theory of weak interactions? Or is the ν_τ a complicated particle with non-zero mass, perhaps with mixing properties, perhaps with decays? All confirmed experimental results agree with the first alternative. In this section I summarize that date, but I also outline some speculations on the ν_τ being a complicated particle.

7.1. ν_τ Mass Limits

Present upper limits from terrestrial experiments on the ν_τ mass, m_{ν_τ} are derived from the decay modes

$$\tau^- \to \nu_\tau + 3\pi^- + 2\pi^+ \qquad (7.1)$$

$$\tau^- \to \nu_\tau + 2\pi^- + \pi^+ + 2\pi^0. \qquad (7.2)$$

Cerutti (1995) has reviewed the recent 95% CL upper limits. In alphabetical order

$$\begin{array}{ll} \text{ALEPH} & m_{\nu_\tau} < 23.8 \text{ MeV}/c^2 \\ \text{ARGUS} & m_{\nu_\tau} < 31.0 \text{ MeV}/c^2 \\ \text{CLEO} & m_{\nu_\tau} < 32.7 \text{ MeV}/c^2 \\ \text{OPAL} & m_{\nu_\tau} < 74.0 \text{ MeV}/c^2. \end{array} \qquad (7.3)$$

The decay mode (Gomez-Cadenas *et al.* 1990)

$$\tau^- \to \nu_\tau + K^- + K^+ + \pi^- \tag{7.4}$$

and the decay mode (Gomez-Cadenas and Gonzales-Garcia 1989, Mendel *et al.* 1986)

$$\tau^- \to \nu_\tau + e^- + \bar{\nu}_e \tag{7.5}$$

can also be used to probe m_{ν_τ}, but are probably less sensitive.

Thus the upper limits on the masses of the three neutrinos are (Cerutti 1995, Montenat *et al.* 1994)

$$\begin{aligned} m_{\nu_\tau} &< 24 \text{ MeV}/c^2, & 95\% \text{ CL} \\ m_{\nu_\mu} &< 0.27 \text{ MeV}/c^2, & 90\% \text{ CL} \\ m_{\nu_e} &< 7 \text{ eV}/c^2, & 95\% \text{ CL} \end{aligned} \tag{7.6}$$

To compare these limits people sometimes use the assumption

$$\frac{m_{\nu_1}}{m_{\nu_2}} = \frac{m_1^2}{m_2^2}. \tag{7.7}$$

Using Eq. (7.9)

$$m_{\nu_\mu} \left(\frac{m_e}{m_\mu}\right)^2 < 6.3 \text{ eV}/c^2 \tag{7.8a}$$

$$m_{\nu_\tau} \left(\frac{m_e}{m_\tau}\right)^2 < 2.0 \text{ eV}/c^2 \tag{7.8b}$$

to be compared to $m_{\nu_e} \lesssim 7 \text{ eV}/c^2$.

There are also astrophysical and cosmological limits on m_{ν_τ} (Harari and Nir 1987, Turner 1995). As discussed recently by Harari (1995), if the ν_τ is stable and $m_{\nu_\tau} \sim$ few MeV/c^2 then

$$m_{\nu_\tau} \lesssim 65 \text{ eV}/c^2. \tag{7.9}$$

7.2. ν_τ as Dark Matter

There have been many papers considering the possibility that the ν_τ is the hypothetical dark matter of the universe (Harari 1989, Bergström and Rubinstein 1991, Langacker 1988, Ciudice 1991). For example, Harari (1995) has discussed the possibility that m_{ν_τ} lies in the range of 15-65 eV/c^2, and the use of $\nu_\mu - \nu_\tau$ oscillations to detect such a mass. Also, Ellis *et al.* (1992) has suggested $m_{\nu_\tau} \sim 10$ eV/c^2.

7.3. ν_τ Lifetime Limits

There is <u>no</u> evidence that the ν_τ is unstable. However, if $m_{\nu_\tau} > 0$, then ν_τ might decay in a variety of ways:

$$\nu_\tau \to \gamma + \nu_x \qquad (7.10a)$$

$$\nu_\tau \to e^+ + e^- + \nu_x \qquad (7.10b)$$

$$\nu_\tau \to \nu_x + \bar{\nu}_x + \nu_y \qquad (7.10c)$$

$$\nu_\tau \to b^0 + \nu_x \,. \qquad (7.10d)$$

In Eq. (7.10d) b^0 would be a boson. If the ν_τ decayed through the processes in Eqs. (7.10a) or (7.10b), then with a sufficiently short ν_τ lifetime, T_{ν_τ}, these decays would have been seen in $e^+e^- \to \tau^+\tau^-$ events. None have been reported and I estimate this leads to a lower limit

$$T_{\nu_\tau}/m_{\nu_\tau} \gtrsim 1 \text{ sec/eV} \,. \qquad (7.11)$$

There are much more stringent lower limits from astrophysical and cosmological consideration as summarized by Montenat et al. (1995). These lower limits depend upon assumptions for m_{ν_τ}. Lower limits of the order of $T_{\nu_\tau}/m_{\nu_\tau} \sim 10^{15}$ sec/eV have been calculated. The subject of instability of the ν_τ remains speculative.

7.4. ν_τ Weak Interactions

In earlier sections I have discussed the $\tau - W - \nu_\tau$ vertex and pointed out that all evidence agrees with that vertex being conventional.

Precise studies of the invisible width and leptonic widths in Z^0 decays (Montenat et al. 1994) give the number of light neutrinos as

$$N_\nu = 2.983 \pm 0.025 \,. \qquad (7.12)$$

Assuming that the ν_e and ν_μ couplings to the Z are conventional, from Eq. (7.12)

$$g_{\nu_\tau z \nu_\tau}/g_{\nu_e z \nu_e} = 0.991 \pm 0.013 \,. \qquad (7.13)$$

Hence within present errors the $\nu_\tau - Z - \nu_\tau$ vertex is conventional.

7.5. The ν_τ and Neutrino Mixing

At present there is no confirmed evidence for the mixing of ν_τ with other neutrinos or for any neutrino mixing (Montenat et al. 1994, Loverre 1995). The theory of neutrino mixing and oscillation is recounted well by Böehm and Vogel (1987).

The present upper limits on $v_e \to \nu_\tau$ and $\nu_\mu \to \nu_\tau$ mixing come from the oscillation search experiment of Ushida et al. (1986). At present searches are going on at CERN for $\nu_\mu \leftrightarrow \nu_\tau$ oscillations using the NOMAD experiment (Astier et al. 1991, Rubbia 1995) and the CHORUS experiment (Armenise et al. 1990, Rosa 1995). Searches will also take place at Fermilab (Reay 1995, Michael 1995).

7.6. ν_τ-Nucleon Interactions

As yet there are no experiments on the interaction of the ν_τ with matter. The study of ν_τ interactions would be directed first to the weak charged current reaction

$$\nu_\tau + N \to \tau^- + \text{hadrons} \tag{7.14}$$

where N is a nucleon. Eventually the weak neutral current reaction

$$\nu_\tau + N \to \nu_\tau + \text{hadrons} \tag{7.15}$$

and the weak leptonic reaction

$$\nu_\tau + e^- \to \nu_\tau + e^- \tag{7.16}$$

might be studied. However, at present just studying Eq. (7.14) is very difficult because: (a) it is necessary to produce a neutrino beam with sufficient ν_τ intensity and (b) it is difficult to identify the $\nu_\tau - N$ interaction.

The best known method for producing a neutrino beam containing ν_τ's begins with the reactions

$$p + N \to D_s^\pm + \text{hadrons}. \tag{7.17}$$

Here N means p, n or nucleus. These reactions are followed by the meson decays

$$\begin{aligned} D_s^- &\to \tau^- + \bar{\nu}_\tau \\ D_s^+ &\to \tau^+ + \nu_\tau \end{aligned} \tag{7.18}$$

and then the τ decays

$$\begin{aligned} \tau^- &\to \nu_\tau + \text{other particles} \\ \tau^+ &\to \bar{\nu}_\tau + \text{other particles}. \end{aligned} \tag{7.19}$$

This beam of ν_τ's and $\bar{\nu}_\tau$'s would also contain the other neutrinos: $\nu_e, \bar{\nu}_e, \nu_\mu, \bar{\nu}_\mu$.

The reactions

$$\begin{aligned} \nu_\tau + N &\to \tau^- + \text{hadrons} \\ \bar{\nu}_\tau + N &\to \tau^+ + \text{hadrons} \end{aligned} \tag{7.20}$$

would then be studied using a neutrino interaction detector with properties which allowed separation of Eq. (7.20) from non-ν_τ reactions such as

$$\begin{aligned} \nu_e + N &\to e^- + \text{hadrons} \\ \nu_e + N &\to \nu_e + \text{hadrons} \end{aligned} \tag{7.21}$$

and so forth.

A bubble chamber experiment (Talebzadeh et al. 1987) used this method with 400 GeV protons interaction in a Cu target and beam dump. No ν_τ or $\bar{\nu}_\tau$ interactions were found, but the upper limit was consistent with the expected rate of such interactions assuming conventional weak interaction theory.

Experiment E-872, scheduled to run at Fermilab in the second half of the 1990's is designed to measure the charged current interaction of τ neutrinos (Ludberg *et al.* 1994, Reay 1995).

As discussed by De Rújula and Rückl (1984), Isaev and Tsarev (1989), and De Rújula *et al.* (1992), the higher energies of future proton-proton colliders bring two substantial benefits. First the cross section for the D_s production reactions (Eq. (7.18)) increase with energy. Second, the principle proposed method for detecting

$$\nu_\tau + N \to \tau^- + \text{hadrons}$$

and

$$\bar{\nu}_\tau + N \to \tau^+ + \text{hadrons}$$

uses the spatial separation between the primary ν_τ or $\bar{\nu}_\tau$ interaction vertex and the secondary decay vertex of the τ^- or τ^+. The larger the initial proton energy in Eq. (7.17) the larger the average ν_τ and $\bar{\nu}_\tau$ energies, and hence the larger the separation between the vertices.

References

Abe K. *et al.* (1992). *Phys. Rev. Lett.* **68** 3398.
Abe K. *et al.* (1994). SLAC-PUB-6605, submitted to *Phys. Rev. Lett.*
Acosta D. *et al.* (1994). *Phys. Rev.* **D44** 5690.
Adeva B. *et al.* (1986). *Phys. Lett.* **B179** 177.
Adeva B. *et al.* (1990). *Phys. Lett.* **B250** 205.
Aguilar-Benitez M. *et al.* (1990). *Phys. Lett.* **B239** 1.
Akrawy M. Z. *et al.* (1990). *Phys. Lett.* **B244** 135.
Albrecht H. *et al.* (1990). *Phys. Lett.* **B246** 278.
Albrecht H. *et al.* (1992a). *Phys. Lett.* **B292** 221.
Albrecht H. *et al.* (1992b). *Z. Phys.* **C55** 25.
Albrecht H. *et al.* (1992c). *Z. Phys.* **C55** 179.
Alexander *et al.* (1994). CLEO CONF 94-5.
Alitti J. *et al.* (1992). *Phys. Lett.* **B280** 137.
Almeida L. D. *et al.* (1991). *Phys. Rev.* **D44** 118.
Altarelli G. (1995). *Proc. Third Workshop on Tau Lepton Physics* (Montreux, 1994) *Nucl. Phys. B, Proc. Supp.* **40**, ed G. Rolandi.
Amaglobeli N. S. *et al.* (1991). *Massive Neutrinos, Tests of Funcdamental Symmetries* (Editions Frontiéres, Gif-sur-Yvette) Ed. J. Tran Thanh Van p. 335.
Aoki S. *et al.* (1993). *Prog. Theor. Phys.* **89** 131.
Armenise N. *et al.* (1990). CERN Proposal CERN-SPSC/90-42, unpublished.
Artuso M. *et al.* (1992). *Phys. Rev. Lett.* **69** 3278.
Astier P. *et al.* (1991). CERN Proposal CERN-SPSC/91-21, unpublished.
Avilez C. *et al.* (1978). *Lett. Nuovo Cimento* **21** 301.
Avilez C. *et al.* (1979). *Phys. Rev.* **D19** 2214.
Bacino W. *et al.* (1978). *Phys. Rev. Lett.* **42** 749.
Balest R. *et al.* (1993). *Phys. Rev.* **D47** 3671.
Barkov L. M. *et al.* (1985). Nucl. Phys. **B256** 365.

Bartel W. et al. (1986). *Z. Phys.* **C30** 371.
Bartett J. et al. (1994). *Phys. Rev. Lett.* **73** 1890.
Bean A. et al. (1993). *Phys. Rev. Lett.* **70** 138.
Beers L. (1989). *Proc. Tau-Charm Factory Workshop* (SLAC, Stanford) SLAC-REPORT-343, Ed. L. Beer.
Behrend H. J. et al. (1989). *Phys. Lett.* **B222** 163.
Bella G. (1995). *Proc. Third Workshop on Tau Lepton Physics* (Montreux, 1994) *Nucl. Phys. B, Proc. Supp.* **40**, Ed. G. Rolandi.
Berger E. L. and Lipkin H. J. (1987). *Phys. Lett.* **B189** 226.
Bergström L. and Rubinstein H. R. (1991). *Phys. Lett.* **B253** 168.
Bernreuther W. et al. (1993). *Phys. Rev.* **D48** 78 and references contained therein.
Boehm F. and Vogel P. (1987). *Physics of Massive Newtrinos* (Cambridge: Cambridge Univ. Press).
Bottcher C. and Strayer M. R. (1990). *J. Phys. G: Nucl. Part. Phys.* **16** 975.
Braaten E. (1988). *Phys. Rev. Lett.* **60** 1606.
Bramon A. et al. (1987). *Phys. Lett.* **B196** 543.
Buskulic D. et al. (1993). *Phys. Lett.* **B298** 479.
Cerutti F. (1995). *Proc. Third Workshop on Tau Lepton Physics* (Montreux, 1994) *Nucl. Phys. B, Proc. Supp.* **40**, Ed. G. Rolandi.
Ching C. H. and Oset E. (1991). *Phys. Lett.* **B259** 239.
Decamp D. et al. (1990). *Phys. Lett.* **B236** 501.
Decker R. and Finkemeir M. (1993). *Phys. Lett.* **B316** 403.
Decker R. and Finkemeir M. (1993). *Phys. Rev.* **D48** 4203.
del Aguila F. et al. (1991). *Phys. Lett.* **B271** 256.
De Rújula A. and Rückl R. (1984). *Proc. Large Hadron Collider in the LEP Tunnel* CERN 84-10, **Vol. II**, Ed. M. Jacob (CERN, Geneva) p 573.
De Rújula A. et al. (1992). CERN-TH-6452, IFAE-92/001.
Drell P. (1992). *Proc. 26th Int. Conf. High Energy Physics*, (American Inst. Physics, New York, 1993), Ed. J. R. Sanford, p. 3.
Drell S. D. (1958). *Ann. Phys.* **4** 75.
Duflot L. (1995). *Proc. Third Workshop on Tau Lepton Physics* (Montreux, 1994) *Nucl. Phys. B, Proc. Supp.* **40**, Ed. G. Rolandi.
Eicher R. (1987). *Proc. 1987 Int. Symp. Lepton and Photon Interactions at High Energies*, (Amsterdam: North Holland), Eds. W. Bartel and R. Rückl, p. 389.
Eicher E., Lane K. and Peskin M. (1983). *Phys. Rev. Lett.* **50** 811.
Eidelman S. (1995). *Proc. Third Workshop on Tau Lepton Physics* (Montreux, 1994), *Nucl. Phys. B, Proc. Supp.* **40**, Ed. G. Rolandi.
Eigen G. (1995). *Proc. Third Workshop on Tau lepton Physics* (Montreux, 1994) *Nucl. Phys. B, Proc. Supp.* **40**, ed G. Rolandi.
Ellis J. et al. (1992). CERN-TH-6569/92, CTP-TAMU-53/92.
Escribano R. and Masso E. (1993). *Phys. Lett.* **B301** 414.
Feynman R. P. (1949). *Phys. Rev.* **76** 769.
Gan K. K. (1993). *Proc. Second Workshop on Tau Lepton Physics* (World Scientific, Singapore, 1993), Ed. K. K. Gan.
Gilman F. J. and Rhie S. H. (1985). *Phys. Lett.* **D31** 1066.
Giudice G. F. (1991). *Mod. Phys. Lett.* **A6** 851.
Gomez-Cadenas J. J. Heusch C. A. and Seiden A. (1989). *Particle World* **1** 10.
Gomez-Cadenas J. J. and Gonzales-Garcia M. C. (1989). *Phys. Rev.* **D39** 1370.
Gomez-Cadenas J. J. et al. (1990). *Phys. Rev.* **D42** 3093.
Gomez-Cadenas J. J. et al. (1991). *Phys. Rev. Lett.* **66** 1007.
Harari H. (1995). *Proc. Third Workshop on Tau Lepton Physics* (Montreux, 1994) *Nucl. Phys. B, Proc. Supp.* **40**, Ed. G. Rolandi.

Harari H. and Nir Y. (1987). *Nucl. Phys.* **B292** 251.
Harari H. (1989). *Phys. Lett.* **B216** 413.
Harton J. (1995). *Proc. Third Workshop on Tau Lepton Physics* (Montreux, 1994) *Nucl. Phys. B, Proc. Supp.* **40**, Ed. G. Rolandi.
Hayes K. G. and Perl M. L. (1988). *Phys. Rev.* **D38** 3351.
Hayes K. G. et al. (1989). *Phys. Rev.* **D39** 274.
Hayes K. G. (1995). *Proc. Third Workshop on Tau Lepton Physics* (Montreux, 1994), *Nucl. Phys. B, Proc. Supp.* **40**, Ed. G. Rolandi.
Heiliger P. and Sahgal L. M. (1989) *Phys. Lett.* **B229** 409.
Heltsley B. (1995). *Proc. Third Workshop on Tau Lepton Physics* (Montreux, 1994) *Neul. Phys. B, Proc. Supp.* **40**, Ed. G. Rolandi.
Isaev P. S. and Tsarev V. A. (1989). *Sov. J. Part. Nucl.* **20** 419.
Jowett J. M. (1987). CERN LEP-TH/87-56, unpublished.
Kane G. (1987). "Modern Elementary Particle Physics" (Addison-Wesley, Redwood City, 1987) p. 231.
Kinoshita T. and Sirlin A. (1959). *Phys. Rev. Lett.* **2** 177.
Kirkby J. (1987). *Spectroscopy of Light and Heavy Quarks*, Eds. U. Gastaldi, R. Klapisch and F. Close (New York: Plenum), p 401.
Kirkby J. and Kirkby R. (1994). *Proc. Third Workshop on the Tau-Charm Factory* (Editions Frontieres, Gif-sur-Yvette, 1994), Eds. J. Kirkby and R. Kirkby.
Kirkby J. and Rubio J. A. (1992). *Particle World* **3** 77.
Kounine A. (1995) *Proc. Third Workshop on Tau Lepton Physics* (Montreux, 1994) *Nucl. Phys. B, Proc. Supp.* **40**, Ed. G. Rolandi.
Kühn J. H. and Santamaria A. (1990). *Z. Phys.* **C48** 445.
Landau L. D. and Lifshitz E. M. (1958) *Quantum Mechanics - Non-Relativistic Theory* (Reading: Addison-Wesley) section 113.
Langacker P. (1988). *New Directions in Neutrino Physics at Fermilab* (Fermi Nat. Accel. Lab., Batavia) p. 95.
Laursen M. L. et al. (1984). *Phys. Rev.* **D29** 2652.
Lee T. D. (1994). *Remarks at the Workshop on the Tau-Charm Factory in the Era of B-Factories and CESR.*
Leroy C. and Pestieau J. (1978). *Phys. Lett.* **B72** 398.
Loverre P. F. (1995). *Proc. Third Workshop on Tau Lepton Physics* (Montreux, 1994) *Nucl. Phys. b, proc. Supp.* **40**, Ed. G. Rolandi.
Lundberg B. et al. (1994). Proposal P872 to Fermilab, unpublished.
Marciano W. J. and Sirlin A. (1988). *Phys. Rev. Lett.* **61** 1815.
Matorras F. (1995). *Proc. Third Workshop on Tau Lepton Physics* (Montreux, 1994) *Nucl. Phys. b, Proc. Supp.* **40**, Ed. G. Rolandi.
Mendel R. R. et al. (1986). *Z. Phys.* **C32** 517.
Michael D. (1995). *Proc. Third Workshop on Tau Lepton Physics* (Montreux, 1994) *Nucl. Phys. b, Proc. Supp.* **40**, Ed. G. Rolandi.
Mistry N. (1992). *Proc. Second Workshop on Tau Lepton Physics* (World Scientific, Singapore 1993) Ed. K. K. Gan, p. 84.
Moffat J. W. (1975). *Phys. Rev. Lett.* **35** 1605.
Montanet L. et al. (1994). "Review of Particle Properties", *Phys. Rev.* **D50** 1173.
Morley P. D. (1992). Preprint-92-0080, Dept. of Physics, Univ. of Texas-Austin.
Narison S. (1994). *Proc. Third Workshop on the Tau-Charm Factory* (Editions Frontieres, Gif-suf-Yvette, 1994), Eds. J. Kirkby and R. Kirkby, p. 183.
Narison S. and Pich A. (1993). *Phys. Lett.* **B304** 359.
Nelson C. A. (1995). *Proc. Third Workshop on Tau Lepton Physics* (Montreux, 1994) *Nucl. Phys. B, Proc. Supp.* **40**, Ed. G. Rolandi.

Patterson J. R. (1994). "Weak and Rare Decays", *Proc. 27th Int. Conf. High Energy Physics* (Inst. of Physics, London, 1995) Eds. P. J. Bussay and I. G. Knowles, p. 149.

Perl M. L. (1992). *Reports Progress Phys.* **55** 653.

Perl M. L. (1993). *Proc. Twentieth SLAC Summer Inst. Part. Phys.*, SLAC-REPORT-412, Ed. L. Vassilian, p. 213.

Perl M. L. and Beers L. V. (1994). *Proc. Tau-Charm Factory in the Era of B-Factories and CESR* SLAC-451, Eds. L. V. Beers and M. L. Perl.

Perrotte M. (1994). *Proc. Third Workshop on the Tau-Charm Factory* (Editions Frontieres, Gif-sur-Yvette, 1994), Eds. J. Kirkby and R. Kirkby, p. 89.

Pich A. (1987). *Phys. Lett.* **B196** 561.

Qi N. (1995). *Proc. Third Workshop on Tau Lepton Physics* (Montreux, 1994) *Nucl. Phys. B, Proc. Supp.* **40**, Ed. G. Rolandi.

Reay N. W. (1995). *Proc. Third Workshop on Tau Lepton Physics, Nucl. Phys. B. Proc. Supp.* **40**, Ed. G. Rolandi.

Rolandi G. (1995). *Proc. Third Workshop on Tau Lepton Physics, Nucl. Phys. B. Proc. Supp.* **40**, Ed. G. Rolandi.

Rosa G. (1995). *Proc. Third Workshop on Tau Lepton Physics, Nucl. Phys. B. Proc. Supp.* **40**, Ed. G. Rolandi.

Rubbia A. (1995). *Proc. Third Workshop on Tau Lepton Physics, Nucl. Phys. B. Proc. Supp.* **40**, Ed. G. Rolandi.

Samuel M. A. *et al.* (1991). *Phys. Rev. Lett.* **67** 668.

Savoy-Navarro A. (1991). *Proc. Workshop on Tau Lepton Physics* (Editions Prontiéres, Gif-sur-Yvette), Eds. M. Davier and B. Jean-Marie.

Smith B. H. and Voloshin M. B. (1994). *Phys. Lett.* **B324** 117, erata **B333** 564.

Snow S. (1992). *Proc. Second Workshop on Tau Lepton Physics* (World Scientific, Singapore, 1993) K. K. Gan, p. 308.

Sobie R. (1995). *Proc. Third Workshop on Tau Lepton Physics* (Montreux, 1994) *Nucl. Phys. B, Proc. Supp.* **40**, Ed. G. Rolandi.

Stahl A. (1995). *Proc. Third Workshop on Tau Lepton Physics* (Montreux, 1994) *Nucl. Phys. B, Proc. Supp.* **40**, Ed. G. Rolandi.

Strobel G. L. and Wills E. L. (1983). *Phys. Rev.* **D28** 2191.

Stroynowski R. (1995). *Proc. Third Workshop on Tau Lepton Physics* (Montreux, 1994) *nucl. Phys. B, Proc. Supp.* **40**, Ed. G. Rolandi.

Stugu B. (1995). *Proc. Third Workshop on Tau Lepton Physics* (Montreux, 1994) *Nucl. Phys. B, Proc. Supp.* **40**, Ed. G. Rolandi.

Talebzadeh M. *et al.* (1987). *Nucl. Phys.* **B291** 503.

Trischuk W. (1992). *Proc. Second Workshop on Tau Lepton Physics* (World Scientific, Singapore 1993), Ed. K. K. Gan, p. 290.

Truong T. N. (1984). *Phys. Rev.* **D30** 1509.

Tsai Y. S. (1971). *Phys. Rev.* **D4** 2821.

Tsai Y. S. (1979). SLAC-PUB-2356.

Tsai Y. S. (1989a). *Proc. Tau-Charm Factory Workshop*, Ed. L. Beers, SLAC-REPORT-343 (SLAC, Stanford), p. 387.

Tsai Y. S. (1989b). *Proc. Tau-Charm Factory Workshop*, Ed. L. Beers, SLAC-REPORT-343 (SLAC, Stanford), p. 394.

Tsai Y. S. (1992a). private communication.

Tsai Y. S. (1992b). private communication.

Tsai Y. S. (1994). SLAC-PUB-6685, submitted to Phys. Rev., also 1994 *Proc. Tau-Charm Factory in the Era of B-Factories and CESR*, SLAC-451, Eds. L. V. Beers and M. L. Perl, p. 28.

Turner M. (1995). *Proc. Third Workshop on Tau Lepton Physics* (Montreux, 1994) *Nucl. Phys. B, Proc. Supp.* **40**, Ed. G. Rolandi.

Ushida N. *et al.* (1986). *Phys. Rev. Lett.* **57** 2897.
Vorobiev I. (1995). *Proc. Third Workshop on Tau Lepton Physics* (Montreux, 1994) *Nucl. Phys. B, Proc. Supp.* **40**, Ed. G. Rolandi.
Weinstein A. J. and Stroynowski R. (1993). *Annu. Rev. Nucl. Part. Sci.* **43** 457.
Wu D. Y. (1990a). Ph.D. Thesis, California Inst. TEch. CALT-68-1638, unpublished.
Wu D. Y. *et al.* (1990b). *Phys. Rev.* **D41** 2339.
Zachos C. K. and Meurice Y. (1987). *Mod. Phys. Lett.* **A2** 247.

COMMENTS ON B2
"The Future of Tau Physics and Tau-Charm Detector and Factory Design"

I wrote this paper in 1991 about three years after Jasper Kirkby and John Jowett conceived the idea of a tau-charm factory, a high luminosity electron-positron collider which would operate in the threshold region for the production of tau lepton pairs and charm meson pairs. In 1995 there is no tau-charm factory built or even under construction, although there is a strong interest in such a research instrument at the Institute for High Energy Physics in Beijing.

Beginning in 1988, Juan Antonio Rubio, Jasper Kirkby, John Jowett, Rafe Schindler, I, and many other tried hard to get a tau-charm factory built. We carried out studies of the potential physics, we designed the collider and the detector, we write proposals, we lobbied European and United States high energy physics laboratories, we lobbied funding agencies in Europe and the United States, we organized workshops and issued proceedings, we tried to get a tau-charm factory built at CERN in Geneva, at SLAC in the San Francisco Bay Area, in the south of Spain near Seville. Other physicists considered sites in Russia and in France.

But a tau-charm factory has not been built. Why? The immediate barrier was and still is money: 150 million dollars for constructing the collider, 80 million dollars for constructing the detector, at least 25 million dollars a year in operating expenses. However in the past five years the construction of similar cost B-factories has begun, one in Japan and one in the United States. The B-factory is a high luminosity electron-positron collider operating at 10 GeV whose primary purpose is to provide beauty (B) meson pairs. And in Europe, construction has begun of a very large proton-proton collider, called the LHC (Large Hadron Collider). The LHC construction and operating costs are ten times that of a tau-charm factory or B-factory.

There are strong physics arguments for the construction priorities given to the Large Hadron Collider and the B-factory compared to the tau-charm factory. The Large Hadron Collider is the next step in the exploration of very high energy phenomena in elementary particle physics and in the search for new particles of very large mass. The B-factory will permit the discovery and the study of the violation of time reversal invariance in the decays of B mesons, a rare but very important phenomenon. However these arguments do not deal with the question of why the high energy physics world was not able to make a more comprehensive plan in which one B-factory, one tau-charm factory, and the Large Hadron Collider were all to be built. Or if there were insistence on two B-factories, then the construction of the Large Hadron Collider in Europe could be stretched out for an additional year to provide the funds to build a tau-charm factory in Europe.

The answer to this last question is that there is not and never was any world wide plan for priorities in the construction of new high energy physics facilites; this in spite of the best efforts of committees of physicists and individual physicists. The process by which new high energy physics facilities come into being is too complicated to describe here, but I will comment on the somewhat irrational and obsessive elements which are often part of that process. The community which proposes big science projects can become obsessive about what is perceived to be the most *glamorous* project with the largest *projected* discovery payoff, the so-called *best* project. Reacting to the ever increasing pressure on research funds for big science projects, the community develops great anxieties about getting the *best* project built. And individual scientists and groups develop anxieties that they will not be able to participate in that *best* project if they don't join the project early. As a result other projects are ignored, other projects are often denigrated to weaken opposition, and sometimes two examples of the same *best* projects are built. There is nothing unusual in finding some irrationality, obsessiveness, and anxiety in human affairs, particularly when the stakes are so large.

The tau-charm factory is not a *glamorous* project; it's discovery potential does not include the potential for a *glamorous* and *irresistible* discovery such as the study of violation of time reversal invariance for the B-factory or the possibility of finding the Higgs particle using the LHC. I have already granted this, my point is that there could have been a more comprehensive set of construction projects that included one tau-charm factory.

Perhaps I have granted too much. We use our present knowledge and speculations to try to guess the physics that might be done using a new accelerator. But this ignores much of the history of discoveries in elementary particle physics using accelerators and colliders. For example the building of the Brookhaven AGS proton accelerator and the SLAC SPEAR electron-positron collider led to great discoveries, although the proposals for the construction of these instruments did not contain identified *glamorous and irresistible* discovery plans.

Opponents of the tau-charm factory have said that much of the physics which would be done using this instrument will also be done using existing instruments such as the CESR electron-positron collider at Cornell University and the LEP electron-positron collider at CERN, or will be done at the B-factories now under construction. This has some truth. Indeed I have grown weary of the studies of the potential physics for a tau-charm factory, of the designing of the detector, of the proposals and lobbying, of the tau-charm factory workshops and proceedings. I have recently joined the CLEO Collaboration which uses the CESR collider, and I will try to do some of the tau lepton physics which I once proposed for the tau-charm factory.

However if a tau-charm factory is not built we will not see what the special properties of the tau-charm factory collider and detector might bring to particle physics. Single-minded devotion to tau lepton and charm meson physics would produce a dynamics of interactions between the improving of the collider over time, the improving of the detector over time, the conceiving of new things to measure, and the stimulation of new hypotheses and theories. There is a magic in such interactions in science. Experiments may be done and discoveries may be made that are now only in our dreams. Perhaps a tau-charm factory in China will make our dreams come true.

I have recounted the history of the tau-charm factory concept in this comment, I have no solution to the funding problems which bring some irrational and obsessive pressures and so much anxiety into the choosing of big science projects.

182

THE FUTURE OF TAU PHYSICS AND TAU-CHARM DETECTOR AND FACTORY DESIGN*

Martin L. Perl

Stanford Linear Accelerator Center, Stanford University
Stanford, California 94309

ABSTRACT

Future research on the tau lepton requires large statistics, thorough investigation of systematic errors, and direct experimental knowledge of backgrounds. Only a tau-charm factory with a specially designed detector can provide all the experimental conditions to meet these requirements. This paper is a summary of three lectures delivered at the 1991 Lake Louise Winter Institute.

A. FUTURE TAU RESEARCH

A.1 Introduction

In Part A of this paper I outline the future research which should be carried out on the tau lepton; I show that much of this research requires large statistics, thorough investigation of systematic errors, and direct experimental knowledge of backgrounds. Only a tau-charm factory (Kirkby 1987, Jowett 1988, Kirkby 1989a) can provide all the experimental conditions to meet these requirements. A tau-charm factory is a very high luminosity electron-positron circular collider operating in the total energy range of 3 to 4.5 GeV. Parts B and C are introductions to the design principles for the tau-charm detector and a tau-charm factory.

As described in this paper, a tau-charm factory with a well-designed tau-charm detector offers great prospects for future research in tau physics. Similarly these instruments offer great prospects in charm physics research: D, D_s, ψ/J, ψ'. I did not have the time in these lectures to discuss this charm physics, I must refer

---

* Work supported by Department of Energy contract DE–AC03–76SF00515.

you to others: Schindler (1989, 1990a, 1990b), Barish *et al* (1989), Kirkby (1990) and the Proceedings of the Tau-Charm Factory Workshop (Beers 1989).

But the prospects for tau and charm physics which we can foresee will probably only be part of what is learned about particle physics at the tau-charm factory. The most important discoveries at many accelerators have not been foreseen: CP violation and the J at the Brookhaven AGS, neutral currents and the W and Z at CERN, the ψ and the τ itself at SPEAR. And it may be the same with the tau-charm factory. It will be a powerful new instrument for studying particle physics. There are very strong reasons based on the physics we know for building it, and it may bring us discoveries we cannot now conceive.

These lectures are in that spirit. I tell very generally what we know now and what we expect to learn about the τ. I also show how little we know about the τ and that the future of τ research is very open. I encourage speculation and dreams.

The subjects of future τ research and tau-charm detector and factory design are large, but I want to keep this paper brief and introductory. Therefore I use a summary style: equations, tables, brief comments, a few figures.

General references to τ physics are Barish and Stroynowski (1988), Burchat (1988), Gan and Perl (1988), Kiesling (1988), Pich (1990a, 1990b), and Perl (1991). In addition a comprehensive Workshop on Tau Lepton Physics was held in Orsay, France in September 1990. The proceedings of the Workshop, soon to be published, will provide a most valuable summary of our present knowledge of the τ and ν_τ.

In these lectures all measured values are quoted from the Particle Data Group's Review of Particle Properties (Aguilar-Benitez 1990) unless otherwise noted.

A.2 Overview of τ Decays: Present Data

The average measured values (Aguilar-Benitez *et al* 1990) of the branching ratios of the major decay modes of the τ are:

Pure leptonic modes:

$$B_e = B(\tau^- \to \nu_\tau e^- \bar{\nu}_e) = (17.7 \pm 0.4)\% \qquad (A1a)$$

$$B_\mu = B(\tau^- \to \nu_\tau \mu^- \bar{\nu}_\mu = (17.8 \pm 0.4)\% \qquad (A1b)$$

1-charged particle, non-strange, hadronic modes:

$$B_\pi = B(\tau^- \to \nu_\tau \pi^-) = (11.0 \pm 0.5)\% \qquad (A2a)$$

$$B_\rho = B_{\pi^-\pi^0,\text{resonant}} = B(\tau^- \to \nu_\tau \rho^- \to \nu_\tau \pi^- \pi^0) = (22.7 \pm 0.8)\% \quad (A2b)$$

$$B_{\pi^-\pi^0,\text{non-resonant}} = \left(0.37^{+0.30}_{-0.22}\right)\% \qquad (A2c)$$

$$B_{\pi^-2\pi^0} = B(\tau^- \to \nu_\tau \pi^- \pi^0 \pi^0) = (7.5 \pm 0.9)\% \text{ (See Sec. A.6)} \quad (A2d)$$

3-charged-particle, non-strange, hadronic modes:

$$B_{2\pi^-\pi^+} = B(\tau^- \to \nu_\tau \pi^- \pi^- \pi^+) = (7.1 \pm 0.6)\% \qquad (A3a)$$

$$B_{2\pi^-\pi^+\pi^0} = B(\tau^- \to \nu_\tau \pi^- \pi^- \pi^+ \pi^0) = (4.4 \pm 1.6)\% \qquad (A3b)$$

In addition there are the strange, hadronic modes such as:

$$B_K = B(\tau^- \to \nu_\tau K^-) = (0.68 \pm 0.19)\% \qquad (A4a)$$

$$B_{K^*} = B\left(\tau^- \to \nu_\tau K^{*-}(892)\right) = \left(1.39^{+0.18}_{-0.20}\right)\% \qquad (A4b)$$

which are all small, of order $\sin^2 \theta_c = 0.05$ times the corresponding non-strange hadronic modes.

Finally there are a great number of large multiplicity modes such as

$$\tau^- \to \nu_\tau + \pi^- + n\pi^0, \ n > 2 \qquad (A5a)$$

$$\tau^- \to \nu_\tau + \pi^- + \eta^0 + n\pi^0, \ n \geq 1 \qquad (A5b)$$

$$\tau^- \to \nu_\tau + 2\pi^- + \pi^+ + n\pi^0, \ n > 1 \qquad (A5c)$$

$$\tau^- \to \nu_\tau + 3\pi^- + 2\pi^+ + n\pi^0, \ n \geq 0 \qquad (A5d)$$

The branching ratios for very few of these modes have been measured, two of the few examples are

$$B(\tau^- \to \nu_\tau 3\pi^- 2\pi^+) = (0.056 \pm 0.016)\% \qquad (A6a)$$

$$B(\tau^- \to \nu_\tau 3\pi^- 2\pi^+ \pi^0) = (0.051 \pm 0.022)\% \qquad (A6b)$$

and there is the 90% CL upper limit

$$B(\tau^- \to \nu_\tau 4\pi^- 3\pi^+ n\pi^0, \ n \geq 0) \leq 0.019\% \qquad (A7)$$

We know very little about these large multiplicity modes because present techniques are inadequate for isolating modes with three or more neutral mesons:

π^0's and η's. One of the purposes of a tau-charm factory and proper associated detector is to provide the instruments for isolating and studying these modes.

Measurement of a branching ratio is only the first step in studying a τ decay mode if there are three or more particles in the mode. Once there are three or more particles the dynamics of the decay is revealed through the spectra and correlations of momenta, angles and spins. But present statistics and techniques have limited such studies to a few cases: In

$$\tau^- \to \nu_\tau + \pi^- + \pi^0 \qquad (A8)$$

the $\pi^-\pi^0$ mass spectrum shows the ρ resonance. In

$$\tau^- \to \nu_\tau + e^- + \bar{\nu}_e$$
$$\tau^- \to \nu_\tau + \mu^- + \bar{\nu}_\mu \qquad (A9)$$

the charged lepton spectrum has been extensively studied. In

$$\tau^- \to \nu_\tau + \pi^- + \pi^+ + \pi^- \qquad (A10)$$

the spectra of the 3π's has been studied and show the $a_1(1260)$ resonance and its decay mode

$$a_1 \to \rho + \pi \qquad (A11)$$

But that is about all that has been done. Present techniques are very restrictive. For example

$$\tau^- \to \nu_\tau + \pi^- + \pi^0 + \pi^0 \qquad (A12)$$

has barely been isolated, the various 3π spectra have not yet been studied with care.

A.3 Overview of τ Decays: Standard Model Theory

In the Standard Model of elementary particle physics the decay of the τ takes place through the W exchange processes of Fig. 1. The τ and ν_τ are taken to be spin 1/2 point particles with masses

$$m_\tau = 1784.1^{+2.7}_{-3.6} \text{ MeV} \qquad (A13a)$$
$$0 \leq m_{\nu_\tau} \leq 35 \text{ MeV} \qquad (A13b)$$

The $\tau - W - \nu_\tau$ vertex is taken to have the V-A form and standard coupling strength. And perfect τ lepton number conservation is assumed.

The goal of τ decay theory is to enable calculation of all the spectra of each decay mode, and integrating over those spectra, to calculate the decay width $\Gamma(i)$ of each mode i.

This goal was achieved for the pure leptonic modes twenty years ago by Tsai (1971) before the τ was discovered. From the diagram in Fig. 1a, the well-known result is

$$\Gamma_\ell = \Gamma(\tau^- \to \nu_\tau \ell^- \nu_\ell) = \frac{G_F^2 m_\tau^5}{192\pi^3} F_\ell(y) \; ; \; \ell = e, \mu \qquad (A14)$$

where

$$F_\ell(y) = 1 - 8y + 8y^3 - y^4 - 12y^2 \ln y$$
$$y = m_\ell^2/m_\tau^2$$
$$m_{\nu_\ell} = 0$$
$$m_{\nu_\tau} = 0$$

and G_F is the Fermi coupling constant 1.166×10^{-5} GeV^{-2}. Electromagnetic radiative corrections and a correction for the W mass are given by Marciano and Sirlin (1988), Wu (1990), and Perl (1991).

But at present there is no equivalent usable theory for the hadronic modes which constitute over 60% of the branching ratios. The theory of the W-hadron vertex in Fig. 1b is the theory of quantum chromodynamics *in the 1 GeV energy region*. There is no general and precise way to calculate from this theory.

This serious deficiency is sometimes overlooked because it has become commonplace in discussions of hadronic τ decays to write about the comparison of measured branching ratios with theory. What is meant is the prediction of a τ decay width *using other data and a general theoretical connection*. There are two common examples. The $\tau^- \to \nu_\tau \pi^-$ decay width, Fig. 2a, is given by

$$\Gamma(\tau^- \to \nu_\tau \pi^-) = \frac{G_F^2 f_\pi^2 \cos^2 \theta_c m_\tau^3}{16\pi} \left[1 - \frac{m_\pi^2}{m_\tau^2}\right]^2 \qquad (A15)$$

where f_π is the pion decay constant. To calculate $\Gamma(\tau^- \to \nu_\tau \pi^-)$ we do not calculate f_π from quantum chromodynamics, we obtain it from the *measured width* for $\pi^- \to \mu^- \bar{\nu}_\mu$ decay:

$$\Gamma(\pi^- \to \mu^- \bar{\nu}_\mu) = \frac{G_F^2 f_\pi^2 \cos^2 \theta_c m_\pi m_\mu^2}{8\pi} \left[1 - \frac{m_\mu^2}{m_\pi^2}\right]^2 \qquad (A16)$$

The factor f_π appears in both equations because the $W-\pi$ vertex at the energy of the π mass appears in both decay diagrams, Fig. 2. The other common example is that $\Gamma(\tau^- \to \nu_\tau \rho^-)$ is calculated from the cross section for

$$e^+ + e^- \to \rho^0 \; ; \tag{A17}$$

we have no way at present to precisely calculate from quantum chromodynamics the $W-\rho$ vertex contribution to $\Gamma(\tau^- \to \nu_\tau \rho^-)$.

A major area in future τ research will be the symbiosis between precise methods for low energy quantum chromodynamics calculations and precise measurements of the properties of the hadronic decay modes of the τ. Some references to present applications of quantum chromodynamics to hadronic τ decay are Braaten (1988, 1989a, 1989b), Pumplin (1989, 1990), Pich (1989), Narison and Pich (1988), Maxwell and Nicholls (1990), Ghose and Kumbhakar (1981).

Table 1. Comparison of B_i/B_e from measurements with predictions from theory and other data. I use $B_e = B(\tau^- \to \nu_\tau e^- \bar{\nu}_e) = (17.7 \pm 0.4)\%$ from (Aguilar-Benitez 1990).

B_i/B_e	Measurement	Prediction	Reference and notes
$B(\tau^- \to \nu_\tau \mu^- \bar{\nu}_\mu)/B_e$	1.006 ± 0.032	0.973	use Eqs. A1, A14
$B(\tau^- \to \nu_\tau \pi^-)/B_e$	0.63 ± 0.03	0.607	Smith 1990 use Eqs. A15, A16 and $\Gamma_{meas}(\pi^- \to \mu^- \bar{\nu}_\mu)$
$B(\tau^- \to \nu_\tau \rho^-)/B_e$	1.26 ± 0.05	1.32 ± 0.05	Smith 1990 use $\sigma_{meas}(e^+e^- \to \rho^0)$ and CVC
$B(\tau^- \to \nu_\tau 2\pi^- \pi^+)/B_e$	0.28 ± 0.02	0.3 to 0.7	Smith 1990 very approximate theory
$B(\tau^- \to \nu_\tau 2\pi^- \pi^+ \pi^0)/B_e$	0.28 ± 0.02	0.28 ± 0.06	Smith 1990 use $\sigma_{meas}(e^+e^- \to 4\pi)$ and CVC
$B(\tau^- \to \nu_\tau K^-)/B_e$	0.038 ± 0.01	0.041 ± 0.002	Jain 1990 use $\Gamma_{meas}(K^- \to \mu^- \bar{\nu}_\mu)$

I conclude this section with, Table 1, a comparison of measured branching ratios with predictions from Standard Model theory *and* other data. In these comparisons, the predictions are ratios of widths, for example

$$\left(\frac{B_\pi}{B_e}\right)_{predicted} = \left(\frac{\Gamma_\pi}{\Gamma_e}\right)_{predicted} = \frac{12\pi^2 f_\pi^2 \cos^2\theta_c}{m_\tau^2}\left[1 - \frac{m_\pi^2}{m_\tau^2}\right]^2 \quad (A18)$$

from (A15) and (A16). Given the errors and uncertainties there are no discrepancies.

A.4 τ Pair Production and Backgrounds at a Tau-Charm Factory

Tau pair production

$$e^+ + e^- \rightarrow \tau^+ + \tau^- \quad (A19)$$

occurs through γ and Z^0 exchange. All measurements of the cross section, $\sigma_{\tau\tau}$, for (A19) are in agreement with these mechanisms; more precise tests should still be made.

In the tau-charm factory energy region with $E_{tot} \leq 5$ GeV, only γ exchange contributes to $\sigma_{\tau\tau}$. Ignoring electromagnetic radiative corrections.

$$\sigma_{\tau\tau} = \frac{4\pi\alpha^2(\hbar c)^2}{3 E_{tot}^2}\left[\frac{\beta(3-\beta^2)}{2}\right] \quad (A20a)$$

where $\alpha, \hbar$, and c are the fine structure constant, Planck constant, and velocity of light, and $\beta = v/c$ where v is the τ velocity. Inserting numerical values

$$\sigma_{\tau\tau} = \frac{87.8}{E_{tot}^2}\left[\frac{\beta(3-\beta^2)}{2}\right] \text{nb} \quad (A20b)$$

where E_{tot} is in GeV.

Voloshin (1989) has pointed out that close to threshold, $E_{tot} = 2m_\tau$, the Coulombic attraction between the τ^+ and τ^- changes (A20a) to

$$\sigma_{\tau\tau} \text{ (near threshold)} = \frac{4\pi\alpha^2(\hbar c)^2}{3 E_{tot}^2}\left[\frac{\pi\alpha}{1-\exp(-\pi\alpha/\beta)}\right]\left[\frac{3-\beta^2}{2}\right] \quad (A21a)$$

so that at threshold $\sigma_{\tau\tau}$ is not 0 but is

$$\sigma_{\tau\tau} \text{ (threshold)} = \frac{\pi^2\alpha^3(\hbar c)^2}{2m_\tau^2} = 0.23 \text{ nb} \quad (A21b)$$

This is important for τ research at the tau-charm factory.

Figure 3 shows $\sigma_{\tau\tau}$ and

$$R_{\tau\tau} = \sigma_{\tau\tau}/\sigma_{point} \qquad (A22)$$

where

$$\sigma_{point} = \frac{4\pi\alpha^2(\hbar c)^2}{3\, E_{tot}^2} \qquad (A23)$$

Also shown in Fig. 3 is $R_{hadrons}$ for

$$e^+ + e^- \to \text{hadrons} \qquad (A24)$$

showing the J/ψ, ψ', ψ'' resonances as well as the hadronic continuum. Figure 3 illustrates three important advantages of studying the τ at a tau-charm factory:

(i) The most serious and difficult hadronic backgrounds in τ pair data come from D, D_s and B meson production and decay. By obtaining the τ pair data below the ψ'' energy, these backgrounds are avoided.

(ii) Below the ψ' energy, the nature of the $e^+e^- \to$ hadrons continuum changes very slowly with energy as the energy goes below the $e^+e^- \to \tau^+\tau^-$ threshold. Therefore for τ data obtained below the ψ' energy, the hadronic contamination can be directly measured by operating the collider just below the τ threshold.

(iii) At τ threshold $\sigma_{\tau\tau} = 0.23$ nb and 2 MeV above threshold $\sigma_{\tau\tau} = 0.4$ nb. Thus τ pair data can be obtained with the τ's produced almost at rest, an important condition for some τ studies (Gomez-Cadenas, Heusch, and Seiden 1989).

A.5 The Future of τ Physics

No one knows how to break out of the Standard Model. Historically most fundamental advances in particle physics have been made by going to higher energies and more violent collisions; and many particle physicists are following that path through LEP II, HERA, the SSC, or the LHC. But history also has a counterexample. The fundamental discoveries in e^+e^- annihilation of the ψ and ψ', the τ, charmed mesons, and hadron jets from quark were made at energies below 7 GeV. From 7 GeV to the Z^0 there has been much wonderful physics in e^+e^- annihilation such as the identification of hadron jets from gluons and the measurements of τ, D, and B lifetimes; but the discovery record below 7 GeV has not been equaled. Nature was not kind above 7 GeV. At the Z^0 there may yet be major discoveries, the full story is not yet known.

Extensive and precise experiments on the τ, and where possible on the τ neutrino, offer another direction for breaking out of the standard model, another direction for finding new physics beyond the Standard Model. We don't know if nature will be kind to the τ researcher, we don't know what experiments will be crucial; but we do know that there is a world of τ and ν_τ experiments to do. That world is summarized in Table 2 (Perl, 1990). In the remainder of the first part of these lectures I discuss some of the experiments which are best carried out at a tau-charm factory, commenting on the significance of the experiments for testing the Standard Model and on what new physics might be found.

Table 2. Experiments on τ and ν_τ.

Type of Experiment	Search For New Physics	Test Standard Model	Tau-Charm Factory
Understand 1-charged particle modes puzzle	✓	✓	✓
Untangle multiple π^0 and η modes		✓	✓
Precise measurement of B_e, B_μ, B_π and B_ρ,	✓	✓	✓
Precise measurement of Cabibbo-suppressed modes	✓	✓	✓
Full study of dynamics of $\tau \to e\nu\nu$, $\tau \to \mu\nu\nu$ analogous to $\mu \to e\nu\nu$ in detail	✓	✓	✓
Detailed study of 3, 5, 7-charged particle modes		✓	✓
Find and study rare allowed modes such as radiative decays and second class currents	✓	✓	✓
Explore forbidden decay modes	✓		✓
Search for anomalous moments of τ	✓	✓	✓
Precise measurement of τ lifetime		✓	
Explore ν_τ mass to a few MeV/c^2	✓	✓	
Detect ν_τ	✓	✓	
Study interactions of ν_τ	✓	✓	
Search for ν_τ decays	✓		
Precise low energy study of $e^+e^- \to \tau^+\tau^-$, $\tau^+\tau^-\gamma$	✓	✓	✓
Precise high energy study of $e^+e^- \to \tau^+\tau^-$, $\tau^+\tau^-\gamma$	✓	✓	
Study of $Z^0 \to \tau^+\tau^-$	✓	✓	
Study of $W^- \to \tau^-\bar{\nu}_\tau$	✓	✓	
Measure $B(D^- \to \tau^-\bar{\nu}_\tau)$	?	✓	✓
Measure $B(D_s^- \to \tau^-\bar{\nu}\tau)$	?	✓	✓
Measure $B(B^- \to \tau^-\bar{\nu}_\tau)$	?	✓	
Make and study $\tau^+\tau^-$ atom	✓	✓	?

A.6 The 1-Charged Particle Modes Puzzle

This is the only anomaly in elementary particle physics which is discussed in detail in the Review of Particle Properties (Aguilar-Benitez et al 1990). I take the text and Table 3, the work of K. G. Hayes, directly from Aguilar-Benitez et al (1990), only changing how the references are noted.

Table 3. The 1-charged particle modes puzzle (Aguilar-Benitez 1990.)

1-Prong Branching Fractions of the τ(%)		
Decay Mode	Experiment	Theory[a]
$e^-\bar{\nu}_e\nu_\tau$	17.7 ± 0.4	18.0
$\mu^-\bar{\nu}_\mu\nu_\tau$	17.8 ± 0.4	17.5
$\rho^-\nu_\tau$	22.7 ± 0.8	22.7
$\pi^-\nu_\tau$	11.0 ± 0.5	10.8
$K^-(\geq 0 \text{ neutrals})\nu_\tau$	1.71 ± 0.23	
$K^{*-}\nu, K^{*-} \to \pi^-(2\pi^0 \text{ or } K_L)$	0.6 ± 0.1	
$\pi^-(2\pi^0)\nu_\tau$	7.5 ± 0.9	$\leq 6.7 \pm 0.4$
$\pi^-(\geq 3\pi^0)\nu_\tau$		< 1.4[b]
$\pi^-(\geq 1\eta)(\geq 0\pi^0)\nu_\tau$[c]	< 1.3	< 0.8
Sum of measured modes	79.0 ± 1.4	
Theoretical limits of unmeasured modes		< 2.2
Sum of exclusive modes		$< 80.2 \pm 1.4$
Measured 1-prong branching fraction	86.1 ± 0.3	
Difference		$> 5.8 \pm 1.4$

[a] Normalized to constrained fit to $e\nu\nu$ and $\mu\nu\nu$ measurements assuming $BF_\mu = 0.973\, BF_e$.
[b] Assumes 15% systematic error on the measured cross section for $e^+e^- \to 2\pi^+2\pi^-$.
[c] Contribution to 1-prong mode only.

"There exists a problem in understanding the 1-charged-particle decay modes of the τ. The problem, first discussed by Truong (1984) and Gilman and Rhie (1985), is that the measured inclusive branching fraction to 1-charged prong is larger than the sum of exclusive 1-charged-particle modes. Since the measurement of exclusive modes with 2 or more neutral hadrons is difficult given the limitations of present detectors, the inequality between the sum of exclusive modes and the inclusive measurements is significant only if theoretical predictions are used to put limits on unmeasured or poorly measured modes."

"The current status of the 1-prong modes is summarized in Table 3. For the theoretical estimates, we use the results of Gilman and Rhie (1985) and Gilman (1987) updated to include new experimental data and electroweak radiative corrections (Marciano and Sirlin 1988)."

"The discrepancy is due to errors in the experimental measurements or theoretical limits, or to the existence of one or more modes not included in Table 3. Early measurements of the inclusive one-prong branching fraction reported significantly lower values but suffered from large backgrounds not present in more recent experiments."

"Systematic errors dominate most measurements, particularly for the $\tau^- \to \pi^- \nu_\tau$ and $\tau^- \to \rho^- \nu_\tau$ modes. The technique used to obtain the experimental averages ignores correlated errors, which can be specially important when systematic errors are dominant. There is a tendency for multiple experimental measurements of a given mode to be more consistent than expected from their quoted errors (Hayes and Perl 1988). This indicates either the existence of systematic errors accounted for by the experimenters which are correlated and should not be averaged, or inflation or experimental errors, or a bias in the experimental measurements. The $\tau^- \to \rho^- \nu_\tau$ measurements show this tendency even if the systematic errors are ignored and only the statistical errors are used."

"Resolution of the missing one-prong puzzle will require either new measurements with much reduced systematic and statistical errors, or an explicit measurement of a mode which is presently unmeasured or very poorly measured."

The measurements of Behrend *et al* (1990) give a possible solution to this puzzle. Compared to the world average values in Table 3, they find larger values for $B(\tau^- \to \nu_\tau e^- \bar{\nu}_e)$, $B(\tau^- \to \nu_\tau \pi^- 2\pi^0)$ and B_3, eliminating the 5% discrepancy in Table 3. This has been discussed by Kiesling (1988, 1989, 1990).

A.7 Untangling Multiple π^0 and η Modes

As discussed in Sec.A.2, we have scanty information on τ decay modes with two or more neutral mesons: π^0's and η's. The untangling and study of these decay modes requires that τ pair data be acquired at low E_{tot} where γ's from π^0 and η's are well separated in angle so that the π^0's and η's can be efficiently reconstructed. Furthermore the backgrounds from $e^+e^- \to$ hadrons must be measured directly. These requirements can only be jointly met at a tau-charm factory. In addition a special detector is required with close-to-4π γ detection efficiency even for low energy γ's (Seiden 1989, Kirkby 1989b, Gan 1989).

A.8 Precise Measurements of B_e, B_μ, B_π and B_ρ

The measured values of B_e, B_μ, B_π and B_ρ, Sec. A.2 have fractional errors

$$0.02 \lesssim \Delta B_i / B_i \lesssim 0.04 \qquad (A25)$$

These are average measured values, the individual experiments have larger fractional errors; in addition we don't understand how to average the systematic errors over the experiments (Hayes and Perl 1988). Thus for at least some of these modes the $\Delta B/B$ may be 0.05.

At a tau-charm factory a single, high-statistics experiment can reduce the fractional errors for B_e, B_μ, B_π and B_ρ to

$$\Delta B_i / B_i \approx 0.005 \quad , \qquad (A26)$$

an improvement by a factor of 10! The method due to Gomez-Cadenas, Heusch and Seiden (1989) collects the τ pair sample at a few MeV above τ pair threshold using the mode.

$$\tau^- \to \nu_\tau + \pi^- \qquad (A27)$$

to tag the τ pairs. Since the τ's are produced almost at rest the τ is almost monochromatic in energy. This combined with efficient $e-\pi$, $\mu-\pi$, and $K-\pi$ separation gives very clean τ pair selection. Backgrounds from $e^+e^- \to$ hadrons will be measured directly by going below τ threshold.

The measurement of B_e, B_μ B_π, and B_ρ with a precision of

$$\Delta B_i / B_i \approx 0.005$$

will be compared with the theoretical predictions for B_μ/B_e, $1 - B_\mu/B_e$, B_π/B_e and B_ρ/B_e. Gomez-Cadenas, Heusch and Seiden (1989), Tsai (1989a), Tsai (1989b), Heusch (1989a) and others have discussed how such precise studies can uncover new physics such as a Higgs-like particle or a leptoquark.

194

A.8 Full Study of Dynamics of $\tau^- \to \nu_\tau e^- \bar{\nu}_e$ and $\tau^- \to \nu_\tau \mu^- \bar{\nu}_\mu$

In the Standard Model the matrix element for the decay

$$\tau^- \to \nu_\tau + e^- + \bar{\nu}_e \qquad (A28)$$

has the form

$$M = \frac{G}{\sqrt{2}} \left[\bar{u}_e \gamma^\mu (1 - \gamma_5) v_{\bar{\nu}_e} \right] \left[\bar{u}_{\nu_\tau} \gamma_\mu (1 - \gamma_5) u_\tau \right] \qquad (A29)$$

where the u's and v's are Dirac spinors of particle and antiparticles. If we want to allow some deviation in the $\tau - W - \nu_\tau$ vertex from the Standard Model than we write

$$M = \frac{G}{\sqrt{2}} \left[\bar{u}_e \gamma^\mu (1 - \gamma_5) v_{\bar{\nu}_e} \right] \left[\bar{u}_{\nu_\tau} \gamma_\mu (v_\tau + a_\tau \gamma_5) u_\tau \right] \qquad (A30)$$

This leads to a formula for the e energy spectrum known since the first theoretical studies of μ decay:

$$\frac{d\Gamma_e}{\Gamma_e dx} = 4 \left[3(x^2 - x^3) + 2\rho \left(\frac{4}{3}x^3 - x^2 \right) \right] \qquad (A31a)$$

$$\rho = \frac{3}{4} \frac{(v_\tau - a_\tau)^2}{(v_\tau - a_\tau)^2 + (v_\tau + a_\tau)^2} \qquad (A31b)$$

Here $x = 2E_e/m_\tau$, E_e is the electron energy, ρ is the Michel parameter, and the ν_e, ν_τ, and e masses have been set to zero. The same formula holds for

$$\tau^- \to \nu_\tau + \mu^- + \bar{\nu}_\mu \qquad (A32)$$

with $m_\mu = 0$.

In the Standard Model $v_\tau = 1$, $a_\tau = -1$ and $\rho = 0.75$. Stroynowski (1990) gives the average measured values

$$\begin{aligned}
\rho_e &= 0.705 \pm 0.041 \\
\rho_\mu &= 0.763 \pm 0.051 \\
\rho_e \text{ and } \rho_\mu &= 0.727 \pm 0.033
\end{aligned} \qquad (A33)$$

to be compared with the 0.75 value. So far, so good.

But as discussed in a beautiful paper by Fetscher (1990), the $\tau - W - \nu_\tau$ vertex can be much more general than allowed by (A30). Indeed, this has been known for μ decay for four decades (Scheck 1978) and was discussed for τ decay in the 1970's. In (A30)

$$(v_\tau + a_\tau \gamma_5) = \left(\frac{v_\tau - a_\tau}{2}\right)(1 - \gamma_5) + \left(\frac{v_\tau + a_\tau}{2}\right)(1 + \gamma_5)$$

Also use the notation

$$\frac{1}{2}(1 - \gamma_5)u = u_L$$
$$\frac{1}{2}(1 + \gamma_5)u = u_R$$

to denote left-handed (L) and right-handed (R) spinors. Then (A30) is rewritten

$$M = \frac{2G}{\sqrt{2}}\left\{(v_\tau - a_\tau)\left[\bar{u}_{eL}\gamma^\mu v_{\bar{\nu}_e}\right]\left[\bar{u}_{\nu_\tau}\gamma_\mu u_{\tau L}\right] \right. \\ \left. + (v_\tau + a_\tau)\left[\bar{u}_{eL}\gamma^\mu v_{\bar{\nu}_e}\right]\left[\bar{u}_{\nu_\tau}\gamma_\mu u_{\tau R}\right]\right\} \quad (A34)$$

This is now easily generalized. Let $\ell = e$ or μ and let the $\ell - W - \nu_\ell$ vertex also be non-standard with $1 - \gamma_5 \to v_\ell + a_\ell \gamma_5$. Then, following Fetscher (1990) and Mursula and Scheck (1985) define

$$g_{LL}^V = (v_\ell - a_\ell)(v_\tau - a_\tau)/4$$
$$g_{LR}^V = (v_\ell - a_\ell)(v_\tau + a_\tau)/4 \quad (A35)$$

and so forth with the superscript V denoting the vector γ^μ coupling. Then (A34) is more generally

$$M = \frac{4G}{\sqrt{2}} \sum_{ij} g_{ij}^V \left[\bar{u}_{\ell i}\gamma^\mu v_{\bar{\nu}_\ell}\right]\left[\bar{u}_{\nu_\tau}\gamma_\mu u_{\tau j}\right] \quad (A36)$$

with $i = L, R$ and $j = L, R$. The final generalization adds scalar and tensor coupling with γ^μ in (A36) replaced by 1 and $\sigma^{\mu\nu} = i\left[\gamma^\mu\gamma^\nu - \gamma^\nu\gamma^\mu\right]/2$ respectively. Denoting the coupling operators by Γ^s, Γ^v, and Γ^T.

$$M = \frac{4G}{\sqrt{2}} \sum_{\substack{i,j=L,R \\ N=S,V,T}} g_{ij}^N <\bar{u}_{\ell i}|\Gamma^N|v_{\bar{\nu}_\ell}> <\bar{u}_{\nu_\tau}|\Gamma_N|u_{\tau j}> \quad (A37)$$

Of the 12 g_{ij}^N's, g_{LL}^T and g_{RR}^T are identically zero. Since the g_{ij}^N's can be complex, there are 19 independent parameters ignoring an overall phase. This is in contrast to the Standard Model where

$$g_{LL}^V = 1 \qquad (A38a)$$
$$\text{all other } g_{ij}^N = 0 \qquad (A38b)$$

In μ decay

$$\mu^- \rightarrow \nu_\mu + e^- + \bar{\nu}_e \qquad (A39)$$

a tremendous amount of work has been done to set upper limits on (A38b) (Fetscher, Gerber and Johnson 1986). As discussed by Fetscher (1990), Pich (1990) and others a great deal of work needs to be done to carry out similar investigations of the τ leptonic decays. Evidence in $\tau \rightarrow \nu_\tau \ell^- \bar{\nu}_\ell$ for any $g_{ij}^N \neq 0$ except g_{LL}^V means the discovery of new physics. The detailed study of τ leptonic decays will make use of the correlated spins of the τ's produced in pairs through correlations of the momenta and angles of the e's and μ's. Many of these studies are best carried out at a tau-charm factory. As noted by Stroynowski in a private communication, it is even possible at a tau-charm factory to study the sequence

$$\begin{aligned} \tau^- &\rightarrow \nu_\tau + \mu^- + \bar{\nu}_\mu \\ \mu^- &\rightarrow \nu_\mu + e^- + \bar{\nu}_e \end{aligned} \qquad (A40a)$$

so that the polarization of the μ^- is measured, an important aspect of studying the g_{ij}^N's.

A.9. Exploring the ν_τ Mass to a Few MeV

The present upper limit on the ν_τ mass (Albrecht *et al* 1988) is

$$m_{\nu_\tau} < 35 \; MeV \;, \; 95\% \; \text{CL} \qquad (A41a)$$

compared to

$$m_{\nu_e} < 17 \; \text{eV} \;, \; 95\% \; \text{CL} \qquad (A41b)$$
$$m_{\nu_\mu} < 0.27 \; \text{MeV} \;, \; 90\% \; \text{CL} \qquad (A41c)$$

Since we have no understanding of the masses of the leptons, indeed since we do not know if neutrinos have non-zero mass, we do not know the significance of the three upper limits in (A41). For example, if one believes in the hypothesis

$$\frac{m_{\nu_1}}{m_{\nu_2}} = \frac{m_1^2}{m_2^2} \qquad (A42)$$

then

$$m_{\nu_\mu} \left(\frac{m_e}{m_\mu}\right)^2 < 5.9 \text{ eV} \qquad (A43a)$$

$$m_{\nu_\tau} \left(\frac{m_e}{m_\tau}\right)^2 < 2.9 \text{ eV} \qquad (A43b)$$

and the limits have comparable significance.

The m_{ν_τ} upper limit (A41a) was obtained by studying the spectrum of the invariant mass of the 5π's in the decay

$$\tau^- \to \nu_\tau + 3\pi^- + 2\pi^+ \qquad (A44)$$

As discussed by Gomez-Cadenas et al (1990) this is still the most promising method for exploring for m_{ν_τ} values in the few MeV range, perhaps down to 3 MeV. A tau-charm factory is required to probe m_{ν_τ} down to this level for several reasons: the contamination in (A44) from $e^+e^- \to$ hadrons must be directly measured, the π momentum measurements can be directly calibrated using the decay $D^- \to 3\pi^- + 2\pi^+$ at the ψ', and large statistics are needed.

A.10. Rare Decay Modes of the τ

Some hadronic decay modes will have small branching fractions because of large multiplicity, for example

$$B(\tau^- \to \nu_\tau \, 4\pi^- 3\pi^+ n\pi^0, n \geq 0) \leq 1.9 \times 10^{-4}, \qquad (A45)$$

or because the modes have moderate multiplicity but include K's or η's. At present we don't expect any unusual physics to be associated with such modes as long as they obey the first-class hadronic current rules (Tsai 1971, Barish and Stroynowski 1988, Burchat 1988, Pich 1990). Namely, for non-strange hadronic states the G-parity is $G = +1$ for the weak vector current and $G = -1$ for the weak axial vector current.

On the other hand, second-class weak currents have

$$\text{Vector:} \quad G = -1 \,, \quad J^P = 1^- \quad (A46a)$$
$$\text{Axial vector:} \quad G = +1 \,, \quad J^P = 0^-, 1^+ \quad (A46b)$$

Decays with such properties have never been seen in nuclear or elementary particle physics because they have very small branching fractions. The τ offers the best possibility to observe decays through the second class current. Possibilities are (Leroy and Pestieau 1978, Pich 1987, Zachos and Meurice 1987)

$$\tau^- \to \nu_\tau + \pi^- + \eta \quad (A47a)$$
$$\tau^- \to \nu_\tau + a_0\,(980) \quad (A47b)$$
$$\tau^- \to \nu_\tau + b_1\,(1235) \quad (A47c)$$

The $a_0(980)$ has $G = -1, J^P = 1$ so (A47b) obeys (A46a), similarly the b_1 (1235) has $G = +1$, $J^P = 1^+$ so (A47c) obeys (A46b). In (A47a) η has $G = +1$, $J^P = 0^-$, therefore $G(\pi\eta) = -1$ and $J^P(\pi\eta) = 0^-$ or 1^+, so (A47a) obeys (A46b).

In the Standard Model, second-class current decays do not occur if one ignores the electromagnetic corrections to isospin symmetry in the strong interaction. Therefore there are two interests in observing and studying second-class current decays. First, what is the strength of a second-class current decay due to the electromagnetic correction, that is a decay within the Standard Model? Second, are there second-class current decays whose properties cannot be explained by the Standard Model? Interesting discussions are given by Berger and Lipkin (1987) and by Bramon, Narison and Pich (1987).

There are two ways to estimate the strength of a second-class current decay due to the electromagnetic correction. That correction introduces the fine structure constant α in a second-class current decay amplitude. Then

$$K^2 = \frac{\Gamma(\text{second-class current})}{\Gamma(\text{first-class current})} \sim \alpha^2 \sim 10^{-4} \quad (A48)$$

Alternately, the second-class current decay may be thought of as due to the difference of the d quark and u quark current masses: $\Delta m = m_d - m_u \sim 1$ MeV.

$$K^2 = \frac{\Gamma\,(\text{second-class current})}{\Gamma\,(\text{first-class current})} \sim \left(\frac{m_d - m_u}{m_\pi}\right) \sim 10^{-4} \quad (A49)$$

More generally the range of such crude estimates is

$$10^{-3} \lesssim K^2 \leq 10^{-5} \tag{A50}$$

Thus for the second-class current decay

$$\tau^- \to \nu_\tau + \pi^+ + \eta$$

the estimates are (Pich 1987, Zachos and Meurice 1987)

$$B(\tau^- \to \nu_\tau \pi^- \eta) \sim K^2 \, B(\tau^- \to \nu_\tau \pi^-) \sim 10^{-4} \text{ to } 10^{-6} \tag{A51}$$

The observation and study of such a small decay mode requires the experimental conditions of a tau-charm factory: large statistics, good control of errors, and direct knowledge of backgrounds.

The radiative decay modes such as

$$\tau^- \to \nu_\tau + e^- + \bar{\nu}_e + \gamma \tag{A52a}$$
$$\tau^- \to \nu_\tau + \mu^- + \bar{\nu}_\mu + \gamma \tag{A52b}$$
$$\tau^- \to \nu_\tau + \pi^- + \gamma \tag{A52c}$$
$$\tau^- \to \nu_\tau + \rho^- + \gamma \tag{A52d}$$

have been extensively discussed theoretically. Wu (1990a) has surveyed the theory for leptonic decays. Hadronic decays are discussed by **Queijeiro** and Garcia (1988), Dominguez and Sola 1988, Banerjee (1986), Garcia and Riviera-Robolledo (1981), and Kim and Resnick (1979). But there is only one experimental study by Wu *et al* (1990b), a study of the decay.

$$\tau^- \to \nu_\tau + \mu^- + \bar{\nu}_\mu + \gamma$$

For small values of E_γ

$$\frac{d\Gamma(\tau^- \to \nu_\tau \mu^- \bar{\nu}_\mu \gamma)}{dE_\gamma} \approx \left[\frac{\alpha}{\pi}\left(2\ln\frac{m_\tau}{m_\mu} - \frac{17}{6}\right)\right] \frac{\Gamma(\tau^- \to \nu_\tau \mu^- \bar{\nu}_\mu)}{E_\gamma} \tag{A53}$$

This equation is derived from the treatment of radiative muon decay, $\mu^- \to \nu_\mu e^- \bar{\nu}_e \gamma$, by Kinoshita and Sirlin (1959). Thus in the simplest approximation there is in the bremsstrahlung spectrum $1/E_\gamma$. But precise studies of the γ energy and angle spectra will provide interesting ways to examine the properties of the τ and its decay dynamics as discussed in the references just given. Finally, an interesting use of the decays in (A52a) and (A52b) is to look for an anomalous value of the g-value of the τ using the radiation zero idea (Laursen, Samuel, and Sen 1984).

A.11   Forbidden Decay Modes of the Tau

Since the early days of τ research, there have been searches (Hayes *et al* 1982) for decays which violate the conservation of τ lepton number. Examples of such proposed decays are

$$\begin{aligned}
\tau^- &\to e^- + \gamma \\
\tau^- &\to \mu^- + \gamma \\
\tau^- &\to e^- + \pi^0 \\
\tau^- &\to \mu^- + \pi^0 \\
\tau^- &\to e^- + e^+ + e^- \\
\tau^- &\to e^- + \mu^+ + \mu^-
\end{aligned} \quad (A54)$$

and so forth. The interest is the same as searches for lepton number non-conservation in decays such as $\mu^- \to e^- + \gamma$ and $K^0 \to \mu^\pm + e^\mp$: the desire to find connections between the leptons, the desire to break out of the Standard Model of elementary particle physics.

No violations of τ lepton number conservation have been found. Table 4 from Aguilar-Benitez (1990) gives the upper limits on the branching ratios. Most of these limits are from Albrecht *et al* (1987), Keh *et al* (1988) and Hayes *et al* (1982). Searches for these modes are straightforward because all the particles in the final state can be detected, and the mass of the τ reconstructed if there indeed is such a τ decay. Thus in a sample of 10^n τ pairs with one identified τ decay, the upper limit is of order 3×10^{-n} if an event with the unconventional decay is not found.

Baltrusaitis *et al* (1985) have carried out an interesting but null search for a hypothetical light boson G by looking for the decays

$$\begin{aligned}
\tau^- &\to e^- + G \\
\tau^- &\to \mu^- + G
\end{aligned} \quad (A55)$$

Table 4. Upper limits on branching ratios for forbidden decay modes of τ with 90% CL (Aguilar-Benitez 1990).

Mode	Upper Limit on Branching Ratio
$\mu^-\gamma$	5.5×10^{-4}
$e^-\gamma$	2.0×10^{-4}
$\mu^-\pi^0$	8.2×10^{-4}
$e^-\pi^0$	1.4×10^{-4}
$\mu^-\mu^+\mu^-$	2.9×10^{-5}
$e^-\mu^+\mu^-$	3.3×10^{-5}
$\mu^-e^+e^-$	3.3×10^{-5}
$e^-e^+e^-$	3.8×10^{-5}
μ^-K^0	1.0×10^{-3}
e^-K^0	1.3×10^{-3}
$\mu^-\rho^0$	3.8×10^{-5}
$e^-\rho^0$	3.9×10^{-5}
$e^-\pi^+\pi^-$	4.2×10^{-5}
$e^+\pi^-\pi^-$	6.3×10^{-5}
$\mu^-\pi^+\pi^-$	4.0×10^{-5}
$\mu^+\pi^-\pi^-$	6.3×10^{-5}
$e^-\pi^+K^-$	4.2×10^{-5}
$e^+\pi^-K^-$	1.2×10^{-4}
$\mu^-\pi^+K^-$	1.2×10^{-4}
$\mu^+\pi^-K^-$	1.2×10^{-4}
$e^-K^*(892)^0$	5.4×10^{-5}
$\mu^-K^*(892)^0$	5.9×10^{-5}
$e^+\mu^-\mu^-$	3.8×10^{-5}
$\mu^+e^-e^-$	3.8×10^{-5}
$e^-\eta$	2.4×10^{-4}

Compared to the decay modes in (A54) it is much more difficult to conduct a sensitive search for this decay since the τ cannot be reconstructed.

Interesting discussions of the possibilities for violation of lepton number conservation in τ decays are given by Pich (1990a), Masiero (1990), Stroynowski (1990), Barish (1989), Heusch (1989b).

B. DETECTOR DESIGN FOR TAU AND CHARM PHYSICS

B.1 Introduction

In the first part of this paper I described the τ research to be done at a tau-charm factory. This research requires not only the luminosity and energy range of a tau-charm factory, but also requires that the detector have certain properties. The ψ/J, ψ', and charm research to be done at a tau-charm factory (Schindler 1990a, Schindler 1990b, Kirkby 1990, Izen 1990, Heusch 1990, Barish et al 1989, Beers 1989) also requires certain properties of the detector. Table 5 from Kirkby (1990) summarizes the connections between the research goals and the overall requirements. I now go on to briefly discuss the requirements using Fig. 4 as the schematic design of the detector.

Table 5. Special detector requirements for some measurements at a tau-charm factory (Kirkby 1990).

Experiment	Detector emphasis				
	Charged particles	Photons	$\pi K p$ i.d.	$e\mu$ i.d.	Hermeticity
$\tau^{\pm}$ physics:					
ν_τ, $\tau^{\pm}$ masses	•	•	•	•	•
$\tau \to \ell \nu_\ell \nu_\tau$ spectra	•			•	•
Precise branching ratios		•	•	•	•
Second class currents			•	•	
Weak hadronic current	•	•	•	•	•
$\tau^{\pm}$ electric dipole moment				•	
Rare decays	•	•	•	•	•
D, D_s physics:					
V_{cs}, V_{cd} (semileptonic decays)		•		•	•
f_D (pure leptonic decays)		•		•	•
Hadronic decays (CA, CS, DCS)	•	•	•		
$D^0 \bar{D}^0$ mixing, CP violation	•			•	•
Rare decays	•			•	•
$J/\psi(3.10)$, $\psi(3.69)$ physics:					
Spectroscopy ($c\bar{c}$, gg, hybrid, uds)	•	•	•		
Rare decays			•	•	•

B.2 Beam Pipe, Focussing Quadrupoles and Masking

The emphasis in the research to be done at a tau-charm factory is on production of τ pairs and D pairs near or at threshold. Therefore secondary vertex detection is not used and the beam pipe is relatively large, 10 or 12 cm in diameter. This greatly simplifies the construction of the pipe and the masking of the detector from synchrotron radiation.

On the other hand, the tight focussing of the bunches requires that the quadrupoles start 80 to 100 cm from the interaction point. This complicates the design of the detector end sections because most of the detector subsystems must extend to small angles.

B.3 Charged Particle Tracking, Momentum Measurement and Magnet

Precise momentum measurement is required for a variety of tau and charm physics studies, the goal is

$$\sigma_p/p = \left[(0.4\%p)^2) + (0.3\%/\beta)^2\right]^{1/2} \qquad (B1)$$

where p is in GeV/c and $\beta = v/c$. The small multiple Coulomb scattering term assumes that the beam pipe and the tracking chamber cause minimum multiple scattering. This in turn requires a 1 to 1.5 T solenoid magnet, probably with a superconducting coil.

The tracking chamber might be a drift chamber or a TPC. In either case the goal is to track charged particles over 90% of 4π.

Some of the tau and charm physics studies which use the momentum precision of (B1) are: probing the ν_τ mass, studying narrow mass D decays, and searching for rare narrow mass processes such as Higgs particles produced in radiative J/ψ decays.

B.4 Electromagnetic Calorimeter

Several strict requirements are imposed by tau and charm physics goals on the electromagnetic calorimeter. For example, in tau physics when using the τ **tag modes** (Kirkby 1990, Gomez-Cadenas, Heusch, and Seiden 1989, Barish *et al* 1989)

$$\tau^- \to \nu_\tau + e^- + \bar{\nu}_e$$
$$\tau^- \to \nu_\tau + \mu^- + \bar{\nu}_\mu \qquad (B2)$$
$$\tau^- \to \nu_\tau + \pi^-$$

it is important to know that the *missing energy is carried off by ν's, not by undetected γ's*. Also highly efficient detection and measurement of all γ's are important in untangling multiple π^0 and η decay modes. Similarly in charm physics, highly efficient photon detection and measurement is necessary for clean studies of pure leptonic and semi-leptonic D decays. And of course for both tau and charm physics, it is important to have efficient e identification and excellent e–hadron separation.

Thus the measurement of e and γ energies must be precise, the goal is

$$\sigma_E/E = \left[\left(2\%/\sqrt{E}\right)^2 + (1\%)^2 \right]^{1/2} \qquad (B3)$$

where E is in GeV. Hence a crystal electromagnetic shower detector such as CsI is required. The noise must be sufficiently low so that γ's with energies as low as 10 MeV can be detected. It is also important that photon angles be well measured; therefore silicon pad shower positron detectors may be used in conjunction with the crystal shower energy detectors.

Another electromagnetic calorimeter design goal is that the photon detection inefficiency be less than 1%, the calorimeter covering almost all of 4π. This is part of the hermeticity requirement of the detector noted in connection with (B2).

B.5 Muon Detector and Hadron Veto Calorimeter.

A unique aspect of the detector is the combined muon identification and hadron veto calorimeter system outside the solenoid and on both ends. This system would consist of thin iron plates, 1. to 2.5 cm thick, separated by tracking devices such as drift chambers with long drift regions. The separation of μ's from π's and K's would be done by the lack of nuclear interactions by the μ and careful range measurement.

This system would also detect hadrons by their nuclear interactions, in particular the neutral hadrons: K^0's and n's. The K^0 and n detection would mostly be used as a signal that energy was being carried off by a neutral particle, although that energy would be poorly measured. Such a signal would veto the use of a τ pair event which had been tagged by the decays in (B2) since the missing energy might be due to a K^0 or n, not due to ν's. Like the electromagnetic calorimeter the muon detector and hadron veto system must cover almost all of 4π.

B.6 Particle Identification and Time-of-Flight Counters

I have already discussed the design goals for e, μ, and γ identification. It is also important, very important for charm physics, to separate π's and K's. Excellent $\pi - K$ separation is required for advanced studies of D decays and D^0 mixing. In the present detector concept π and K identification is accomplished with time-of-flight scintillation counters just outside the tracking chamber and by dE/dx measurements in the tracking chamber. The design goal for the time-of-flight counters is a time resolution of 120 ps.

B.7 Detector Summary

Past and current design work on the detector for tau-charm physics shows that all the requirements on the detector can be met by conventional and well developed detector technology.

C. THE TAU-CHARM FACTORY

C.1 Requirements for a Tau-Charm Factory

In these talks I have only discussed the τ physics to be done at a tau-charm factory, the principle operating part being

$$E_{tot} = 3.57 \text{ GeV (Several MeV above threshold)} : \sigma = 0.4 \text{ nb}$$
$$E_{tot} = 3.67 \text{ GeV (just below } \psi) : \sigma = 2.3 \text{ nb} \tag{C1}$$

There is also a tremendous amount of charm quark physics to do at a tau-charm factory as described in the Proceedings of the Tau-Charm Factory Workshop (Beers, 1989) and by Schindler (1989), Schindler (1990a), Schindler (1990b), Kirkby (1989a), Kirkby (1990), Barish *et al* (1989). The principle operating points are the $c\bar{c}$ resonances

$$J/\psi \text{ at } E_{tot} = 3.10$$
$$\psi' \text{ at } E_{tot} = 3.69 \tag{C2}$$

and the D pair production points

$$D^+D^- \text{ at the } \psi'', \; E_{tot} = 3.77 \text{ GeV} : \; \sigma = 4.2 \text{ nb}$$
$$D^0\bar{D}^0 \text{ at the } \psi'', \; E_{tot} = 3.77 \text{ GeV} : \; \sigma = 5.8 \text{ nb} \tag{C3}$$

$$D_s \bar{D}_s \text{ at } E_{tot} = 4.03 \text{ GeV}: \quad \sigma = 0.7 \text{ nb}$$
$$D_s \bar{D}_s^* \text{ at } E_{tot} = 4.14 \text{ GeV}: \quad \sigma = 0.9 \text{ nb} \quad (C4)$$

Finally Klein (1989) has suggested

$$\Lambda_c \bar{\Lambda}_c \text{ at } E_{tot} = 4.6 \text{ GeV}: \quad \sigma \sim 0.2 \text{ nb} \quad (C5)$$

Thus the energy range of a tau-charm factory is

$$3.0 < E_{Tot} \lesssim 5.0 \text{ GeV} \quad (C6)$$

The next question is luminosity. Since the maximum luminosity cannot be constant over the E_{tot} range in (C6), the energy of maximum luminosity must be picked. The two operating parts where maximum luminosity is desired are τ pairs at $E_{tot} = 3.67$ GeV and D^+D^-, $D^0\bar{D}^0$ pairs at 3.77 GeV. Then

$$E_{tot}(L_{max}) \approx 3.7 \text{ to } 3.8 \text{ GeV} \quad (C7)$$

The physics to be done at these two operating points requires 10^6 to 10^8 pairs per year of τ's, D^0's or $D^\pm$'s. The design luminosity should be set as close to the 10^8 requirement as can be achieved with reliable and conservative collider technology.

A tau-charm factory would be operated 8 months per year for particle physics, the other months would be used for accelerator physics studies, maintenance and upgrading. And unlike other types of accelerators and colliders, a tau-charm factory could achieve 80% efficiency if built with sufficient reliability. This is because the operating mode of the machine is fixed for long-time periods. This means 1.7×10^7 s/year. Then the 10^8 events goal requires for example

$$N_{yr} L_{max} = 2.6 \times 10^{33} \text{ yr cm}^{-2} \text{ s}^{-1} \text{ for } 10^8 \tau \text{ pairs}$$
$$N_{yr} L_{max} = 1.4 \times 10^{33} \text{ yr cm}^{-2} \text{ s}^{-1} \text{ for } 10^8 D^\pm \text{ pairs} \quad (C8)$$

where N_{yr} is the number of operating years.

Now remembering that 10^8 events is not a rigorous goal but just a rough goal, we see that an L_{max} of 5×10^{33} cm^{-2} s^{-1} would be wonderful, would enable the physics goals to be accomplished in half a year. Eventually $L_{max} \sim 5 \times 10^{33}$

cm^{-2} s^{-1} may be possible using a "crab-crossing" technique (Voss, Paterson, and Kheifets 1989). But as discussed in the next section

$$L_{max} \approx 10^{33} \text{ cm}^{-2} \text{ s}^{-1} \quad (C9)$$

is the best that can be reliably designed with our present knowledge of collider physics and technology. And this L_{max} is certainly adequate for the 10^8 events goal. Another requirement on a tau-charm factory is that all the particles in the beams have the same energy within an MeV or so:

$$\Delta E_{tot} \sim \text{ few MeV}$$

This is very important for the τ physics studies carried out a few MeV above τ pair threshold, and is useful at other energies, for example at the ψ/J and ψ'.

Summarizing, there are four required properties of a tau-charm factory:

(i) $3.0 \leq E_{tot} \lesssim 5.0$ GeV

(ii) $L_{max} \approx 10^{33}$ cm^{-2} s^{-1}

(iii) Highly reliable operation

(iv) $\Delta E_{tot} \sim$ few MeV

It is the second and third property which will enable experiments at a tau-charm factory to make great advances. At $E_{tot} \approx 3.8$ GeV (C3) the L_{max} of the new BEPC storage ring will be about 10^{31} cm^{-2} s^{-1} and for SPEAR L_{max} was about 10^{30} cm^{-2} s^{-1}. Thus in full operation the tau-charm factory will have 100 times the data rate of BEPC and 1000 times that of SPEAR. In addition the improved reliability made possible by new technology will increase these factors on a yearly bases. Even in the first period of tau-charm factory operation when $L \sim 10^{32}$ to 3×10^{32} cm^{-2} s^{-1}, the data rate will be very, very good.

C.2 Obtaining High Luminosity in a Tau-Charm Factory

I now give a qualitative discussion, following the seminal work of Jowett (1987), on how $L_{max} = 10^{33}$ cm^{-2} s^{-1} is obtained in a tau-charm factory. I start with some basic formulas for just 1 e^+ and 1 e^- bunch, each bunch having N_b particles. At the collision point each bunch has a cross sectional area

$$A = 4\pi \sigma_x^* \sigma_y^* \quad (C10)$$

where σ_x^* and σ_y^* are the sigmas of the width and height of the bunch. Let ℓ_c be the circumference of the ring and c the velocity of light. Then

$$L = \frac{cN_B^2}{\ell_c A} \qquad (C11)$$

The obvious first stage in maximizing L is to use *large* N_b, many particles in a bunch. And to use *small* A by tight focussing of the bunches at the interaction point. But there is a *beam-beam interaction* between the e^+ and e^- beams when they collide which perturbs the orbits of the particles. The magnitude of the perturbation increases on the average as N_b/A increases, therefore there is a practical upper limit to N_b/A. There is also a *single bunch instability* upper limit on N_b because a bunch interacts electromagnetically with the walls of the beam pipe as it moves around the ring. This interaction if too strong will disrupt the bunch. Finally, there is a *dynamic aperture* upper limit on A.

Looking back at (C10) with these limits on N_b and A the next step in increasing L is to decrease ℓ_c. But here again there is a limit, a lower limit on ℓ_c. The energy lost per turn by synchrotron radiation in a ring of effective radius ρ is

$$V \propto E^4/\rho \qquad (C12)$$

This power must be absorbed by the walls of the beam pipe, it is important that this power be kept moderate so that a conventional and simple beam pipe can be used. (A simple beam pipe reduces single-beam instabilities.) For this and other reasons there is a lower limit on ℓ_c.

The way out of these limits is to use multiple bunches, k_b e^+ bunches and $k_b e^-$ bunches. Then (C10) becomes

$$L = \frac{ck_b N_b^2}{\ell_c A} \qquad (C13)$$

But now, in a simple, single ring collider, a bunch will collide $2k_b$ times during one revolution around the ring and the multiple beam-beam interactions per revolution put a more severe upper limit on N_b/A. To avoid this, bunch collisions must be limited to one or two interaction regions where there are experiments; at $2k_b - 1$ or $2k_b - 2$ potential collision points the bunches must pass by each other without touching. In the tau-charm energy region, Jowett (1987, 1988) found the best solution to this problem is to use separate e^+ and e^- rings with only 1 or 2 places where the ring intersect. Figure 5 from Jowett (1988) is a schematic of the design for 1 interaction point.

Jowett's original design was for L_{max} at $E_{tot} = 5$ GeV with the following parameters

$$\begin{aligned} \ell_c &= 377 \text{ m} \\ N_b &= 1.63 \times 10^{11} \,,\ k_b = 24 \\ \sigma_x^* &= 443\ \mu\text{m}\,,\ \sigma_y^* = 8\ \mu\text{m} \\ L_{max} &= 1.6 \times 10^{33}\ \text{cm}^{-2}\text{s}^{-1} \end{aligned} \qquad (C14)$$

Another important bunch shape parameter is the rms length σ_z. In Jowett's original design $\sigma_z = 6.2$ mm. The bunch must be kept so short in order to make use of the tight focussing of the bunches at the collision point. This small σ_z imposes two more requirements on the design of a tau-charm factory. The radio frequency cavities must have a large overvoltage. And once again, the beam pipe must be simple so that it has a small impedance for the electromagnetic interaction of a bunch with the pipe walls.

C.3 History of Tau-Charm Factory Design

The first design of a tau-charm factory was done by Jowett (1987, 1988, 1989) who worked out the basic principles:

- Separate e^+ and e^- rings.
- One interaction region. Later designs allow two.
- Tight focussing of the bunches at the interaction point, called microbeta insertion.
- Multiple bunches, about 20 to 30 in each ring.
- Rings have a large radius to keep synchrotron radiation moderate and allow a conventional beam pipe.
- Substantial RF overvoltage and low beam pipe impedance to produce short bunches.
- Feedback systems to control multibunch instabilities.
- A high intensity e^+ and e^- injector to maintain luminosity by "top-off" of the circulating bunches.

Further design work was carried out by many accelerator physicists at the 1989 Tau-Charm Factory Workshop (Beers 1989). This group confirmed that $L_{max} \approx 10^{33}$ cm^{-2} s^{-1} was feasible with present technology (Brown, Fieguth, and Jowett 1989).

A separate conceptual design was carried out by Gonichen, Le Duff, Mouton, and Travier (1990). This report discusses the accelerator physics in very useful detail, for example comparing flat beams with round beams.

Another conceptual design based more closely on the original Jowet design was prepared by Barish *et al* (1990). Both this design and the Gonichon *et al* design attained $L_{max} \approx 10^{33}$ cm^{-2} s^{-1}.

Danilov *et al* (1990) have also discussed tau-charm factory design.

The most recent conceptual design (Baconnier, 1990) was carried out by physicists from CERN, LAL in France, and CIEMAT in Spain and is intended for a tau-charm factory laboratory in Spain. These lectures conclude with a summary of this design.

C.4 The Design of a Tau-Charm Factory for Spain

Figure 6 is the title page of the Baconnier *et al* (1990) design for a tau-charm factory in Spain, Fig. 7. The powerful, high-rate e^+ and e^- injector, Fig. 8, also allows a separate synchrotron light source ring. The collider is designed for L_{max} at $E_{tot} = 4$ GeV. Some of the luminosity connected parameters are

$$\begin{aligned}
\ell_c &= 360 \ m \\
N_b &= 1.6 \times 10^{11} \quad k_b = 30 \\
\sigma_y^* &= 280 \ \mu m \quad \sigma_y^* = 14 \ \mu m \\
\sigma_z &= 6 \ mm \\
L_{max} &= 1 \times 10^{33} \ \text{cm}^{-2}\text{s}
\end{aligned} \qquad (C15)$$

where N_b and k_b are the particles per bunch and the number of bunches in a beam, σ_x^* and σ_y^* are the horizontal and vertical rms bunch sizes at the collision point, and σ_z is the bunch length. Figure 9 shows the collider tunnel and magnets.

The total current per beam, I, is 0.6 A, a large current for an e^+e^- collider which in turn requires care in the beam pipe, interaction region, and RF cavity design. But the maximum synchrotron radiation power dissipated in the beam pipe is 1.9 kW/m which is moderate and can be managed by a conventional beam pipe. The total power dissipated per ring is 120 kW which is also moderate. Since I = 0.6 A, this power loss requires an average accelerating voltage per ring of 0.2 MW. However to keep σ_z small the RF cavities produce a large overvoltage, 19 MV. Superconducting RF cavities with a resonant frequency of 500 MHz are proposed.

C.5 Summary

Thus design has reached the stage that we can build a powerful and very general new electron-positron collider – the tau-charm factory. In these lectures I have described the tremendous amount of research which can be foreseen in tau physics. Tremendous amounts of charm research will also be done. And as I said in the introduction, the tau-charm factory may bring us discoveries we cannot now conceive.

ACKNOWLEDGEMENT

It is evident from the references that a very large number of people have contributed to the particle and accelerator physics associated with a tau-charm factory. I am particularly grateful for many conversations with Jasper Kirkby, John Jowett, Juan Antonio Rubio, and Rafe Schindler.

REFERENCES

Aguilar-Benitez *et al* 1990 *Phys. Lett.* **B239** 1

Albrecht H *et al* 1987 *Phys. Lett.* **B185** 228

Albrecht H *et al* 1988 *Phys. Lett.* **B202** 142

Baconnier Y *et al* 1990 CERN/AC/90-07 (unpublished)

Baltrusaitis R M *et al* 1985 Phys. Rev. Lett. **55** 1842

Banerjee S 1986 *Phys. Rev.* **D34** 2080

Barish B and Stroynowski R 1988 *Phys. Reports* **157** 1

Barish B *et al* 1989 SLAC-PUB 5053 (unpublished)

Barish B *et al* 1990, SLAC-PUB-5180 (unpublished)

Barish B 1989 *Proc. Tau-Charm Factory Workshop* ed. Beers L (SLAC, Stanford) p 113

Beers L 1989 *Proc. Tau-Charm Factory Workshop* ed. Beers L (SLAC, Stanford)

Behrend H J *et al* 1990 *Z. Physics* **C46** 537

Berger E L and Lipkin H J 1987 *Phys. Lett* **B189** 226
Braaten E 1988 *Phys. Rev. Lett.* **60** 1606
Braaten E 1989a *Phys. Rev. Lett.* **63** 577
Braaten E 1989b *Proc. Tau-Charm Factory Workshop* ed. Beers L (SLAC, Stanford) p 408
Bramon A, Narison S and Pich A 1987 *Phys. Lett.* **B196** 543
Brown K L, Fieguth T, and Jowett J M 1989 *TFC* ed. Beers L (SLAC, Stanford) p 244
Burchat P 1988 *Proc. SIN Spring School on Heavy Flavour Physics* (Zuoz, Switzerland) p 125
Danilov M V et al 1990 ITEP-90-67 (unpublished)
Dominguez C A and Sola J 1988 *Phys. Lett.* **B208** 131
Fetscher W 1990 *Phys. Rev.* **D42** 1544
Fetscher W, Gerber H-J and Johnson K F 1986 *Phys. Lett.* **B173** 102
Gan K K and Perl M L 1988 *Int. J. Mod. Phys.* **A3** 531
Gan K K 1989 *Proc. Tau-Charm Factory Workshop* ed. Beers L (SLAC, Stanford) p 554
Gan K K 1990 *Proc. Workshop on Tau Lepton Physics*, Orsay (to be published)
Garcia A and Rivera Rebelledo J M 1981 *Nucl. Phys.* **B189** 500
Ghose P and Kumbhakar D 1981 *Z. Phys.* **C8** 49
Gilman F J and Rhie S H 1985 *Phys. Rev.* **D31** 1066
Gilman F J 1987 *Phys. Rev.* **D35** 3541
Gomez-Cadenas J J et al 1990 *Phys. Rev.* **D41** 2179
Gomez-Cadenas J J, Heusch C A, and Seiden A 1989 *Particle World* **1** 10
Gonichon J, Le Duff J, Mouton B, and Travier C 1990 LAL/RT 90-02 (unpublished)
Hayes K G et al 1982 *Phys. Rev.* **D25** 2869
Hayes K G and Perl M L 1988 *phys. Rev.* **D38** 3351
Heusch C A 1989a *Les Rencontres de Physique de la Vallée d'Aoste*, 1989, ed. Greco M (*Editions Frontières*, Gif-sur-Yvette) p 399
Heusch C A 1989b *Proc. Tau-Charm Factory Workshop* ed. Beers L (SLAC, Stanford) p 528
Heusch C A 1990 SCIPP-90/14, to be published in *Proc. 1990 Rencontre de Physique de la Vallée d'Aoste*
Izen J M 1990 UIUC-HEPG-90-58, to be published in *Proc. 1990 of APS Div. of Particles and Fields*
Jain V 1990 *Proc. Workshop on Tau Lepton Physics*, Orsay (to be published)

Jowett J M 1987 CERN LEP-TH/87-56

Jowett J M 1988 *Proc. European Particle Accelerator Conf.*, ed. Tazzari S (*World Scientific*, Singapore) p 368

Jowett J M 1989 *Proc. Tau-Charm Factory Workshop*, ed. Beers L (SLAC, Stanford) p 7

Keh S *et al* 1988 *Phys. Lett.* **B212** 123

Kiesling C 1988 *High Energy Electron-Positron Physics* eds. Ali A and Söding P (World Scientific, Singapore) p 177

Kiesling C 1989 *Proc. XXIV Rencontre De Moriond* ed. Tran Thanh Van J (*Editions Frontières*, Gif-sur-Yvette) p 323

Kiesling C 1990 *Proc. Workshop on Tau Lepton Physics*, Orsay (to be published)

Kim J H and Resnick L 1980 *Phys. Rev.* **D21** 1330

Kinoshita T and Sirlin A 1959 *Phys. Rev. Lett.* **2** 177

Kirkby J 1987 CERN EP/87-210 and *Spectroscopy of Light and Heavy Quarks* eds. Gastaldi U, Klapisch R, and Close F (Plenum, New York) p 401

Kirkby, J 1989a *Particle World*, **1** 27

Kirkby J 1989b *Proc. Tau-Charm Factory Workshop* ed. Beers L (SLAC, Stanford) p 294

Kirkby J 1990 *Proc. 17th Int. Meeting Fundamental Physics* eds. Aguilar-Benitez M and Cerrada M (*World Scientific*, Singapore) p 181

Klein S 19889 *Proc. Tau-Charm Factory Workshop* ed. Beers L (SLAC, Stanford) p 909

Laursen M L, Samuel M A, and Sen A 1984 *Phys. Rev.* **D29** 2652

Leroy C and Pestieau J 1978 *Phys. Lett.* **B72** 398

Marciano W J and Sirlin A 1988 *Phys. Rev. Lett.* **61** 1815

Masiero A 1990 DFPD/90/TH732 (unpublished)

Maxwell C J and Nicholls J A 1990 *Phys. Lett.* **B236** 63

Mursula K and Scheck F 1985 *Nucl. Phys.* **B253** 189

Narison S and Pich A 1988 *Phys. Lett.* **B211** 183

Perl M L 1990 *Proc. Int. Symposium Lepton and Photon Interactions at High Energies* ed. Riordan M (*World Scientific*, Teaneck)

Perl M L 1991 *Reports Progress Physics* (to be published)

Pich A 1987 *Phys. Lett.* **B196** 561

Pich A 1989 *Proc. Tau-Charm Factory Workshop* ed. Beers L (SLAC, Stanford) p 416

Pich A 1990a *Proc. 17th Int. Meeting Fundamental Physics* eds. Aguilar-Benitez M and Cerrada M (*World Scientific*, Singapore) p 323

Pich, A 1990b *Mod. Phys. Lett. A* (to be published)
Pumplin J 1988 *Phys. Rev. Lett.* **63** 576
Pumplin J 1990 *Phys. Rev.* **D41** 900
Queijeiro A and Garcia H 1988 *Phys. Rev.* **D38** 2218
Scheck F 1978 *Phys. Reports* **D44** 87
Schindler R H 1989 *Les Rencontres de Physique de la Vallée d'Aoste, 1989*, ed. Greco M (*Editions Frontières*, Gif-sur-Yvette) p 89
Schindler R H 1990a *12th Inst. Workshop on Weak Interactions and Neutrinos* eds. Singer P and Eilam G (North-Holland, Amsterdam)
Schindler, R H 1990b *Proc. Lake Louise Winter Inst. on the Standard Model and Beyond* ed. Astbury A et al (*World Scientific*, Teaneck)
Seiden A 1989 *Proc. Tau-Charm Factory Workshop* ed. Beers L (SLAC, Stanford) p 252
Smith J G 1990 *Proc. Workshop on Tau Lepton Physics*, Orsay (to be published)
Stroynowski R 1990a CALT-68-1092 (unpublished)
Stroynowski R 1990b CALT-68-1683, to be published in *Proc. Workshop on Tau Lepton Physics*, Orsay
Truong T N 1984 *Phys. Rev.* **D30** 1509
Tsai Y S 1971 *Phys. Rev.* **D4** 2821
Tsai Y S 1989a *Proc. Tau-Charm Factory Workshop* ed. Beers L (SLAC, Stanford) p 387, submitted to *Phys. Rev.*
Tsai Y S 1989b *Proc. Tau-Charm Factory Workshop* ed. Beers L (SLAC, Stanford) p 394, submitted to *Phys. Rev.*
Voloshin M 1989 TPI-MINN-89/33-T (unpublished)
Voss G A, Paterson J M, and Kheifets S A 1989 *Proc. Tau-Charm Factory Workshop* ed. Beers L (SLAC, Stanford) p 31
Wu D Y 1990a Ph.D. Thesis, Cal. Inst. Tech., CALT-68-1638 (unpublished)
Wu D Y et al 1990b Phys. Rev. **D41** 1990
Zachos C K and Meurice Y 1987 *Mod. Phys. Lett.* **A2** 247

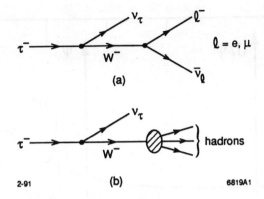

Fig. 1. Feynman diagram for τ decays for (a) pure leptonic decays, (b) hadronic decays.

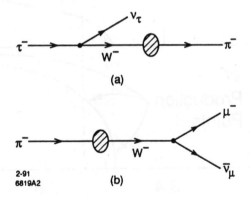

Fig. 2. Feynman diagram for (a) $\tau^- \to \nu_\tau \pi^-$ and (b) $\pi^- \to \mu^- \bar{\nu}_\mu$.

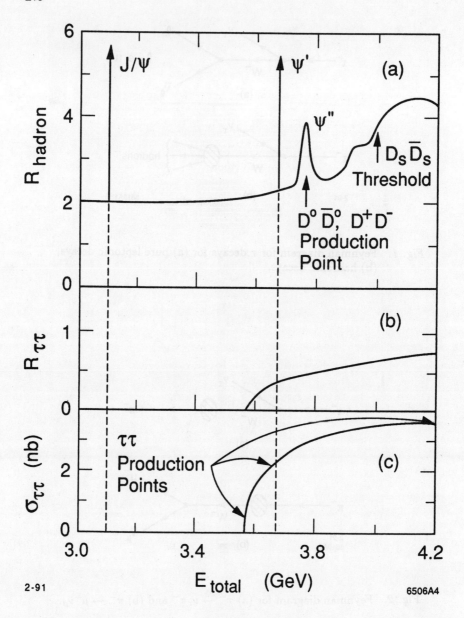

Fig. 3. $\sigma_{\tau\tau}$, $R_{\tau\tau}$, and $R_{hadrons}$ for the main part of the tau-charm factory energy range.

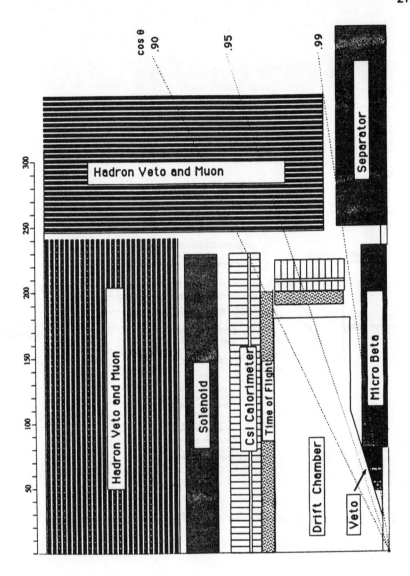

Fig. 4. Schematic design of detector for tau and charm physics research at a tau-charm factory.

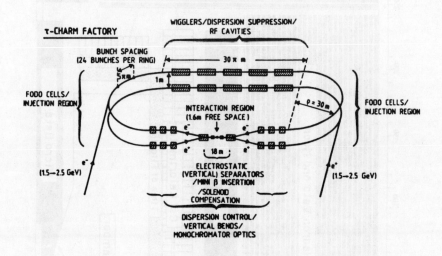

Fig. 5. Schematic design of a tau-charm factory collider from Jowett (1988).

EUROPEAN LABORATORY FOR PARTICLE PHYSICS

CERN/AC/90-07

A TAU-CHARM FACTORY LABORATORY IN SPAIN
combined with a
SYNCHROTRON LIGHT SOURCE
(A conceptual study)

Y. Baconnier, J.-L. Baldy, J.-P. Delahaye, R. Dobinson,
A. De Rújula, F. Ferger, A. Hofmann, J.M. Jowett,
J. Kirkby, P. Lefèvre, D. Möhl, G. Plass, L. Robertson,
J. A. Rubio, T.M. Taylor and E.J.N. Wilson
CERN, Geneva, Switzerland

F. Dupont and J. le Duff
LAL, Orsay, France

C. Willmott
CIEMAT, Madrid, Spain

Abstract

A conceptual design for a τ-charm factory and its associated laboratory is given. It includes the physics interest, a description of the scope and layout of the new laboratory in Spain, the τ-charm factory collider and detector, the injector system and a synchrotron light source, together with estimates of the time-scale and necessary resources.

Geneva, Switzerland
20 November 1990

Fig. 6. Title page from design for a Tau-Charm Factory in Spain (Baconnier 1990).

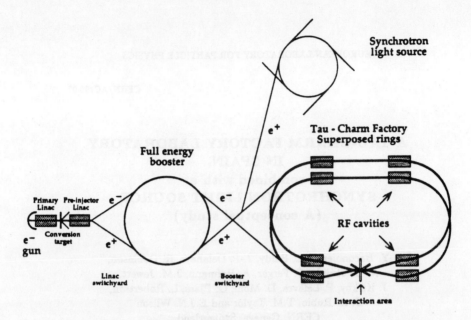

Fig. 7. Schematic representation of the injector, collider, and synchrotron light source from Baconnier (1990). The collider may be designed to allow a second interaction area to be installed later.

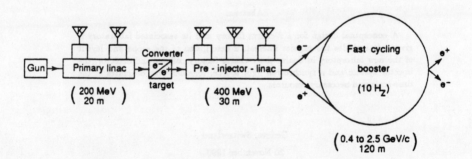

Fig. 8. Injector for tau-charm collider and synchrotron light source from Baconnier (1990).

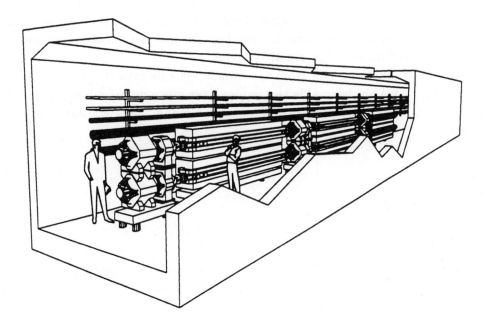

Fig. 9. Collider tunnel and magnets showing the vertically separated rings from Baconnier (1990).

COMMENTS ON B3
"Beyond the Tau: Other Directions in Tau Physics"

A tremendous amount has been learned about quantum mechanics, quantum electrodynamics, and elementary particles by the experimental study of atoms, that is electron-nucleus systems, and the experimental study of positronium, the electron-positron atom e^+e^-. Similarly a great deal has been learned from mu mesic atoms, the system consisting of a negative muon and a nucleus usually with additional electrons.

This paper concentrates mainly on the properties of systems in which a tau lepton replaces an electron or positron, namely an atom containing a negative tau lepton or a system consisting of a negative tau lepton and a positive tau lepton $\tau^+\tau^-$. These thoughts came out of my work on the tau-charm factory because that is the only instrument where we might hope to produce such tau atomic systems.

But can we learn anything new from tau atomic systems even if we could make them and study them? I have not been able to propose clear research objectives. But in physics studying a system for it's own sake can be pleasurable and there is always the dream of the unexpected.

BEYOND THE TAU:
OTHER DIRECTIONS IN TAU PHYSICS*

MARTIN L. PERL
Stanford Linear Accelerator Center
Stanford University, Stanford, California 94309

ABSTRACT

This paper calls attention to four topics in tau lepton physics which are outside our present areas of tau physics research: $\tau^+\tau^-$ atoms, τ^- nucleus atoms, photoproduction of τ's, and heavy ion production of τ's.

---

* This work was supported by the Department of Energy, contract DE-AC03-76SF00515.

Copyright © 1993 by World Scientific Publishing Co. Pte. Ltd.

A. Introduction

This paper is based on a talk delivered at the *Second Workshop on Tau Lepton Physics* held at The Ohio State University, September 8–11, 1992. In that talk I called attention to four out-of-the-way topics in tau physics: $\tau^+\tau^-$ atoms, τ^- – nucleus atoms, photoproduction of τ's, and heavy ion production of τ's; and these are the areas covered in this paper. Two other topics from that talk will not be discussed in this paper: future searches for heavy leptons and speculations on missing modes in tau decay (Perl 1992).

B. The $\tau^-\tau^+$ Atom

B.1 Introduction

In the early years of the discovery of the τ there was some discussion of the physics of an atom that would consist of the Coulombic bound state of a τ^+ and a τ^- (Moffat 1975, Avilez *et al.* 1978, Avilez *et al.* 1979), an entity analagous to the e^+e^- atom positronium (Rich 1978). The $\tau^+\tau^-$ atom can be made in e^+e^- annihilation just below τ pair threshold.

$$e^+ + e^- \to \tau^+\tau^- \text{ atom} , \qquad (1)$$

hence the tau-charm factory offers the best route for making these atoms as discussed in Sec. B.5.

B.2 Static Properties

The energy levels of the $\tau^+\tau^-$ atom are shown in Fig. 1 where the atomic spectroscopy notation

$$n^{2S+1}L_J$$

is used. Here n is the principle quantum number; S is the total spin quantum number and is 0 or 1, L is the orbital angular momentum quantum number with L = S, P, D ... for L = 0, 1, 2 ..., and J is the total angular momentum quantum number. Ignoring fine structure, the energy levels are given by

$$E_n = -\frac{m_\tau c^2 \alpha^2}{4n^2} = -\frac{23.7 \text{ keV}}{n^2} \qquad (2)$$

The 4 in the denominator comes from the usual 2 in the denominator and and $m_{reduced}(\tau^+\tau^- \text{ atom}) = m_\tau/2$ in the numerator. I use $m_\tau = 1777$ MeV/c^2.

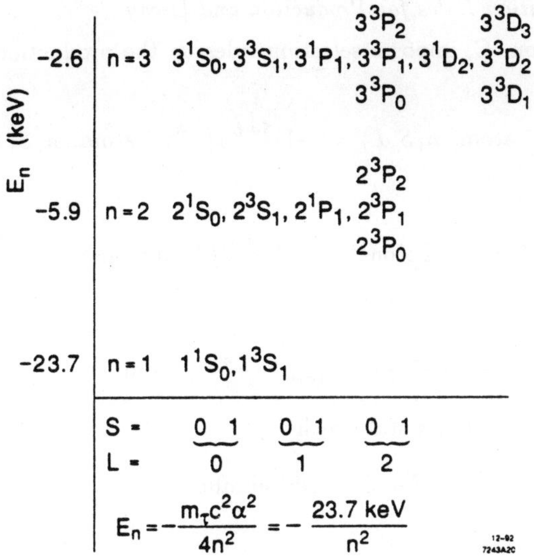
3^3P_2 3^3D_3
-2.6 n = 3 $3^1S_0, 3^3S_1, 3^1P_1, 3^3P_1, 3^1D_2, 3^3D_2$
3^3P_0 3^3D_1
E_n (keV)
2^3P_2
-5.9 n = 2 $2^1S_0, 2^3S_1, 2^1P_1, 2^3P_1$
2^3P_0
-23.7 n = 1 $1^1S_0, 1^3S_1$
S = 0 1 0 1 0 1
L = 0 1 2
$E_n = -\frac{m_\tau c^2 \alpha^2}{4n^2} = -\frac{23.7\text{ keV}}{n^2}$
12-92
7243A20

Figure 1

The Bohr radius is given by

$$a_0 = \frac{2\hbar^2}{m_\tau e^2} = 3.04 \times 10^{-12} \text{ cm} \quad , \tag{3}$$

which is three orders of magnitude smaller than the Bohr radius for hydrogen of 5.29×10^{-9} cm.

The $n = 1, 2$ wave functions are given by

$$\psi_{n\ell} = R_{n\ell}(r) Y_{\ell m}(\theta, \phi) \tag{4a}$$

where $Y_{\ell m}$ is a normalized spherical harmonic and

$$\begin{aligned} R_{10} &= \frac{1}{a_0^{3/2}} \, 2\, e^{-r/a_0} \\ R_{20} &= \frac{1}{a_0^{3/2}} \frac{1}{\sqrt{2}} \left(1 - \frac{r}{2a_0}\right) e^{-r/2a_0} \\ R_{21} &= \frac{1}{a_0^{3/2}} \frac{1}{\sqrt{6}} \frac{r}{2a_0} e^{-r/2a_0} \end{aligned} \tag{4b}$$

B.3 Charge Conjugation Rules for Production and Decay

Charge conjugation, C, imposes selection rules on the production and decay of the $\tau^+\tau^-$ atom

$$C\psi(\tau^+\tau^- \text{ atom}, n, S, L) = (-1)^{S+L}\psi(\tau^+\tau^- \text{ atom}, n, S, L) \quad (5)$$

and for a state of N photons

$$C\,\psi(N \text{ photons}) = (-1)^N\,\psi(N \text{ photons}) \quad (6)$$

Therefore in production

$$e^+ + e^- \to \gamma_{virtual} \to \tau^+\tau^- \text{ atom} \quad (7a)$$

the atom must be produced in a state with

$$S + L = \text{ odd number} \quad (7b)$$

The decay

$$\tau^+\tau^- \text{ atom} \to \gamma + \gamma \quad (8a)$$

requires

$$S + L = \text{ even number} \quad (8b)$$

and the decay

$$\tau^+\tau^- \text{ atom} \to \gamma + \gamma + \gamma \quad (9a)$$

requires

$$S + L = \text{ odd number} \quad (9b)$$

B.4 Decay channels of the $\tau^+\tau^-$ atom

Next I discuss the decay of the $\tau^+\tau^-$ atom. There are two classes of decay channel. In the first class the τ^+ or τ^- decay through the weak interaction in the normal way and the atomic state disappears. The decay width is

$$\Gamma(\text{atom, } \tau \text{ decay}) = 2\hbar/\tau_{lifetime} = 4.4 \times 10^{-3} \text{ eV} \quad (10a)$$

where the 2 occurs because the decay of either τ breaks up the atomic state. I have used the τ lifetime (Trischuk 1992) of

$$T_\tau = (2.96 \pm 0.03) \times 10^{-13} \text{ s} \quad (10b)$$

In the second class of decay channels the τ^+ and τ^- annihilate. The annihilation requires that the atomic wave function $\psi(r)$ be unequal to 0 at r = 0

$$\psi(0) \neq 0$$

Here r is the distance between the τ^+ and τ^-. Therefore in lowest order annihilation only occurs in L = 0 states, that is, S states. This is illustrated in Eq. 4b. There are five annihilation channels:

$$\tau^+\tau^- \text{ atom} \to \gamma + \gamma \qquad (11a)$$
$$\tau^+\tau^- \text{atom} \to \gamma + \gamma + \gamma \qquad (11b)$$
$$\tau^+\tau^- \text{atom} \to e^+ + e^- \qquad (11c)$$
$$\tau^+\tau^- \text{atom} \to \mu^+ + \mu^- \qquad (11d)$$
$$\tau^+\tau^- \text{atom} \to \text{hadrons} \qquad (11e)$$

The annihilation channel

$$\tau^+\tau^- \text{ atom} \to \gamma + \gamma \qquad (12a)$$

is even under charge conjugation, therefore

$$\text{atomic state} = n\ ^1S_0 \qquad (12b)$$

The decay width is

$$\Gamma(\text{atom} \to 2\gamma) = \frac{\alpha^5 m_\tau c^2}{2n^3}$$
$$= \frac{1.8 \times 10^{-2} \text{ eV}}{n^3} \qquad (12c)$$

The four other annihilation channels have odd charge conjugation, therefore

$$\text{atomic state} = n\ ^3S_1 \qquad (13)$$

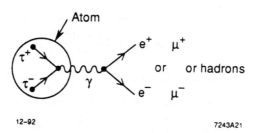

Figure 2

The channel

$$\tau^+\tau^- \text{atom} \rightarrow \gamma + \gamma + \gamma \tag{14a}$$

has the width

$$\Gamma(\text{atom} \rightarrow 3\gamma) = \frac{2(\pi^2 - 9)\alpha^6 m_\tau c^2}{9\pi n^3}$$

$$= \frac{1.7 \times 10^{-5} \text{ eV}}{n^3} \tag{14b}$$

The two channels, Fig. 2,

$$\tau^+\tau^- \text{ atom} \rightarrow e^+ + e^- \tag{15a}$$
$$\tau^+\tau^- \text{ atom} \rightarrow \mu^+ + \mu^- \tag{15b}$$

have the same width

$$\Gamma(\text{atom} \rightarrow e^+e^-) = \Gamma(\text{atom} \rightarrow \mu^+\mu^-) = \frac{\alpha^5 m_\tau c^2}{6n^3}$$

$$= \frac{6.1 \times 10^{-3} \text{ eV}}{n^3} \tag{15c}$$

when we neglect the masses of the e and μ. Finally there is the channel, Fig. 2,

$$\tau^+\tau^- \text{ atom} \rightarrow \text{hadrons} \tag{16a}$$

The width cannot be calculated from first principles, however from colliding beams e^+e^- annihilation data at $E_{tot} \sim 2m_\tau$ we know

$$\sigma(e^+ + e^- \rightarrow \text{hadrons}) \approx 2\sigma(e^+ + e^- \rightarrow \mu^+ + \mu^-) \tag{16b}$$

Therefore

$$\Gamma(\text{atom} \rightarrow \text{ hadrons}) \approx 2\,\Gamma_{\mu\mu} \tag{16c}$$

Collecting all this together, for $n\ ^1S_0$ states

$$\Gamma_{tot}(n\ ^1S_0) = \Gamma(\text{atom}, \tau \text{ decay}) + \Gamma(\text{atom} \rightarrow 2\gamma)$$

$$= \left(4.4 \times 10^{-3} + \frac{3.7 \times 10^{-2}}{n^3}\right) \text{eV} \tag{17}$$

For the $n\ ^3S_1$ states we can neglect $\Gamma(\text{atom} \rightarrow 3\gamma)$, Eq. 14b, and set

$$\Gamma_{tot}(n\ ^3S_1) \approx \Gamma(\text{atom}, \tau \text{ decay}) + 4\Gamma(\text{atom} \rightarrow e^+e^-)$$

$$\approx \left(4.4 \times 10^{-3} + \frac{2.44 \times 10^{-2}}{n^3}\right) \text{ eV} \tag{18}$$

Table I gives the widths and lifetimes for various S states.

Table I. Widths and lifetimes of $^3S_1$ states of the $\tau^+\tau^-$ atom due to τ decay and $\tau^+\tau^-$ annihilations.

n	Width (eV)	Lifetimes (s)
1	29×10^{-3}	2.3×10^{-14}
2	7.5×10^{-3}	8.8×10^{-14}
3	5.3×10^{-3}	12×10^{-14}
4	4.8×10^{-3}	14×10^{-14}

I remind the reader that in addition to the decays which destroy the $\tau^+\tau^-$ atom there are electromagnetic decays within the atom from an upper level to a lower level (Sec. B6)

$$\psi(\tau^+\tau^- \text{ atom}, n') \to \psi(\tau^+\tau^- \text{ atom}, n) + \gamma \, , \quad n' > n \tag{19}$$

B.5 Production of the $\tau^+\tau^-$ Atom

As noted in Sec. B.1 the production process

$$e^+ + e^- \to \gamma_{virtual} \to \tau^+\tau^- \text{ atom} \tag{20}$$

requires S + L = odd number. Furthermore, the produced state must have $\psi(0) \neq 0$ and hence L = 0. Therefore, S = 1 and the produced state must be $n\,^3S_1$.

The production cross section for the process in Eq. 20 is

$$\sigma(e^+e^- \to \tau^+\tau^- \text{ atom}) = \frac{3\pi(\hbar c)^2}{4m_\tau^2} \frac{\Gamma_{ee}\Gamma_{tot}}{(E_{tot} - 2m_\tau)^2 + \Gamma_{tot}^2/4} \tag{21}$$

Here Γ_{ee} means $\Gamma(\text{atom} \to e^+e^-)$ and is given by Eq. 15c. Γ_{tot} is given by Eq. 18. Thus the production cross section is given by the Breit-Wigner equation with full width at half-height of Γ_{tot} and peak cross section

$$\sigma(e^+e^- \to \tau^+\tau^- \text{ atom, peak}) = \frac{3\pi(\hbar c)^2}{m_\tau^2} \frac{\Gamma_{ee}}{\Gamma_{tot}} \tag{22}$$

As an example consider $\tau^+\tau^-$ atom production into the ground state $1\,^3S_1$. Then

$$\Gamma_{ee} = 6.1 \times 10^{-3} \text{ eV} \tag{23a}$$

$$\Gamma_{tot} \approx 2.9 \times 10^{-2} \text{ eV} \tag{23b}$$

$$\Gamma_{ee}/\Gamma_{tot} = 0.21 \tag{23c}$$

and

$$\sigma(e^+e^- \to \tau^+\tau^- \text{ atom, peak}) \approx 2.4 \times 10^{-28} \text{ cm}^2 \qquad (24)$$

This is a large cross section, <u>but</u> the energy spread of the e^+ and e^- beams, ΔE, is much larger than Γ_{tot}. Thus in a tau-charm factory we expect

$$\Delta E \sim 1 \text{ MeV} \qquad (F.25)$$

and the effective cross section is

$$\sigma(e^+e^- \to \tau^+\tau^- \text{ atom, effective}) \sim$$
$$2.4 \times 10^{-28} \text{ cm} \times \frac{2.9 \times 10^{-2}}{10^6} \sim 10^{-35} \text{ cm}^2 \qquad (26)$$

Therefore for a tau-charm factory luminosity of 10^{33} cm^{-2} s^{-1} we expect

$$\tau^+\tau^- \text{ atoms produced per sec.} \sim 10^{-2} \qquad (27)$$

B.6 Detecting $\tau^+\tau^-$ Atoms?

Equation 27 shows that $\tau^+\tau^-$ atoms can be produced at a reasonable rate at a tau-charm factory. However, we don't know how to detect $\tau^+\tau^-$ atoms in the ground state. One difficulty, Table I, is the small width, 2.9×10^{-2} eV, compared to the 1 MeV energy spread of the beams. The other difficulty is the short lifetime, 2.3×10^{-14} s.

Another approach discussed by Moffat (1975) and Avilez *et al.* (1979) is to look for atoms produced in an excited state and look for photons produced in the transition to the ground state. First, suppose $\tau^+\tau^-$ atoms are produced in the $2\,^3S_1$ state. This is a metastable state and will decay by annihilation

$$\tau^+\tau^- \text{ atom} \to e^+ + e^-,\ \mu^+ + \mu^-,\ \text{hadrons} \qquad (28)$$

before it decays to an $n = 1$, S state of the atom. The next possibility is to produce the $\tau^+\tau^-$ atom in the $3\,^3S_1$ state (Avilez *et al.* 1979) and look for the x-ray photon emitted in the transition

$$3\,^3S_1 \to 2\,^3P_J + \gamma\,,\ E_\gamma = 3.3 \text{ kev} \qquad (29)$$

where J = 0, 1, 2. E_γ is the energy of the x-ray.

The width for $\tau^+\tau^-$ atom decay from the atomic state a to the atomic state b is

$$\Gamma_{ab} = \frac{4e^2 w^2 \hbar}{m_\tau c^3} f_{ab} \tag{30}$$

where

$$E_\gamma = \hbar w$$

and f_{ab} is the oscillator strength, a number of order 0.1 or less. For our purpose it is useful to rewrite Eq. 30 as

$$\Gamma_{ab} = \frac{4\alpha E_\gamma^2}{m_\tau c^2} f_{ab} \tag{31}$$

and to use Eq. 2 to obtain

$$\Gamma_{ab} = \frac{\alpha^5 m_\tau c^2}{4} \left[\frac{1}{n_b^2} - \frac{1}{n_a^2}\right]^2 f_{ab} \tag{32}$$

If a is an $n\ ^3S_1$ state, then comparing Eq. 32 with Eq. 15c and then Eq. 18

$$\Gamma_{ab} < \Gamma_{tot}(n\ ^3S_1) \tag{33}$$

Hence in $\tau^+\tau^-$ atoms an $n\ ^3S_1$ state is more likely to decay by annihilation than make an x-ray transition to a lower atomic state.

For example, in the specific case of Eq. 29

$$f_{ab} = 0.42 \tag{34}$$

from Table 4[5] of Condon and Shortley (1959). Hence from Eq. 32

$$\Gamma(3\ ^3S_1 \to 2\ ^3P_J) = 7.4 \times 10^{-6}\ \text{eV} \tag{35}$$

and from Table 1

$$\Gamma_{tot}(3\ ^3S_1) = 5.3 \times 10^{-3} \tag{36}$$

Dividing Eq. 35 by Eq. 36

$$\frac{\text{Probability}\ (3\ ^3S_1 \to 2\ ^3P_J)}{\text{Probability}\ (3\ ^3S_1\ \text{annihilation})} = 1.4 \times 10^{-3} \tag{37}$$

Therefore, if we made $\tau^+\tau^-$ atoms in the $3\ ^3S_1$ state only 1.4×10^{-3} of them will make an x-ray transition before decaying. Furthermore, the production rate in Eq. 27

is reduced because for the $3\,^3S_1$ state

$$\sigma(e^+e^- \to \tau^+\tau^-\text{ atom, effective}) \sim 3 \times 10^{-37} \qquad (38)$$

For a luminosity of 10^{33} cm^{-2} s^{-1}, there will be about 4×10^{-7} transitions per second of the form

$$3\,^3S_1 \to 2\,^3P_J + \gamma \ ,$$

a rate much too small to detect.

Finally, as pointed out by Avilez *et al.* (1979) the transition

$$2\,^3P_J \to 1\,^3S_1 + \gamma\ ,\ E_\gamma = 17.8\text{ keV} \qquad (39)$$

has the much more favorable ratio

$$\frac{\text{Probability }(2\,^3P_J \to 1\,^3S_1)}{\text{Probability }(2\,^3P_J\text{ annihilation})} = 0.16 \qquad (40)$$

But $2\,^3P_J$ states cannot be produced directly by

$$e^+ + e^- \to \tau^+\tau^-\text{ atom}$$

as discussed in Sec. B5.

Summarizing, with a tau-charm factory we can make $\tau^+\tau^-$ atoms but we don't know how to detect their production. Beyond that problem, is the yet deeper question of what physics we can do with $\tau^+\tau^-$ atoms.

C. The τ^-–Nucleus Atom

C.1 Static Properties

The τ^-–Nucleus atom in analogy to the μ^-–Nucleus atom consists of a τ^- and $Z-1$ e^-'s around a nucleus of charge Z and atomic number A. In the $\tau - N$ atom the reduced mass of the τ is

$$m = \frac{m_\tau m_N}{m_\tau + m_N} \qquad (41)$$

and ignoring the fine structure and effects of the non-zero nuclear radius, the nth energy level is

$$\begin{aligned}E_n &= -\frac{m_\tau c^2 \alpha^2 Z^2}{2n^2}\left(\frac{m_N}{m_\tau + m_N}\right) \\ &= -\frac{47.4 Z^2}{n^2}\left(\frac{m_N}{m_\tau + m_N}\right)\text{ keV}\end{aligned} \qquad (42)$$

The Bohr radius is given by

$$a_0 = \frac{\hbar^2}{m_\tau e^2} \left(\frac{m_\tau + m_N}{m_N}\right)$$
$$= 1.52 \times 10^{-12} \left(\frac{m_\tau + m_N}{m_N}\right) \quad (43)$$

The average value of the radius of the τ^- orbit is

$$\bar{r} = \frac{a_0}{2Z} [3n^2 - \ell(\ell+1)] \quad (44)$$

ignoring the effect of the non-zero nuclear radius. Thus for $Z \gtrsim 4$ and small n, $\bar{r}$ is of the order of 10^{-13} cm or less. Then particularly for S states, the τ^- is inside the nucleus part of the time. This effect reduces the magnitude of E_n. This is illustrated in Table II taken from Strobel and Wills (1983) who limit their calculations to $Z \leq 12$.

Table II. Energy levels of the 1S and 2P states of a τ^- nucleus atom in keV. E_p is for a point nucleus and E_{ex} is for an extended size nucleus. The proton is always taken as a point. These calculations are from Strobel and Wills (1983) and are corrected for the τ mass of 1777 MeV/c^2.

Nucleus	1S		2P	
	E_p	E_{ex}	E_p	E_{ex}
$^1_1$H	−16.3	−16.3	−4.1	−4.1
$^4_2$He	−128	−118	−32	−32
$^9_4$Be	−625	−474	−156	−155
$^{24}_{12}$Mg	−6310	−2940	−1580	−1460

C.2 Atomic Transitions

Table III, also from Strobel and Wills (1983) gives the energy of the emitted x-ray and the lifetime for the transition

$$2P \to 1S + \gamma \quad (45)$$

We see that the lifetime of the 2P-1S transition is much shorter than the τ lifetime of 3.0×10^{-13} s, Eq. 10b. Therefore, once the τ^- is in the 2P state, the τ^- will make the transition to the 1S state before it decays. Of course, the experimental question is how to get the τ^- into that state or other low lying states.

Table III. Transition energy and lifetime for $2P \to 1S$ in a τ^-–nucleus atom. From Strobel and Wills (1983) corrected for τ mass of 1777 MeV/c^2.

Nucleus	$E(2P \to 1S)$ keV	Lifetime $(2P \to 1S)$ s
$^1_1$H	12.2	5.0×10^{-14}
$^4_2$He	86	2.1×10^{-15}
$^9_4$Be	319	2.3×10^{-16}
$^{24}_{12}$Mg	1480	2.6×10^{-17}

C.3 τ^- Capture in the Nucleus

An interesting result of the τ^- orbit passing through the nucleus is that the τ^- can interact with the protons in the nucleus

$$\tau^- + p \to \nu_\tau + n \qquad (46)$$

in analogy to e^- and μ^- capture. Ching and Oset (1991) have studied the process for heavy nuclei where the capture rate is greatest. They find for $^{208}_{82}Pb$ the following captive rates

$$\begin{aligned}\Gamma(\tau \text{ capture from 1S}) &= 2.5 \times 10^9 \text{ s}^{-1} \\ \Gamma(\tau \text{ capture from 2S}) &= 2.3 \times 10^9 \text{ s}^{-1} \\ \Gamma(\tau \text{ capture from 2P}) &= 5.2 \times 10^9 \text{ s}^{-1}\end{aligned} \qquad (47)$$

However from Eq. 10b

$$\Gamma(\tau \text{ decay}) = 1/T_\tau = 3.4 \times 10^{12} \text{ s}^{-1}$$

Therefore, even in the best case in Eq. 47 there is only a 10^{-3} chance that a τ will be captured with $\tau^- + p \to \nu_\tau + n$ compared to the chance that the τ^- decays.

Morley (1992) has given an interesting discussion of the $\tau^- - U$ atom. He discusses in some detail the process of the τ^- slowing down in solid uranium, the τ^- being captured in a high atomic orbit, and then cascading down to a low orbit.

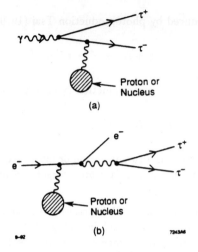

Figure 3

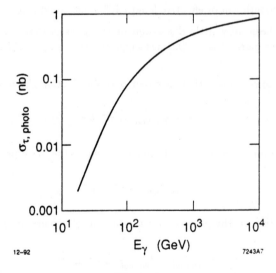

Figure 4

D. Photoproduction of τ's

τ pairs can be produced by photoproduction Tsai (1979)

$$\gamma + N \to \tau^+ + \tau^- + N' \tag{48}$$

as shown in Fig. 3a and by electroproduction (virtual photoproduction)

$$e^- + N \to e^- + \tau^+ + \tau^- + N' \tag{49}$$

as shown in Fig. 3b. Here N is a target proton or nucleus and N' is the final hadronic state. The cross section, $\sigma_{\tau,photo}$, for a proton target is given in Fig. 4 as a function of energy.

As an example, suppose that at SLAC one photoproduces τ pairs with a photon beam of maximum energy 40 GeV and intensity 10^{12} γ/s. Then in a 1 radiation length hydrogen target using an average cross section of 3×10^{-36} cm^{-2}, the τ pair production rate would be

$$\tau \text{ pairs}/s \sim 100 \tag{50}$$

Thus in a one month run of effective length 10^6 s one could produce 10^8 τ pairs.

There has been very little discussion of the physics that might be done with photoproduced τ pairs. Tsai (1992) has suggested that a ν_τ, $\bar{\nu}_\tau$ beam could be made this way.

It is useful to remember that in τ pair photoproduction the basic process is

$$\gamma + \gamma_{virtual} \to \tau^+ + \tau^- \tag{51}$$

in contrast to production by e^+e^- annihilation where the basic process is

$$\gamma_{virtual} \to \tau^+ + \tau^- \tag{52}$$

In the next section on the proposal for production of τ pairs in heavy ion collisions the basic process is

$$\gamma_{virtual} + \gamma_{virtual} \to \tau^+ + \tau^- \tag{53}$$

Therefore some of the goals of heavy ion tau physics may be applicable to photoproduction τ physics.

Returning to the first topic in this paper, $\tau^+\tau^-$ atoms, consider

$$\gamma + N \to \tau^+\tau^- \text{ atom} + N' \qquad (54)$$

Olsen (1986) has discussed the relativistic production of positronium

$$\gamma + N \to e^+e^- \text{ atom} + N' \qquad (55)$$

He shows that at high energy there is the crude relationship

$$\sigma(\gamma + N \to \ell^+\ell^- \text{ atom} + N') \sim \alpha^3 \, \sigma(\gamma + N \to \ell^+ + \ell^- + N') \qquad (56)$$

The α^3 comes from a_0^{-3} (Eq. 3) involved in the phase space factor for the atom relative to the phase space factor for the unbound pair. Applying Eq. 56 to the unbound τ pair cross section in Fig. 4 we see that the cross section for photoproduction of a $\tau^+\tau^-$ atom is in the range of 10^{-39} to 10^{-41} cm^2, much too small to use.

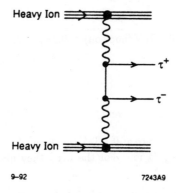

Figure 5

E. τ Pair Production in Heavy Ion Collisions

There have been a number of papers on the production of μ pairs and τ pairs in relativistic collisions of heavy ions (Bottcher and Strayer 1990, del Aquila *et al.* 1991, Almeida *et al.* 1991, Amaglobeli *et al.* 1991). The overall process is

$$\text{ion} + \text{ion} \to \tau^+ + \tau^- + \text{ion} + \text{ion} \qquad (57)$$

as shown in Fig. 5. And the basic process is given in Eq. 53.

498

At sufficiently high energies the cross section will be of the order of

$$\sigma_0(\text{coherent}) = \frac{\alpha^4(\hbar c)^2 Z^4}{m_\tau^2 c^4} \tag{58}$$

The charge Ze at each ion$-\gamma-$ion vertex entering the amplitude as Ze. At lower energies the momentum transfer to the ions becomes large and the process has an incoherent cross section of the order of

$$\sigma_0(\text{incoherent}) = \frac{\alpha^4(\hbar c)^2 Z^2}{m_\tau^2 c^2} \tag{59}$$

Bottcher and Strayer (1990) have studied the production cross section when the ion is $^{197}_{79}A_u$. First consider $Au + Au$ at the LHC with 7.5 TeV per proton which is 3.0 TeV per nucleon. Extrapolating the Bottcher and Strayer calculation

$$\sigma(3.0 \text{ TeV/nucleon}) \approx 40\sigma(\text{coherent}) \approx 0.5 \text{ mb} \tag{60}$$

On the other hand, at a RHIC energy of 0.25 TeV per proton which is 0.1 TeV per nucleon, they obtain

$$\sigma(0.1 \text{ TeV/nucleon}) \approx 0.2\sigma_0 \text{ (coherent}$$
$$\approx 2.8 \times 10^{-3} \text{ mb} \tag{61}$$

This is still larger than

$$\sigma_0(\text{incoherent}) = 2.0 \times 10^{-5} \text{ mb} \tag{62}$$

hence there is still some coherence at 0.1 TeV/nucleon. As another example del Aguila et al. (1991), consider the $^{20}_{82}P_b$ ion. For the LHC they find a cross section of 1 mb, similar to Eq. 60.

If we take the proposed LHC heavy ion luminosity as 10^{28} cm^{-2} s^{-1}, a 1 mb cross section for 10^7 s/year gives a yield of 10^8 τ pairs per year, comparable to a tau-charm factory. Can these pairs be used to do τ physics? This has been partially discussed by del Aguila et al. (1991). They point out that most of the τ pair events will be clean with the ions themselves proceeding along the beam pipe and no additional particles produced. But I think there is a problem in non$-\tau$ events contaminating the data sample, since the cross section for non$-\tau$ events is so much larger. It may be that the only clean samples are the old faithful

$$\tau^+ + \tau^- \to e^\pm + \mu^\mp + \text{missing energy} \tag{63}$$

events.

There have been two suggestions for the tau physics that might be done with τ pairs produced in heavy ion collisions. The suggestion of del Aguila *et al.* (1991) is that one can measure the anomalous magnetic moment of the τ.

$$\mu_\tau(\text{anom}) = a_\tau \frac{e\hbar}{2m_\tau c} \qquad (64a)$$

$$a_\tau = \frac{\alpha}{2\pi} + \sum_{n>1} c_n \alpha^n \qquad (64b)$$

to about 1%. And one can also look for unconventional behavior of the $\tau - \gamma - \tau$ vertex such as an electric dipole moment.

Amaglobeli *et al.* (1991) have suggested using high rate τ production to look for the unconventional decay

$$\tau^- \rightarrow \mu^- + \mu^+ + \mu^- \qquad (65)$$

A τ pair event with one such decay would stand out in the data sample. It would have 4 or 6 charged particles, with 3 of the particles being μ's whose invariant mass is the τ mass.

F. Acknowledgement

This paper is based on the work of those authors given in the references. I am very grateful to them for this work.

References

Almeida L D *et al.* 1991 *Phys. Rev.* **D44** 118

Amaglobeli N S *et al.* 1991 *Massive Neutrinos, Tests of Fundamental Symmetries* (Editions Frontiéres, Gif-sur-Yvette) ed J Tran Thanh Van p 335

Avilez C *et al.* 1978 *Lett. Nuovo Cimento* **21** 301

Avilez C *et al.* 1979 *Phys. Rev.* **D19** 2214

Bottcher C and Strayer M R 1990 *J. Phys. G: Nucl. Part. Phys.* **16** 975

Ching C H and Oset E 1991 Phys. Lett. B **259** 239

Condon E V and Shortley G H 1959 The Theory of Atomic Spectra (Cambridge Univ. Press, Cambridge) p 136

del Aguila F *et al.*1991 *Phys. Lett.* **B271** 256

Moffat J W 1975 *Phys. Rev. Lett.* **35** 1605

Morley P D 1992 Univ. of Texas-Austin Preprint-92-0080

Olsen H A 1986 Phys. Rev. **D33** 2033

Perl M L 1992 *Proc. Trieste Workshop on Search for New Elementary Particles: Status and Prospects* (World Scientific) eds L Beers, G Herten, M L Perl, to be published

Rich A 1981 *Rev. Mod. Phys.* **53** 127

Strobel G L and Wills E L 1983 Phys. Rev. **D28** 2191

Trischuk W 1992 *Proc. Second Workshop on Tau Lepton Physics* (The Ohio State Univ., Columbus) ed K K Gan, to be published, and CERN PPE 92-190

Tsai Y S 1979 SLAC-PUB-2356

Tsai Y S 1992 private communication

COMMENTS ON B4
"Tau Physics at Future Facilities"

As of Spring, 1995 this was my most recent paper on tau lepton physics. There has been a tremendous amount of experimental research on the properties of the tau since my discovery of the tau in 1975. There have been a thousand experimental and theoretical papers and probably a hundred Ph.D. theses. Research techniques have become more and more refined and there has been a steady increase in the amount of data on tau production and decay from experiments at electron-positron colliders.

There are two summarizing observations to be made about our present knowledge of the physics of the tau lepton. First, contrary to earlier hopes that the tau possessed some non-observed decay modes (the one-prong problem) all existing measurements confirm that the tau obeys conventional theories of weak and electromagnetic interactions. Second, as the amount of data increases and statistical measurement errors decrease in the study of the tau decay modes, measurement uncertainties are becoming dominated by systematic errors.

In this paper I try to project how much these systematic errors might decrease in the future and this illustrates a general problem in projecting scientific research. I can guess, sometimes quite well, how much present particle detector technology will improve. But I cannot guess what new technology might be applied to the study of the tau. If I had an idea for a new and powerful technology for studying the tau or other elementary particles, I would be writing papers about it and looking for funds to realize it. I would do this in spite of my discouragement about the time I spent on the tau-charm factory.

Often research technologies are stagnant for years, the only improvements being the building of larger or more precise versions of old technology instruments. This is illustrated by the modern history of research using optical technology. Starting in the middle of the nineteenth century and extending into the second decade of the twentieth century the study of optical spectra made great progress owing to inventions such as diffraction gratings and vacuum pumps. This culminated in the development of quantum mechanics. But then research using optical techniques stagnated for three decades until the laser was invented. We are still in this renaissance of optics.

Returning to elementary particle physics research we are in a period in which we are mostly building larger or more precise versions of old technology instruments. The only new ideas on the accelerator front are electron-positron linear colliders and muon-muon circular colliders. There are no new technology ideas on the detector front.

Perhaps the demonstration for which I hope, that the tau is not a conventional lepton, will require new detector technology. Meanwhile we will have to do the best we can with improving the old detector technology. In my experience new experimental ideas do not come from sitting in the office; I must be doing an experiment to get a new idea.

Tau Physics at Future Facilities*

Martin L. Perl

Stanford Linear Accelerator Center, Stanford University, Stanford, CA 94309 USA

This paper discusses and projects the tau research which may be carried out at CESR, at BEPC, at the SLC, in the next few years at LEP I, at the asymmetric B-factories under construction in Japan and the United States and, if built, a tau-charm factory. As the size of tau data sets increases, there is an increasing need to reduce the effects of systematic errors on the precision and search range of experiments. In most areas of tau physics there is a large amount of progress to be made, but in a few areas it will be difficult to substantially improve the precision of present measurements.

CONTENTS

1. Introduction
2. Future Facilities
3. Detectors, Efficiencies and Systematic Errors
4. Precisely Calculable Decay Modes
5. Dynamics of Lepton Decays
6. Semileptonic Decay Modes
7. Rare Decays, Forbidden Decays and Lepton Conservation in Tau Production
8. CP Violation in Tau Production and Decay
9. Tau Neutrino Mass
10. Electromagnetic Properties of the Tau

1. INTRODUCTION

The τ was discovered two decades ago. In this paper I look ahead to the next two decades of τ research. I begin in Section 1.1 by estimating the number of τ pairs, $N_{\tau\tau}$, which may be produced at ongoing and future τ research facilities, and I note special properties of these facilities. But $N_{\tau\tau}$ by itself is not sufficient for forecasting future research, and so in Section 2 I discuss detector properties and systematic errors, σ_{sys}. In the next eight sections I apply the discussion of $N_{\tau\tau}$ and σ_{sys} to the newer areas of τ physics, comparing where we are now to where we might go in τ research. In the course of this comparison, I note possible new directions

---

* Work supported by Department of Energy contract DE-AC03-76SF00515.

Elsevier Science B.V.
SSDI 0920-5632(95)00178-6

in physics and techniques. After the year 2000 the facilities with large $N_{\tau\tau}$ production per year will be CESR, the asymmetric B-Factories (ABF) at KEK and SLAC, and perhaps a tau-charm factory (TCF). Therefore I particularly compare CESR, the ABFs and a possible TCF.

There is always uncertainty in predicting human activities. The list

- Roulette wheel
- Stock market
- Technology trends
- Economic trends
- Population growth

goes from the unpredictable to something which can be predicted a decade ahead. Where is forecasting high energy physics research on this list? It is probably at the level of forecasting economic trends. But there is also an analogy between forecasting technological trends and forecasting τ research. A new technique that drastically reduces σ_{sys} or enables the finding of a deviation from the standard model in τ physics will change τ research, just as fiber optics changed communications technology. Therefore the most useful way to use this paper is that I provide information on $N_{\tau\tau}$ and σ_{sys} and the like, and let the reader do the forecasting.

The data used in this paper is taken from this Workshop [1], from the 1994 Review of Particle Properties [2], and from the review talk of R. Patterson [3]. I do not discuss tau neutrino physics other than the ν_τ mass.

Reprinted with permission from *Proc. Third Workshop Tau Lepton Physics, Nucl. Phys. B (Proc. Suppl.)* **40**, ed. L. Rolandi, pp. 541–556.
© 1995 Elsevier Science B. V.

Some conventions used in this paper are: B means decay fraction, $\delta B/B$ means the fractional error in B, and a data acquisition year is 10^7's.

2. FUTURE FACILITIES

I begin with the low energy facilities.

2.1 CESR

If almost all future data acquisition at CESR is at or close to the $\Upsilon(4S)$, the τ pair cross section is

$$\sigma_{\tau\tau} = 0.78 \text{ nb} \quad . \quad (1a)$$

At present, the CLEO collaboration has accumulated [4]

$$N_{\tau\tau}(1994) \approx 2.5 \times 10^6 \quad , \quad (1b)$$

and by the year 2000 it is expected that

$$N_{\tau\tau}(2000) \approx 2 \times 10^7 \quad (1c)$$

$$\mathcal{L}(\text{CESR}) \approx 1 \times 10^{33} \text{cm}^{-2}\text{s}^{-1} \quad (1d)$$

$$N_{\tau\tau}/\text{yr} \approx 8 \times 10^6/\text{yr} \quad . \quad (1e)$$

Beyond 2000, $\mathcal{L}$(CESR) may increase above $3 \times 10^{33} \text{cm}^{-2}\text{s}^{-1}$ and then $N_{\tau\tau}/\text{yr}$ will exceed 2×10^7.

2.2 Asymmetric B-Factories (ABF)

Two asymmetric B-factories are under construction: the KEK B-Factory in Japan with $3.5 \oplus 8.0$ GeV and the SLAC B-Factory, PEP-II, with $3.1 \oplus 9.0$ GeV. There is one experiment at each facility. These colliders will begin operation for data acquisition about the year 2000 with:

$$\mathcal{L} \approx 1 \times 10^{33} \text{cm}^{-2}\text{s}^{-1} \to 3 \times 10^{33} \text{cm}^{-2}\text{s}^{-1} \quad (2a)$$

$$N_{\tau\tau}/\text{yr} \approx 8 \times 10^6/\text{yr} \to 2.3 \times 10^7/\text{yr} \quad . \quad (2b)$$

The second $\mathcal{L}$ value is $\mathcal{L}$ design. Eventually these facilities might attain

$$\mathcal{L} \approx 1 \times 10^{34} \text{ cm}^{-2}\text{s}^{-1} \quad (2c)$$

$$N_{\tau\tau}/\text{yr} \approx 10^8/\text{yr} \quad . \quad (2d)$$

2.3 BEPC

The BES Collaboration at the Beijing Electron Position Collider (BEPC) has collected [5]

$$N_{\tau\tau}(1994) \approx 90,000 \quad . \quad (3a)$$

The luminosity of BEPC is being upgraded [5,6] to

$$\mathcal{L} \approx 1.5 \times 10^{31} \text{ cm}^{-2}\text{s}^{-1} \quad (3b)$$

which will yield

$$N_{\tau\tau}/\text{yr} \approx 4 \times 10^5/\text{yr} \quad . \quad (3c)$$

2.4 Tau-Charm Factory (TCF) [7,8]

In August 1994 a Workshop [9] jointly organized by physicists from the Stanford Linear Accelerator Center (SLAC), the Institute for High Energy Physics (IHEP) in Beijing, and the BEC Collaboration and entitled "The Tau-Charm Factory in the Era of B-Factories and CESR" was held at SLAC.

The participants discussed the tau, charm, and charmonium physics which would be studied at a TCF that began operation on or after the year 2000. TCF designs were presented for sites at IHEP in Beijing [5,10], Argonne National Laboratory [11], the Budker Institute in Novosibirsk [12] and at IHEP in Dubna [13]. In all these presentations the design luminosity was

$$\mathcal{L} \approx 1 \times 10^{33} \text{ cm}^{-2}\text{s}^{-1} \quad . \quad (4a)$$

At the usual proposed three operating points for τ research [7]

$E_{cm} = 3.56$ GeV :
$$\sigma_{\tau\tau} = 0.5 \text{ nb}, N_{\tau\tau}/\text{yr} = 0.5 \times 10^6/\text{yr} \quad (4b)$$

$E_{cm} = 3.67$ GeV :
$$\sigma_{\tau\tau} = 2.4 \text{ nb}, N_{\tau\tau}/\text{yr} = 2.4 \times 10^7/\text{yr} \quad (4c)$$

$E_{cm} = 4.25$ GeV :
$$\sigma_{\tau\tau} = 3.5 \text{ nb}, N_{\tau\tau}/\text{yr} = 3.4 \times 10^7/\text{yr} \quad . \quad (4d)$$

An advanced upgraded TCF might achieve [14] $\mathcal{L} = 4 \times 10^{33} \text{cm}^{-2}\text{s}^{-1}$.

Next I consider high energy facilities.

2.5 TRISTAN

At the present TRISTAN energy [15] of 58 GeV

$$\mathcal{L} \approx 4 \times 10^{31} \text{cm}^{-2}\text{s}^{-1} \quad (5a)$$

$$\sigma_{\tau\tau} \approx 20 \text{ pb} \quad (5b)$$

$$N_{\tau\tau}/\text{yr, experiment} \approx 8 \times 10^3/\text{yr} \quad . \quad (5c)$$

Thus the main τ research is the study [16] of $\gamma - Z^0$ interference.

2.6 LEP I

At present [17], each of the four LEP experiments has

$$N_{\tau\tau}(1994)/\text{experiment} \approx 9 \times 10^4 \quad , \quad (6a)$$

and at the conclusion of LEP I

$$N_{\tau\tau}(1996)/\text{experiment}$$
$$\approx 1.9 \times 10^5 \text{ to } 2.4 \times 10^5 \quad . \quad (6b)$$

2.7 SLC

By the end of the present data acquisition period the SLD Collaboration using the SLAC Linear Collider (SLC) will have acquired

$$N_{\tau\tau}(1994) \approx 5 \times 10^3 \quad . \quad (7a)$$

If a total of $10^6 Z^0$'s are produced

$$N_{\tau\tau} \rightarrow 2 \times 10^4 \quad . \quad (7b)$$

These τ pairs have the special property that they are produced using an e^- beam that is 70% to 80% longitudinally polarized [18].

2.8 LEP II

At $E_{cm} = 180$ GeV

$$\sigma_{\tau\tau} \approx 8 \text{ pb} \quad , \quad (8a)$$

and it is expected that LEP II will give an integrated luminosity

$$\int \mathcal{L} dt \approx 500 \text{ pb}^{-1} \quad . \quad (8b)$$

Then

$$N_{\tau\tau}/\text{experiment} = 4 \times 10^3 \quad . \quad (8c)$$

About half of these events will be from the radiative tail of the Z^0 and may not be useful.

Thus at LEP II and at an e^-e^+ linear collider, $N_{\tau\tau}$ is small and τ research is restricted (see Section 10).

2.9 Electron-Positron Linear Collider

Far above the Z^0 energy

$$\sigma_{\tau\tau} \approx \frac{0.1}{s} \text{ pb, s in TeV}^2 \quad . \quad (9a)$$

At $E_{cm} = 0.5$ TeV and

$$\mathcal{L} = 10^{33} \text{cm}^{-2}\text{s}^{-1} \quad , \quad (9b)$$

$$N_{\tau\tau}/\text{yr} = 4 \times 10^3/\text{yr} \quad . \quad (9c)$$

2.10 Longitudinal Polarization of e^-e^+ Beams

A substantial amount of present day τ research makes use of τ spin distributions through τ spin $-\tau$ spin correlations .[19-22].

The sensitivity of such research can be substantially increased by using a longitudinally polarized e^- or e^+ beam in the collider [23], but it is not necessary to polarize both beams.

There are already longitudinally polarized e^- beams at the SLC [18] and HERA [24]. Longitudinally polarized beams are under discussion for LEP, and such beams might be considered for CESR and the ABFs. A very suitable candidate is a tau-charm factory [12,14,26].

Radiative transverse polarization [25] cannot be used at a TCF unless wigglers are inserted in the ring [12]. Otherwise the polarization time is too long. A separate small radius ring may however be used with the polarized e^- injected into the TCF. An attractive alternative is to use a linear accelerator with a polarized e^- source as the TCF injector. In all cases spin rotators must be used before and after the interaction point so that the e^- beam is longitudinally polarized at the interaction point, but transversely polarized in the ring.

2.11 Tau production at Hadron Colliders

There are two ways in which τ's can be produced at hadron colliders [27]. At a pp collider

$$p + p \to D_s \text{ or } B + \ldots$$
$$D_s \text{ or } B \to \tau + \nu_\tau \quad . \quad (10)$$

A more practical way is to use a heavy ion collider, RHIC or LHC, and the two virtual photon reactions [28-31]

$$\text{ion} + \text{ion} \to \text{ion} + \text{ion} + \gamma_{\text{virtual}} + \gamma_{\text{virtual}}$$
$$\gamma_{\text{virtual}} + \gamma_{\text{virtual}} \to \tau^+ + \tau^- \quad . \quad (11)$$

The ions would not be disrupted and the event would be quite clean.

A Pb-Pb collision at the LHC gives

$$\sigma_{\text{PbPb}\tau\tau} \approx 1 \text{ mb} \quad , \quad (12a)$$

and with

$$\mathcal{L} \approx 10^{28} \text{cm}^{-2}\text{s}^{-1} \quad (12b)$$

$$N_{\tau\tau}/\text{yr} = 10^8/\text{yr} \quad . \quad (12c)$$

However event detections may be difficult because the transverse momentum of the τ's is less than m_τ.

3. DETECTORS, EFFICIENCIES AND SYSTEMATIC ERRORS

A large $N_{\tau\tau}$ is not enough, the significance of the measurement depends upon many properties of the experiment: event selection efficiency, backgrounds, detector simulation quality, systematic errors. Valuable comparisons of many of these experiment properties have been carried out by Weinstein [32] and Burchat [33].

3.1 Efficiencies and Backgrounds

It has been known for quite [4,32,33] a while and was emphasized again at this meeting [1] that τ data analyses at the $\Gamma(4S)$ compared to τ data analyses at the Z^0 involve smaller efficiencies, ϵ, for event acceptance and larger fractional backgrounds f_b. The ϵ's and f_b's will be about the same at ABFs as they are at CESR. One of the goals of a TCF project is to design a detector so that at the smaller τ physics operating points [7,8], 3.56 and 3.67, the f_b's are smaller, and then the ϵ's can be larger.

3.2 Systematic Errors

In the past few years (and even more so at this meeting [1]) in many measurements the systematic errors, σ_{sys}, are larger than the statistical error, σ_{stat}. The determination of a systematic error is often a complicated process, and there is always some nervousness in the way we combine them quadratically

$$\sigma_{\text{sys,tot}} = \left[\sum_i \sigma_{\text{sys},i}^2\right]^{1/2} \quad . \quad (13)$$

As $N_{\tau\tau}$ increases, the future of τ research depends upon reducing systematic errors such as $\sigma_{\text{sys},\epsilon}$ and σ_{sys,f_b}. Reduction of σ_{sys,f_b} requires in part improvements in particle identification as sketched next.

3.3 The μ/π Separation

The separation of μ's from π's becomes difficult for momentum below about 0.5 GeV/c. This is not a problem at the Z^0 energy and above, it is a problem at CESR and the ABFs, and is even more of a problem at the TCF [34].

3.4 The π/K Separation

The problem of π/K separation behaves in the opposite way versus energy. At LEP only the DELPHI experiment [35] permits event-by-event separation of π's from K's, the other experiments will continue with statistical π/K separation. On the other hand, there will be powerful event-by-event π/K separation in the CLEO III detector [4,36] and the ABF detectors [37,38]. It is easier to achieve π/K separation over most of the momentum range at CESR compared to the ABFs, and it is easiest at a TCF (Table 1).

Table 1
Maximum π and K momentum from τ decays

Collider	Maximum π and K momentum (GeV/c)
TCF at 3.56 GeV	0.8 to 0.9
TCF at 3.67 GeV	1.1
CESR	5.1
PEP II	8.7

3.5 The γ and π^0 Detection

Substantial reduction of systematic errors in the detection of γ's and π^0's will be necessary for substantial improvement in research on semileptonic decay modes, radiative decay modes and rare decay modes. The LEP I experiments will conclude in a few years, hence the reduction of $\sigma_{sys,\gamma}$ and σ_{sys,π^0} must be carried out at CESR, the ABF's and perhaps a TCF. Detection problems include (a) the rejection of false γ's from "split-offs" from hadronic interactions in the electromagnetic calorimeter, (b) the efficiency for detecting low energy γ's, and (c) the efficiency for reconstructing π^0's. The CsI calorimeter of the CLEO II detector gives

$$\sigma_{sys,\pi^0}/B_i \approx .01 - .02 \qquad (14)$$

per π^0 in decay modes with π^0's and reduction in σ_{sys},π^0 requires further tuning of the π^0 simulation programs using data from known, non-τ events containing π^0's.

The ABF detectors will have about the same $\sigma_{sys,\gamma}$ and σ_{sys,π^0} as CLEO II unless the electromagnetic calorimeters are improved by using longitudinal segmentation of the CsI crystals.

At a TCF the electromagnetic calorimeter must detect smaller energy γ's compared to CESR and the ABF's. As a compensation the ψ and $\psi\prime$ decays such as

$$\psi \to \pi^+\pi^-\pi^0 \,,\; B = 1.5\%$$
$$\psi\prime \to 2(\pi^+\pi^-)\pi^0 \,,\; B = 3.4\% \qquad (15)$$

are a copious source of calibration π^0's.

4. PRECISELY CALCULABLE DECAY MODES

The decay widths and dynamics of the modes

$$\tau^- \to \nu_\tau + e^- + \bar{\nu}_e \qquad (16a)$$
$$\tau^- \to \nu_\tau + \mu^- + \bar{\nu}_\mu \qquad (16b)$$
$$\tau^- \to \nu_\tau + \pi^- \qquad (16c)$$
$$\tau^- \to \nu_\tau + K^- \qquad (16d)$$

are predicted precisely from weak interaction theory and well measured quantities such as the π^- lifetime and the $K^- \to \mu^- \bar{\nu}_\mu$ decay width. How well can we expect to compare prediction with measurement?

4.1 Ratio of B_i's

How well can and will we be able to measure

$$B_\mu/B_e \,,\; B_\pi/B_e \,,\; B_K/B_\pi \quad ? \qquad (17)$$

As the first example consider B_e. Including the new data presented at this meeting the world average value is [39,40]

$$B_e(wa) = (17.79 \pm 0.09)\% \qquad (18a)$$

which is heavily weighted by the B_e's from the LEP experiment, particularly

$$B_e(\text{ALEPH}) = (17.76 \pm 0.13)\% \quad . \qquad (18b)$$

According to Harton [17], by the end of LEP I we might expect that the average of the LEP experiments has the errors on B_e

$$\sigma_{stat} = 0.05\% \,,\; \sigma_{sys} = 0.06\% \sigma_{tot} = 0.08\% \quad (19c)$$

which is not much better than the present value. Thus from the LEP experiments the final fractional error will be

$$\delta B_e/B_e \approx 0.005 \quad . \qquad (18d)$$

The fractional error on B_μ will be similar. Hence

$$\delta\left(\frac{B_\mu}{B_e}\right) / \left(\frac{B_\mu}{B_e}\right) \approx 0.007 \quad . \qquad (18e)$$

Improvements in precision will have to come from CESR, the ABF's and, if built, a TCF. A two-year-old CLEO II measurement [41] gives

$$B_e(\text{CLEO II}) = (17.97 \pm 0.14 \pm 0.23)\% \quad . \qquad (18f)$$

As $N_{\tau\tau}$ increases, σ_{stat} can certainly be reduced to $\sigma_{stat}/B_e \approx 0.001$; the question is how much σ_{sys} can be reduced at CESR or the ABF's? Can experimenters at these colliders attain $\sigma_{sys}/B_e \to 0.002$? Can they attain $\sigma_{sys}/B_\mu \to 0.002$?

Initial studies for a TCF [42,43] indicate that 0.002 might be attained for σ_{sys}/B_e and σ_{sys}/B_μ by using the special properties of the E_{cm} = 3.56 operating point.

Heltsley [44] gives the new world average values

$$B(\tau^- \to \nu_\tau h^-, wa) = (11.76 \pm 0.14)\% \quad (19a)$$

$$B_K(wa) = (0.68 \pm 0.04)\% \, . \quad (19b)$$

By subtraction

$$B_\pi(wa) = (11.08 \pm 0.15)\% \, . \quad (19c)$$

Thus at present

$$\delta B_\pi/B_\pi \approx 0.014 \quad (19d)$$

$$\delta B_K/B_K \approx 0.06 \, . \quad (19e)$$

Weinstein [32] predicts that CESR and the ABF's will reduce σ_{sys} in B_π to give

$$\delta B_\pi/B_\pi \to 0.01 \, . \quad (19f)$$

This is not a brick wall limit, as $N_{\tau\tau}$ increases $\delta B_\pi/B_\pi$ could decrease further.

Thus future reductions in $\delta B_e/B_e$, $\delta B_\mu/B_\mu$, $\delta B_\pi/B_\pi$ and probably $\delta B_K/B_K$ will be by factors of 2 to 4, but not by a factor of 10. One or more radically new techniques will be needed to reduce the fractional errors by a factor of 10.

There is an additional reason for new measurements of B_e, B_μ, B_π, and B_K. At this meeting [1] Hayes [45] and Smith [46] have shown that τ branching fractions such as B_ρ and B_1 have changed over time beyond the range of the world average σ_{sys} errors. Might the same happen for B_e, B_μ, B_π, or B_K?

4.2 Comparison of τ_τ, B_ℓ and M_τ

At present the world average value of the tau lifetime [40] is

$$\tau_\tau(wa) = 291.6 \pm 1.6 \text{ fs} \, , \quad (20a)$$

based primarily on

$$\tau_\tau(CLEO) = 291 \pm 7.6 \text{ fs}$$

$$\tau_\tau(ALEPH) = 292.5 \pm 3.2 \text{ fs}$$

$$\tau_\tau(DELPHI) = 295.2 \pm 4.2 \text{ fs} \quad (20b)$$

$$\tau_\tau(L3) = 296.4 \pm 7.8 \text{ fs}$$

$$\tau_\tau(OPAL) = 288.8 \pm 2.6 \text{ fs} \, .$$

Thus at present

$$\delta\tau_\tau/\tau_\tau \approx 0.005 \, . \quad (20c)$$

This represents amazing improvement in the last four years, but I do not think that the LEP I experiments can improve much more.

Reduction of δ_τ will have to come from the CLEO experiment when the new vertex detector is introduced and from the ABF detectors. In the CLEO τ_τ measurement [47] the largest σ_{sys} are from (a) vertexing and tracking and (b) background. Certainly (a) will be drastically reduced.

Using

$$B_e = \left(\frac{m_\tau}{m_\mu}\right)^5 \left(\frac{\tau_\tau}{\tau_\mu}\right)(1+c) \quad (21)$$

with c a small correction term [19,48]. Stroynowski [39] finds Eq. 21 confirmed within one standard deviation.

As is the case with B_e and B_μ, the sensitivity of this comparison cannot be much improved unless radically new techniques are used to decrease $\delta\tau_\tau/\tau_\tau$ in Eq. 20c.

In this section I have taken some space to show how we try to forecast future precision in τ measurements from our knowledge of present errors. From now on I will be more concise.

Table 2
The τ Michel parameters ρ, η, δ, ξ. The first row gives the expected values for a Dirac charged lepton. The second row gives the world average values. The third row gives the projected total errors from LEP I experiments. The next two rows give the projected statistical errors for CESR, ABF's and a TCF. The bottom row gives the projected total error for a TCF. The ρ, δ and ξ values and errors are averaged over the e and μ decay modes. The η values and errors are only for μ decay mode.

	ρ	η	ξ	δ	Ref
Dirac charged lepton	3/4	0	1	3/4	
World average	0.732 ± 0.024	-0.01 ± 0.14	$1.04 \pm .010$	0.70 ± 0.15	40
Projected LEP I total errors	≤ 0.025	≈ 0.07	≤ 0.10	≤ 0.10	17
Projected CESR or ABF statistical errors	≈ 0.002	≈ 0.03	≈ 0.01	≈ 0.01	21
Projected TCF statistical errors at 4 GeV	≈ 0.002	≈ 0.03	≈ 0.02	≈ 0.02	21
Projected TCF total errors at 3.56 GeV	≈ 0.003		≈ 0.001	≈ 0.01	43

5. DYNAMICS OF LEPTONIC DECAYS

Table 2 gives the τ Michel parameters ρ, η, δ, ξ (a) the world average values, (b) the final projected errors from LEP I experiments, and (c) projected errors for CESR, ABF's, and a TCF. For simplicity, $e - \mu$ universality has been used. In contrast to the discussion in the previous section, the future will bring substantial reductions in the errors on the Michel parameters.

All the measurements and projections in Table 2 are for unpolarized e^- and e^+ beams and use τ spin $- \tau$ spin correlations [21,49]. We expect even further reduction in the errors on δ and ξ when a longitudinal polarized e^- or e^+ beam is used in a collider (Section 2.10).

6. SEMILEPTONIC DECAY MODES

Table 3 lists world averages for B_i and $\delta B_i / B_i$ for semileptonic decay modes as given by Patterson [3].

Once the LEP I experiments are concluded the burden of reducing $\delta B_i / B_i$ falls on the CLEO, ABF and BES collaborations, and perhaps on a TCF collaboration. I look at three examples from CLEO II analysis to illustrate the larger sources of σ_{sys}.

First consider [32,50]

$$\tau^- \to \nu_\tau + h^- + \pi^0 \quad . \quad (22a)$$

Using three different topologies, l vs. ρ, ρ vs. ρ, and 3-prong vs. ρ

$$B(\tau \to \nu_\tau h \pi^0) = (25.87 \pm 0.12 \pm 0.42)\% \quad (22b)$$

$$\sigma_{sys}/B = 0.016 \quad .$$

Table 3
Branching fractions B_i and errors, δB_i, in percent for semileptonic decay modes from Patterson.[3] $\delta B_i / B_i$ is the fractional error.

Mode	$B_i \pm \delta B_i$ in %	$\delta B_i / B_i$
ν_τ h π^0	25.20 ± 0.37	0.015
ν_τ h $2\pi^0$	9.08 ± 0.27	0.030
ν_τ h $3\pi^0$	1.27 ± 0.16	0.13
ν_τ h $4\pi^0$	0.16 ± 0.07	0.4
ν_τ 3h	8.91 ± 0.34	0.038
ν_τ 3h π^0	4.25 ± 0.15	0.035
ν_τ 3h $2\pi^0$	0.48 ± 0.06	0.1
ν_τ 5h	0.07 ± 0.01	0.1
ν_τ 5h π^0	0.02 ± 0.01	0.5

The largest contributions to σ_{sys}/B in Eq. 22b are 0.009 for π^0 reconstruction, 0.009 for extra shower veto and 0.005 to 0.010 for acceptance.

The second example from CLEO II [51] is

$$B(\tau \to \nu_\tau 3h\pi^0) = (4.25 \pm 0.09 \pm 0.26)\% \quad (23)$$

$$\sigma_{sys}/B = 0.06 \quad ,$$

the analysis used e and μ tags. The largest contribution to σ_{sys}/B are .041 to .048 for cuts, .03 for π^0 reconstruction, and .025 for tracking efficiency.

The third example from CLEO II [50] is

$$B(\tau \to \nu_\tau 5h\pi^0) = (0.019 \pm 0.004 \pm 0.004)\%$$

$$\sigma_{sys}/B = 0.21 \quad . \quad (24)$$

The major contributors to σ_{sys}/B are π^0 reconstruction, tracking and backgrounds.

I do not see how to predict the reductions in σ_{sys} for the semileptonic decay modes width which will be accomplished by the CLEO, ABF, and BES collaborations. Certainly experience, larger $N_{\tau\tau}$'s, new analysis ideas, and improved detectors will bring reductions in σ_{sys}, but will $\delta B_i/B_i < 0.01$ be attained?

Conversely, do we need $\delta B_i/B_i < 0.01$ for semileptonic decay modes? Accurate comparisons with theory and other data can only be made for decay modes with even numbers of π's. And these comparisons using CVC and e^+e^- annihilation cross section data are limited in their accuracy by the e^+e^- data. Thus to compare with Eqs. 22a and 23 Eidelman reports [52]

$$B(\tau \to \nu_\tau \pi \pi^0 \text{ , CVC prediction})$$
$$= (24.9 \pm 0.07)\% \quad (24a)$$

$$B(\tau \to \nu_\tau 3\pi\pi^0 \text{ , CVC prediction})$$
$$= (4.20 \pm .29)\% \quad , \quad (24b)$$

and Sobie gives [53]

$$B(\tau \to \nu_\tau \pi \pi^0 \text{ , CVC prediction})$$
$$= 24.3 \pm 1.1)\% \quad . \quad (24c)$$

(Add .5% for $\tau^- \to \nu_\tau K^- \pi^0$ to B in Eqs. 24a and 24c to compare with Eq. 22b.) Reduction of the errors in the predicted B's in Eqs. 24 requires better e^+e^- cross section data in the energy region $2m_\pi < E_{cm} < m_\tau$. Such data can be obtained at the VEPP-2M e^+e^- collider which has $E \le 1.4$ GeV and, if it is built, from a tau-charm factory.

As has been emphasized by Kühn [54] at this meeting, there is much more to semileptonic decays than branching fractions. The hadronic resonances contained in the modes, the kinematic distributions, the measurement of form factors, the great variety of modes containing hadrons; all this data provides the best highway to the study of hadron physics for $E_{cm} < 1.8$ GeV. Large values of $N_{\tau\tau}$ will be of great help in providing precise data.

Improvements will also be required in π/K separation, π^0 reconstruction, and in removal of backgrounds. In particular, it will be important to avoid using cuts which distort kinematic distributions.

Finally, I come to the question of what might be hidden in the semileptonic modes. Since the sum of the exclusive mode B_i's is within 1% of 100% [39,44] there are no mysterious modes with $B \gtrsim 1\%$. But are there mysterious modes with $B \approx 10^{-3}$ or $B \approx 10^{-4}$. Can we begin to use our detectors as bubble chambers which were once used to pick out a few "new physics" events out of thousands of ordinary events? For example, is there a mysterious decay

$$\tau^- \to \nu_\tau + x^- + 3\gamma \quad (25)$$

which does not come from

$$\tau^- \to \nu_\tau + h^- + \pi^0 + \text{fake } \gamma$$

or

$$\tau^- \to \nu_\tau + h^- + 2\pi^0 (\gamma \text{ lost})$$

or

$$\tau^- \to \nu_\tau + e^- + \bar{\nu}_e + 3\gamma \quad ?$$

7. RARE DECAYS, FORBIDDEN DECAYS, AND LEPTON NONCONSERVATION IN TAU PRODUCTION

There will be tremendous progress in research on rare τ decays, forbidden τ decays, and lepton number nonconservation in τ pair production as

$$N_{\tau\tau} \to 10^8 \quad (26)$$

at CESR, ABFs, and perhaps a TCF.

7.1 Second-Class Current Rare Decays

Most interesting is the search for the second-class current decay mode [19,55]

$$\tau^- \to \nu_\tau + \pi^- + \eta \quad (27a)$$

for which the standard model predicts

$$B(\tau^- \to \nu_\tau \pi^- \eta) \approx 10^{-5} \text{ to } 10^{-6} \quad (27b)$$

and the present upper limit [56] is

$$B(\tau^- \to \nu_\tau \eta^- \eta) < 3.4 \times 10^{-4}, \quad 95\% \text{ CL} \quad (27c)$$

The best signature uses $\eta \to \gamma + \gamma$ hence

$$\tau^- \to \nu_\tau + \pi^- + \gamma + \gamma \quad . \quad (27d)$$

The backgrounds are

$$\tau^- \to \nu_\tau + \pi^- + \pi^0, \ \nu_\tau + \pi^- + 2\pi^0 \quad (27e)$$

so that once again π^0 and γ detection and selection is crucial.

The second-class current decay mode

$$\tau^- \to \nu_\tau + \pi^- + \omega \text{ via } b_1(1235) \quad (28)$$

will be more resistant to demonstration because it must be proven that the $\pi^- \omega$ come from the b_1.

7.2 Other Rare Decays

There is a variety of rare decay modes which will be a challenge to detect and study although they are not of special theoretical interest. Examples are the higher multiplicity Cabibbo suppressed decays and the seven-prong decay

$$\tau^- \to \nu_\tau + 4h^- + 3h^+ + n\pi^0, \ n \geq 0 \quad . \quad (29)$$

Another example is the five particle leptonic decays

$$B(\tau^- \to \nu_\tau e^- e^- e^+ \overline{\nu}_e, \text{ predicted}) = 4 \times 10^{-5}$$

$$B(\tau^- \to \nu_\tau \mu^- e^- e^+ \overline{\nu}_\mu, \text{ predicted}) = 2 \times 10^{-5}$$

$$B(\tau^- \to \nu_\tau \mu^- \mu^- \mu^+ \overline{\nu}_\mu, \text{ predicted}) = 1 \times 10^7, \quad (30)$$

the branching fractions have been calculated by Dicus and Vega [57].

7.3 Forbidden Decays Without Neutrinos

The search for τ decay modes which do not contain neutrinos, such as

$$\tau^- \to \ell^- + \gamma \quad (31a)$$

$$\tau^- \to \ell^- + \ell^+ + \ell^- \quad (31b)$$

$$\tau^- \to \ell^- + (\text{hadrons})^0 \quad (31c)$$

with $\ell = e$ or μ, will of course greatly benefit from very large $N_{\tau\tau}$. The smallest upper limits on the B_i's are [55,58]

$$B_i \leq \text{few} \times 10^{-6} \quad (31d)$$

based on $N_{\tau\tau} \approx 1.5 \times 10^6$ from CLEO II. As $N_{\tau\tau}$ goes to 10^7 and then 10^8 at CESR and the ABF's, experimenters can attain sensitivity 1/10 and then 1/100 of Eq. 31d, if backgrounds can be suppressed. Looking at existing search data [55,58] the search for forbidden modes containing hadrons, Eq. 31c, are most likely to suffer from backgrounds. Table 4 from Alemany et al. [59]. gives projected sensitivity limits.

Table 4
Attainable limits for the branching fractions for forbidden, neutrinoless τ decays. The TCF is assumed to have $N_{\tau\tau} = 2.4 \times 10^7$ at 3.67 GeV. CESR or the ABF is assumed to have $N_{\tau\tau} = 0.9 \times 10^7$.

Mode	Tau-Charm Factory	CESR or B-Factory
$\tau \to e\gamma$	10^{-7}	10^{-6}
$\tau \to \mu\gamma$		
$\tau \to \mu\mu\mu$		
$\tau \to \mu\, ee$	10^{-7}	10^{-7}
$\tau \to e\mu\mu$		
$\tau \to eee$		

7.4 Forbidden Decays with an Undetectable Particle

There is no recent progress in the search for

$$\tau^- \to l^- + x^0 \qquad (32a)$$

with $\ell = e$ or μ and x^0 a weakly interacting particle. In 1990 Albrecht et al. [60] reported with a 95% C.L.

$$B(\tau^- \to e^- x^0) < 0.003\,, \quad m_{x^0} < 100 \text{ MeV} \quad (32b)$$

rising to 0.009 at $m_{x^0} = 500$ MeV. The limits [59] on $B(\tau^- \to \mu^- x^0)$ are similar. The problem is that these forbidden modes cannot be distinguished from the corresponding leptonic modes

$$\tau^- \to \nu_\tau + l^- + \bar{\nu}_l \,, \qquad (32c)$$

if m_x^0 equals the invariant mass of the $\nu_\tau \bar{\nu}_l$ combination. Indeed the search method requires that the x^0 be detected as a bump above the $\nu_\tau \bar{\nu}_l$ mass spectrum. Alemany et al [59]. have shown that a TCF will permit a more sensitive search than CESR or ABF's particularly for the $\tau^- \to e^- x^0$ mode, Table 5.

Table 5
Attainable limits for the branching fractions for forbidden τ decays with a weakly interacting particle. The TCF is assumed to have $N_{\tau\tau} = 2$ to 5×10^6 at 3.56 GeV. CESR or the ABF is assumed to have $N_{\tau\tau} = 9 \times 10^6$ (from Ref. 59).

Mode	Tau-Charm Factory	CESR or B-Factory
$\tau \to ex^0$	10^{-5} to 10^{-6}	5×10^{-3}
$\tau \to \mu x^0$	10^{-3} to 10^{-4}	5×10^{-3}

7.5 Lepton Nonconservation in Tau Pair Production

Vorobiev [61] has reviewed the upper limits

$$\begin{aligned} e^+ + e^- &\to e^\pm + \tau^\mp \\ e^+ + e^- &\to \mu^\pm + \tau^\mp \end{aligned} \qquad (33a)$$

The smallest 95% C.L. upper limit at the Z^0 is from the L3 experiment with

$$\begin{aligned} B(Z^0 \to e\tau) &< 0.9 \times 10^{-5} \\ B(Z^0 \to \mu\tau) &< 1.1 \times 10^{-5} \end{aligned} \qquad (33b)$$

To my knowledge, the smallest upper limit measured below the Z^0 is [62]

$$\sigma(e^+e^- \to e^\pm \tau^\mp)/\sigma(e^+e^- \to \mu^+\mu^-) < 1.2 \times 10^{-3}$$
$$\sigma(e^-e^- \to \mu^\pm \tau^\mp)/\sigma(e^+e^- \to \mu^+\mu^-) < 4.1 \times 10^{-3} \qquad (33c)$$

with 95% C.L. at 29 GeV.

The sensitivity to $B(Z^0 \to e\tau)$ and $B(Z^0 \to \mu\tau)$ can probably be extended to 5×10^{-6} at LEP I [61], but that is not a significant increase in sensitivity. I do not know how much the sensitivity can be improved at CESR, the ABF's or a TCF over that in Eq. 33c.

8. CP VIOLATION IN TAU PRODUCTION AND DECAY

At this meeting Stahl [63] has reviewed the search for CP violation in

$$e^+ + e^- \to Z^0 \to \tau^+ + \tau^- \quad (34a)$$

using τ spin–τ spin correlations. The upper limits on a weak dipole moment, d_τ^Z, from LEP I experiments are [63]

$$|\text{Re}(d_\tau^Z)| < 6.4 \times 10^{-18} \text{ e cm}$$
$$|\text{Im}(d_\tau^Z)| < 4.5 \times 10^{-17} \text{ e cm} \quad . \quad (34b)$$

Table 6 from Bernreuther et al. [22] gives projected 1 σ accuracies for measurement of d_τ^Z and d_τ^γ using τ spin–τ spin correlations. The bottom row shows that as $N_{\tau\tau} \to 2 \times 10^5$ (Eq. 6b), there will be some increase in sensitivity at the Z^0.

Sensitivity to the electric dipole moment, d_τ^γ, is given in the top three rows of Table 6. Weinstein and Stroynowski [19] have reviewed other ways to find d_τ^γ. Present upper limits on d_τ^γ are [2]

$$|d_\tau^\gamma| < \text{few} \times 10^{-16} \text{ ecm} \quad . \quad (35)$$

Table 6
Projected 1σ accuracies for measurement of the CP violating electric dipole moment of d_τ^γ and weak dipole moment d_τ^Z for various E_{cm} and certain $N_{\tau\tau}$. The upper value is for $|\text{Re}(d)|$ and the lower value is for $|\text{Im}(d)|$ from Bernreuther et al.[22]

E_{cm} (GeV)	$N_{\tau\tau}$	d_τ^γ (e cm)	d_τ^Z (e cm)
3.67	2.4×10^7	2×10^{-16} 1×10^{-16}	
4.25	3.5×10^7	4×10^{-17} 2×10^{-17}	
10.58	5×10^7	1×10^{-18} 3×10^{-18}	
91.2	3.3×10^5		2×10^{-18} 3×10^{-17}

Ananthanarayan and Rindani [64] have discussed using a longitudinally polarized e^- beam to search for CP violation in τ pair production.

The τ provides almost the only way to search for CP violation in the decay of leptons. At this meeting Nelson [49] described the theory of using τ spin–τ spin correlations to search for CP violation.

An alternative method of searching for CP violation in τ decay is to use a longitudinally polarized e^- beam or e^+ beam as discussed by Tsai [23] at the Workshop [9] on "The Tau-Charm Factory in the Era of B-Factories and CESR." There are two advantages. First, the search will be more sensitive by a factor of 10 or more. Second, the experimenter will be able to reverse the beam polarization or set it to zero, thus obtaining better control of the systematic errors in the required asymmetry measurements.

9. TAU NEUTRINO MASS

As reviewed by Cerutti [65] the present upper limits on m_{ν_τ} with 95% C.L. are

ALEPH : 23.8 MeV

ARGUS : 31.0 MeV

CLEO : 32.6 MeV

OPAL : 74.0 MeV .

There have been numerous projections of the smallest m_{ν_τ} which could be explored at CESR, at a B-factory or a tau-charm factory. A comparative discussion has been given by Gomez-Cadenas [66]. He discusses the use of the different decay modes:

$$\tau^- \to \nu_\tau + \pi^- + K^+ + K^-$$
$$\tau^- \to \nu_\tau + 3\pi^- + 2\pi^+ \quad (36)$$
$$\tau^- \to \nu_\tau + 2\pi^- + \pi^+ + 2\pi^0 \quad .$$

He finds that the sensitivity to m_{ν_τ} in tau-charm factory experiments is 2.0 MeV/c^2 and in CESR or B-factory experiments is 2.5 MeV/c^2, assuming in both cases the data set contains 10^8 tau pairs. These projections may be optimistic, for example, Weinstein [32] predicts a sensitivity of about 15 MeV for CESR. On the other hand, the new two-dimensional search technique introduced by ALEPH experimenters [65] may also be helpful at CESR and ABF's.

10. ELECTROMAGNETIC PROPERTIES OF THE TAU

10.1 Radiative Decays

There is much work to be done on the radiative decays of the τ such as:

$$\tau^- \to \nu_\tau + l^- + \bar{\nu}_l + \gamma \,, \ l = e, \mu \quad (37a)$$

$$\tau^- \to \nu_\tau + \pi^- + \gamma \quad (37b)$$

$$\tau^- \to \nu_\tau + \rho^- + \gamma \quad . \quad (37c)$$

There are three physics issues. First, precise comparisons of the measured ratios B_π/B_e zand B_K/B_e with theory require calculation of radiative corrections [67]. Second, as discussed by Decker and Finkemeir [68] and the references they give, we can learn about internal bremsstrahlung and structure-dependent radiation from distribution such as the γ energy spectrum and the $\pi\gamma$ invariant mass spectrum in Eq. 37b. Third, can there be "new physics" in radiative decays?

To my knowledge, there are only two experiments on radiative tau decays [69,70].

10.2 Tau Magnetic Moment

If the τ is a conventional Dirac charged particle, its magnetic moment is given [71] by

$$\mu_\tau = g_\tau \frac{e\hbar}{2m_\tau c} \quad (38a)$$

$$\frac{g_\tau - 2}{2} = \frac{\alpha}{2\pi} + O(\alpha^2) = a_\tau \,, \quad (38b)$$

where

$$\frac{\alpha}{2\pi} = 1.16 \times 10^{-3} \quad (39a)$$

is the Schwinger term. In Eq. 38b[71]

$$a_\tau = 1.177 \times 10^{-3} \quad (39b)$$

As calculated by Escribano and Masso [72] from LEP I experimental data

$$-8 \times 10^{-3} \leq a_\tau(\text{measured}) \leq 10 \times 10^{-3} \,. \quad (39c)$$

Thus measured limits are ten times larger than the expected value. Can we eventually measured a_τ so as to test the τ? Laursen et al. [73] have suggested a method using the leptonic radiative decays in Eq. 37a.

10.3 Tau Cross Section Near Threshold

The last measurement of the behavior of the τ pair production cross section, $\sigma_{\tau\tau}$, from threshold to $E_{cm} = 4$ GeV was made 16 years ago in the DELCO experiment at SPEAR [74]. The theory of $\sigma_{\tau\tau}$ in this threshold region is now well understood.[75] I believe it will be interesting to make a precision study of the ratio $\sigma_{\tau\tau}(\text{measured})/\sigma_{\tau\tau}(\text{theory})$ as a function of E_{cm}.

10.4 $\tau^+\tau^-$ Atom

I have reviewed [27] the atomic structure and decay process of $\tau^+\tau^-$ atoms, as well as the cross section for

$$e^+ + e^- \to \gamma \to \tau^+\tau^- \text{atom} \quad . \quad (40)$$

The 1^3S_1 ground state which is 24 KeV below threshold has a peak cross section and width

$$\sigma_{\tau\tau\text{atom}}(\text{peak}) \approx 2.4 \times 10^{-28} \text{cm}^2$$
$$\Gamma = 2.9 \times 10^{-2} \text{eV} \quad . \quad (41)$$

The observed peak cross section depends upon $\sigma_{E_{cm}}$, the spread in E_{cm}, as follows

$$\sigma_{E_{cm}} = 1 \text{ MeV} \,, \ \sigma_{\tau\tau\text{atom}}(\text{peak}) \approx 0.003 \text{ mb}$$
$$\sigma_{E_{cm}} = 100 \text{ KeV} \,, \ \sigma_{\tau\tau\text{atom}}(\text{peak}) \approx 0.03 \text{ mb} \quad . \quad (42)$$

Skrinsky [12] has shown the fascinating behavior of $\sigma_{\tau\tau\text{atom}}$ if $\sigma_{E_{cm}}$ can be reduced to 20 KeV or 5 KeV in a future upgrade of a tau-charm factory. If $\sigma_{E_{cm}} = 20$ KeV $\sigma_{\tau\tau\text{atom}}(\text{peak}) \approx 0.1$ mb, and is $\sigma_{E_{cm}} = 5$ KeV $\sigma_{\tau\tau\text{atom}}(\text{peak}) \approx 0.5$ mb. There is still the deeper question of what physics can we do with $\tau^+\tau^-$ atoms?

ACKNOWLEDGEMENT

I am greatly indebted to many members of the CLEO, ARGUS, BES, ALEPH, DELPHI, L3 and OPAL collaborations for their kindness and patience in discussing with me the details of how they carry out their measurements. I want to thank L. Rolandi and the other organizers and staff of the Third Workshop on Tau Lepton Physics for having set up and conducted a very valuable meeting.

REFERENCES

1. See Proc. Third Workshop on Tau Lepton Physics, Montreux, 1994, Nucl. Phys. B, Proc. Supp., ed. G. Rolandi.
2. Review of Particle Properties, Phys. Rev. D50, (1994) 1173.
3. R. Patterson, "Weak and Rare Decays," Proc. 27th Int. Conf. High Energy Physics (Glasgow, 1994), to be published.
4. D. Besson, Proc. Tau-Charm Factory in the Era of CESR and B-Factories, SLAC, August 15-16, 1994, SLAC Report-451, eds. L. V. Beers and M. L. Perl.
5. Z. Zhong, ibid.
6. N. Qi, in Proc. Third Workshop on Tau Lepton Physics, Montreux, 1994, Nucl. Phys. B Proc. Supp., ed. G. Rolandi.
7. J. Kirkby and J. A. Rubio, Part. World 3 (1992) 77.
8. For recent reviews, see Proc. Third Workshop on the Tau-Charm Factory (Editions Frontieres, Gif-sur-Yvette, 1994), eds. J. Kirkby and R. Kirkby.
9. See Proc. Tau-Charm Factory in the Era of CESR and B-Factories, SLAC, August 15-16, 1994, SLAC Report-451, eds. L. V. Beers and M. L. Perl.
10. Y. Wu, ibid.
11. J. Repond, ibid.
12. A. Skrinsky, ibid.
13. G. Chelkov, ibid.
14. J. LeDuff, Proc. Third Workshop on the Tau-Charm Factory (Editions Frontieres, Gif-sur-Yvette, 1994), eds. J. Kirkby and R. Kirkby, p. 797.
15. S. Iwata, KEK preprint 93-70 (1993).
16. H. Hanai, in Proc. Third Workshop on Tau Lepton Physics, Montreux, 1994, Nucl. Phys. B, Proc. Supp., ed. G. Rolandi.
17. J. Harton, Proc. Tau-Charm Factory in the Era of CESR and B-Factories, SLAC, August 15-16, 1994, SLAC Report-451, eds. L. V. Beers and M. L. Perl.
18. B. Schumn, in Proc. Third Workshop on Tau Lepton Physics, Montreux, 1994, Nucl. Phys. B, Proc. Supp., ed. G. Rolandi.
19. A. J. Weinstein and R. Stroynowski, Ann. Rev. Nucl. Part. Phys. 43 (1992) 457.
20. Y. S. Tsai, Phys. Rev. D4 (1971) 2821; D13 (1976) 77 .
21. W. Fetscher, Phys. Rev. $D42$, (1990) 1544.
22. W. Bernreuther et al., Phys. Rev. D48 (1993) 78.
23. Y.-S. Tsai, Proc. Tau-Charm Factory in the Era of CESR and B-Factories, SLAC, August 15-16, 1994, SLAC Report-451, eds. L. V. Beers and M. L. Perl.
24. D. P. Barber et al., Nucl. Inst. Meth. A338, (1994) 166, and G. Voss, private communication.
25. R. Prepost, Proc. Tau-Charm Factory in the Era of CESR and B-Factories, SLAC, August 15-16, 1994, SLAC Report-451, eds. L. V. Beers and M. L. Perl.
26. A. Zholents, CERN SL/92-27 (1992).
27. M. L. Perl, Proc. Second Workshop on Tau Lepton Physics (World Scientific, Singapore, 1993), ed. K. K. Gan, p. 483.
28. F. del Aguila et al., Phys. Lett. B271 (1991) 256.
29. C. Bottcher and M. R. Strayer, J. Phys. G.: Nucl. Part. Phys. 16 (1990) 975.
30. N. S. Amaglobel et al., Massive Neutrinos, Tests of Fundamental Symmetries (Editions Frontieres, Gif-sur-Yvette, 1991), ed. J Tran Thanh Van, p. 335.
31. L. D. Almeidi et al., Phys. Rev. D44 (1991) 118 .
32. A. J. Weinstein, Proc. Third Workshop on the Tau-Charm Factory (Editions Frontieres, Gif-sur-Yvette, 1994), eds. J. Kirkby and R. Kirkby, p. 11.
33. P. R. Burchat, Proc. SIN Spring School on Heavy Flavor Physics (Zuoz, 1988).

34. J. Kirkby, Proc. Third Workshop on the Tau-Charm Factory (Editions Frontieres, Gif-sur-Yvette, 1994), eds. J. Kirkby and R. Kirkby, p. 805.
35. W. Ruckstuhl, in Proc. Third Workshop on Tau Lepton Physics, Montreux, 1994, Nucl. Phys. B, Proc. Supp., ed. G. Rolandi.
36. CLEO Collaboration, CLNS 94–1277 (1994).
37. K. Abe et al., KEK 92–3 (1992).
38. BABAR Collaboration, SLAC–443 (1994).
39. R. Stroynowski, in Proc. Third Workshop on Tau Lepton Physics, Montreux, 1994, Nucl. Phys. B, Proc. Supp., ed. G. Rolandi.
40. M. Davier, ibid.
41. D. S. Akerib et al., Phys. Rev. Lett. 69 (1992) 3610; 71 (1993) 3395.
42. J. Kirkby, Proc. Workshop on Tau Lepton Physics (Editions Frontieres, Gif-sur-Yvette, 1991), eds. M. Davier and B. Jean-Marie), p. 453
43. J. J. Gomez-Cadenas, Proc. Tau-Charm Factory Workshop, SLAC REPORT–343, ed. L. V. Beers, p. 48.
44. B. Heltsley, in Proc. Third Workshop on Tau Lepton Physics, Montreux, 1994, Nucl. Phys. B, Proc. Supp., ed. G. Rolandi.
45. K. Hayes, ibid.
46. J. Smith, ibid.
47. C. White, ibid.
48. W. Marciano, ibid.
49. C. Nelson, ibid., and references cited in this paper.
50. D. F. Cowan, CALT–68–1934 (1994).
51. W. Ford, in Proc. Third Workshop on Tau Lepton Physics, Montreux, 1994, Nucl. Phys. B, Proc. Supp., ed. G. Rolandi.
52. S. Eidelman, ibid.
53. R. Sobie, ibid.
54. J. Kühn, ibid.
55. G. Eigen, ibid.
56. M. Artuso et al., Phys. Rev. Lett. 69 (1992) 3278.
57. D. A. Dicus and R. Vega, CPP–93–30 (1993).
58. T. Bowcock et al., Phys. Rev. D41 (1990) 805.
59. R. Alemany et al., CERN–PPE/93-49 (1993).
60. H. Albrecht et al., Phys. Lett. B246 (1990) 278.
61. I. Vorobiev, in Proc. Third Workshop on Tau Lepton Physics, Montreux, 1994, Nucl. Phys. B, Proc. Supp., ed. G. Rolandi.
62. J. J. Gomez-Cadenas et al., Phys. Rev. Lett 66 (1991) 1007.
63. A. Stahl, in Proc. Third Workshop on Tau Lepton Physics, Montreux, 1994, Nucl. Phys. B, Proc. Supp., ed. G. Rolandi.
64. B. Ananthanarayan and S. D. Rindani, Phys. Rev. Lett. 73 (1994) 1215.
65. F. Cerutti, in Proc. Third Workshop on Tau Lepton Physics, Montreux, 1994, Nucl. Phys. B, Proc. Supp., ed. G. Rolandi.
66. J. J. Gomez-Cadenas, Proc. Third Workshop on the Tau-Charm Factory (Editions Frontieres, Gif-sur-Yvette, 1994), eds. J. Kirkby and R. Kirkby, p. 97.
67. R. Decker and M. Finkemeir, Phys. Lett. B316 (1993) 403.
68. R. Decker and M. Finkemeir, Phys. Rev. D48 (1993) 4203.
69. D. Y. Wu et al., Phys. Rev. D41 (1990) 2339.
70. N. Mistry, Proc. Second Workshop on Tau Lepton Physics (World Scientific, Singapore, 1993), ed. K. K. Gan, p. 84.
71. M. A. Samuel et al., Phys. Rev. Lett. 67 (1991) 668.
72. R. Escribano and E. Masso, Phys. Lett. B301 (1993) 419.
73. M. L. Laursen et al., Phys. Rev. D29 (1984) 2652.
74. W. Bacino et al., Phys. Rev. Lett. 42 (1978) 749.
75. B. H. Smith and M. B. Voloshin, Phys. Lett. B324 (1994) 117; B333 (1994) 564.

Discussion

G. Rolandi, CERN
1) Could the τ lifetime be measured better at ABF?
2) Would a τ physics optimized detector at ABF look different?

M. Perl
1) It would not be improved at CESR. At ABF maybe.
2) Yes, but it is CP violation in B that drives the detector design and money.

N. Wermes, Universität Bonn
I have a comment on your conclusion on the reach of CP-violation studies at LEP. I think that – given that the analyses are still statistics dominated and that the results of the four experiments have not yet been combined – values of $d_\tau < 10^{-18}$ ecm are well within reach by the end of 1995.

Part C

Innovations in Experimental Methods and New Directions in Physics

COMMENTS ON C1
"Nuclear Electric Quadrupole Moment of Na^{23}"

This atomic beam experiment was my first physics experiment, my Ph.D. thesis research and my first physics paper. Now, forty years later, I remember little about the significance of the measurement itself, the determination of the electric quadrupole moment of the sodium nucleus, but I remember a great deal about building and running the atomic beam apparatus.

I was lucky that the general nature of this experiment coincided with the aspects of scientific work which gave me the most pleasure. First of all the experiment was grandly mechanical and physical: the large brass vacuum chamber; the vacuum pumps, the physical beam of sodium atoms; the mechanical adjustments for positioning the magnets, lenses, and detector along the atomic beam. Even the power supply for the magnets, submarine batteries charged with a motor-generator set, was more mechanical than electronic.

In most of my later experiments I have preferred the mechanical to the electronic; I love the mechanical, the electronic I tolerate. It is true that mechanical has taken on for me very broad meanings. Thus for me optics is a mechanical subject and my ways of thinking about fundamental physical processes are also mechanical. In this first experiment I saw in my mind the photons hitting the sodium atom in the beam, knocking them up to an excited state, and then the radio frequency electromagnetic field twisting the excited atoms into a different hyperfine structure state. It has been the same in my work on the tau lepton. I see elementary particles as very small mechanical objects which collide and decay.

This love of the mechanical, of building things, started very early in my life and led to my strong desire to be a mechanical or civil engineer. But my father had business connections with chemical companies and so I studied chemical engineering. The fluid mechanics and physical process in the chemical engineering field interested me more than the chemical reactions.

Returning to my PhD. thesis, the second aspect which gave me so much pleasure was that led by my thesis professor, Isadore Rabi, I used two technologies new to atomic beam experiment. First the radio frequency transitions was made in the excited state rather than the traditional ground state. This had to be done because there is no nuclear quadrupole interaction in the $^2S_{1/2}$ ground state of sodium. The sodium lamps used for the excitation bathed the top of the apparatus in a dramatic orange light.

Even more dramatic was the second innovation, the use of a lock-in circuit to measure the weak current produced by the detector through ionization of the sodium beam. When I came to the atomic beam laboratory at Columbia University the standard way to measure small atomic beam currents was to use a highly sensitive mirror galvanometer. The galvanometer was mounted on one wall of the room, a narrow beam of light coming from the opposite wall was reflected by the mirror back onto a scale mounted on that same opposite wall. The mirror responded with agonizing slowness to a change in current and a single reading could take many minutes. As suggested by Rabi, I went to the atomic beam laboratory at the Massachusetts Institute of Technology and learned how to build a lock-in current amplifier system in which the radio frequency applied to the excited atoms and the amplifier itself were both square wave modulated at 34Hz. The rate of data acquisition was increased by a factor of ten and the experiment became possible.

Even the lock-in system was somewhat mechanical, it used a Western Electric mercury relay switch. All this was described in a long paper, Perl, Rabi, and Senitzky, Phys. Rev. **98**, 611 (1955).

A third aspect of experimental physics that started with this first experiment and has always remained with me is the excitement of getting an experiment working, making the measurements, and taking data. This excitement includes the ups-and-downs, the pleasures of a successful data run, the anxiety and despair after an experiment breaks or fails.

In the atomic beam experiment a run lasted a half day or a full day, the run ending when the oven which produced the beam ran out of sodium. I remember the preparations for a run, the loading of the oven with sodium, the pump down, the checking of the submarine batteries, the stabilization of the sodium lamp. And then the start of the run. Would the atomic beam come out of the oven cleanly or would one of those mysterious clogs occur in the oven? And finally, I remember the pleasant tiredness at the end of a good run, the shutting down of the experiment, and the nights data, having been acquired on a Speedomax recorder, now safely stored.

Nuclear Electric Quadrupole Moment of Na²³†

M. L. PERL, I. I. RABI, AND B. SENITZKY

Columbia University, New York, New York

(Received December 7, 1954)

THE hyperfine interaction constants of the $3\,^2P_{\frac{3}{2}}$ and $3\,^2P_{\frac{1}{2}}$ excited states of Na²³ have been measured by the atomic beam resonance method proposed by Rabi.[1] In this method, the Na²³ beam is excited by resonance radiation from a sodium discharge lamp, while a radio-frequency magnetic field is simultaneously applied. The rf changes the magnitude of the refocused beam when the rf corresponds to the separation of two hyperfine levels in the excited state. Since this change is small, high rf power and an intense light source are necessary. For the light source both Philips SO 60 W and General Electric NA-1 sodium lamps were used. From 10 to 20 percent of the atoms in the beam were excited, depending on the way the lamps were operated. Two continuously tunable triode oscillators were used to cover the rf range of 6 to 240 Mc/sec.

The small excited state rf effect also required a more elaborate detection system than is used in the usual atomic beam experiment. The rf field is square-wave modulated at 34 cps so that the refocused beam contains modulated excited state signal and unmodulated background. The atoms are ionized on the usual hot, oxidized, tungsten ribbon and deflected through a simple mass spectrometer onto the first dynode of an Allen type electron multiplier. The modulated portion of the output signal of the Allen tube is then amplified in a narrow band tuned amplifier. The signal is then rectified in a lock-in circuit, averaged in a long time constant RC circuit, and put on a recorder.

A typical set of experimental points for the $3\,^2P$ state at near-zero magnetic field is shown in Fig. 1. The narrow lines of the typical atomic beam resonance experiment are absent in this experiment. This is because the fundamental line width is here determined by the natural lifetime of the excited state, rather than by the atom in the rf field. The experimental points are fitted by means of two overlapping resonance curves of the form

$$\frac{h_\mathrm{I}}{(f-f_\mathrm{I})^2+g^2}+\frac{h_\mathrm{II}}{(f-f_\mathrm{II})^2+g^2},$$

where h_I and h_II are the heights of the individual resonances, f_I and f_II are the centers of the individual resonances, and g is their width. These five constants are determined by trial for a best fit. The results obtained in a series of five runs are given in Table I. The widths are in agreement with the results of Stephenson.[2] In addition, the $F=2$ to $F=1$ hyperfine separation in the

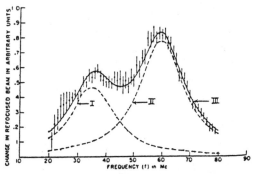

FIG. 1. Typical resonance curve for $3\,^2P_{\frac{3}{2}}$ state of Na²³, taken at 1.84-gauss static field. The dots with vertical probable error lines are the experimental points. Curves I and II are the individual resonance curves and curve III is their sum. The parameters of the curves are selected to give a best fit of curve III to the experimental points.

TABLE I. Constants obtained from the resonance curves of five different runs.

f_I, center of lower peak (Mc/sec)	f_II, center of upper peak (Mc/sec)	g, line width (Mc/sec)
35.0	58.5	9.75
35.5	59.8	9.80
35.4	59.8	9.80
36.0	60.2	9.75
35.8	60.5	9.90

$3\,^2P_{\frac{3}{2}}$ state was measured at near-zero magnetic field. This measurement gives a magnetic hyperfine interaction constant in the $3\,^2P_{\frac{1}{2}}$ state of $a_{\frac{1}{2}} = 94.4 \pm 0.5$ Mc/sec.

In the-$3\,^2P_{\frac{3}{2}}$ state, the three hyperfine level separations are given by

$(3a_{\frac{3}{2}} + b)$ for $F = 3$ to $F = 2$,

$(2a_{\frac{3}{2}} - b)$ for $F = 2$ to $F = 1$,

$(a_{\frac{3}{2}} - b)$ for $F = 1$ to $F = 0$,

where $a_{\frac{3}{2}}$ and b are the magnetic dipole and electric quadrupole hyperfine interaction constants for the $3\,^2P_{\frac{3}{2}}$ state. Since the theoretical ratio of $(a_{\frac{1}{2}})/(a_{\frac{3}{2}})$ is 5 we can uniquely assign the two observed overlapping resonances to the 3–2 and 2–1 transitions. The weak 1–0 transition was not observed, so that there are two possible values of b, each giving a different value for the nuclear electric quadrupole moment, Q. It is difficult to assign a probable error to the values of b which one gets from the kind of curve-fitting one is compelled to use in evaluating this experiment. The average value of the two possible coefficients are $b = +2.58 \pm 0.18$ Mc/sec and $b = -21.6 \pm 0.5$ Mc/sec, where the error is taken to be the extreme of the deviation from the average. However if the peak positions of I and II from different runs are taken the limits become much wider. For this reason we do not trust the resultant values of b to better than about 10 percent. The corresponding values of Q are $Q = +0.11 \times 10^{-24}$ cm^2 and $Q = -0.91 \times 10^{-24}$ cm^2, without corrections of the Sternheimer type.

In conventional optical spectroscopy, further assignment can be made by the ratio of the intensities of lines. In our case a comparison of the experimental ratio with a simplified theoretical calculation of the ratio favors the assignment of the upper transition to the 3–2 transition. This gives a Q of $+0.11 \times 10^{-24}$ cm^2. However we do not feel that the experimental conditions of either the light source or the application of the rf are sufficiently definite to make this assignment with full confidence. These remarks apply equally to the results of Sagalyn[3] obtained by the double-resonance optical method. Current nuclear theory would find it very difficult to accept the value $Q = -0.91 \times 10^{-24}$ cm^2.

Full details will be published in an extended paper in this journal. We are now in the course of making similar measurements on rubidium. We wish to acknowledge the important contributions of Dr. Charles A. Lee and Dr. G. K. Woodgate who participated in the construction and operation of the early form of the apparatus.

† This work was supported in part by the U. S. Office of Naval Research.
[1] I. I. Rabi, Phys. Rev. 87, 379 (1952).
[2] G. Stephenson, Proc. Phys. Soc. (London) A64, 463 (1951).
[3] P. L. Sagalyn, Phys. Rev. 94, 885 (1954).

COMMENTS ON C2
"Scattering of K^+ Mesons by Protons"

After receiving my PhD. in atomic physics at Columbia University in 1955 I went to the University of Michigan to work in elementary particle physics. My changing physics fields was mostly due to the urging of my thesis professor, Isadore Rabi. Although he never worked in elementary particle physics, he saw its great potential; also, he liked to annoy his fellow atomic physicsts at Columbia by telling his students that atomic physics was a dying field.

I chose Michigan because Donald Glaser had just invented the bubble chamber at Michigan. For about the next ten years I worked intermittently in bubble chamber experiments. I enjoyed the first few experiments where we used a non-magnetic propane bubble chamber with a small sensitive volume of 12 cm by 12 cm by 30 cm. We built the entire chamber in the Physics Department of the University of Michgan and the construction involved interesting mechanical and optical work The entire apparatus was small enough so that we could push it into and out of the particle beams of the Cosmotron at Brookhaven.

Runs lasted a few days to a week. I remember that as I would come on shift walking onto the floor of the Cosmotron, I would listen for the periodic expansion noise from the chamber. The same kind of run excitement as I felt during my atomic beam experiment. Was the chamber still working? Would the expansion system last the run?

This reprint describes one of our first experiments using the propane chamber. It is interesting how in the first section we argue the advantages of a bubble chamber over an emulsion. This was the beginning of my interest in the elastic scattering of elementary particles, I thought that this simplest of reactions could reveal the most about the interaction of particles.

Within the next few years I moved into the luminescent chamber technique and then the optical spark chamber technique because I was interested in high energy elastic scattering and needed a technique that was more selective than the bubble chamber. But until 1968 I returned to the bubble chamber from time to time. My final bubble chamber experiment (T. H. Tan *et al.*, *Phys. Lett.* **28B**, 195 (1968)) was an exposure of the Berkeley 72 inch hydrogen bubble chamber to a 6 GeV/c proton beam. My secret hope was that we would find a stable proton-proton particle, analogous to the deuteron. I had the idea that if two protons were pushed together violently, they might stick together. But I couldn't emphasize this in our proposal to the Bevatron Program Committee and so the proposal was mostly to study mechanisms of nucleon isobar formation; and that is what we eventually published.

Why was my primary interest in searching for a proton-proton resonance or particle in my final bubble chamber experiment, when the study of nucleon isobar physics would have been a much more direct and fruitful goal? Indeed a proton-proton resonance or particle has never been found and everything we know now about nuclear forces says that such a resonance or particle could never exist. This contrariness has occurred numerous times in my research work. I'll have the opportunity to do some straightforward and obviously needed research, yet I will do some other research. Several strands of my experimenter's personality lead me this way. First, I don't feel that I am a more competent experimenter than many of my colleagues, and if they are going to work on a piece of research and are not shorthanded then they don't need me.

Second. I don't do well in specialized research areas that already have enough experimenters at work. Many scientists like to attend meetings with colleagues who specialize in their research area; they like to fully discuss the details of the experiments and compare the magnitudes and precision of similar measurements. These meetings are important for the stability of experimental results; the discussions often lead to the understanding and reduction of errors. But I get bored in such meetings, my thoughts wander and my eyes become glassy.

I learned early in my scientific career that I do best and am least annoying to my colleagues when I work in an area where experimenters are scarce, where there are few papers to read, and where there are few meetings to attend.

PHYSICAL REVIEW VOLUME 107, NUMBER 1 JULY 1, 1957

Scattering of K^+ Mesons by Protons*

DONALD I. MEYER, *Brookhaven National Laboratory, Upton, New York and University of Michigan, Ann Arbor, Michigan*

AND

MARTIN L. PERL AND DONALD A. GLASER, *University of Michigan, Ann Arbor, Michigan*

(Received March 15, 1957)

The scattering of 20- to 90-Mev K^+ mesons by protons has been investigated by using a propane bubble chamber. On the basis of 32 events, a total cross section of 9.4 ± 1.7 mb is obtained. The experimental angular distribution is not isotropic.

INTRODUCTION

THIS paper describes a study of the elastic scattering of positive K mesons by hydrogen nuclei at kinetic energies of less than 100 Mev. The experiment was carried out in the University of Michigan 12-inch propane bubble chamber using K^+ mesons from a specially purified beam at the Cosmotron of the Brookhaven National Laboratory.

The study of the scattering of K mesons on nucleons is of basic importance to the elucidation of the nature of K meson-nucleon interaction. One may hope eventually to obtain the same type of information from K meson-nucleon scattering experiments as has been obtained from π meson-nucleon scattering experiments. A step toward this goal is the study of the elastic scattering of positive K mesons on protons at sufficiently low energy so as to drastically restrict the number of angular momentum states involved in the interaction. The information which is required is the total and differential scattering cross sections over a range of K-meson kinetic energies.

Previous to this work all K^+ meson-proton scattering data was collected in nuclear emulsions.[1,2] While the use of emulsions has the advantage of not requiring a particularly pure K^+ beam, the scanning for K^+-proton scatterings is extremely laborious. The low kinetic energy of the K^+ meson also makes the use of counters difficult. However, a propane bubble chamber of reasonable size is suitable for this experiment, because for these low-energy scatterings both the scattered K^+ meson and the recoil proton stop in the chamber. Moreover, the speed of scanning in a bubble chamber is much greater than in an emulsion, and the separation of carbon from hydrogen scatterings is direct at these low energies. The present experiment was undertaken for these reasons with a propane bubble chamber.

EXPERIMENTAL APPARATUS

K^+ mesons were obtained by allowing the external beam of the Cosmotron to bombard a 3-inch-thick copper target. Because of the large number of π^+ mesons of a given momentum produced with the K^+ mesons, it was necessary to find some means of separating the two kinds of particles. For example, for 3-Bev protons producing mesons of momentum around 400 Mev/c, this ratio of π^+/K^+ at the target is about 100 in the forward hemisphere. Allowing a reasonable distance for momentum analysis, in which distance a large fraction of the K^+ mesons decay, one K^+ meson could have been accepted in about every 40 pictures

---

* This work was supported in part by the U. S. Atomic Energy Commission.
[1] N. N. Biswas *et al.*, Nuovo cimento 3, 1481 (1956).
[2] M. Ceccarelli (private communication of a collection of data from Göttingen, Padua, Berkeley, Bristol, Dublin, and Rochester).

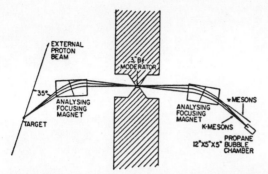

FIG. 1. The external proton beam from the Cosmotron strikes a 1.5-inch-thick copper target. A mixed beam of K and π mesons and protons is momentum-analyzed and focused, their energy degraded in Be, then reanalyzed and focused to obtain a mass separation.

without the accompanying π^+ mesons being so numerous so as to prevent the scanning of the pictures. Protons of the correct momentum have such a short range that they are stopped in the chamber wall and cause no trouble.

The method used to separate the two types of mesons consisted of a magnet to momentum-analyze the beam, followed by a moderator of Be which reduces the K^+ momentum more than that of the π^+, and a second magnet to reanalyze the beam. The system is shown in Fig. 1. In order to attain reasonable K^+-meson intensities, it was necessary to incorporate some focusing in the system. To keep the distances short so that not too many K^+ mesons would decay, momentum analysis was combined with strong focusing in a single magnet.[3] By using a large gradient, both horizontal and vertical focusing were attained with an object-to-image distance of 3 meters for each magnet. The beam of particles focused and analyzed at the chamber was 2.5 cm high by 5 cm wide. A figure of merit for the separation is that the π^+/K^+ ratio at the chamber is a factor of 50 smaller than it would be if the Be were removed and the magnet readjusted for the unseparated beam. With this system about one K^+ meson entered the chamber in every 6 pictures.

The University of Michigan propane bubble chamber has a sensitive volume of 12 by 12 by 30 cm, and was operated at 52.8°C. The K^+ mesons enter the chamber with an energy of about 100 Mev, and the stopping power is sufficiently high so that they come to rest in the chamber and decay. Stereographic pictures were taken at five-second intervals using 70-mm film.

METHOD OF ANALYSIS AND RESULTS

About 25 000 pictures were taken, of which 20 400 pictures were scanned. The remainder were not scanned because of excess beam intensity, no beam being present, or damage in the developing process. The large size of the pictures made it possible to scan with the naked eye directly from the film. The scanning was done by looking along the edge of the picture where K^+ mesons entered the chamber. Tracks having the appropriate number of bubbles per cm were then followed. A decay into a lightly ionizing particle or the characteristic τ^+ decay would then identify the K^+ meson. No attempt was made to distinguish the various decay modes except that of the τ^+. In a few percent of the tracks, where absolute bubble density was not sufficient to distinguish K mesons from π or μ mesons, the criteria involving multiple scattering of the track and rate of change of bubble density along the track were used. Occasionally a K^+ meson would be found by a general area-scanning of the pictures which was done after the above procedure was completed. Finally all tracks which were counted as K mesons were required to enter the chamber within perscribed angles and to have a length equal to at least half the chamber length if they did not scatter in the chamber. The total cross section was corrected for this second requirement which eliminates a known small fraction of K mesons which decay in flight.

3507 K^+ mesons were found, including 189 τ^+ mesons, giving a ratio of the number of τ^+ mesons to the total number of K mesons of 5.4%. This ratio agrees well with other published τ^+ to K^+ ratios and is a direct verification that the method of identification of K^+ mesons was correct.

The separation of K-meson scatterings on hydrogen from those on carbon depends upon the presence of a recoil track in the former case. In the elastic scattering on carbon of K mesons with kinetic energies in the range of this experiment, the velocity of the recoiling carbon nucleus is always much too low to produce a visible track. However, the elastic scattering on free protons gives a visible proton recoil track, if the scattering angle of the K meson is above a minimum angle which depends on the energy. This minimum angle, as shown on Fig. 2, varies from 33° at 30 Mev to 17° at 90 Mev. The minimum angle is sufficiently small so that there is only a small fraction of the solid angle

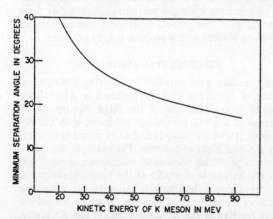

FIG. 2. Hydrogen and carbon events can be distinguished by the recoil in the hydrogen case. The minimum laboratory scattering angle at which this recoil is visible is shown as a function of the kinetic energy of the incoming K meson.

[3] R. M. Sternheimer, Rev. Sci. Instr. 24, 573 (1953).

about the forward direction where the separation of hydrogen from carbon events cannot be made.

All entering tracks which satisfied the bubble density criteria were scanned for scatterings. Since the minimum scattering angle for hydrogen-carbon separation is much larger than the minimum discernable scattering angle, all measurable hydrogen scatterings, that is, all scatterings with a visible recoil, were detected. Scatterings without visible recoils were noted, but will not be discussed in this paper. Measurements from the stereographic photographs provide the angles of the scattered and recoil tracks with respect to the incoming tracks. If either of these tracks stop in the chamber, the range is also known, or if they leave the chamber a minimum range is determined. Since K mesons have almost nonrelativistic energies, the relation between the scattering angle and the recoil angle is almost energy independent. These angles combined with the requirement of coplanarity are the first test as to the genuineness of a K meson-proton collision. When the ranges can also be measured, the kinematics is considerably overdetermined and the genuineness of the event can be ensured.

34 events were found in which the recoil track and scattered track stopped in the chamber and the scattered track had the characteristic K-meson decay. 32 of these events were shown by all of the aforementioned criteria to be K meson-proton scatterings. Of the remaining 2 events, one involved a K meson of less than 20 Mev and could not be satisfactorily analyzed. The other event was an inelastic K meson-carbon scattering, which could be separated from K meson-proton scatterings since the binding energy of the carbon nucleus is important at these low energies.

The chamber was sufficiently large so that no recoil track could leave the chamber, but at the higher energies some of the K mesons which are scattered by protons through certain angles should leave the chamber before decaying, if the K meson is traveling near a chamber wall. While this should not often occur, an extensive search was made for such events, by measuring all scattered incoming tracks which obeyed the bubble density criteria but did not decay in the chamber. To decrease the possibility of such bias, the bubble density criteria were relaxed to include probable π-meson tracks, if the track was too short positively to identify the nature of the incoming particle. One possible K meson-proton scattering was found in 56 such events. However, the recoil proton track was too short to be measured accurately and the nature of the event was not clear.

The distribution of the K meson-proton scattering angles in the laboratory system is shown in Fig. 3. The minimum angles for hydrogen-carbon separation are also shown; line A indicating that angle for 30 Mev and line B that angle for 90 Mev. On the basis of the analysis of the aforementioned 56 events, no more than two or three events have been missed in the range of laboratory angles of 30° to 60° in which the K meson could decay outside of the chamber after scattering on

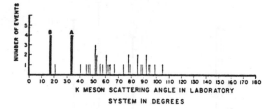

FIG. 3. The distribution of events in the laboratory system. Lines A and B correspond to the minimum angle at which hydrogen recoils can be detected for 30-Mev and 90-Mev K mesons, respectively.

a proton. The small number of events in the backward hemisphere of the laboratory system is believed to be representative of the true angular distribution because a special effort was made not to miss these possibly obscure events. The distribution of the K meson-proton scattering angles in the center-of-mass system is shown in Fig. 4 divided into six groups. The center-of-mass angles of 50° and 26° correspond to the minimum angles for carbon-hydrogen separation for 30-Mev and 90-Mev mesons, respectively.

CONCLUSIONS

The total cross section for the scattering of K^+ mesons on protons is 9.4 mb averaged over the meson kinetic energy range of 20 to 90 Mev. The statistical mean square deviation is 1.7 mb and systematic errors are expected to be of about this magnitude. The number of events is too small to obtain a definite relationship between total cross section and energy, but Fig. 5 indicates that the total cross section may increase with energy in this range.

The differential cross section averaged over the meson kinetic-energy range of 20 to 90 Mev can be derived from Fig. 4. Unfortunately the number of events is not large enough to permit a meaningful determination

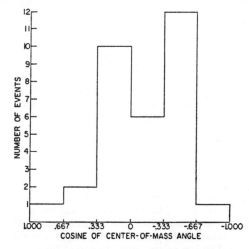

FIG. 4. The center-of-mass distribution of K meson-proton scattering in terms of equal solid angle intervals.

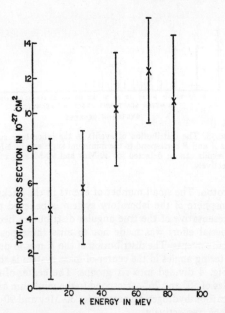

FIG. 5. The total cross section for K meson-proton scattering as a function of the K-meson kinetic energy.

of the parameters in Eq. (1).

$$d\sigma/d(\cos\theta) = 2\pi\lambda^2(A + B\cos\theta + C\cos^2\theta + D\cos^3\theta + E\cos^4\theta). \quad (1)$$

The inclusion of the $\cos^3\theta$ and $\cos^4\theta$ terms in Eq. (1) are necessary because most of the events are at an energy where d-wave effects may be expected. If it is assumed that only s and p waves are involved in the scattering, then Eq. (2) can be fitted by least squares to obtain best values of A, B, and C.

$$d\sigma/d(\cos\theta) = 2\pi\lambda^2(A + B\cos\theta + C\cos^2\theta). \quad (2)$$

There are, however, some general restrictions on the values of A, B, and C from the requirement that the differential cross section, Eq. (2), cannot be negative. This leads to the following restrictions

$$A + B + C \geq 0, \quad (3)$$

$$A - B + C \geq 0. \quad (4)$$

There is a stricter condition on the sum of A, B, and C if the polarization density matrix of the incoming meson beam and of the target protons has only diagonal elements, namely,

$$A + B + C \geq (\sigma/4\pi\lambda^2)^2, \quad (5)$$

where σ is the total cross section.

Table I lists the values of the coefficients, A, B, and C for the unrestricted least-squares fitting, for the least-squares fitting subject to Eq. (3) and Eq. (5), and for the assumption that only s waves are involved. These values are for a value of λ^2 averaged over the 32 events. This last column gives the probability based on the χ^2 test for the experimental distribution to be

TABLE I. The least-squares fit of the experimental data with various restrictions, and the χ^2 probability that these coefficients agree with the experimental data.

Restriction	A	B	C	χ^2 probability
None	0.201	−0.050	−0.245	5%
$A+B+C \geq 0$	0.171	−0.017	−0.154	1.5%
Eq. (5)	0.166	−0.012	−0.139	1%
Only s-wave	0.120	0	0	0.1%

produced by each set of coefficients. The first section in Fig. 4 is not used in this test.

From Table I it is clear that the probability is very small of the experimental distribution coming from either of the restricted sets of coefficients. Therefore, if it is assumed that there have been no substantial systematic errors, it can be concluded that the data cannot be interpreted with only s and p waves. It is believed that the systematic errors are not so substantial as to negate this conclusion, but only a repetition of the experiment can confirm this belief. The presence of d waves would not be surprising. The use of only low-energy events in the angular distribution is not possible because the statistics are too small to permit a division into energy ranges and still give meaningful angular distributions.

Emulsion data which have been published[1] give a higher total cross section and different angular distribution than has been found in this bubble chamber work. But unpublished additions[2] to the published emulsion results have considerably modified those results and brought the total cross section closer to the bubble chamber value.

The results of this experiment indicate both the great need for a substantial increase in K^+ meson-proton scattering statistics and the fruitfulness of the bubble chamber approach to this problem. Therefore we are now working toward the repetition of this experiment in a larger bubble chamber, with an improved K-meson beam. In this way it is hoped that statistics can be increased and bias problems considerably reduced.

We wish to thank Dr. Gustave Zorn for his close collaboration in the design of the separation system and Mr. John L. Brown for his help in the taking of data. We also wish to thank Dr. George Collins, Dr. William Moore, and the staff of the Cosmotron Department for their aid, encouragement, and patience.

Note added in proof.—A compilation containing most of the observed K^+ hydrogen scatterings in nuclear emulsions has been published by Ceolin, Cresti, Dallaporta, Grilli, Gueiriero, Merlin, Saladin, and Zago, Nuovo cimento V, 402 (1957). They find a total cross section of 14.6±3.3 mb. The differential cross section is small in the backward direction in agreement with our results. The emulsion data give many events in the forward hemisphere in contrast to our more backward peak. Our bubble chamber results have an experimental bias against very small scattering angles, so the results cannot be compared in the most forward directions.

COMMENTS ON
C3 "The Use of a Sodium Iodide Luminescent Chamber to Study Elastic and Inelastic Scattering of Pions in Hydrogen" and C4 "Negative Pion–Proton Elastic Scattering at 1.51, 2.01, and 2.53 Bev/c outside the Diffraction Peak"

Although busy with bubble chamber physics in the late 1950's I kept looking for a new particle physics technology which would combine the best features of bubble chambers and counters. The magnetic bubble chamber technology provided large solid angles, precise measurements of vector momenta, good vertex identification, dE/dx information, and sometimes photon detection. But bubble chambers could not be triggered at a high rate on selected events, and the scanning and measuring of bubble chamber pictures had become an industry. On the other hand counters could be used to study selected events at a high rate, but they provided limited information.

When my friend and colleague Lawrence Jones heard about a new idea in the Soviet Union, the luminescent chamber, he and I immediately started to build one. Part of our motivation was to be on our own in particle research and so we wrote to the Office of Naval Research for research support. We were young and unknown, instructors or assistant professors, I don't remember which, but SPUTNIK had just gone up and Washington was worried about a Russian lead in science. What better scientists to support than two young Americans, who wanted to work on a new Russian idea in high energy physics.

Figure 1 in Reprint C3 shows the idea: a visible light image intensifying tube with sufficient amplification allows photography of the trail of a charged particle moving through a scintillator. The volume of scintillator acts as a bubble chamber, the time discrimination of the image intensifier allows high rate triggering on selected events.

We had two problems: how to balance high light collection efficiency against large depth of field, and how to obtain enough light amplification. To ameliorate the first problem we used thallium-doped sodium iodide viewed through $f/5$ aperture lenses. The second problem occurred because there were no image intensifier tubes of sufficient amplification available to us. Most image intensifier tubes were being made for military use and were classified. We had to cascade three tubes to get enough amplification.

The apparatus worked, and as described in Reprint C4 with Kwan Lai we were able to study π^--p elastic scattering using the Bevatron in Berkeley. The entire experiment, the apparatus development and construction in Ann Arbor, the setting up and running of the experiment in Berkeley, and the data analysis back again in Ann Arbor, was carried out by one graduate student, one marvelously skilled technician, Orman Haas, and two faculty members. Mr. Haas was one of those physics technicians who could build anything, just give him a rough sketch or a few sentences to tell him what you wanted. These people were treasures.

What happened to the luminescent chamber? Why did the technique suddenly disappear around 1960? The answer is that the optical spark chamber came along. While we were at the Bevatron carrying out our luminescent chamber experiment we saw our first optical spark chamber, and the optical spark chamber had three advantages over the luminescent chamber: it could be built cheaper and of very large size, there was no optical depth-of-field problem and there was no need for those hard-to-get and often classified image intensifier tubes. Lawrence Jones and I jumped into the technology of optical spark chambers as described in the comment on Reprint C5.

I don't regret my luminescent chamber work, the technology was fascinating and we enjoyed the special pleasure of working in a very small field where the two of us could make many of the inventions and innovations. We even invented a regenerative image intensifier in which using mirrors we passed the light image again and again through a single intensifier tube of moderate gain. The total gain depended upon the number of passes which in turn was controlled by the pulsed on-time of the tube. Quite tricky, too long an on-time and the tube's phosphor screen burnt out. Our long paper in Advances in Electronics and Electron Physics **XII**, 153 (1960) describes this work. We also obtained U.S. Patent 3,,154,687 entitled "OPTICAL FEEDBACK IMAGE INTENSIFIER SYSTEM".

The luminescent chamber was born again in the last decade in design proposals for large particle detectors. The sodium iodide chamber, is replaced by a chamber of scintillating fibers, and the light amplification is accomplished with a mosaic of photodiodes.

IV. 8. THE USE OF A SODIUM IODIDE LUMINESCENT CHAMBER TO STUDY ELASTIC AND INELASTIC SCATTERING OF PIONS IN HYDROGEN*

Martin L. Perl, Lawrence W. Jones, and Kwan Lai

University of Michigan
Ann Arbor, Michigan

Presented by Martin L. Perl

Introduction

For the past year a homogeneous luminescent chamber system has been operated using pion beams from the Bevatron of the Lawrence Radiation Laboratory. In April of this year, a preliminary run on the "large-angle" elastic scattering of 1.5-Bev/c negative pions on hydrogen was made, and currently an extension of this experiment together with an experiment to study pion production from pions on protons is in progress. In this paper the nature of the luminescent chamber system is briefly summarized; the current experiments are described; and comments based on our experience in the design, execution, and interpretation of luminescent chamber systems are presented.

Luminescent Chamber

General Properties

The idea of photographing tracks of particles in scintillators by using a system of image-intensifier tubes has been described and discussed extensively.[1] Assuming that the image tubes have sufficient gain and resolution to record photographically single photoelectrons from the first image-tube cathode, the depth of field realized in a homogeneous scintillator is related to the poorest track resolution and recorded information by the relationships indicated on Fig. 1. Thus with thallium-activated sodium iodide crystals the track resolution in the crystal is 1 to 3 mm if scintillators up to 4 in. thick are used.

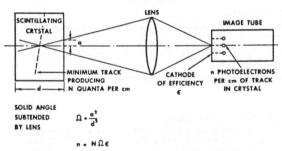

Fig. 1. Relationships between track resolution, a, depth of field, d, and track information, n, for the homogeneous luminescent chamber. For NaI(Tl) in which we have $N=10^5$, $a=1.7$ mm for $d=10$ cm and $n=10$ photoelectrons per cm of track in the crystal.

The relevant properties of image tubes and of typical lenses available for coupling are summarized in Tables I and II. The particular combination of tubes and lenses in use currently at the Bevatron is schematically indicated in Fig. 2. Magnetically focused tubes will provide a further improvement in the total number of resolved image elements when they are incorporated; however, to date they have not been available.

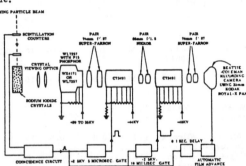

Fig. 2. Schematic diagram of the luminescent chamber system currently in use. The chamber-viewing optics and beam-defining scintillation counters are oversimplified and generalized in this diagram.

Time Resolution

The time resolution of our system is determined by the decay characteristics of the P-15 phosphor on the first image tube. As pointed out earlier,[2] this phosphor has an hyperbolic decay characteristic, so that although 35% of the light is given off in the first microsecond, about 40% of the light comes off slowly after 5 μsec. This has put a practical limit of less than 10^4 to the number of particles we may pass through our chamber per Bevatron pulse (150 milliseconds) without having an excessive background of photon dots from "old" tracks appearing on the film.

Partly as a result of this, our experiments are generally planned with a hydrogen (or polyethylene) target in a particle beam, and the scintillators arranged about the target to detect and study reaction products.

Gating

The fast gating of our second image tube (RCA C73491) is accomplished by applying a positive 2500-volt 5-μsec pulse to the grid (focusing electrode) of the first stage of the tube. With the grid grounded, the tube is cut off; with the grid at +2500 volts, a well-focused image is transmitted. The pulse is derived from a simple thyratron circuit driven by an EFP 60. The total delay from the time the particle

Table I

Image tube parameters

Tube mfr.	Tube type	Focus	Cathode diam. (in.)	Anode diam. (in.)	Axial anode resolution[a]	No. of stages	Total voltage (kv)	Quantum gain[b]	Phosphor type	Noise[c]	Cathode efficiency at 4400Å	Cathode type	Approx. price $
Westinghouse	WL7257	F.S.	5	1	15	1	25-35	10-30 / 80-120	P15 / P11	10^7-10^8 / 10^8-10^9	8%	S11	3,000
	WX4171	F.S.	5	1	15	1	25-35	20-60	P15	--	15%	S20	5,000
RCA	C73458	E.S.	1	1	18[d]	2	20-25	300-2000	P11, P20	10^7-10^9	15%-20%	S20	3,000-5,000
	C73459	F.S.	1 or 2[e]	1	--	2	20-30	≈1000	P11, P20	--	--	S20	6,000
	C73491	F.S.	1	1	12	3	30-45	5000[f] / 30,000	P11, P20	10^8-10^9	15%-20%	S20	10,000
RCA	- (b) C70012	E.M.	1.5, 3	1.5, 3	15	2	20-30	1000	P11, P20	--	--	S20	--
I.T.-T.	FW113[g]	E.M.	1.5	1.5	15	2	--		P11, P20	--	--	S11	--
Westinghouse	WX4342	E.M.	1	1	10-12	4[h]	25-36	>5000	P11	--	5%-7%	S11	--
20th Century-England	Wilcock-Emberson Intensifier	E.M.	0.8	0.8	15	5[h]	35	20,000-30,000[f]	P11	--	--	S11	--

[a] Resolution in line pairs per mm.
[b] Quantum gain defined as (light power out)/(light power in).
[c] Noise in units of quanta per second from anode with no cathode illumination.
[d] Resolution and distortion very poor at edge of field.
[e] Available with 1- or 2-in. first cathode, P11 or P15 first phosphor.
[f] With the best tubes of these types, single photoelectrons from the first cathode may be visually observed by using a 10X eyepiece.
[g] Developmental types. No working tubes yet available.
[h] Transmission secondary-electron emission dynodes.

---

passes through the scintillator to the time the voltage has reached its correct values is less than 0.4 μsec. A positive pulse has been used here, as it was observed that the cathode resistivity was high enough to affect the image quality with these short gate times when conventional cathode-bias gating was used. The second C73491 is gated on by use of cathode biasing for 10 msec as described elsewhere.[1,2] No mechanical camera shutter is used.

Fiducial Light Sources

Earlier Bevatron experience indicated that the scintillator image was shifted on the film because of the Bevatron stray magnetic field (affecting the image tubes) even within the duration of the beam spill. As a result, fiducial lights placed near the corners of the crystals are flashed simultaneously with the fast gate to allow correct orientation of the film for reading track angles and positions relative to the true beam and target positions. These fiducials consist of four Amperex 6977 indicator triodes with their "flash" adjusted to give 5 or 10 photon dots apiece on the film.

Crystal-Viewing Optics

The lenses used to focus the image of the crystal or crystals onto the 5-inch cathode of the WX 4171 are chosen and adjusted for each particular experimental geometry. In general we have subtended an f/5 aperture from the scintillator. For stereo photography of tracks from one 2×2×4-inch rectangular crystal of sodium iodide, two pairs of 15-in. F.L. f/5 lenses were used at about unity magnification, each pair viewing one face. During the April run of the π⁻-p scattering, a configuration of two scintillators, one 2×2×4-in. crystal and the other either a similar crystal or a cylinder 3 in. in diameter and 2 in. deep, was viewed nonstereoscopically by a single lens system consisting of a 15-in. F.L. f/5 achromat in conjunction with a 5-in. F.L. f/1.9 projection television lens. The combination demagnified the 7-in.-diameter field to the central 3 in. of the 5-in. cathode.

In the current geometry for π-p scattering, the scintillators are arranged in two units, each 2×2×8 in. as in Fig. 3. Two sets of lenses, each consisting of an f/5 achromat and an f/1.9 5-in. F.L. lens, are

Table II

Lens properties. Typical properties of lens systems. Each system consists of a pair of identical units combined for unity magnification

Manufacturer	Focal[a] length (mm)	Rated[a] aperture	Theoretical collection efficiency[b]	Transmission[c] P11	Transmission[c] P20	60% vignetting radius[d] (mm)
Farrand	76	f/0.85	0.26	.50	.64	11.0
Taylor-Taylor-Hobson	50	f/0.80	0.28	.49	.72	5.5
Canon	50	f/1.2	0.16	.61	.73	8.0
Zeiss	75	f/1.5	0.10			8.5
Nikkor	85	f/1.5	0.10	.58	.74	10.0
Carl Meyer	60	f/1.6	0.09	.47	.72	7.2
Bausch and Lomb	127	f/1.9	0.07	.45	.67	50.0
Kodak (Aeroektar)	178	f/2.5	0.04	.28	.38	> 25

[a] Focal length and aperture for each element of the pair.
[b] Theoretical collection efficiency from a Lambertian surface given by $\sin^2\theta$, where θ is the half-angle subtended by the lens.
[c] Transmission is ratio of actual transmission of the lens to the theoretical transmission, recorded for light from P11 and P20 phosphors.
[d] The recorded values are the radii (in mm) at which the transmission falls to 60% of its value on axis.

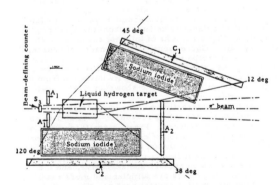

Fig. 3. Geometry of hydrogen target, sodium iodide crystals, and scintillation counters used in elastic π-p scattering experiments. The coplanarity counters subtend ±7 deg from the target off the vertical plane (plane of the drawing).

For the experiment studying pion production by pions, a scintillator array 4 in. high, 8 in. long, and from 2 in. deep (top) to 4 in. deep (bottom) is imaged by a lens combination consisting of a 24-in. F.L. f/5 achromat together with a 12-in. F.L. f/2.5 camera lens (Kodak Aeroektar) onto the 5-in. cathode (Fig. 4). A low-grade stereo image from a 45-deg prism 8 in. long placed below the scintillator is presented through the same lens.

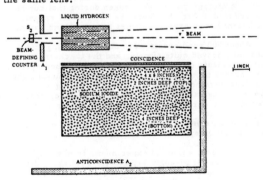

Fig. 4. Geometry of target, sodium iodide, and scintillation counters used in detection of protons from inelastic π-p scattering experiments.

used to bring the two crystal images separately onto the cathode with 2.5-to-1 demagnification. Such a system better accommodates the lens properties and, by presenting the two images close together on the cathode, makes better use of the available system resolution.

Thus far, all our experiments have made use of 2×2×4-in. sodium iodide bricks, completely encased in glass and stacked together in various arrays. Six such bricks are currently in use, and the flexibility provided by such a system has proven very convenient.

It should be noted that the crystal-viewing optics transmit 40% to 80% of the light from sodium iodide entering their aperture; the loss is due primarily to absorption in the glass. Simpler lenses (e.g., achromat combinations) have the highest transmission, while the large, highly corrected Aeroektar is the worst. Special optics using quartz and other ultraviolet-transmitting glasses would allow a closer approach to the numbers given in Fig. 1.

For purposes of orientation, an f/5 aperture view of a sodium iodide crystal (index of refraction 1.8) subtends 1/1300 of the light from a track in the crystal. If the crystal-viewing optics transmit 60% of this, one would expect 45 photons falling on the image tube cathode per centimeter of minimum-ionizing track in the sodium iodide, or 3 to 5 photoelectrons per centimeter, depending on the cathode efficiency.

Luminescent Chamber Experiments

Elastic Pion-Proton Scattering

As our group is, to our knowledge, the first to use a fast chamber of any kind (e.g., spark chamber, scintillation or luminescent chamber) in a physics experiment, comments on the configurations used might be of interest. The April experiment on "large-angle" pion-proton scattering (that is, elastic scattering outside the diffraction peak) used five scintillation counters in coincidence, three in a beam-defining telescope and two coplanarity counters. The latter counters, outside the sodium iodide crystals, required any event triggering the image tubes to have produced at least one particle up and one particle down within ±10 deg of the vertical plane. Anticoincidence counters rejected events due to particles not in the central 1/2×1/2-in. beam.

For simplicity, a polyethylene target was used as a source of hydrogen in this run, and the carbon contribution was determined by making a background run with a graphite target. Analysis of the pictures showed that, of the events triggering the image tubes, only about one in fifteen was a "two-prong" final state satisfying the kinematics of large-angle pion-proton elastic scattering. The graphite pictures showed that the great majority of these were indeed elastic scattering in hydrogen. A photograph of tracks taken during that run is reproduced in Fig. 5, where the beam axis, target, and scintillator-crystal outlines have been indicated.

About 600 good events were obtained during this run. Scanning proved quite simple, and the track-angle measurements could be made to ±2 deg. Since systematic errors of ±2 deg were also present (fiducial light sources were not then in use), tracks falling within ±4 deg of the calculated kinematic pion-proton correlation angle were accepted. Corrections for image-tube distortion were made by photographing a polar-ray test pattern and correlating angles read on film with angles in space at the focal plane.

The current series of experiments is designed to measure the large-angle π-p elastic-scattering differential cross sections behind the diffraction peak at 1.5, 2.0, and 2.55 Bev/c (the total large-angle elastic-scattering cross section is of the order of 2 mb in this region). A liquid hydrogen target 3 in. long by 1-1/2 in. in diameter is used here. The pion beam is again defined to a 1/2-in.-square cross section at the target.

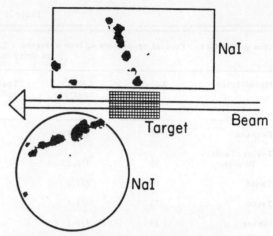

Fig. 5. π-p scattering, April 1960. The large rectangle and circle outline the images of the 2×2×4-in. and 3-in. diam×2-in.-thick sodium iodide scintillators. The beam axis and polyethylene targets are also indicated, with the pion beam direction from left to right.

With 10^{11} protons per pulse on the Bevatron target, more than 10^4 pions pass through the beam-defining telescope. The geometry illustrated in Fig. 3 employs two coplanarity counters as before, together with appropriate anticoincidence counters. The vacuum envelope and heat shield around the hydrogen target necessitate scaling up the dimensions of scintillators used to cover the desired angles over the earlier run with polyethylene. The polar angles subtended by the two coplanarity counters are 12 to 45 deg and 38 to 120 deg, as indicated on Fig. 3. The azimuthal angles subtended about the beam axis are ±7 deg from the vertical plane. The sodium iodide crystals completely shadow the coplanarity counters from the hydrogen target, so that all tracks of interest should be seen in the crystals. Corrections must be made, of course, for pion and proton interactions in the sodium iodide. The range of azimuthal angle accepted is small enough so that corrections arising from its deviation from 0 deg are negligible. Figure 6 shows an elastic scattering in this setup.

The target interaction volume is made as small as possible consistent with a reasonable rate of data collection, for two reasons. First, a small target allows a compact configuration of counters and scintillation crystals. Track resolution can be better if a smaller volume of sodium iodide is viewed; the scintillator crystals are few in number and expensive, and the crystal-viewing optics are simpler and more readily available for smaller crystal configurations. Second, the smaller the target is, relative to the scintillator and counter sizes, the more nearly equal are the angles subtended from different parts of the target volume.

Pion Production With Slow-Proton Recoils

In Fig. 7a, tracks of protons stopping near the center of a 2×4×6-in. array of sodium iodide are shown. Figure 7b shows a minimum-ionizing track in comparison. These tracks, 2 to 4 ins. long, correspond to entering protons of 125 to 200 Mev kinetic energy. The ionization makes these tracks very easy to identify and to differentiate from stopping mesons. Such track photographs encouraged us to use this ability to identify stopping protons in the study of inelastic

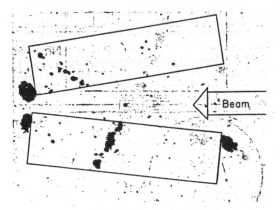

Fig. 6. π-p scattering, August 1960. The two rectangles indicate the two 2×2×8-in. sodium iodide crystals. The liquid hydrogen target lies between these crystals. The three large dots are the fiducial lights. The direction of the incoming beam is also indicated. The NaI crystals are relatively closer together on the photograph than in the real setup because the crystal-viewing lenses are arranged to do this.

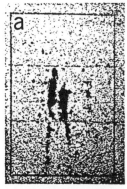 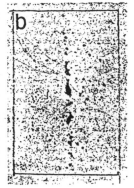

Fig. 7. a. Photographs of protons stopping near the center of a 2×4×6-in. array of sodium iodide crystals, viewed normal to the 4×6-in. face.
b. A minimum-ionizing track for comparison in the same NaI array.

pion scattering when slow protons are produced with small laboratory-system velocities. The experimental setup presented in Fig. 4 employs a coincidence counter below the hydrogen target to indicate the entrace of a particle into the sodium iodide, and an anticoincidence counter below the sodium iodide to indicate that the particle has stopped. If entering pions of 700, 850, and 1000 Mev are used, the number of inelastic pion events other than production of one additional pion is small. If the reaction is $\pi^- + p \rightarrow \pi^- + \pi^0 + p$, the energy of the incident pion and the energy and angle of the outgoing proton determine w, the total energy in the pion-pion-system center of mass. Determining the recoil proton energy in the laboratory system between 20 and 90 deg and from 80 Mev to 200 Mev for the three initial pion energies adequately covers the range of w values suspected to contain the pion-pion resonance, while using a range of incident pion energies allows discrimination against other effects such as the nucleon isobar state and the statistical model. Pictures will be scanned, then, for stopping proton tracks, and the data plotted versus the parameter w. Again the comments on the small target volume, made concerning the elastic-scattering experiment, apply equally well here. However, in this case the information sought from the sodium iodide on the proton range and ionization places a premium on a large volume of sodium iodide.

Discussion of Experimental Techniques

Bias Corrections

In the parlance of classical nuclear physics, most high-energy counter experiments employ "good geometry," while bubble chambers employ very "poor geometry." Our luminescent chamber experiments employ detectors subtending an intermediate range of solid angles, and as a result, it is important to determine just what this solid angle is in order to unambiguously interpret the track information available. These solid-angle bias corrections have proven a very important and sometimes tricky factor in our design of experiments.

We have found it convenient to use beams and hydrogen targets of small dimensions in order to keep the required scintillator volume low and minimize solid-angle biases. For experiments studying interactions of a few millibarns total cross section, 10^4 particles per pulse through a 3-in. liquid hydrogen target are adequate to give an interaction photograph on every beam pulse.

Chamber Comparisons

The track-angle resolution of the various fast chambers (discharge chambers, plastic scintillator filament luminescent chamber, and sodium iodide homogeneous luminescent chamber) are all comparable. The homogeneous chamber is best suited to experiments in which it is useful to know the track ionization and where a moderately large stopping power is desired. Thus the inelastic pion-scattering experiment is better suited to our technique than the elastic-scattering experiment above. In fact, the corrections to the elastic-cross-section data due to particle interactions in the sodium iodide are an annoyance that would not be present in a thin-plate discharge chamber.

Earlier, the ability to take stereo photographs of particle tracks and events was considered important, and an older photograph is reproduced in Fig. 8 for illustration. However, to date, our experiments have centered around two-dimensional presentation of a larger area, with the third dimension roughly defined by scintillation counters.

Resolution

The information content in our photographs, as with fast chambers in general, is restricted by the system resolution. We use a 1.5-cm-diameter area on the film, and the image of a single photon is 0.1 mm in diameter on the film. Thus our final image contains between 10^4 and 2×10^4 resolved elements. Although this restricted resolution may be improved with better image tubes, etc., it is to a certain extent an inherent characteristic of the system. When supplemented by a characteristic signature of particular scintillation counters, and with the fast time resolution of the chamber, this resolution is entirely appropriate for a wide variety of significant high-energy experiments. As these fast chambers may be expected to come into wider us-

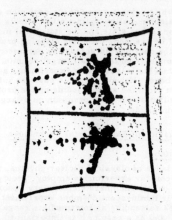

Fig. 8. 1959 stereoscopic photographs of events produced by 1-Bev π⁻ in one 2×2×4-in. sodium iodide crystal. The (distorted) outlines of the two 90-deg stereo images of the crystal are indicated on the film.

age in the future, we believe it would be very appropriate for those concerned with the automatic analysis of track photographs to direct some attention toward this field.

Footnotes and References

*Supported in part by the U.S. Office of Naval Research and by Project Michigan of the U.S. Army.

1. a. Zavoiskii, Butslov, Plakhov, and Smolkin, Atomnaya Energiya 1, 34 (1956); J. Nuclear Energy 4, 340 (1957).
   b. M. L. Perl and L. W. Jones, Phys. Rev. Letters 2, 116 (1959).
   c. K. Lande, A. K. Mann, M. M. Schlacter, D. M. Skyrme, and H. Uto, Rev. Sci. Instr. 30, 496 (1959).
   d. G. T. Reynolds, R. Giacconi, and D. Scarl, Rev. Sci. Instr. 30, 497 (1959).
   e. L. W. Jones and M. L. Perl, in CERN Conference on High-Energy Accelerators and Instrumentation, CERN, 1959, p. 561.
   f. M. L. Perl and L. W. Jones, Nucleonics 18, 5, 92 (1960).
2. Jones, Lai, Newsome, and Perl, Luminescent Chamber Technical Report No. 2, Feb. 1960 (to be published in the Proceedings of the Seventh Scintillation Counter Symposium by the Institute of Radio Engineers).

NEGATIVE PION-PROTON ELASTIC SCATTERING AT 1.51, 2.01, AND 2.53 Bev/c OUTSIDE THE DIFFRACTION PEAK*

Kwan W. Lai,[†] Lawrence W. Jones, and Martin L. Perl
The University of Michigan, Ann Arbor, Michigan
(Received July 27, 1961)

The differential elastic scattering cross sections for negative pions on protons have been measured at incident pion momenta of 1.51, 2.01, and 2.53 Bev/c with emphasis on the angular region outside the diffraction peak. The purpose of the experiment was to examine the behavior of the large angle differential elastic cross section as a function of energy from the energy of the highest known resonance in the pion-nucleon system into the region where the total cross sections appear to be approaching an asymptotic value.[1]

The experiment was performed at the Bevatron using a luminescent chamber system to photograph the tracks of the scattered pion and the recoil proton from a liquid hydrogen target. The chamber configuration used is shown in Fig. 1, where the plane of the figure is normal to the optic axis of the image intensifier tube system. Lenses focused this image of the sodium iodide scintillators onto the first image tube cathode, and the gain of the system was sufficient to record single photoelectrons from this cathode. A coincidence between the coplanarity counters, which subtended $\Delta\Phi = \pm 7.15°$ from the vertical plane, triggered the system on probable elastic events. This post-event triggering enabled the tracks to be subsequently recorded on film with a time resolution of a few microseconds. A description of the apparatus has been reported elsewhere in detail.[2] The film was double scanned, and angles for elastic scattering events were recorded with suitable corrections for distortions in the image tube system. The angles and ionization of the two tracks were used to identify an event as elastic, and subsequent correlations of the track angles with kinematics were consistent

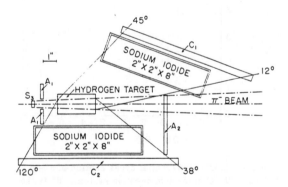

FIG. 1. Geometry of liquid hydrogen target, scintillation counters, and sodium iodide luminescent chambers. The coplanarity scintillation counters C_1 and C_2 subtended ±7.15° from the vertical plane (the plane of the drawing) for scatterings on the beam axis. Besides anticoincidence counters A_1 and A_2, two other anticoincidence counters flanked the target parallel to the plane of the figure to bias against inelastic events. S_3 is the final beam defining counter. The sodium iodide crystals were imaged by lenses onto the image tube system which was oriented perpendicularly to the plane of the figure.

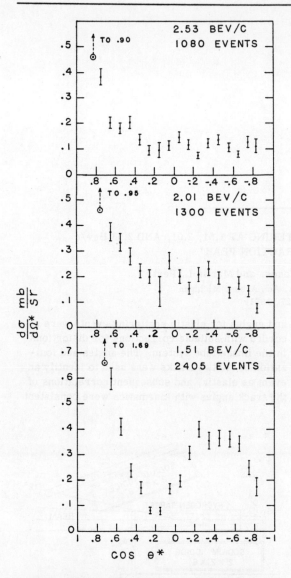

FIG. 2. Elastic differential cross section of $\pi^- + p \to \pi^- + p$ in the center-of-momentum system. θ^* is the angle of the scattered pion in that system. The values of $d\sigma/d\Omega^*$ at the smallest angle measured at each energy are too large to be placed at the correct places on the graphs, but their values are indicated.

2405 elastic scatterings were analyzed at 1.51 Bev/c, 1300 events at 2.01 Bev/c, and 1080 events at 2.53 Bev/c. The results are presented in Fig. 2. The differential cross sections have been corrected for multiple Coulomb scattering and nuclear interactions in sodium iodide, for geometrical and scanning biases, and for pseudo-elastic events from the target vessel (from data taken with the target empty). The absolute value of the differential cross section may be uncertain by as much as ±15% from systematic errors associated with this new technique. The data at 1.51 Bev/c are in good agreement with the bubble chamber result of Chretien et al.[3]

From the data it may be noted that the backward hump, which has a maximum height of 2.1 mb/sr at 900 Mev and 1.1 mb/sr at 1020 Mev,[4] is down to 0.4 mb/sr at 1.51 Bev/c (1.37 Bev) and is not present at 2.01 or 2.53 Bev/c. The angular distributions behind the diffraction peak at 2.01 and 2.53 Bev/c are roughly constant with angle, decreasing from 0.18 mb/sr at 2.01 Bev/c to 0.11 mb/sr at 2.53 Bev/c. Although the data at these two higher momenta can be taken to suggest some oscillatory structure behind the diffraction peak, they are not inconsistent with an isotropic distribution behind the diffraction peak.

We wish to acknowledge our appreciation to Dr. E. J. Lofgren, his associates, and the staff of the Bevatron of the Lawrence Radiation Laboratory for their hospitality and assistance throughout the preparation, execution, and analysis of this experiment.

---

*This work was supported by the Office of Naval Research.

†Submitted by Kwan W. Lai in partial fulfillment of the requirement for the degree of Doctor of Philosophy.

[1]V. S. Barashenkov, Soviet Phys.-Uspekhi 3, 689 (1961).

[2]M. L. Perl, L. W. Jones, and K. Lai, in Proceedings of the 1960 International Conference on Instrumentation for High-Energy Physics (Interscience Publishers, Inc., New York, 1961), p. 186.

[3]M. Chretien, J. Peitner, N. P. Samios, M. Schwartz, and J. Steinberger, Phys. Rev. 108, 383 (1957).

[4]C. D. Wood, T. J. Devlin, J. A. Helland, M. J. Longo, B. J. Meyer, and V. Perez-Mendez, Phys. Rev. Letters 6, 481 (1961).

COMMENTS ON C5
"Pion-Proton Elastic Scattering at 2.00 GeV/c"

I loved the technology of optical spark chamber experiments: the mechanical problems of building a large, rugged, precise chamber; the high speed cameras; the optics and optical paths that extended over meters; the film scanning and measuring tables. For more than a decade I used optical spark chambers to explore pion-proton elastic scattering, neutron-proton elastic scattering, muon-proton inelastic scattering, and finally electron-proton inelastic scattering.

For me the combination of optical spark chamber technology with elastic scattering experiment was particularly satisfying. Lawrence Jones and I were entranced by the early 1960's interest in the Regge pole model of strong interactions, and the cleanest example: elastic scattering. Bubble chamber experiments could not provide enough data and counter experiment only worked with high intensity proton beams. The optical spark chamber was ideal.

The paper reprinted here describes one of several pion-proton elastic scattering studies we carried out at the Bevatron. We had enough statistics to study two burning issues in the Regge pole model of the 1960's. As the energy increases, does the diffraction peak become narrower or, to use the jargon of the period, does the diffraction peak shrink? The second burning issue was the existence and significance of secondary maxima in the elastic differential cross section.

It is hard now to understand the beliefs in the early 1960's that elastic scattering through the Regge pole model could give us a basic understanding of high energy hadronic interactions. For a time I, too, was a believer. After all, what could be better than to use the mechanical collisions of elementary particles to study those particles? But as the Regge pole model grew in complexity with ghosts and daughter trajectories my faith weakened and, as described in Memoir A1 in this volume, I began to turn to leptons.

While compiling this volume, I compared Fig. 1 in Reprint C5, 2.00 GeV/c $\pi^- p$ elastic scattering, with the 2.01 GeV/c $\pi^- p$ luminescent chamber measurement in Fig. 2 of Reprint C4. The latter measurement gave too large a backwards cross section and does not have enough statistics to show the second maximum at $\cos\theta^* = 0.2$. The luminescent chamber was a tough technique.

In 1964 I summed up my understanding of πp elastic scattering in a paper entitled "Empirical Partial Wave Analysis of $\pi + p$ Elastic Scattering Above 1 GeV/c" (M. L. Perl and M. C. Corey, *Phys. Rev.* **136**, B787 (1964)). I obtained neat descriptions of πp elastic scattering, but the connection to the fundamentals of the strong interaction was no better than in Regge pole models.

Except for one more foray into hadron-hadron elastic scattering, the neutron-proton elastic scattering experiment described in Reprint C6, I left hadron-hadron elastic scattering forever.

PION-PROTON ELASTIC SCATTERING AT 2.00 GeV/c[†]

D. E. Damouth,[*] L. W. Jones, and M. L. Perl[‡]
University of Michigan, Ann Arbor, Michigan
(Received 1 July 1963)

In the course of a spark chamber experiment which studied pion-proton elastic scattering up to 5 GeV/c,[1,2] we measured the π^-p elastic differential cross section at 2.01 GeV/c with high statistical accuracy (7000 elastic events) and the π^+p elastic differential cross section at 2.02 GeV/c with moderate statistical accuracy (1400 elastic events). This momentum is of particular interest, as a resonance has recently been found in the π^-p total cross section at 2.08 GeV/c by Longo and Moyer[3] and by Diddens et al.[4] Diddens et al. also found a resonance at 2.51 GeV/c in the π^+p total cross section, so that 2.0 GeV/c lies midway between this new π^+p resonance and the previously known π^+p resonance[5] at 1.5 GeV/c. The data presented below show that there is a second maximum in the π^-p differential cross section at $\cos\theta = 0.2$ (in the barycentric system). This second maximum is less pronounced in the π^+p system. It is also shown that while the width of the π^+p diffraction peak changes considerably in the 1.0 to 3.0 GeV/c momentum interval, no significant change in the width of the π^-p diffraction peak is observed in this interval.

The differential cross sections are plotted in a semilogarithmic form in Figs. 1 and 2. The errors are statistical and do not include an over-all normalization error of ±8% for π^-p and +20%, -10% for π^+p. The total elastic cross sections are 7.94 ± 0.9 mb for π^-p and $9.1^{+2}_{-1.3}$ mb for π^+p.

The second maximum which appears in the π^-p differential cross section, Fig. 1, is not seen at 3.0 GeV/c or above,[1,2] or at 1.6 GeV/c[6]

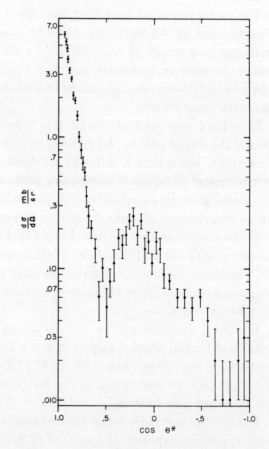

FIG. 1. Differential cross sections in the barycentric system for π^-p elastic scattering. The errors are statistical and do not include an over-all normalization error of ±8%.

main diffraction peak. Third, the model predicts additional peaks which are not seen. These would, however, probably disappear with the use of a more realistic model, e.g., one with a slightly diffuse boundary.

The π^+p differential cross section confirms the previous measurement and conclusions of Cook et al.[14] that the π^+p differential cross section is larger than the π^-p for all regions outside the diffraction peak at 2.0 GeV/c. Cook et al. and Helland et al.[9] showed that in the momentum region of the 1.5-GeV/c π^+p resonance, there is a large bump in the π^+p differential cross section in the backward hemisphere of the barycentric system. The comparatively larger size of the π^+p differential cross section outside the diffraction peak at 2.0 GeV/c may be the remains of this bump and therefore related to the 1.5-GeV/c resonance. As seen in Fig. 2, in the π^+p system there is evidence for a second maximum at $\cos\theta = 0.2$ rising out of this background, but it is considerably less pronounced than in the π^-p case. Therefore our tentative conclusion on the basis of existing data is that the second maximum at $\cos\theta = 0.2$ is the strongest in the π^-p system and is related to the 2.08-GeV/c π^-p total cross-section resonance.

As another way to look for relations between the peak in the total cross section and the shape of the elastic differential cross section, we have made use of the diffraction peak parametrization used at higher energies:

$$d\sigma(\theta)/d\Omega = [d\sigma(\theta)/d\Omega]_0 e^{At}, \quad (1)$$

where $d\sigma(\theta)/d\Omega$ is the differential cross section in the barycentric system in mb/sr, and t is the square of the four-momentum transfer in $(\text{GeV}/c)^2$. In this momentum range, and with measurements of high statistical accuracy, this simple exponential is not a very good fit, but it is a very useful way to measure the width of the diffraction peak, because the peaks are roughly exponential out to $t = 0.4$ $(\text{GeV}/c)^2$. In Table I we have listed the values of A obtained by a least-squares fit to this and other experiments for the interval $0.0 \le t \le 0.4$ $(\text{GeV}/c)^2$. $P(\chi^2)$, the probability of obtaining a χ^2 as large as given by the fit, is also listed. For comparison, it should be noted[2] that at momenta above 3 GeV/c the A's of the $\pi^\pm p$ diffraction peaks have a constant value of 7.6 to 7.9 $(\text{GeV}/c)^{-2}$. Table I

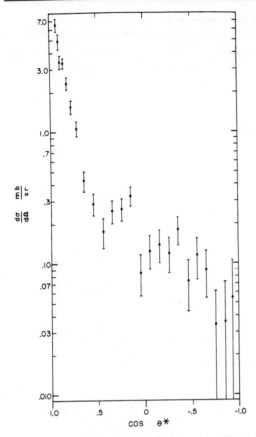

FIG. 2. Differential cross section in the barycentric system for π^+p elastic scattering. The errors are statistical and do not include an over-all normalization error of +10%, -20%.

or below.[7-11] Unfortunately, other measurements at 2.5[8] and 2.8[12] GeV/c have insufficient accuracy to observe this second maximum. However, an unpublished differential cross-section measurement at about 1.85 GeV/c by Erwin and Walker[13] gives some evidence for this second maximum.

The interpretation of this second maximum as simply the second maximum in a diffraction pattern meets with several difficulties. First, if such a diffraction effect exists, one would expect it to be seen over a very large range of energies. Second, one cannot simultaneously fit the width of the first diffraction peak and the position of the second peak. If one adjusts the interaction radius to fit the position of the second maximum, it is then about 30% smaller than one would obtain by fitting the width of the

Table I. Exponential fits to diffraction peaks.

Incident pion laboratory momentum (GeV/c)	A (GeV/c)^2	$P(\chi^2)$	Reference
π^-p elastic scattering			
1.34	7.5 ± 0.4	0.40	a
1.48	7.5 ± 0.4	0.20	b, c
1.59	7.1 ± 0.2	0.01	d
1.85	9.3 ± 1.7	0.30	e
2.01	7.8 ± 0.2	0.50	This experiment
2.5	8.5 ± 0.8	0.20	c
3.15	7.9 ± 0.3	0.02	f
π^+p elastic scattering			
1.12	4.1 ± 0.2	0.25	g
1.45	7.4 ± 0.6	0.30	g
1.50	8.2 ± 0.3	0.15	h
1.69	6.4 ± 0.2	0.02	g
2.00	5.0 ± 0.4	0.70	h
2.02	5.7 ± 0.4	0.40	This experiment
2.50	6.9 ± 0.5	0.02	h
2.92	7.6 ± 0.3	0.20	f

a See reference 10.
b See reference 7.
c See reference 8.
d See reference 6.
e R. C. Whitten and M. M. Block, Phys. Rev. 111, 1676 (1958).
f See reference 1.
g See reference 9.
h See reference 14.

shows that for the π^-p system the A values rise rather smoothly from 7.0 (GeV/c)$^{-2}$ at 1.34 GeV/c to 7.9 (GeV/c)$^{-2}$ at 3.15 GeV/c. That is, the π^-p diffraction peak simply narrows slightly over this momentum range. On the other hand, the A values of the π^+p system increase from 4.0 (GeV/c)$^{-2}$ at 1.12 GeV/c to a peak of about 8.2 (GeV/c)$^{-2}$ at 1.5 GeV/c, then decrease to 5.0 (GeV/c)$^{-2}$ and finally rise again at 2.92 GeV/c to 7.6 (GeV/c)$^{-2}$ which is close to the π^-p value at that momentum. This behavior can be thought of as a considerable narrowing of the diffraction peak over the 1- to 3-GeV/c interval combined with a sudden and temporary narrowing at 1.5 GeV/c, possibly associated with the resonance at that momentum.

Finally, we point out that the high statistics in the π^-p data make evident some structure in the diffraction peak. In particular, the $0.0 \leq t \leq 0.2$ (GeV/c)^2 interval has a steeper slope than the $0.0 \leq t \leq 0.4$ interval, 9.6 ± 0.9 as compared to 7.8 ± 0.2 (GeV/c)$^{-2}$. That is, on the semilogarithmic plot there is a definite concave upward slope to the diffraction peak.

We would like to suggest that the detailed structure and energy dependence of the elastic diffraction peak parameters might prove to be a useful approach to studying properties of resonances at higher energies where the interaction is mostly inelastic.

Thus it would be particularly interesting to compare accurate π^-p diffraction data at several energies about the 2.08-GeV/c resonance to see if the diffraction peak has structure at this point analagous to the narrowing of the π^+p diffraction peak at the 1.5-GeV/c resonance.

We wish to acknowledge the help and hospitality of E. J. Lofgren, his colleagues, and the staff of the Bevatron in supporting this experiment; the help of W. J. Holley in taking the π^+p data; and the assistance of C. C. Ting, K. W. Lai, and O. Haas in conducting the experiment.

---

†Work supported in part by the U. S. Office of Naval Research.
*Present address: Xerox Corporation, Rochester, New York.
‡Present address: Stanford University, Stanford, California.
[1] C. C. Ting, L. W. Jones, and M. L. Perl, Phys. Rev. Letters 9, 468 (1962).
[2] M. L. Perl, L. W. Jones, and C. C. Ting (to be published).

[3]M. J. Longo and B. J. Moyer, University of California Lawrence Radiation Report No. UCRL-10174, 1962 (unpublished); Phys. Rev. Letters 9, 466 (1962).

[4]A. N. Diddens, E. W. Jenkins, T. F. Kycia, and K. F. Riley, Phys. Rev. Letters 10, 262 (1963).

[5]T. J. Devlin, B. J. Moyer, and V. Perez-Mendez, Phys. Rev. 125, 690 (1962); J. C. Brisson et al., Nuovo Cimento 19, 210 (1961).

[6]J. Alitti et al., Nuovo Cimento 22, 1310 (1961); F. Shively (private communication).

[7]M. Chretien et al., Phys. Rev. 108, 383 (1957).

[8]K. W. Lai, L. W. Jones, and M. L. Perl, Phys. Rev. Letters 7, 125 (1961).

[9]J. A. Helland et al., Phys. Rev. Letters 10, 27 (1963); J. A. Helland, University of California Lawrence Radiation Laboratory Report No. UCRL-10378, 1962 (unpublished).

[10]L. Bertanza et al., Nuovo Cimento 19, 467 (1961).

[11]C. D. Wood et al., Phys Rev. Letters 6, 481 (1961).

[12]L. P. Kotenko, E. P. Kusnetsov, G. I. Merzon, and Yu. B. Sharov, Zh. Eksperim. i Teor. Fiz. 42, 1158 (1962) [translation: Soviet Phys.—JETP 15, 800 (1962)].

[13]W. D. Walker (private communication).

[14]V. Cook, B. Cork, W. R. Holley, and M. L. Perl, Phys. Rev. 130, 762 (1963).

COMMENTS ON C6
"Neutron-Proton Elastic Scattering from 1 to 6 Ge/V"

This middle 1960's experiment was designed in an afternoon. I had already moved from the University of Michigan to the Stanford Linear Accelerator Center. Michael Longo from Michigan and I had asked the Bevatron Program Committee for time to study very small angle $\pi^- - p$ elastic scattering in the angular region when coulomb scattering and diffractive scattering interfere.

On that afternoon we were talking about this proposed experiment but we were dissatisfied with it. I don't remember the precise sources of our dissatisfaction. I think we were worried that our experimental design was faulty and we were not sure we could reach the accuracy we wanted. We began to look for another idea and realized that we could do an accurate and broad study of high energy neutron-proton elastic scattering using a new technique.

There were no previous measurements of high energy $n - p$ scattering because no one knew how to make a single energy neutron beam. We didn't know how to make one either, but we realized we didn't need one! We proposed to use an incident neutron beam of all energies, scattering on a hydrogen target. We would measure the angles and momentum of the scattered proton and we would measure the angles and energy of the scattered proton and we would measure the angles and energy of the scattered neutron. The scattered neutron energy would be obtained from the interaction of the neutron and a set of thick plate optical spark chambers, an early form of a hadron calorimeter. We went before the Bevatron Program Committee and they accepted our new proposal. Proposing experiments was simpler in those days.

The experiment was led by Michael Kreisler as his Stanford PhD. thesis. As you can see from the author list there were just five of us on the experiment, but now we had help from four technicians including, once again, Orman Haas from Michigan.

I was very happy with the efficiency of the experiment and the precision of this data. This was my only experiment in which we accumulated more data than we analyzed. Figure 2 is beautiful; in one pioneer experiment we covered broad ranges in energy and scattering angle and we found the "Regge pole shrinkage" in the diffraction peak. I liked and still like our discovery of a flat $d\sigma/dt$ cross section at 90°.

With a larger group of physicists, eight authors, we went on to repeat the experiment at the AGS at Brookhaven for neutrons in the momentum range of 8 to 30 GeV/c (B. G. Gibbard *et al.*, *Phys. Rev. Lett.* **24**, 22 (1970)). Again beautiful data, but now we did not have the statistics to go to large angles. The largest value of $|t|$ was about 1.2 $(GeV/c)^2$.

My colleagues, Michael Longo and Lawrence Jones, went on to extend this technique to larger angles and higher energies. In their research at Fermilab reported in C. E. DeHaven, Jr., *et al.*, *Phys. Rev. Lett.* **41**, 669 (1978), they attained 360 GeV/c neutron energies and demonstrated interesting structure in the neutron-proton differential cross section.

This was as high as the technique could read in energy. Much higher energy proton-proton and proton-antiproton elastic cross sections have been studied but only by using colliders.

NEUTRON-PROTON ELASTIC SCATTERING FROM 1 TO 6 GeV*

M. N. Kreisler,†‡ F. Martin, and M. L. Perl
Stanford University, Stanford, California

and

M. J. Longo and S. T. Powell, III
University of Michigan, Ann Arbor, Michigan
(Received 19 May 1966)

In this Letter we are reporting the first high-energy measurements (1- to 6.3-GeV kinetic energy) of neutron-proton elastic scattering extending from the small-angle, diffraction-peak region to the region beyond 90° in the center-of-mass system. Previous high-energy measurements[1,2] have concerned only elastic neutron-proton scattering near 180° in the so-called charge-exchange backward-peak region. This experiment was carried out at the Bevatron of the Lawrence Radiation Laboratory and used a neutron beam, spark chambers, and a liquid-hydrogen target. There were three objectives in this experiment: (1) to verify the existence of the expected but hitherto unobserved diffraction peak, to determine its parameters, and to investigate possible shrinkage; (2) to examine the differential cross section at and beyond 90° in the center-of-mass system, a region inaccessible in proton-proton scattering; (3) to look for the secondary forward peak which appears in pion-proton elastic scattering[3,4] but not in proton-proton elastic scattering.

The experiment involved a new technique using a neutron beam containing neutrons of all energies up to 6.3-GeV kinetic energy. Neutrons, produced by the external proton beam of the Bevatron hitting a beryllium target, were formed into a beam by a 15-foot-long lead collimator set at 1° to the proton-beam direction. Bending magnets removed charged particles from the beam and lead plates reduced gamma-ray contamination. From analysis of the elastic events, the neutron spectrum was found to peak at 5.0 GeV and two-thirds of the neutrons which gave events had energies above 4.0 GeV. Thus, this is a high-energy beam and, in fact, the spectrum was more favorable than expected.

The neutron beam with a diameter of 1.25 inches interacted in a 12-inch-long hydrogen target as shown in Fig. 1. A system of thin-plate spark chambers and a magnet were used to detect the recoil proton from the elastic scattering and to measure its angle and momentum. A set of seven spark chambers with $\frac{3}{16}$-inch-thick stainless-steel plates was used to detect the scattered neutron by its interactions. The interaction or conversion of the neutron appeared as a neutron star of one or more prongs. The proton-detecting system and the neutron-detecting chambers were both on a circular rail centered on the hydrogen target. With seven different settings of their positions, all scattering angles at all energies above 1 GeV were covered. The spark chambers were triggered when a set of long, horizontal, scintillation counters interspaced among the neutron chambers and two long, horizontal, scintillation counters, P_1 and P_2, in the proton system indicated that an approximately coplanar event had occurred.

The angle of the incident neutron was known to ±0.2 deg and the angle of the scattered neutron, determined by the line from the interaction point in the target to the conversion point in the neutron spark chambers, was known to ±0.5 deg. These angles, combined with the angle and momentum of the recoil proton, overdetermined an elastic scattering, and yielded the incident neutron energy. The energy dependence of the conversion efficiency of the neutron spark chambers was measured by setting up the apparatus for small-angle scattering and triggering only with the proton-system counters. At small angles the recoil proton angle and momentum was sufficient by itself to determine an elastic scattering. The fraction of events which showed neutron conversion in the neutron chambers then gave directly the conversion efficiency. This efficiency was

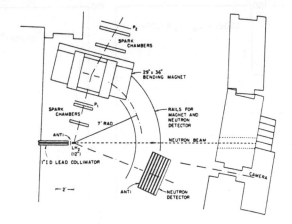

FIG. 1. Layout of experimental apparatus.

62% at 6 GeV and dropped to 45% at 2 GeV.

Corrections have been applied to the data for angular bias in the spark chambers, multiple scattering in the hydrogen target, small-angle cutoffs, and inelastic contamination. The relative normalization between the different settings was obtained by two sets of scintillation counters which measured the scattered charged-particle flux from the hydrogen target. No absolute normalization was available from the experiment itself. We have normalized the data by fitting the small-angle regions with an exponential in t, the square of the four-momentum transfer from the incident to the scattered neutron, and by using the optical theorem and the neutron-proton total cross sections.[5] We took the real part of the forward-scattering amplitude to be zero.[6]

For presentation we have grouped the events into ranges of incident-neutron kinetic energy. The data presented are based on 6219 elastic events which represent about 15% of our available data. Figure 2 is a semilogarithmic plot of the differential cross sections $d\sigma/d|t|$ vs $|t|$ [$|t|$ is expressed in $(\text{GeV}/c)^2$].

In terms of the center-of-mass scattering angle θ^* and the center-of-mass momentum p^*,

$$|t| = 2p^{*2}(1-\cos\theta^*)$$

and

$$d\sigma/d|t| = (\pi/p^{*2})d\sigma/d\Omega^*.$$

The cross sections as a function of $\cos\theta^*$ may be computed from $d\sigma/d|t|$ using the average values of p^* for each incident energy range

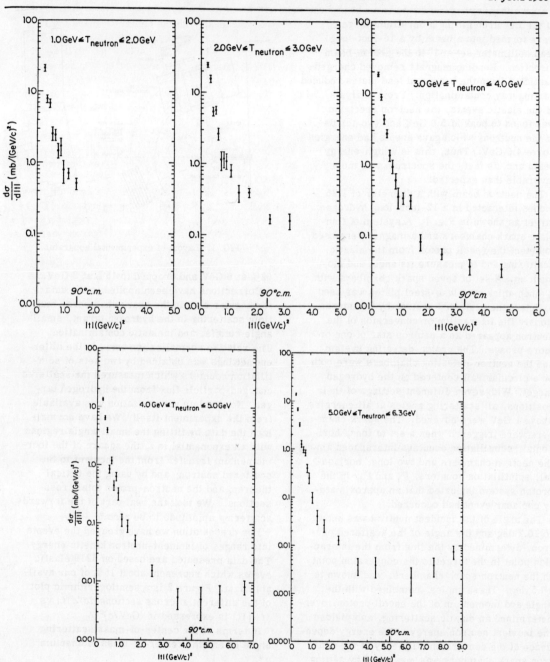

FIG. 2. Neutron-proton differential cross sections versus $|t|$.

shown in Table I.

We first observe that, as expected by our general understanding of high-energy elastic scattering in the presence of a large inelastic cross section, there is a strong diffraction peak at all energies. The peak has a roughly exponential behavior.

We have fit the region $0.2 < |t| < 0.6$ $(GeV/c)^2$ with the form $d\sigma/d|t| = A \exp(b|t|)$, and b is given in Table I. According to recent data of

Table I. Values of the slope of the diffraction peak for each energy interval.

Incident kinetic energy range (GeV)	p^* average (Momentum in center of mass) (GeV/c)	b [(GeV/c)$^{-2}$]
1.0–2.0	0.851	−6.321 ± 0.647
2.0–3.0	1.096	−5.527 ± 0.463
3.0–4.0	1.287	−6.655 ± 0.432
4.0–5.0	1.460	−7.720 ± 0.411
5.0–6.3	1.612	−7.562 ± 0.391

Clyde et al.,[7] the values of $-b$ for proton-proton scattering at 2.2, 4.1, and 6.2 GeV are 6.50 ± 0.03, 7.44 ± 0.04, and 7.69 ± 0.04 (GeV/c)$^{-2}$. The neutron-proton and proton-proton diffraction-peak slopes have about the same values except perhaps at the lower energies. The slopes in the energy region from 2 to 6 GeV indicate a shrinkage of the diffraction peak quite similar to proton-proton scattering.

The following observations may be made on the large-angle region. The differential cross section deviates from exponential and begins to flatten out as $\theta^* = 90°$ is approached. It is roughly flat, that is, isotropic near 90°, and the minimum in the differential cross section is at or just beyond 90°. The isotropy near 90° is predicted both by the statistical model[8] and by the model of Wu and Yang.[9] Beyond 90°, $d\sigma/d|t|$ increases even though the values of $|t|$ are very large. This leads to the idea that t is no longer the meaningful parameter because the neutron and proton are exchanging their charges, and u (the square of four-momentum transfer from the incident neutron to the recoil proton) is the relevant parameter. Since $|u| = 4p^{*2} - |t|$, $|u|$ is decreasing as θ^* approaches 180°. In this experiment we used a slightly different technique to measure the region near 180°, but that data are not completely analyzed yet. Comparison with other data[1,2] near 180° indicates that our differential cross section will rise roughly monotonically into the charge-exchange peak. However, in the backward-angle regions presented in Fig. 2, $d[\ln(d\sigma/d|u|)]/d|u|$ is about 0.6 (GeV/c)$^{-2}$ compared to 40 or 50 (GeV/c)$^{-2}$ at 180°.

The statistical model predicts an exponential decrease with the center-of-mass total energy W^* of the form $(d\sigma/d\Omega^*)_{90°} = A\exp(-gW^*)$. We found g to be 3.73 ± 0.38 (GeV)$^{-1}$. If we fit the decrease with a power of W^*, namely $(d\sigma/d\Omega^*)_{90°} = CW^{*-N}$, then $N = 11.04 ± 1.15$. Finally, Clyde et al.[7] give the proton-proton differential cross section at 90° as 0.45 ± 0.01, 0.016 ± 0.0009, and 0.000 78 ± 0.000 04 mb/(GeV/c)^2 at 2.2, 4.1, and 6.2 GeV.[10] Interpolation of our data yields 0.32 ± 0.05, 0.017 ± 0.006, and 0.0014 ± 0.0008 mb/(GeV/c)^2 at these energies.

We find no clear evidence for a second forward peak in the neutron-proton system. There are some ambiguous indications as can be seen from Fig. 2, which should be resolved when the remainder of our data are analyzed.

We are very grateful to Mr. T. L. R. Elder, Mr. Orman Haas, Mr. James Moss, Mr. Ronald Seefred, and to the Bevatron staff for their very valuable assistance during the setup and execution of the experiment.

―――――

*Work supported in part by the U. S. Atomic Energy Commission and by the U. S. Office of Naval Research Contract No. NONR 1224(23).

†National Science Foundation Predoctoral Fellow.

‡This paper constitutes part of the thesis of M. N. Kreisler to be submitted in partial fulfillment of the requirements for the degree of Doctor of Philosophy, Stanford University.

[1]J. L. Friedes, H. Palevsky, R. L. Stearns, and R. J. Sutter, Phys. Rev. Letters 15, 38 (1965).

[2]G. Manning et al., Nuovo Cimento 41, 167 (1966).

[3]D. E. Damouth, L. W. Jones, and M. L. Perl, Phys. Rev. Letters 11, 287 (1963).

[4]C. T. Coffin et al., Phys. Rev. Letters 15, 838 (1965).

[5]D. V. Bugg et al., Rutherford High-Energy Laboratory Report No. RPP/H/13, 1966 (unpublished).

[6]L. Kirillova et al., abstract presented at Proceedings of the Oxford International Conference on Elementary Particles, Oxford, England, 1965 (Rutherford High-Energy Laboratory, Chilton, Berkshire, England, 1966); Kh. Chernev et al., Joint Institute for Nuclear Research Report No. JINR-E-2413, 1965 (unpublished). The data are consistent with the real part being zero.

[7]A. R. Clyde et al., University of California Radiation Laboratory Report No. UCRL-11441, 1964 (unpublished); A. R. Clyde, private communication.

[8]G. Fast, R. Hagedorn, and L. W. Jones, Nuovo Cimento 27, 856 (1963).

[9]T. T. Wu and C. N. Yang, Phys. Rev. 137, B708 (1965).

[10]The 6.2-GeV value was actually at $\theta^* = 82.3°$.

COMMENTS ON C7
"A High Energy, Small Phase-Space Volume Muon Beam"

In Part A of this book I described my work on muon-proton inelastic scattering. This search for differences between the muon and the electron, by its failure, led to my discovery of the tau lepton. Here I have reprinted the article describing the muon beam itself. It was a neat design.

A 17 GeV primary electron beam from the SLAC linear accelerator hit a water-cooled copper target producing muon pairs via bremsstrahlung and pair production. The target was immediately followed by a 5.5 m long beryllium filter to remove pions. The filter was made of 30 cm long sections, each section had an 18 cm diameter. This was, and still is, a lot of beryllium. We obtained it from the Rocky Flats National Laboratory and after the muon experiments were done, we lent it to other high-energy laboratories.

The size and complexity of the muon beam made these muon-proton experiments my first big engineering experiment. It was a taste of the even bigger engineering jobs that would be required to build electron-positron annihilation detectors. Since my first career love was engineering, this was a pleasant return to earlier interests. More than pleasant, there is a great thrill in bringing a large apparatus from the first sketches through to construction and operation.

NUCLEAR INSTRUMENTS AND METHODS 69 (1969) 77-88; © NORTH-HOLLAND PUBLISHING CO.

A HIGH ENERGY, SMALL PHASE-SPACE VOLUME MUON BEAM*

J. COX, F. MARTIN, M. L. PERL, T. H. TAN, W. T. TONER and T. F. ZIPF

Stanford Linear Accelerator Center, Stanford University, Stanford, California, U.S.A.

and

W. L. LAKIN

High Energy Physics Laboratory, Stanford University, Stanford, California, U.S.A.

Received 15 July 1968

The design and performance of a high-momentum, small phase-space volume muon beam is described. The muons are photoproduced by the bremsstrahlung of electrons incident on a thick target. Pion contamination is reduced to 3×10^{-6} of the muon flux by a 5.5 m beryllium filter placed immediately after the target. The optical system has two stages of momentum analysis to give an almost dispersion free beam, and a final imaging stage to clean up the tails of the beam distribution. Standard BNL 8Q48 and 18D72 magnets are used throughout. At 10 GeV/c, the beam yields 1.0×10^5 $\mu^\pm$/sec in a phase-space volume of 3×10^{-3} cm$^2 \cdot$ sterad and a momentum band of $\pm 1.5\%$.

1. Introduction

At the Stanford Linear Electron Accelerator, muons are produced copiously when an electron beam strikes a thick target. The production process is electromagnetic pair production by the bremsstrahlung of the electrons in the target. The bulk of the production takes place in the first four radiation lengths, and the source of muons has a cross section similar to that of the incident electron beam, of the order of 5 mm × 5 mm. It is therefore possible to make a muon beam with optical properties similar to those of the high energy beams of strongly interacting particles common at proton accelerators. This may be contrasted with the situation at a proton synchrotron, where the muons are the product of the decay in flight of π-mesons contained in a beam. There, the source of muons has the dimensions of the pion beam, of the order of 10 cm × 10 cm, and extends over a length of several tens of meters.

In this paper, we describe a beam which has been built at SLAC for a high energy muon scattering experiment. The muons are produced in a water-cooled copper target and then pass through a beryllium filter 5.5 m long, placed immediately after the target. The π/μ ratio in the beam is reduced to 3×10^{-6} by the filter. Multiple Coulomb scattering of the muons in the filter has the effect of making them appear to come from a source with a diameter of about 2.5 cm close to the beginning of the filter. The beam transport system which follows has two stages of momentum analysis to give an almost dispersion free beam, and a final imaging section to clean up the tails of the beam

distribution and to shape the beam at the experimental target. With 100 kW of 17 GeV electrons incident on the production target, the beam yields 1.0×10^5 μ/sec at 10 GeV/c in a momentum band of $\pm 1.5\%$. 90% of the beam is contained within an area of 5 cm × 10 cm, 99% within an area of 10 cm × 10 cm.

In section 2 we discuss muon production in the target and the target design, in section 3 the effects of the filter needed to remove strongly interacting particles, in section 4 the optics of the beam transport system, and in section 5 the parameters of the beam and its uses. In section 6 we discuss some possible improvements to the beam.

2. Muon production and target design

The principal process by which electrons produce secondary particles in a thick target takes place in two steps. First, an electron radiates in the field of a nucleus. The secondary particles are then photoproduced at another nucleus in the target by the bremsstrahlung. The direct electroproduction reaction $e^- + \text{nucleus} \rightarrow e^- + \text{nucleus} + \mu^+ + \mu^-$ can be described as photoproduction by virtual photons whose spectrum is equivalent to the real bremsstrahlung which would be produced by the electron in a target of 0.02 radiation lengths[1]. It can therefore be neglected. The calculation of the muon yield from a thick target involves the integration of the Bethe-Heitler cross section for the pair production over the thick target bremsstrahlung spectrum, taking account of the nuclear form factor and of pair production from individual nucleons. These calculations have been made by Tsai and Whitis[2,3]. Some results of their calculations for 18 GeV electrons incident on a 10 radiation length

---

* Supported by the U.S. Atomic Energy Commission.

Fig. 1. Muon production by 18 GeV electrons in a 10 radiation length target as a function of angle. The production angle is in units of $\alpha = m_\mu/E_\mu$. Common curve for $p_\mu \leq 12$ GeV/c (a); $p_\mu = 15$ GeV/c (b); $p_\mu = 17$ GeV/c (c); $[1+(\theta/\alpha)^2]^{-2}$ (d) and $\exp\{-\tfrac{1}{2}[\theta/(0.57\,\alpha)]^2\}$ (e).

copper target are shown in figs. 1 and 2. In fig. 1, the muon flux relative to the flux at 0° is plotted as a function of angle, with the angle in units of m_μ/E_μ. Multiple Coulomb scattering in the target has not been included. In the case of 4, 8 and 12 GeV/c muons, the points lie very closely on a common curve. For 15 and 17 GeV/c muons, the fluxes at large angles are suppressed. Also shown in the figure is the function $\{1+(\theta/\alpha)^2\}^{-2}$, and a gaussian, $\exp\{-\tfrac{1}{2}[\theta/(0.57\,\alpha)]^2\}$, chosen so as to fit the common curve at 0° and at the half maximum point. The first gives an overestimate, the second an underestimate of the large angle production. For the $\{1+(\theta/\alpha)^2\}^{-2}$ distribution, the total flux is $\pi(m_\mu/E_\mu)^2$ times the flux at 0°, and half of this is contained within an angle of m_μ/E_μ. For a 10.5 GeV/c muon, m_μ/E_μ is 10 mrad, an angle typical of the acceptance of a high energy beam transport system using conventional 8" bore quadrupoles. We can therefore expect to capture a significant fraction of all muons produced into the beam. In fig. 2 the muon flux at 0° and the total production, integrated over the angle, are shown, plotted versus muon momentum. The total muon production at 10 GeV/c for 3.5×10^{13} electrons/sec (100 kW of beam power) is 8.7×10^5 μ/sec/% · · $\delta p/p$. Therefore, the production target must be designed to dissipate large amounts of power, if we are to have an intense beam of muons. The muon yield from a thick target depends very little on the target material. However, pions are also produced. The pion production process is a nuclear interaction of the bremsstrahlung with a cross section which is roughly proportional to A. Therefore, the number of g/cm² per radiation length should be kept small, particularly in the first few radiation lengths, where particle production is a maximum. These considerations make a water-cooled copper target an obvious choice. Gold would be better still. A simplified diagram of the target is shown in fig. 3. It is 11.6 radiation lengths thick. This is sufficient to absorb about 70% of the power in the electron beam, and to diffuse the cascade so that its diameter when it leaves the target is ≈ 3 cm. The beam power per unit area is then sufficiently small to be handled in the beryllium filter which follows.

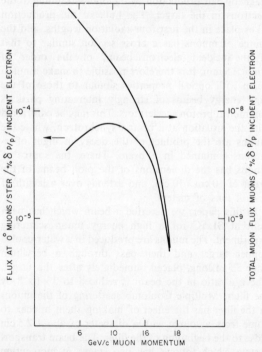

Fig. 2. Muon production by 18 GeV electrons in a 10 radiation length target as a function of momentum. The lower curve (left hand scale) is the flux at 0°. The upper curve (right hand scale) is the lower curve multiplied by $\pi(m_\mu/E_\mu)^2$, and represents approximately the total muon flux.

A HIGH ENERGY, SMALL PHASE-SPACE VOLUME MUON BEAM

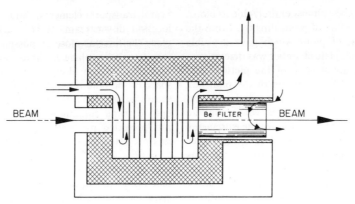

Fig. 3. Schematic diagram of water-cooled production target.

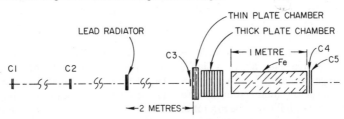

Fig. 4. Schematic diagram of arrangement used to measure π/μ ratio.

3. Pion filter

3.1. Pion contamination

Some preliminary measurements showed that strongly interacting particles, mostly pions, are produced in the copper at the rate of approximately one for every two muons. The interaction cross sections for pions and muons differ by more than three orders of magnitude. To reduce the contamination of pion induced events in an experiment to less than one percent, it is, therefore, necessary to attenuate the strongly interacting particles by a factor of more than $\times 10^5$, the precise value depending upon the particular experiment. More than 11 attenuation lengths of filter are required. The filter is placed immediately after the production target so as to keep the effects of Coulomb scattering of the muons, discussed later, to a minimum. All forward produced strongly interacting particles make a cascade in the filter and we must estimate the number which emerge relative to muons of the same energy.

Studies have been made[4] of the attenuation of monoenergetic protons in shielding material. Measurements of the particle flux as a function of depth show that, at first, the flux of particles increases. Then, after about two mean free paths, it begins to be attenuated exponentially. The conditions in the present case are quite different, although the qualitative behaviour should be similar, and so measurements of the pion flux and attenuation were made in a region where it was expected to be already exponential. A schematic diagram of the apparatus used is shown in fig. 4. The incident beam is defined by counters 1, 2 and 3. They are followed by a four gap thin plate spark chamber, a spark chamber with nine 2.5 cm thick iron plates, an iron absorber 1 m thick and anticoincidence counters 4 and 5. The anticoincidence counters were made big enough so that multiply-scattered muons, or muons which had interacted in the iron would still be registered. About 2 m upstream of the spark chambers, four radiation lengths of lead were placed in the beam, so that electrons from muon decay would produce a characteristically spread out shower in the first spark chamber. Electrons from the target were reduced to a negligibly low level by placing a filter of two radiation lengths of lead at the first focus in the beam. This technique has been described already[5]. The signal 1 2 3 $\bar{4}$ $\bar{5}$ was used to trigger the spark chambers. Almost all pions would give a trigger signal and some definite proportion of them would give a visible interaction in the thick plate spark chamber. The rest would look like muons. Measurements were made with filter lengths of 2.14, 3.35 and 4.88 m of Be. With 2.14 m of Be, the trigger rate was almost entirely due to pions, and this rate gave the flux of pions directly. From the pictures, the fraction of pions which gave a visible interaction in the thick plate chamber was determined. With 3.35 m of Be, about 30% of the pictures had electron-like events. The attenuation of the pion flux produced by the additional 1.2 m of Be was determined by the relative number of pictures with pion-like interactions in the two cases. Counter measurements were made with a 4.88 m filter. In this case, the trigger rate was almost entirely due to electrons, and was consistent with the number of electron-like events observed with 3.35 m of Be. The results of this study are as follows: With 3.35 m of Be the π/μ ratio is $(3.2\pm0.4)\,10^{-4}$, and the attenuation length of Be is $0.47^{+0.03}_{-0.02}$ m. In the geometry we used, $(2.6\pm0.2)\,10^{-4}$ electrons/μ from μ decay were counted. These measurements were made with 8 GeV/c μ^- from 16 GeV/c e^-. With the 2.14 m filter, measurements were made at other momenta. They showed that above 4 GeV/c the π/μ ratio did not change significantly with momentum. Therefore, the change in momentum due to dE/dx loss in the filter does not influence the π/μ ratio. Our present filter is 5.5 m of Be, giving a value of $(0.30^{+0.13}_{-0.09})\times 10^{-5}$ π/μ in the beam. This is adequate for the experiments which are at present being carried out, since they incorporate a steel shield through which the scattered muons must pass, which gives a further rejection against pion-induced events.

3.2. Electron contamination

The primary electrons incident on the target very rapidly lose their energy by bremsstrahlung. The bremsstrahlung is then attenuated at a rate of $\exp(-\tfrac{7}{9}t/x_0)$. Conversion of these photons in the last radiation length of the filter leads to a contamination of electrons in the beam. This may be estimated from the results of[2]. With a total thickness of target and filter of 27.1 radiation lengths, the electron contamination is approximately 0.3% of the muon flux.

3.3. Multiple Coulomb scattering. Choice of beryllium

The source of muons is characterized by the size of the incident electron beam and the production angular distribution folded in with the multiple Coulomb scattering in the ten radiation length target. The muons then undergo further multiple Coulomb scattering in the filter. Since it extends over a finite distance from the source, the scattering in the filter broadens the apparent source seen by the subsequent beam transport elements, and shifts its apparent location downstream. It is clearly advantageous to place the filter as close as possible to the production target, or to a plane in which the target is imaged. The effect of the filter on the distribution in phase space of the muons can be calculated in a straightforward manner. We project all angles and displacements onto a plane containing the beam. Let θ_prod be the rms projected production angle (rad) and θ_scatt the rms projected angle of scattering in the target. Then $\theta_S = (\theta^2_\text{prod}+\theta^2_\text{scatt})^{\frac{1}{2}} =$ rms projected angle of muons emerging from the target. We have seen in section 2 that a gaussian approximation to the flux at production gives an underestimate of the large angle flux, and $\{1+(\theta/\alpha)^2\}^{-2}$ gives an overestimate. We take the mean of the values of the rms projected angle from these two distributions as a reasonable approximation for the treatment which follows. Its value is $(\theta^2_\text{prod})^{\frac{1}{2}} = 0.8\,m_\mu/E_\mu$ rad. The energy dependence is the same as for Coulomb scattering, and it is instructive to calculate the thickness of scatterer which would give an rms projected scattering angle of the same size. It is $(0.084/0.015)^2 = 31$ radiation lengths. The 8 or so radiation lengths of copper traversed by the muons after they are produced do not, therefore, seriously reduce the fraction of muons which can be captured in the beam transport system. We can write $\theta_S = (0.015/p)\,(39)^{\frac{1}{2}}$ rad, with p in GeV/c. The target is compact enough so that it is effectively located at the source and does not appreciably broaden the beam. The filter is much less compact and does have a serious effect on the phase space occupied by the muons. Let θ_R be the rms projected angle of scattering in the filter, and Z the length of the filter (in m). Starting from the distribution function $P(Z, y, \theta)^{6)}$, which is the probability that a particle incident on the filter along the Z-axis has a displacement y and angle θ at depth Z, and folding in a gaussian source distribution and emittance, it can be shown (after some tedious algebra) that the muons appear to come from a source a distance

$$L = Z/\{2(1+\theta^2_S/\theta^2_R)\},$$

downstream of the target, with an rms projected angle

$$\theta_\text{rms} = (\theta^2_S+\theta^2_R)^{\frac{1}{2}}.$$

The apparent source has a width

$$\sigma = [\sigma^2_S + \tfrac{1}{12}\theta^2_R Z^2\{(\theta^2_R+4\theta^2_S)/(\theta^2_R+\theta^2_S)\}]^{\frac{1}{2}},$$

where σ_S is the rms projected width of the incident electron beam. The density of particles per unit area

TABLE 1
Properties of filters made of different materials.

Material	Length*			Values for 10 GeV/c muons			
	Collision lengths	Radiation lengths	(m)	Energy loss (GeV)	Displacement of apparent source	σ (cm)	D^2
Li	18	11.5	17.0	1.5	2.3	4.5	108.0
Be	18	15.5	5.4	1.6	0.9	1.7	16.4
C	18	25.6	7.2	1.9	1.7	2.6	50.0
Al	18	59.0	5.3	2.3	1.9	2.6	75.0
Cu	18	158.0	2.1	2.7	1.0	1.5	55.0
U	18	390.0	1.4	3.2	0.7	1.4	111.0

σ_S = rms width of incident electron beam, taken as 0.5 cm.
σ is the rms width of the apparent source of muons.
D^2 is the factor by which the density in 4-dimensional phase space of the muons is reduced by the filter.
* Data from UCRL-8030.

TABLE 2
Properties of Be filters as function of length and muon momentum.

Momentum of muons	Length*			π/μ ratio	Energy loss (GeV)	σ (cm)	D^2
	Collision lengths	Rad. lengths	(m)				
12	18	15.5	5.4	3×10^{-6}	1.6	1.4	8.9
8						2.1	25.0
4						3.9	95.0
10	22	19.0	6.6	0.24×10^{-6}	1.9	2.2	30.0
	18	15.5	5.4	3×10^{-6}	1.6	1.7	16.4
	14	12.0	4.2	38×10^{-6}	1.2	1.2	8.2

* Data from UCRL-8030.

of (θ, y) phase space before the filter is proportional to $(\theta_S \sigma_S)^{-1}$ and after the filter to $(\theta_{\rm rms} \sigma)^{-1}$.

The dilution due to the filter is thus

$$D = \{1 + (\theta_R^2/\theta_S^2)\}^{\frac{1}{2}} \cdot$$
$$\cdot [1 + (\tfrac{1}{12} \theta_R^2 Z^2/\sigma_S^2) \{(\theta_R^2 + 4\theta_S^2)/(\theta_R^2 + \theta_S^2)\}]^{\frac{1}{2}}.$$

In the four-dimensional phase-space y, θ, x, ϕ, the dilution is just D^2. In the derivation of these expressions, the energy loss of the muons in the filter has been neglected. This may be taken into account exactly in the case of the angles by using the momentum at production to calculate θ_S and by using the square root of the product of the momenta before and after the filter as the momentum in the expression

$$\theta_R = (0.015/p)(Z/x_0)^{\frac{1}{2}}.$$

Substitution of these values in the formulae for σ, L and D^2 still does not allow for the shift downstream of the scattering. To take this into account approximately, we observe that the point in the filter at which the mean square angle of scattering reaches half its final value is $\tfrac{1}{2} Z$ in the case of no energy loss, and $\tfrac{1}{2} Z \{1 - (\tfrac{1}{2} \Delta p/p_0)\}^{-1}$ when the initial momentum is p_0 and the final momentum $p_0 - \Delta p$. We use this value for $\tfrac{1}{2} Z$ in the formula for L, and $Z \{1 - (\tfrac{1}{2} \Delta p/p_0)\}^{-1}$ for Z in the expressions for σ and D^2, together with the corrected angles θ_S and θ_R. In table 1, values of D^2 are given for various materials with a length containing the same number of collision lengths as our 5.5 m Be filter. Beryllium is clearly favored. Values are also given for θ_S, $\theta_{\rm rms}$, and L. Table 2 gives values of D^2 and σ for Be as a function of the length of the filter.

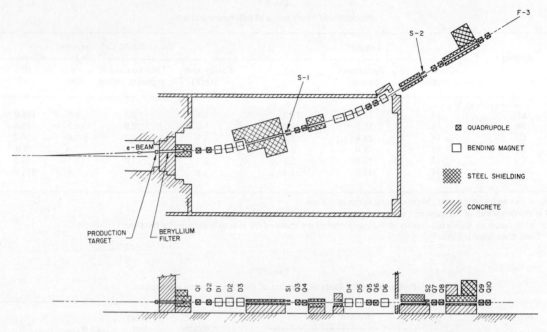

Fig. 5. Floor layout of muon beam.

The advantages in keeping the length of the filter to a minimum are obvious.

3.4. MECHANICAL DESIGN OF THE FILTER

The filter is made of vacuum cast cylinders of Be, each 18 cm dia. and 30 cm long. They are encased in 30 cm × 30 cm square castings of lead, made up in 0.6 m, 1.2 m and 1.8 m sections. End windows of aluminum seal the beryllium. The first 0.3 m of the filter is attached to the target. The beryllium is water-cooled to remove the power of up to 30 kW contained in the electron-photon cascade emerging from the target.

4. Optical design

The floor layout of the beam and particle trajectories through the system are shown in figs. 5 and 6. The effective source of the muons is a short distance downstream of the beginning of the target, as discussed in section 3 above. The beam is at $0°$ to the direction of the incident electrons. The first two lenses, Q1 and Q2, form an image of the source in both planes at the first slit, S1. Lead collimators in these lenses define the aperture of the beam. The magnification of the source at S1 is ×1.08 in the horizontal and ×4.1 in the vertical plane. Bending magnets D1, D2, and D3 deflect the beam through a total angle of 12.9°, to give a dispersion at S1 of 3.2 cm per percent. Q3 and Q4 are field lenses. Q5 and Q6, together with D4, D5 and D6, give an almost dispersion free image of the source in both planes at S2. The magnification of the source at S2 is ×0.93 in the horizontal and ×3.66 in the vertical plane. Q7 and Q8 are field lenses. Q9 and Q10 can be varied to produce different beam configurations in the experimental area. They are usually set to make an image in both planes of the source at the downstream end of the hydrogen target used in the muon scattering experiment. The basic optical parameters of the beam are listed in table 3. The effects of chromatic aber-

TABLE 3
Optical parameters of the beam.

Position	Distance from target (m)	Horizontal plane		Vertical plane
		Magnification	Dispersion	Magnification
S1: momentum slit	30	1.08	3.25 cm/%	4.1
S2: second focus	65	0.93	0.45 cm/%	3.7
F3: third focus	88	0.72	0.35 cm/%	3.4

A HIGH ENERGY, SMALL PHASE-SPACE VOLUME MUON BEAM

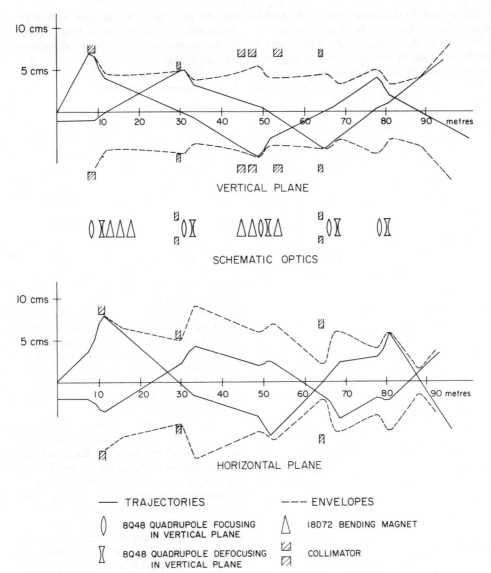

Fig. 6. Particle trajectories and beam envelopes. In the horizontal plane the beam envelope is approximately that of the 10 GeV/c beam with a $\pm 1.5\%$ momentum bite.

rations on the beam are negligible. The most important features of the beam design are as follows:

1. The beryllium filter is placed as close to the production target as possible.
2. The aperture of the beam is defined very early, at the first lenses. The beam thus defined is kept well inside the aperture of all subsequent magnets, lenses and collimators.
3. The vertical extent of the effective source is redefined by S1. It is 10 cm high. The vertical opening of S2 is made significantly bigger than the image of S1 at S2, so that the main beam is not touched by S2.
4. The horizontal extent of the effective source is determined by the source dimensions, modified by Coulomb scattering in the Be filter, and by S2. S2 is made wider than the beam image at this point so that, again, only the tail of the beam can be scattered here.
5. The final beam is re-formed after S2 by Q7, Q8,

Q9 and Q10 so that the scattering target of the experiment is well removed from any heavy collimators.

The definition of the beam aperture and images is determined exclusively by multiple Coulomb scattering and a small amount of dE/dx loss in the collimators. The heavy shielding shown around the beam is to protect against muons which do not pass through the optical channel by bringing them to rest. The small volume in phase-space occupied by the beam, and the fact that the collimators are placed where the angular spread of particles in the beam is small, contribute to the effectiveness of the Coulomb scattering. Slits S1 and S2 are made of steel and are each 1 m long. They form closed iron circuits around the beam and can be magnetized so as to deflect any particle entering the steel further from the beam. In 1 m of iron at 15 kG, particles are deflected by the field through 4 times the rms projected angle of multiple scattering. Tests showed that the effect of magnetizing S1 was not measurable. On the other hand, magnetizing S2 reduced the background trigger rate in the muon experiment by a factor of two.

5. Performance of the beam

5.1. Beam profiles; Momentum bite

The intensity distribution of muons at the effective source, discussed in section 3, is approximately gaussian. The width of this distribution will determine the momentum resolution of the beam. If we neglect chromatic aberrations, the displacement in the horizontal plane of a trajectory at S1 is given by

$$x_1 = M_{01} x_0 + D \delta p/p,$$

where x_0 is the displacement at the source, M_{01} is the magnification of the source at S1, D is the dispersion and δp is the departure from the nominal beam momentum, p. For an infinitesimal slit, $x_1 = $ constant, and the momentum distribution of the transmitted beam will have the shape of the intensity distribution at the source, suitably scaled. As the width of the slit increases, the width of the momentum distribution increases. When S1 is wide enough to pass the whole of the source image at the reference momentum, the full width at half maximum intensity of the momentum distribution becomes W/D. Typical distributions are shown in fig. 7b.

At S2, the beam is sufficiently dispersion free that we can write $x_2 = M_{02} x_0$, independently of the momentum of the particles. Some reflection should be sufficient to convince the reader that the shape of the intensity distribution in the horizontal plane at S2 is independent of the width of the momentum slit, and is the same as that of the source distribution, scaled by the magnification. We may therefore deduce the shape of the effective source distribution and of the momentum distribution from horizontal scans of the beam image at S2. A point at S2 corresponds to a point at the source. The momentum distribution of particles at one point in the image of S2, therefore, corresponds to the momentum distribution which would be obtained with a point source. It is a square distribution with a full width of W/D. At the third focus, F3, the intensity distribution and momentum distribution will be the same as at S2, when account is taken of the magnification, except that the particles in the tails of the distribution are scattered and bent out of the beam by S2. Fig. 7a shows horizontal beam profiles at S2 taken with S1 set at ± 5 cm wide ($\pm 1.5\%$ $\delta p/p$). Momentum distributions deduced from these profiles

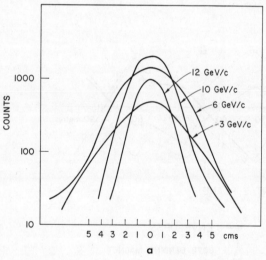

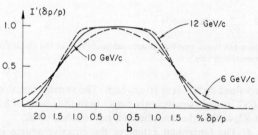

Fig. 7. a. Horizontal beam profiles at the second focus. The ordinate is counts in a 0.63 cm × 0.63 cm counter normalized to the same flux of electrons on the production target for each curve;
b. Momentum distribution in muon beam deduced from the profiles of fig. 7a.

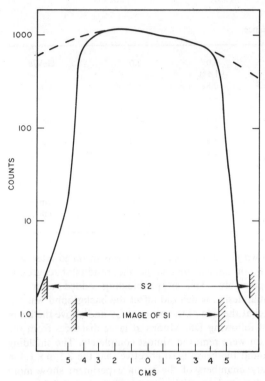

Fig. 8. Vertical beam profile at the second focus for 12 GeV/c muons (solid curve). The 0.63 cm counter width has not been unfolded. The dashed curve is a horizontal beam profile taken with the same beam conditions, with the abscissa expanded by a factor of 4. The limits set by S1 and S2 are indicated.

are shown in fig. 7b. In the vertical plane, the magnification at S1 is $\approx \times 4$, and there is no dispersion, so S1 re-defines the source for the subsequent beam system. Fig. 8 is a vertical scan of the beam at S2. The solid line represents the data. The dashed curve is a horizontal beam profile taken under the same conditions, scaled to the appropriate magnification. The limits set by S1 and S2 are also shown. The effectiveness of the 1 m of iron in S1 in removing muons from the beam is clear. The beam shape in the vertical scan is almost independent of momentum, although the horizontal width increases significantly as the beam momentum is reduced. Fig. 9a, b are horizontal and vertical beam profiles at the third focus for 12 GeV/c muons.

5.2. Background particles and shielding

5.2.1. *Shielding from low energy photons, neutrons, etc.*

The production target is in the B target room at

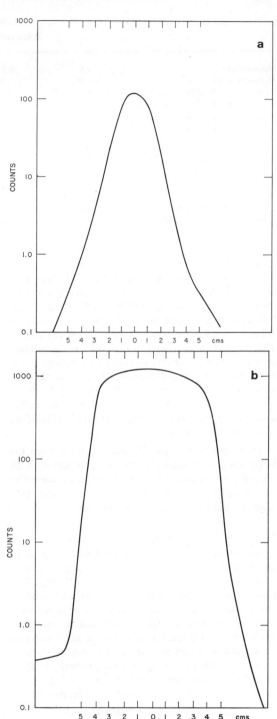

Fig. 9. Horizontal (a) and vertical (b) beam profile at the third focus for 12 GeV/c muons.

TABLE 4
Beam performance.

Momentum	12.0	10.0	6.0	3.0	GeV/c
Momentum bite	±1.5%	±1.5%	±1.5%	~±1.5%	
Flux for 100 kW of 16 GeV electrons	0.31×10^5	0.8×10^5	0.82×10^5	0.29×10^5	μ/sec
Flux for 100 kW of 17 GeV electrons	0.59×10^5	1.03×10^5	—	—	μ/sec
Vertical beam width at S2					
Full width at 10% intensity	9.5	9.8	9.8	10.6	cm
Horizontal beam width at S2					
Full width at 50% intensity	2.9	3.2	5.4	4.8	cm
Full width at 10% intensity	5.3	5.9	8.8	10.9	cm
Vertical beam width at F3					
Full width at 10% intensity	9.8	10.0	—	—	cm
Horizontal beam width at F3					
Full width at 50% intensity	2.4	3.0	—	—	cm
Full width at 10% intensity	4.4	5.1	—	—	cm

SLAC. Along the beam line it is followed by the 5.5 m beryllium filter. This is set in a steel wall, the upstream part of which is 3.7 m thick and completely shuts off the target room from the End Station Building. This is followed by a further 3.7 m of steel 1.8 m wide by 1.2 m high with the beam channel down the center. The radiation background in the end station building is low enough to operate scintillation counters in the beam about 10 m downstream of this shield quite satisfactorily. In view of the 5×10^{-4} duty cycle, this is quite a low level. The experimental area is further shielded by the walls of the End Station Building which are of concrete.

At the position of our experiment, 90 m from the target room, we operate a bank of counters with an area of 2 m^2 without difficulty. We do see evidence of slow neutrons in the few MeV range, but these come after the end of the 1.4 μsec beam pulse.

5.2.2. Shielding against muons

Muon production is so sharply peaked in the forward direction that the bulk of all muons produced pass down the beam channel through Q1 and Q2 to the first bend. There, off-momentum muons are deflected to either side. These bending magnets are followed by a heavy shield long enough to stop muons of up to 12 GeV/c momentum. Muons produced at large angles are of low momentum. The angle-energy relation of muon production works with the range-energy relation to make the maximum lateral excursion of muons in the shielding small and independent of momentum.

In the last leg of the beam there is another 12 m of steel around the beam to stop or scatter any muons which remain outside the optical channel. The shield shown just downstream of Q4 was installed later, and did not appear to change the residual background appreciably. The only shielding added after the initial design which did affect the background was the steel at the side of the beam at Q9 and above the beam just following D6. Muons at large distances from the beam were removed almost completely. The shielding around the beam is so effective that the 2.5 m × 1.4 m spark chambers of the muon experiment show more background particles coming from a collimator 900 m upstream of the muon production target at an angle of the order of 10° to the direction of the muon beam than that they show particles produced in the target room.

5.3. BEAM HALO

The combination of the very low cross section for muon interactions ($\approx$ a few μb for the inelastic cross section) and the 5×10^{-4} duty cycle makes the beam halo one of the most critical features of the performance. At present, the beam halo is sufficiently low for the experiments which are being carried out. About 1% of the total muon flux is in a halo lying just outside the beam envelope, but within a 9 cm radius circle, another 1% or so lies beyond this, extending to a radius of about 75 cm.

With the evidence of fig. 8 of how well a short collimator can define a muon beam, it is clear that the addition of a further stage of beam transport – or possibly just a bend in the last section – could effect a significant reduction in the halo. Ultimately, a muon beam should be better in this respect than a pion beam, since the muons decay only infrequently, and into electrons which are easily distinguishable.

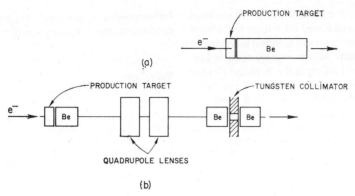

Fig. 10. a. Present arrangement of production target and beryllium filter; b. Possible arrangement of production target and filter using a quadrupole doublet to allow the filter to be concentrated at an image plane and to enable the source of muons to be re-defined

5.4. Summary of beam parameters

The most important features of the beam performance are summarized in table 4. The muon flux peaks at between 8 and 10 GeV/c for 16 GeV electrons incident. The size of the beam shrinks as the momentum is increased. The 0° beam transport system can also be used to give a pion beam, as described in [5]), if the copper target and beryllium filter are removed and replaced by a thin ($\approx$ 1 to 2 radiation lengths) beryllium target.

6. Possible improvements

In the muon-proton scattering experiment which we are performing at present, we are not limited by the beam quality per se, but by knock-on electrons produced in the hydrogen target by the beam muons themselves. The first spark chamber in the experiment is 1.3 m downstream from the end of the liquid hydrogen target. The beam passes through a 12 cm dia. hole in the chamber plates. With the hydrogen target empty, we can run up to 250 muons/1.4 μsec pulse before stray tracks in the chamber become a serious worry. With the target full, the limit is in the region of 150 muons/pulse, despite the fact that we have a 2 kG-meter magnet between the target and chambers to sweep out the δ-ray background.

It is nevertheless interesting to consider what improvements could be made to increase the flux of the beam and decrease the halo, since future experiments might not be limited in the same way.

1. The present beam is limited in flux by power dissipation in the production target to 100 kW. The Stanford Linear Electron Accelerator can run at present with an average power of 300 kW, and up to 1 MW may be attained in the future. Using more power would entail a redesign of the beam dump and of the Be filter for more efficient cooling. The muon flux would, of course, increase linearly with beam current.

2. The addition of another focusing stage at the beginning of the beam, as shown in fig. 10, would bring the following improvements: a. The beam source could be re-defined in both planes at a slit after it had passed through the bulk of the Be filter. The horizontal beam profile would then fall off as steeply as the vertical; b. The beryllium filter could be split in three, so that the scattering would all take place very close to the target, or its image. The effective source size, and therefore the phase-space volume occupied by the muons, would be smaller by a factor of about 9 at 10 GeV/c. A factor of about 2 increase in muon flux would be obtained at 10 GeV/c, somewhat more at lower momenta; c. With the smaller phase-space volume occupied by the beam muons, collimation and removal of halo muons should be more effective.

With a 300 kW beam on the target and the improvements in (2) above, we estimate that the muon flux would increase by a factor of about 6 at 10 GeV/c, and the phase-space volume of the beam decrease by a factor of about 9. The beam halo would also improve, but the amount is difficult to estimate. Adding another stage at the end of the beam, or incorporating a small bend in the last stage, would cut it by an order of magnitude.

The muon beam was the first high energy particle beam to be constructed at SLAC; all magnets, power supplies, control systems and installation techniques were new and untried. The success with which the beam was installed and commissioned is a tribute to the work of many people too numerous to mention individually. We would like to thank all the engineers, surveyors, riggers, electricians and others in the technical division whose work went into the construction of the beam. We would like in particular to thank Mr. R. Vetterlein who did the engineering design and supervised the installation of the beam and Mr. C. A. Harris and his technicians for their tireless work with the power supplies.

We are grateful to Dr. Y. S. Tsai and Mr. V. Whitis for permission to use their unpublished data on muon production and for many useful conversations.

References

[1] W. K. H. Panofsky, C. M. Newton and G. B. Yodh, Phys. Rev. **98** (1955) 751.
[2] Y. S. Tsai and V. Whitis, Phys. Rev. **149** (1966) 1248.
[3] Y. S. Tsai and V. Whitis, SLAC Users Handbook, pt. D (unpublished) and private communication.
[4] CERN shielding studies, Nucl. Instr. and Meth. **32** (1965) 45.
[5] A. Barna et al., Phys. Rev. Letters **18** (1967) 360.
[6] B. Rossi, *High energy particles* (Prentice-Hall, 1952) p. 71.

COMMENTS ON C8
"Measurement of the Inclusive Electroproduction of Hadrons"

In the middle and late 1960's my research concentrated more and more on the nature of leptons and the muon-electron problem. But this research itself led us back into the hadron world in the following way.

My primary interest in muon-proton inelastic scattering was to search for muon electron differences as described in Memoir A1 and Reprint A3. In the muon-proton inelastic scattering experiments we simply measured the angles and momentum of the scattered muon, ignoring the hadrons, mostly pions, that were produced. These hadrons were produced by a virtual photon passing from the muon to the proton. Again my mechanical view of particle physics. We wondered how the degree of virtuality of the photon would affect production of hadrons. How would the multiplicities and kinematic distributions depend on q^2, the square of the four-momentum transferred to the proton?

The muon beam had a large diameter which made it difficult to detect the produced hadrons. Also we wanted higher incident beam energy. The ideal experiment would use a high energy, small diameter electron beam. And so we designed and built an experiment to measure the scattered e^- and forward produced π's and K's in

$$e^- + p \rightarrow e^- + \text{forward } \pi\text{'s and } K\text{'s} + \text{other hadrons.}$$

This experiment was only possible because of a very clever device developed by my colleagues, Frederick Martin, Steven St. Lorant, and William Toner. When a high energy electron beam passes through a hydrogen target a large number of low energy electrons and positions are produced in the beam direction. Since the linear accelerator produced the electron beam in 1.5 μs long pulses, the magnet and spark chambers downstream of the target would be filled with stray electron and positron tracks. The clever device was a superconducting tube which surrounded the beam as it passed through the magnet after exiting the target. The swarm of low energy electrons and positrons simply stayed inside the beam pipe. It worked amazingly well.

Equally amazing was that the superconducting tube was built by our technician, Acie Newton, who had never before built a superconducting device. He was one of the greatest technicians I have ever met. Newton also built the high speed 70 mm cameras we used for our many optical spark chamber experiments. He built them from scratch, only the magazines were commercial. Newton also fixed automobiles, household appliances and machine tools. My group's mechanical shop was filled with boxes of valves and motors and machine parts. He threw nothing away. We are more sophisticated these days in our mechanical design, but I miss Newton and his boxes of parts.

Our research on the electroproduction of hadrons continued into the middle 1970's. We were joined by a group of physicists from the Massachusetts Institute of Technology led by Louis Osborne, and they took on the leadership as our SLAC group became more and more involved in electron-positron annihilation physics at SPEAR.

If the opportunity to study electron-positron annihilation physics had not occurred, would I have become immersed in the study of the electroproduction of hadrons? Along with its analog, muoproduction of hadrons, it has become a major area of particle research. Louis Osborne went on to carry out hadron muoproduction experiments at Fermilab into the late 1980's, and muoproduction experiments have been a major part of the CERN program. At present very high energy electroproduction of hadrons is being studied using two detectors, H1 and ZEUS, at the Deutsches Elektronen-Synchrotron laboratory's HERA electron-positron collider.

It is interesting how the technology of these studies has alternated between muons and electrons. We began with muoproduction, then changed to electroproduction to obtain higher energies and larger data rates. The experimenters at Fermilab and CERN changed back to muoproduction to obtain much higher energies. And now the pioneer electron-positron collider, HERA, has taken experimenters back to electroproduction of hadrons.

Returning to my question to myself, if there had been no SPEAR electron-positron collider and no tau, would I have followed that electroproduction, muoproduction, electroproduction trail? I don't think so, not unless I had thought that the trail would lead me through the electron-muon problem.

Measurement of the Inclusive Electroproduction of Hadrons*

J. T. Dakin, G. J. Feldman, W. L. Lakin,† F. Martin, M. L. Perl, E. W. Petraske,‡ and W. T. Toner§

Stanford Linear Accelerator Center, Stanford University, Stanford, California 94305
(Received 31 July 1972)

> The electroproduction of hadrons was studied with a wide-aperture spectrometer. Inclusive data are presented for the electron-scattering region $-0.5 > q^2 > -2.5$ (GeV/c)^2, $4 < \nu < 14$ GeV. Distributions of the electroproduced hadrons in the three inclusive variables φ, $p_\perp^2$, and x are studied in the region $x > 0$. A striking difference from photoproduction is observed in the excess of positive over negative hadrons at high x and high q^2.

With the recent work in deep inelastic electron-nucleon scattering and its subsequent theoretical interpretations, there has been increasing interest in the hadronic final states produced in such interactions.[1] Here we report some preliminary results on the inclusive electroproduction of hadrons.

In this experiment, we detected in coincidence an electron scattered from a hydrogen target and one or more electroproduced hadrons. Taking each combination of a scattered electron and an electroproduced hadron as an independent inclusive event, the cross section is a function of six variables. Three of them are determined by the electron system: E, the incident electron energy in the laboratory (fixed at 19.5 GeV for all of our data); q^2, the invariant momentum transfer squared to the scattered electron; and ν, the electron energy loss in the laboratory. The remaining three, which concern the detected hadron, are calculated relative to the direction of the electron three-momentum transfer: x, the ratio of the longitudinal momentum in the virtual-photoproduction center-of-mass system to the maximum possible; $p_\perp^2$, the transverse momentum squared; and φ, the hadron azimuthal angle.

Virtual photoproduction cross sections can be derived from experimental cross sections by

$$\frac{d\sigma}{dq^2\,d\nu\,d^3p} = \Gamma(E, q^2, \nu) \frac{d\sigma(q^2, \nu)}{d^3p}, \qquad (1)$$

where the function Γ contains the electrodynamic factors describing the electron-photon vertex. We will report here ratios of differential virtual-photon cross sections to the total virtual-photon cross section. These ratios are derived directly from our data since we had no requirement for a hadron in our trigger.

The differential cross sections will be given in a Lorentz-invariant form:

$$E\frac{d\sigma(q^2, \nu)}{d^3p} = 2\frac{E^*}{p_{max}^*}\frac{d\sigma(q^2, \nu)}{dx\,dp_\perp^2\,d\varphi}, \qquad (2)$$

where E^* is the energy of the hadron in the center-of-mass system and p_{max}^* is the maximum possible center-of-mass momentum.

The experimental apparatus consisted of a 19.5-GeV electron beam incident on a target and a large-aperture spectrometer to detect a large fraction of the forward final-state particles with lab momenta greater than ~ 1 GeV/c. These elements are shown in Fig. 1. The electron beam at the Stanford Linear Accelerator Center (SLAC) contained typically 10^4 e^- per 1.5-μsec-long pulse. The momentum band was 0.2% at 19.5 GeV/c. The target was 4 cm long and was filled with either hydrogen or deuterium. Only the hydrogen data are reported here. The spectrometer magnet had a field integral of 17 kG m. The unscattered beam and the forward electromagnetic backgrounds passed through the magnet in a field-free region created by a cylindrical supercon-

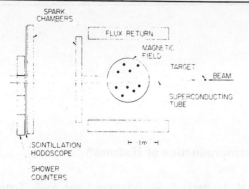

FIG. 1. Schematic view of the apparatus.

ducting tube.[2] Beyond the magnet were two optical spark chambers separated by 1.7 m. Each chamber had a mirror system which allowed tracks to be seen in three views. The apparatus was triggered on the detection of a scattered electron by a hodoscope of twenty scintillation counters and eleven shower counters[3] behind the second spark chamber. The shower counter thresholds were set to ~ 5 GeV.

The pictures taken with hydrogen in the target were scanned and measured, and the particle trajectories were reconstructed in space using two independent systems. The first was Hummingbird II, a flying-spot digitizer, with which we measured all tracks in all pictures.[4] The second was a conventional hand system with which we measured only events with two or more tracks in 20% of the pictures. Each system had only ~ 50% efficiency for fully reconstructing events with two or more tracks, largely because of confusion introduced by spurious tracks. We have not as yet thoroughly studied biases introduced by these inefficiencies. Biases, if any, should be different for the two systems since the Hummingbird system identifies tracks on the basis of the stereo reconstruction while the hand system relies on pattern recognition of spark densities. We have verified that both systems give the same physics results within statistics. The data reported here are solely from the Hummingbird system.

In each reconstructed picture, the electron was identified by matching the positions and momenta of tracks to the position and pulse height of the shower counter which triggered the event. The remaining tracks which were consistent with the scintillator hodoscope pattern for an event were assumed to be hadrons. In a picture in which an electron was identified, if a hadron was present it was identified with $(70 \pm 10)\%$ efficiency. The rms momentum resolution was 2% at 10 GeV.

We have no means of distinguishing pions, kaons, and protons from each other. Hence we will refer only to positively and negatively charged hadrons and use the symbol h to describe them. In computing x, each h is assumed to be a π.

To prepare the physics distribution, each event was weighted inversely as its detection probability, a function of q^2, ν, φ, $p_\perp^2$, x, and h charge. To study the dependence of the data on one of these variables, we summed over some range of the other five variables. When such summations passed over small regions of zero acceptance, corrections were made for the missing events.

The cross sections thus obtained were normalized to the total virtual photoproduction cross section in the same q^2, ν region $[\sigma(q^2, \nu)]$ obtained from the data. The data presented here are 6244 events from the region $4 < \nu < 14$ GeV, $-0.5 > q^2 > -2.5$ $(GeV/c)^2$, the region where the virtual-photon direction is well within the geometric acceptance.

We have made a preliminary study of radiative and spatial-resolution effects and have found that they produce no noticeable changes in the shapes of the $p_\perp^2$ and x distributions at the statistical level of this experiment. No corrections for these effects have been made in the data presented here.

The errors given in figures in this report represent statistical errors only. There is, in addition, a possible 15% error due primarily to uncertainty in the level of the Hummingbird-system efficiency.

We observe no significant φ dependence in the data, although the negative-hadron data are consistent with π^- photoproduction measurements which show transverse polarization effects at high x.[5] As an aid in extracting $p_\perp^2$ and x distributions, the subsequent analysis assumed that the φ distribution was flat.

The $p_\perp^2$ dependence of the data in the range $0 < p_\perp^2 < 0.7$ $(GeV/c)^2$ was studied as a function of q^2, ν, x, and h charge and was found not to depend on any of these variables in a statistically significant way. The data are fitted well with a function of the form $A \exp(-bp_\perp^2)$ with $b = 4.7 \pm 0.5$ $(GeV/c)^{-2}$. We have extracted a value of $b = 5.9 \pm 0.2$ $(GeV/c)^{-2}$ for the same x and ν (E_γ) range from the π^- photoproduction data.[5]

The dependence of the invariant cross section on x is shown in Fig. 2 for positive and negative hadrons. The curve for negative hadrons agrees

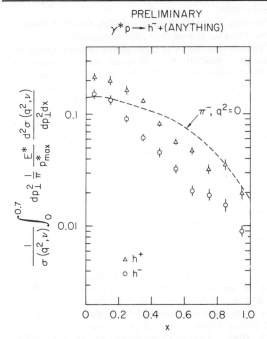

FIG. 2. Dependence of the invariant cross section on x for $-0.5 > q^2 > -2.5$ $(GeV/c)^2$, $4 < \nu < 14$ GeV. A line representing the data in π^- photoproduction (Ref. 5) is included.

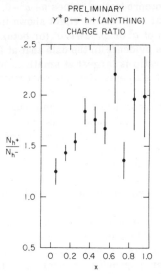

FIG. 3. The ratio of the invariant cross section for positive to negative hadrons at each x for $-0.5 > q^2 > -2.5$ $(GeV/c)^2$ and $4 < \nu < 14$ GeV.

with the analogous photoproduction[5] curve at low x, but falls a factor of 2 to 3 below it at higher x. The slow fall of the photoproduction distribution at moderate x values is due largely to the production of ρ mesons. If proportionally fewer ρ mesons were electroproduced than photoproduced, one would expect the distribution to fall more sharply.

Figure 3 shows the ratio of the invariant cross sections for positive hadrons to those for negative hadrons as a function of x. There is a striking increase in the ratio as x increases.

There are no published photoproduction data for this ratio. However, from available information we estimate the photoproduction ratio to be 1.20 ± 0.10 throughout this x range. The SLAC-Berkeley-Tufts bubble-chamber data[6] give a ratio between 1.00 and 1.10 in this range. These data exclude two classes of events, one-prong events and visible strange-particle production, which together compose ~15% of the cross section and which predominantly yield positive hadrons. The ratio of 1.20 ± 0.10 is also consistent with low-$p_\perp^2$ SLAC spectrometer data.[7] Our data appear to approach the photoproduction value as $x \to 0$.

In Fig. 4 we show the charge ratio as a function of q^2 for two ranges of x. The ratio increases markedly as $|q^2|$ increases, and appears to ap-

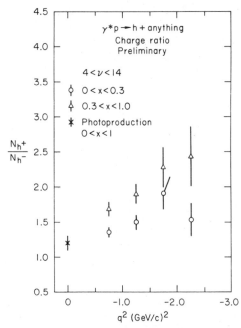

FIG. 4. The ratios of the invariant cross section for positive to negative hadrons at each q^2 for two different x ranges. A point at $q^2 = 0$ from photoproduction (Refs. 6 and 7) is included.

proach the photoproduction values as $q^2 \to 0$.

It is not clear whether the effect shown in Fig. 4 is a function of q^2 or $\omega = 2M\nu/q^2$ (or both). There is some evidence in the data that at fixed q^2 the charge ratio is largest at small ω, but we have been unable to establish this on a statistically significant level for all of our data.

We consider these data to be significant and surprising. As $|q^2|$ increases or as ω decreases, by some mechanism, part of the charge of the proton is being projected forward. We note that quark-parton models can yield a charge ratio as large as 8 at small ω and large x.[8]

We wish to acknowledge the technical support contributed by numerous SLAC groups, particularly the cooperation from Steve St. Lorant and the Low Temperature group, and John Brown and the Data Analysis group. Byron Dieterle and Benson T. Chertok provided valuable assistance in the early stages of the experiment.

---

*Work supported by the U. S. Atomic Energy Commission.

†Present address: Birmingham Radiation Centre, University of Birmingham, P. O. Box 363, Birmingham B15 2TT, England.

‡Present address: Vanderbilt University, Nashville, Tenn. 37203.

§Present address: Rutherford High Energy Laboratory, Chilton, Didcot, Berkshire, England.

[1]The current status of these topics is summarized in talks by H. Kendall, K. Berkelman, and J. Bjorken, in *Proceedings of the Fifth International Symposium on Electron and Photon Interactions at High Energies*, edited by N. B. Mistry (Cornell Univ. Press, Ithaca, N.Y., 1971).

[2]F. Martin, S. J. St. Lorant, and W. T. Toner, to be published.

[3]W. L. Lakin, E. W. Petraske, and W. T. Toner, to be published.

[4]J. L. Brown, SLAC Report No. SLAC-PUB-752, 1970 (unpublished).

[5]K. C. Moffeit, J. Ballam, G. B. Chadwick, M. Della-Negra, R. Gearhart, J. J. Murray, P. Seyboth, C. K. Sinclair, I. O. Skillicorn, H. Spitzer, G. Wolf, H. H. Bingham, W. B. Fretter, W. J. Podolsky, M. S. Rabin, A. H. Rosenfeld, R. Windmolders, C. P. Yost, and R. H. Milburn, Phys. Rev. D 5, 1603 (1972).

[6]K. C. Moffeit, private communication.

[7]A. M. Boyarski, D. Coward, S. Ecklund, B. Richter, D. Sherden, R. Siemann, and C. Sinclair, in *Proceedings of the Fifth International Symposium on Electron and Photon Interactions at High Energies*, edited by N. B. Mistry (Cornell Univ. Press, Ithaca, N.Y., 1971), and private communication.

[8]For example, see C. F. A. Pantin, University of Cambridge Report No. DAMTP 7219, 1972 (to be published).

COMMENTS ON C9
"Small Multiplicity Events in $e^+ + e^- \to Z^0$ and Unconventional Phenomena"

There is a fifteen year gap between the time we did the research described in Reprint C8 and the period of this paper. Almost all my research in that fifteen year period was immersed in the discovery and exploration of the tau lepton; described in Part A of this volume.

This reprint raises the question of why should any individual experimenter labor on an experiment, why not wait until someone else does the experiment? Let me explain.

In the mid-1980's two Z^0 facilities were under construction, the LEP circular electron-positron collider at CERN and the SLC linear electron-positron collider at SLAC. During the proposal and construction of these colliders and the concurrent construction of the needed detectors, the involved high energy physicists organized extensive study groups and workshops on the potential physics of the Z^0. Since then, not only the construction, but even the proposal of a large high energy facility, has brought forth extensive workshops on the potential physics, workshop with voluminous proceedings.

As I wrote in the comment on Reprint C2, I don't do well in specialized research areas where there are already enough experimenters at work. The potential for new physics discoveries in the study of the Z^0 attracted many able experimenters, I felt not needed. I wonder how many other elementary particle experimenters felt that way or feel that way today.

And I wonder how many experimenters feel as I do that there are an excesive number of workshops on the physics which may be done at the large accelerators either proposed or under construction. Of course, substantial thought and planning must be devoted to the design of new colliders and detectors. But do other high energy physicists sometimes think, as I do, that there is no need for endless Monte Carlo studies of signals for new particles. Isn't the actual carrying out of a new particle search the best stimulus for new ideas about how to make the search?

I know that an important purpose, perhaps the main purpose, of many of these workshops is to maintain the morale of the scientists and engineers during the long years of funding requests, design, and construction. Indeed a second purpose is to show the funding agencies that the project is moving along. But think of all the workshops and studies done for the Superconducting Super Collider! Some of these studies have been useful for the design of the Large Hadron Collider and its detectors, but many of these studies have passed into obscurity. As it gets harder to bring large science projects to reality, perhaps these workshops and proceedings are our virtual reality.

To return to this reprint, as my contribution to potential physics at the Z^0 I decided to write on an out of the way topic. I wrote about searching for unconventional particle physics in small multiplicity Z^0 decays. At the energy of the Z^0, decays into hadrons which are always messy have large multiplicities, but the known small multiplicity events such as

$$e^+ + e^- \to e^+ + e^-$$

$$e^+ + e^- \to e^+ + e^- + e^+ + e^-$$

$$e^+ + e^- \to \tau^+ + \tau^-$$

are completely understood. So why not sort through small multiplicity Z^0 decays looking for events which cannot be explained by known decay processes?

I never carried out this search because our experiment, the Mark II experiment, at the SLC did not get enough luminosity. With my graduate student I made a start, but all we could do with the small luminosity was to study the expected processes.

Since then experimenters using the four detectors at LEP have searched for all sorts of unconventional phenomenon in Z^0 decays and found nothing unexpected!

Thus the experimental answer to my 1986 question is that there are no detected unconventional phenomena, small multiplicity or not, in Z^0 decays. Am I better off not having spent three or four years in a fruitless search? Am I fortunate that the circumstance of low luminosity forced me to wait for some other experimenters to do all the work? Yes, I am fortunate, but I'm sorry to have missed the excitement of the Z^0 chase after new physics. I'm sorry to have missed the excitement on a Friday afternoon when the first look at a new batch of data shows an unexpected number of very rare events. I'm not sorry to have missed the Monday morning after a weekend of more data processing when the very rare events turn out to very ordinary events which have been misidentified.

I hope there is no history lesson for elementary particle physicists in the failure to find new physics in all the ingenious, beautiful and careful experimental studies of the Z^0. Are our chances of unexpected discoveries going to be smaller in the future compared to the past? In the past half century, unexpected discoveries in particle physics, the J/ψ and the tau lepton for example, involved dramatic phenomena. Thus the J/ψ is a startling resonance and the tau is produced 20% of the time in low energy electron-positron annihilation. Of course these discoveries were not easy, the experiments were crude and our ignorance was large. However, once the first traces were found of these unexpected phenomena, experimenters could change their apparatus or their analysis to amplify and clarify the unexpected signal.

Suppose that lying somewhere in the masses of data accumulated at CERN and Fermilab and SLAC there is an 0.1% or an 0.01% unexpected signal of new physics. Suppose that signal looks quite a bit like the signal from normal processes. Then the chance is very small for the discovery of this new physics. This may be what we face in particle physics today.

SLAC – PUB – 4165
December 1986
(E)

SMALL MULTIPLICITY EVENTS IN $e^+ + e^- \rightarrow Z^0$ AND UNCONVENTIONAL PHENOMENA[*]

Martin L. Perl
Stanford Linear Accelerator Center
Stanford University, Stanford, California 94305

ABSTRACT

Events with two–, four– or six–charged particles and no photons produced through the process $e^+ + e^- \rightarrow Z^0$ provide an opportunity to search for unconventional phenomena at the SLC and LEP electron-positron colliders. Examples of unconventional processes are compared with the expected background from electromagnetic processes and from charged lepton pair production.

Invited paper to be presented at the XXII Recontres de Moriond:
Electroweak Interactions and Unified Theories Conference, Les Arcs, France, March 8-15, 1987

---

[*] Work supported by the Department of Energy, contract DE–AC03–76SF00515.

A. INTRODUCTION

At the Z^0, the process

$$e^+ + e^- \to Z^0 \to n\text{-charged-particles} + 0\text{-photons} \tag{A1}$$

where $n = 2$, 4 or 6; provides an opportunity to search for unconventional processes. The behavior and rate of background events from conventional processes can be calculated, and the events occupy limited regions in the space of kinematic variables. This paper provides a concise description of this opportunity.

There are several unconventional processes which can yield small multiplicity, 0–photon events. The signatures for some of these processes have been fully discussed, other processes are less known. The discussion here is based on a classification by general production mechanisms and event topology. For example, there are similarities in the signatures for the charged sequential lepton ($L^\pm$) process

$$\begin{aligned} e^+ + e^- &\to Z^0 \to L^+ + L^- \\ L^+ &\to \ell^+ + \nu_\ell + \bar{\nu}_L \\ L^- &\to \ell^- + \bar{\nu}_\ell + \nu_L \end{aligned} \tag{A2}$$

and the supersymmetric scalar lepton ($\tilde{\ell}^\pm$) process

$$\begin{aligned} e^+ + e^- &\to Z^0 \to \tilde{\ell}^+ + \tilde{\ell}^- \\ \tilde{\ell}^+ &\to \ell^+ + \tilde{\gamma} \\ \tilde{\ell}^- &\to \ell^- + \tilde{\gamma} \end{aligned} \tag{A3}$$

if

$$\begin{aligned} m_L &\gg m_\ell, & m_L &\gg m_{\nu_L} \\ m_{\tilde{\ell}} &\gg m_\ell, & m_{\tilde{\ell}} &\gg m_{\tilde{\gamma}} \end{aligned} \tag{A4}$$

Here, ℓ means e or μ, ν is a neutrino, $\tilde{\gamma}$ is a photino and m means mass. When m_L and $m_{\tilde{\ell}}$ are greater than about 20 GeV/c^2, these processes yield acollinear two–charged–particle events with substantial energy and missing momentum. There is little background to such events from conventional processes.

On the other hand, suppose $m_L - m_{\nu_L}$ or $m_{\tilde{\ell}} - m_{\tilde{\gamma}}$ is small, of the order of 1 GeV/c^2. Then, depending on m_L or $m_{\tilde{\ell}}$, such events may have small visible energy and two–virtual–photon processes may cause a troublesome background.

I also discuss unconventional processes which might produce four-charged leptons. A well-known example is a neutral lepton L° which mixes with the e or μ:

$$e^+ e^- \to Z^0 \to L^\circ + \bar{L}^\circ$$
$$L^\circ \to \ell^- + \ell'^+ + \nu_{\ell'} \qquad (A5)$$
$$\bar{L}^\circ \to \ell^+ + \ell'^- + \bar{\nu}_{\ell'}$$

An instructive example is to suppose that an unknown, very high energy, interaction has a low energy residual interaction at the Z^0 mass which yields directly

$$e^+ + e^- \to Z^0 \to \ell^+ + \ell^- + \ell'^+ + \ell'^- \qquad (A6)$$

As I show in sec. E, if the ℓ and ℓ' are required to be μ's, one can search to very small cross sections in the process in eq. A5 or A6.

The comparison of signatures for unconventional processes with the backgrounds from conventional processes depends upon the particle detector. For this paper I use a simplified model of the Mark II detector (sec. B) as it has been upgraded by my colleagues and myself for use at the SLAC Linear Collider (SLC).

Having made many searches for new particles and been successful only once, I know that one cannot precisely set search criteria until the experiment is working and the data is in hand. Usually one does not achieve the expected search sensitivity, unexplained background and imperfect equipment usually intervene first. Therfore, I shall limit myself to indicating general directions for signature selection, and proceed by example.

The plan of the paper is that backgrounds for two-charged-particle events are described in sec. C and compared in sec. D with examples of such events from unconventional processes. In secs. E and F, I discuss the background and unconventional process examples for four-charged- and six-charged-particle events, respectively.

B. SCHEMATIC DETECTOR, ACCEPTANCES AND CROSS SECTIONS

1. Schematic Detector

I discuss and calculate backgrounds and unconventional process systems using a schematic magnetic detector based on the upgraded Mark II detector.[1] In the following list θ is the smallest angle (0°–90°) between the direction of motion of a particle and the e^+e^- beam line, p is the magnitude of a particle momentum and E is its energy.

$$\text{charged particle momentum measured}: \cos\theta < 0.85 \qquad (B1a)$$

$$e \text{ identified}: \cos\theta < 0.85, \ E > 1. \text{ GeV} \qquad (B1b)$$

$$\mu \text{ identified}: \cos\theta < 0.85, \ E > 1.5 \text{ GeV} \qquad (B1c)$$

$$\gamma \text{ detected and energy measured}: \cos\theta < 0.95 \qquad (B1d)$$

e detected for veto: $\theta > 15$ mrad, $E > 0.2$ GeV (B1e)

γ detected for veto: $\theta > 15$ mrad, $E > 0.2$ GeV (B1f)

2. Acceptances

In the background and signature calculations, the acceptance for charged particles is set by

$$\cos\theta < 0.85 \tag{B2a}$$

$$p > 1.\text{ GeV/c} \tag{B2b}$$

3. Cross Sections

The approximate, radiatively corrected, cross section for

$$e^+ + e^- \to Z^0 \to f + \bar{f} \tag{B3}$$

is

$$\sigma_{f\bar{f}} \approx 1400\, T_f(\beta)\text{ pb} \tag{B4}$$

T_f depends upon the f-Z^0-$\bar{f}$ coupling. For conventional charged leptons

$$T_{L^-} \approx \beta(3-\beta^2)/2 \tag{B5a}$$

For conventional neutral leptons

$$T_{L^0} \approx \beta(3+\beta^2)/2 \tag{B5b}$$

C. BACKGROUNDS FOR TWO–CHARGED PARTICLE, 0–PHOTON EVENTS

1. $e^+e^- \to \ell^+\ell^-$

The reaction

$$e^+ + e^- \to \ell^+ + \ell^- \tag{C1}$$

where ℓ is an e or μ, gives a pair of collinear particles to the extent allowed by radiative corrections and instrumental errors. When ℓ is a τ, the one–charged–particle, 0–photon decay modes

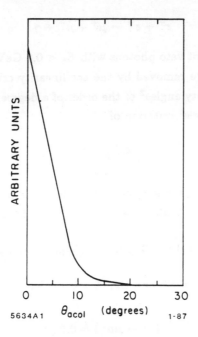

Figure 1

$$\tau^- \to \nu_\tau e^- \bar{\nu}_e \ , \quad \nu_\tau \mu^- \bar{\nu}_\mu \ , \quad \nu_\tau \pi^- \ , \quad \nu_\tau K^- \quad (C2)$$

yield the acollinearity angle distribution of fig. 1. Here, θ_{acol} is defined[2] to be 0 when the particles have exactly oppposite momenta. When

$$\theta_{acol} > 15° \quad (C3)$$

is required, about 0.5% of the decays are accepted under the conditions of eqs. B1 and B2. Hence

$$\sigma(ee \to \tau\tau, 2\text{-prong}, 0\text{-photon} \approx 1.4 \text{ pb}, \qquad \theta_{acol} > 15° \quad (C4)$$

The efficacy of increasing the lower limit on θ_{acol} to reduce this σ depends upon the level of mistracking in a particular detector.

2. $e^+e^- \to \ell^+\ell^-\gamma, \ell^+\ell^-\gamma\gamma$

The radiative corrections to the reactions in eq. C1 lead to a continuum between lepton pair production and

$$e^+ + e^- \to \ell^+ + \ell^- + \gamma \quad (C5a)$$

$$e^+ + e^- \to \ell^+ - \ell^- + \gamma + \gamma \quad (C5b)$$

The completeness of the photon veto (eq. B1f) for the schematic detector controls the level of background from these reactions. For example, in

5

$$e^+ + e^- \rightarrow \mu^+ + \mu^- + \gamma \tag{C6}$$

the schematic detector does not veto photons with $E_\gamma < 0.2$ GeV or with $\theta_\gamma < 15$ mrad. But events with $E_\gamma < 0.2$ GeV are removed by the acollinearity criterion, eq. C3. Events with $\theta_\gamma < 15$ mrad have acoplanarity angles[2] of the order of a degree, hence they can be removed by a nominal acoplanarity angle[2] criterion of

$$\theta_{\text{acop}} > 5° \tag{C7}$$

The practical question is the degree of perfection of the photon veto system. Consider $ee \rightarrow \mu\mu\gamma$ again and suppose no photon veto, only $\theta_{\text{acol}} > 15°$ and $\theta_{\text{acop}} > 5°$, then

$$\sigma(ee \rightarrow \mu\mu\gamma) = 23. \text{ pb}, \qquad \theta_{\text{acop}} > 15°, \theta_{\text{acol}} > 5° \tag{C8}$$

under all other conditions of eqs. B1 and B2. A 1% inefficiency in the photon veto will then leave

$$\sigma(ee \rightarrow \mu\mu\gamma) = 0.2 \text{ pb} \tag{C9}$$

Similar considerations apply to the other reactions in eq. C5.

3. $e^+e^- \rightarrow e^+e^-e^+e^-, e^+e^-\mu^+\mu^-, e^+e^-\pi^+\pi^-$

The two–virtual–photon processes

$$e^+ + e^- \rightarrow e^+ + e^- + e^+ + e^- \tag{C10a}$$

$$e^+ + e^- \rightarrow e^+ + e^- + \mu^+ + \mu^- \tag{C10b}$$

$$e^+ + e^- \rightarrow e^+ + e^- + \pi^+ + \pi^- \tag{C10c}$$

give two–charged–particle, 0–photon events when one e^+ and one e^- are not detected because their angles with the beamline, θ_{e^+} and θ_{e^-}, are very small. This kinematic situation has been studied in several experiments[3,4] and the results confirm the calculation methods developed by Berands, Daverveldt and Kleiss.[5]

With the acceptance conditions of eq. B2, the cross section in the pion pair process (eq. C10c) is a small fraction of the cross sections for the lepton pair processes (eqs. C10a and C10b), hence the former is ignored here. We use the mnemonic $ee \rightarrow (ee)\ell\ell$ to represent the sum of the processes in eqs. C10a and C10b when one e^+ and one e^-, represented by the symbols (ee), are not detected by tracking or veto devices. The observed cross section,

$\sigma(ee \to (ee)\ell\ell)$ depends upon the charged particle acceptance criteria and the angular extent of the $e^\pm$ veto devices. For example, with $\theta_{acol} > 15°, \theta_{acop} > 5°, p > 1$ GeV/c:

$$\sigma(ee \to (ee)\ell\ell) \approx 60 \text{ pb} \quad , \quad \theta_{e^\pm,\text{veto}} > 15 \text{ mrad} \quad (C11a)$$

$$\sigma(ee \to (ee)\ell\ell) \approx 150 \text{ pb} \quad , \quad \theta_{e^\pm,\text{veto}} > 555 \text{ mrad} \quad (C11b)$$

Here, $\theta_{e^\pm,\text{veto}}$ is measured from the beamline. Figures 2 and 3 give the E_{vis} and p_T distributions when the $\theta_{acol} > 15°, \theta_{acop} > 5°, p > 1.$ GeV/c criteria are applied. Here E_{vis} is the total energy of the two observed charged particles and p_T is the vector sum of their momenta transverse to the beamline.

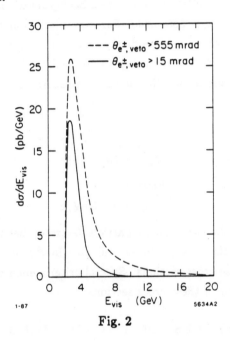

Fig. 2

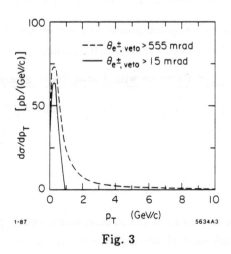

Fig. 3

The $ee \to (ee)\ell\ell$ cross section can be a serious background when one is searching for processes with small E_{vis}, such as the close-mass lepton pair model in sec. D2. This background can also be a problem in searches involving very small cross sections, such as occurred in the search of Perl et al.,[6] for neutral leptons in e^+e^- annihilation events produced at 29 GeV total energy.

When an $e^{\pm}$ is detected with $\theta_{e,\text{veto}} > 15$ mrad with perfect efficiency (eq. B1f), the remaining events with $\sigma = 60$ pb (eq. C11a) can be removed with a p_T criterion such as

$$p_T > 0.8 \text{ GeV/c} \tag{C12}$$

However, if the $e^{\pm}$ veto is inefficient there will be a background from a fraction of the dN/dp_T distribution in fig. 3 for $\theta_{e,\text{veto}} > 555$ mrad. For example, 1% inefficiency would give $\theta(ee \to (ee)\ell\ell) \approx 0.7$ pb for events where p_T exceeds the criterion in eq. C12. Such events would have p_T values up to 8 GeV/c and E_{vis} values up to 20 GeV/c. Of course, a larger p_T criterion could be used, but that reduces the efficiency of some searches for unconventional processes.

4. Summary

With criteria

$$\begin{aligned} \theta_{\text{acol}} &> 15° \\ \theta_{\text{acop}} &> 5° \\ p_T &> 0.8 \text{ GeV/c} \end{aligned} \tag{C13}$$

in the schematic detector, the two-charged particle, 0-proton backgrounds have cross sections of the order of a few tenths of a pb to several pb. Inefficiencies in γ and $e^{\pm}$ vetoes can substantially increase these cross sections. Special criteria such as requiring an $e\mu$ pair can substantially decrease some of these cross sections.

D. SIGNATURES FOR TWO-CHARGED PARTICLE, 0-PHOTON EVENTS FROM UNCONVENTIONAL PROCESSES

In this discussion the unconventional processes are classified according to the effect of the production mechanism on the kinetic variable distributions.

1. Pair Production of Charged Particles with Large Decay Energies

The general process is production of an x^+x^- pair followed by the decays of x^+ and x^- through the weak interaction:

$$e^+ + e^- \to x^+ + x^- \tag{D1}$$

$$x^{\pm} \to y^{\pm} + n_1 + n_2 + \ldots \tag{D2}$$

with the energy released in the decay of the x, large compared to the masses of the y and the $n_1, n_2 \ldots$. Here, $n_1, n_2 \ldots$ are neutral, weakly interacting particles; there may be one or more in the decay. And y^- means e^-, μ^-, π^- or K^-. It can also indicate τ^- when the τ decays into a one–charged–particle, 0–photon mode. The general kinematics are determined by two parameters: (i) the mass of x, called m_x and (ii) the difference, called δ, between m_x and the sum of all the masses of the particles on the right side of the reaction in eq. D2. Explicitly

$$\delta = m_x - \left(m_y + \sum m_n\right) \qquad (D3)$$

The case usually discussed is

$$m_x \approx \delta \qquad (D4)$$

The best known example[7] is a heavy sequential lepton, L^-, with a near-zero-mass neutrino partner ν_L. Then the decay process in eq. D2 is

$$L^- \to \ell^- + \bar{\nu}_\ell + \nu_L \quad ; \quad \ell = e, \mu \qquad (D5)$$

A similar example is the chargino, χ^-, proposed in supersymmetric models, when the χ^- decays to a near-zero-mass photino, $\tilde{\gamma}$:

$$\chi^- \to \ell^- + \nu_\ell + \tilde{\gamma} \quad , \quad \ell = e, \mu \qquad (D6)$$

A two–body example, also from supersymmetric models, is the pair production of scalar leptons

$$\begin{aligned} e^+ + e^- &\to Z^0 \to \tilde{\ell}^+ + \tilde{\ell}^- \\ \tilde{\ell}^- &\to \ell^- + \tilde{\gamma} \end{aligned} \qquad (D7)$$

To illustrate the case of a three–body decay

$$x^- \to \ell^- + n_1 + n_2 \qquad (D8)$$

I use the following simplified model: (i) the production process in eq. D1 is isotropic, the decay process in eq. D8 is calculated using relativistic phase space, and the masses of the final particles are set to zero. The θ_{acol} distribution is given in fig. 4 for $m_x = 20, 30$ and 40 GeV/c^2. Replacing the phase space calculation by one using some combination of V and A couplings changes these distributions slightly. A feeling for the observed cross section can be obtained by using eq. B4 with $\beta = 1$, using the branching fractions

$$B(x^- \to e^- n_1 n_2) = B(x^- \to \mu^- n_1 n_2) = 0.1 \quad , \qquad (D9)$$

and using an acceptance of 0.7 for the criteria of eqs. B2 and C11:

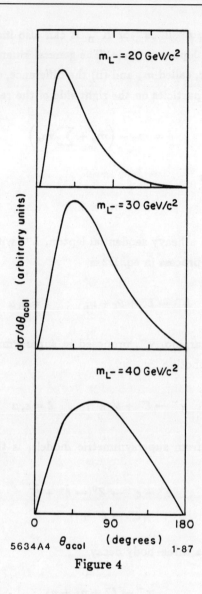

Figure 4

$$|\cos\theta| < .85$$
$$p > 1.\ \text{GeV/c}$$
$$\theta_{\text{acol}} > 15°$$
$$\theta_{\text{acop}} > 5°$$
(D10)

Then
$$\sigma = (ee \to L^+L^- \to \ell^+\ell^-, \text{observed}) = 39\ \text{pb} \qquad (D11)$$

As is well known this is much larger than the background examples given in eqs. C4 and C9, hence such searches are straightforward. Incidently, the lower limit from the UA1 collaboration[8] of

$$m_{L^-} > 41\ \text{GeV/c}^2 \qquad (90\%\ \text{CL})$$

10

on a charged heavy lepton with a near-zero-mass neutrino partner limits this search using $Z^°$ decay to a small mass range.

Summarizing, the events produced by processes defined by eqs. D1, D2 and D4 have the following properties:

1. As m_x approaches $m_Z/2$, the acollinearity increases. Acollinearity and acoplanary criteria such as $\theta_{acol} > 15°, \theta_{acop} > 5°$ separate most of the events from the $ee \to \ell\ell$ and $ee \to \ell\ell\gamma$ backgrounds.

2. For all m_x values the events have large E_{vis} values, hence they separate from $ee \to (ee)\ell\ell$ events.

3. For all m_x values the events have large p_T values.

There have been several detailed discussions[9,10] of how to search for new particles produced by the processes defined by eqs. D1, D2 and D4. I turn to a less known case.

2. Pair Production of Charged Particles with Small Decay Energies

We have recently begun to study models[11] where the production and decay processes are given by eqs. D1 and D2 but

$$m_x \gg \delta \tag{D12}$$

I will concentrate on what I call the close-mass lepton pair model[11,12] The reader can easily extend the discussion to other hypothetical particles, for example, charginos and photinos.

Consider the lepton pair $L^-, L^°$ with masses m_- and m_0, respectively. Suppose $m_- > m_0$ but

$$m_- - m_0 = \delta \ll m_- \tag{D13}$$

The charged particle in the decay modes

$$L^- \to \ell^- + \bar{\nu}_\ell + L^° \quad ; \quad \ell = e, \mu \tag{D14a}$$

$$L^- \to \pi^- + L^° \tag{D14b}$$

has maximum laboratory momentum

$$P_{max} = E_b \left[1 - \left(\frac{m_0}{m_-}\right)^2\right]\left[1 + \left(1 - \left(\frac{m_-}{E_b}\right)^2\right)^{1/2}\right]/2 \tag{D15}$$

Here E_b is the beam energy $m_Z/2$. The ℓ, ν_ℓ and π masses have been set to zero. Using eq. D12

$$E_{vis} < (\delta/m_-)m_Z \tag{D16}$$

Here E_{vis} is the total useable energy.

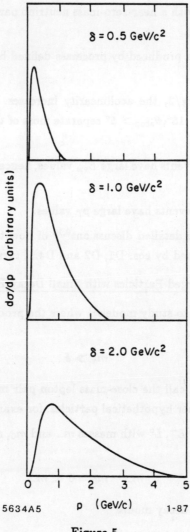

Figure 5

If δ is of the order of a few GeV/c^2 or less, and m_- is of the order of tens of GeV/c^2, E_{vis} is small. Then the $ee \to (ee)\ell\ell$ background (sec. C3) becomes important. A general discussion is unwieldy because there are now two parameters, m_- and δ; I proceed by example.

Suppose $m_- = 30$ GeV/c^2; consider the decay modes of eq. D14a; set $m_\ell = 0, m_{\nu_\ell} = 0$, and let $\delta = m_- - m_0$ have the values 0.5, 1.0 and 2.0 GeV/c^2. Using the pair production cross section in eqs. B4 and B5a and conventional weak interaction theory with V-A coupling, we calculate kinematic distributions, branching fractions and the observed cross sections for the classic signature

$$e^+ + e^- \to L^+ + L^- \to e^\pm + \mu^\mp + \text{ missing energy} \tag{D17}$$

The e and μ momentum distributions are given in fig. 5. Table 1 gives the observed cross sections under the usual criteria

Table 1. Branching fractions and observed cross sections for $e^+e^- \to L^+L^- \to e^\pm\mu^\mp +$ missing energy via $L^- \to L^0 e^- \bar{\nu}_e$, $L^+ \to \bar{L}^0 \mu^+ \nu_\mu$, with $m_- = 30$ GeV/c^2.

$m_- - m_0$ (GeV/c^2)	$B(L^- \to L^0 e^- \bar{\nu}_e)$	$B(L^+ \to \bar{L}^0 \mu^+ \nu_\mu)$	Observed σ (pb)	
			$p > 0.$ GeV/c	$p > 1.$ GeV/c
2.0	.19	.19	71.	.29
1.0	.17	.16	50.	4.
0.5	.18	.15	51.	0.

$$|\cos\theta| < 0.85$$
$$\theta_{\text{acol}} > 15°$$ \hfill (D18)
$$\theta_{\text{acop}} > 5°$$

but with

$$p > 0. \text{ GeV/c}$$
$$\text{or} \hfill \text{(D19)}$$
$$p > 1. \text{ GeV/c}$$

Figure 5 and Table 1 lead to several comments:

1. As $\delta = m_- - m_0$ decreases below 2. GeV/c^2, the $p > 1.$ GeV/c criterion must be abandoned. But then, according to eqs. B1b and B1c, the e and μ can no longer be identified.

2. Without e and μ identification the observed cross sections are the same size as the $ee \to (ee)\ell\ell$ background cross sections in eq. C11.

3. Comparing fig. 6, the p_T distributions for this model, with fig. 3, one sees that for small values of δ this signature will be submerged by the $ee \to (ee)\ell\ell$ background.

There are, of course, other signatures for the process under discussion: $\ell^\pm \pi^\mp$, $\ell^\pm \rho^\mp$ and four-charged particles, depending upon δ. And the Z^0 width, when carefully measured, would reflect the existence of an additional L^- and L^0. But when $\delta \lesssim 1.5$ GeV/c^2, it could be very difficult to elucidate the type of process discussed in this section. Incidently, when $\delta \lesssim m_\pi$, the charged particle lifetime becomes sufficiently long for the L^- to appear stable.

3. Production of Two Neutral Particles

Various types of hypothetical leptons[11,13,14] illustrate how two-charged-particle, 0-photon events could come from the decay of the Z^0. If there is a heavy neutral lepton, L^0, which mixes with the e, μ or τ generation then the following could occur

$$e^+ + e^- \to L^0 + \bar{\nu}_1$$
$$L^0 \to \ell_1^- + \ell_2^+ + \nu_2 \hfill \text{(D20)}$$

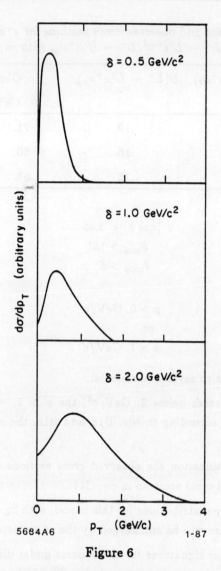

Figure 6

Here ℓ means e, μ or τ and ν is the corresponding neutrino. In a more exotic scheme, consider a pair of neutral leptons L^0, ν_L with the unconventional decay $L^0 \to \nu_L + \ldots$, then

$$e^+ + e^- \to L^0 + \bar{\nu}_L$$
$$L^0 \to \nu_L + \ell^+ + \ell^- \tag{D21}$$

or

$$e^+ + e^- \to L^0 + \bar{L}^0$$
$$L^0 \to \nu_L + \ell^+ + \ell^- \tag{D22}$$
$$\bar{L}^0 \to \bar{\nu}_L + \nu + \bar{\nu}$$

will give two–charged-particle, 0–photon events.

Another possibility is a weakly decaying neutral boson[6], N^0, with

$$e^+ + e^- \to N^0 + \bar{N}^0$$
$$N^0 \to \ell^+ + \ell^-$$
$$\bar{N}^0 \to \nu + \bar{\nu}$$
(D23)

In all these processes, as in the process in sec. D1, the events will have

1. large values of θ_{acol}
2. large values of E_{vis}
3. large values of p_T

Although here the large values of θ_{acol} occur for a different reason than the processes in sec. D1. Indeed in contrast to the latter processes, these processes give events which become more collinear as m_0 approaches m_Z or $m_Z/2$. Figure 7 illustrates this for the reactions in eq. D22. Here the masses of ν_L, ℓ and ν are set to zero, and once again relativistic phase space is used for the decay process. The L^0 mass is given for each curve.

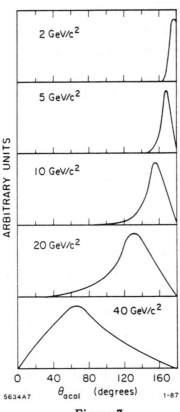

Figure 7

To illustrate what an observed cross section might be, I use eqs. B4 and B5b with $\beta = 1$, and I take the branching fractions from fig. 3 in ref. 6.

$$B(L^0 \to \nu_L \nu_i \bar\nu_i) = 0.25$$
$$B(L^0 \to \nu_L \ell\ell) = 0.07 \qquad \text{(D24)}$$

Before any θ_{acol} or θ_{acop} cuts, the acceptance is 0.7 for

$$|\cos| < .85$$
$$p > 1.\,\text{GeV/c} \qquad \text{(D25)}$$

Then

$$\sigma(ee \to L^0 \bar L^0 \to \ell^+\ell^-,\ \text{observed}) = 69\ \text{pb} \qquad \text{(D26)}$$

Like the result in eq. D11, this σ allows a straightforward search with respect to the background discussed in sec. C, providing the production cross section and branching fractions are as assumed here.

4. Production through a Central Process

As a final example I consider a central process, perhaps the low energy residual of some much higher energy interaction, where the Z^0 decays to four fermions

$$Z^0 \to f_1 + \bar f_1 + f_2 + \bar f_2 \qquad \text{(D27)}$$

If f_1 is a neutrino, and f_2 is a lepton

$$Z^0 \to \nu + \bar\nu + \ell^+ + \ell^- \qquad \text{(D28)}$$

yields two-charged particle, 0-photon events.

The question is how small a cross section could be found in view of the background described in sec. C: e^+e^- and $\mu^+\mu^-$ pairs from $ee \to \tau\tau, ee \to \ell\ell\gamma$ and $e^+e^- \to (ee)\ell\ell$. Important separation criteria are: (i) the lower limit on θ_{acol}, called $\theta_{\text{acol,min}}$ and (ii) the lower limit on E_{vis}, called $E_{\text{vis,min}}$. The former discriminates against all backgrounds, the latter against $ee \to (ee)\ell\ell$. Figure 8 gives the acceptance as a function of these criteria. Relativistic phase space is used and all lepton masses are zero. Useable acceptances can be obtained with large values of $\theta_{\text{acol,min}}$ and $E_{\text{vis,min}}$ and such large values discriminate against the background discussed in sec. C. The lower limit on the detectable cross section from the reaction in eq. D28 will probably be set by detector inefficiencies and malfunctions.

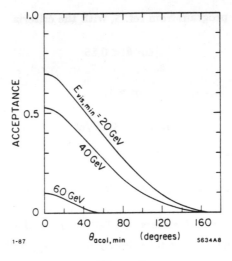

Figure 8

E. BACKGROUND FOR FOUR–CHARGED AND SIX–CHARGED PARTICLE, 0–PHOTON EVENTS

1. $\underline{e^+e^- \to \tau^+\tau^-}$

The process

$$e^+ + e^- \to \tau^+ + \tau^- \tag{E1}$$

gives four–charged or six–charged particles and 0–photons when one or both τ's decay

$$\tau^- \to \pi^- + \pi^+ + \pi^- \tag{E2}$$

But these events will be obvious and easily separated out.

2. $\underline{e^+e^- \to e^+e^-\ell^+\ell^-}$

The two–virtual–photon process

$$e^+ + e^- \to e^+ + e^- + \ell^+ + \ell^- \tag{E3}$$

with all particles detected is the main known background from conventional processes for four–charged particle, 0–photon events. The process in eq. E3 has been studied at PETRA[15,16] and compared with calculations using the Monte Carlo methods of ref. 5. Measurement and calculations agreed in the use of ref. 15, but not in that of ref. 16. The latter discrepancy has not been confirmed.

Using the Monte Carlo program from ref. 5, with the criteria

$$|\cos\theta| < 0.85 \qquad (E4a)$$

$$p > 1. \qquad (E4b)$$

$$m_{\ell\ell} > 1.\ \text{GeV}/c^2 \qquad (E4c)$$

the observed cross section is

$$\sigma(ee \to eeee, ee\mu\mu; \text{observed}) = 0.084 \pm 0.019\ \text{pb} \qquad (E5)$$

Here, $m_{\ell\ell}$ represents the invariant masses of $e^+e^-, \mu^+\mu^-$ and $e^\pm\mu^\mp$ pairs; the lower limit eliminates uninteresting events. The uncertainty in σ is from the statistics of the Monte Carlo calculation.

All the $ee \to ee\mu\mu$ events contributing to σ in eq. E5 have one $m_{\ell\ell}$ close to the lower limit in eq. E4c, the other $m_{\ell\ell}$ is usually close to m_Z^0, an expected distribution. For example, if

$$m_{\ell\ell} > 5.\ \text{GeV}/c^2 \qquad (E6)$$

is required, the observed cross section is reduced to

$$\sigma(ee \to eeee, ee\mu\mu, \text{observed}) = 0.022 \pm 0.008\ \text{pb} \qquad (E7)$$

Thus, the criteria in eqs. E4a, E4b and E6 limit $\sigma(ee \to ee\ell\ell, \text{observed})$ to very small values.

3. $e^+e^- \to$ hadrons

A possible, albeit very small, background is four-charged or six-charged particle, 0-photon hadronic events from quark-antiquark pair production. I do not know how to calculate this. The cross section for such events in the PETRA–PEP region is not measured to my knowledge. And one cannot depend upon the empirical quantum chromodynamic calculational methods used in the several Monte Carlo programs currently applied to $e^+e^- \to$ hadrons. Such programs are designed and adjusted to fit the behavior of the bulk of hadronic events; not the rare events of interest here.

F. FOUR–CHARGED OR SIX–CHARGED PARTICLE, 0–PHOTON EVENTS FROM UNCONVENTIONAL PROCESSES

Here, as in sec. D, the unconventional processes are classified according to the affect of the production mechanism on the kinematic variable distribution. The background has been discussed in sec. E. Excepting the easily recognized $\tau\tau$ background, the known background is 0.1 to 0.01 pb, eqs. E5 and E7. The limitations on search sensitivity will probably come from a combination of the unknown hadronic background (sec. E) and detector malfunctions and inefficiencies.

1. **Four-Charged Particle, 0-Photon Events from Two Neutral Particles**

Using the L^0, ν_L model of eq. D22

$$e^+e^- \to Z^0 \to L^0 + \bar{L}^0$$
$$L^0 \to \nu_L + \ell^+ + \ell^- \quad\quad\quad (F1)$$
$$\bar{L}^0 \to \bar{\nu}_L + \ell^+ + \ell^-$$

or the neutral boson model of eq. D23

$$e^+ + e^- \to Z^0 \to N^0 + N^0$$
$$N^0 \to \ell^+ + \ell^- \quad \text{(both } N^0\text{)} \quad\quad\quad (F2)$$

four-charged particle, 0-photon events can be produced. Here ℓ means e, μ or the one-charged-particle, 0-photon decay modes of the τ. Of course the ℓ's could be replaced by π or K mesons, but such decay modes would probably have very small branching fractions for the L^0 or N^0 masses of interest—above several GeV/c^2.

The L^0, ν_L model with the $\nu_L \ell^+ \ell^-$ branching fraction of eq. D24, with the production cross section of eqs. B4 and B5b with $\beta = 1$, and with an acceptance of 0.4 for

$$|\cos\theta| < .85$$
$$p > 1. \text{ GeV}/c \quad\quad\quad (F3)$$

gives

$$\sigma(ee \to L^0\bar{L}^0 \to \ell^+\ell^-\ell^+\ell^-, \text{ observed}) = 5 \text{ pb} \quad\quad\quad (F4)$$

This is a relatively small cross section, but still much larger than the known background cross sections.

The N^0 model would give obvious events with $E_{vis} = m_Z$ within detector precision and radiative corrections. The distributions of pair masses would indicate m_{N^0}, and be very different from the $ee \to ee\ell\ell$ distributions (Sec. E2).

A generalization of the L^0 and N^0 models would add additional neutral, weakly-interacting, particles to the decay modes in eqs. F1 or F2.

2. **Production through a Central Process**

The central process model in sec. D4 would also give

$$e^+ + e^- \to \ell_1^+ + \ell_1^- + \ell_2^+ + \ell_2^- \quad\quad\quad (F5)$$

Using relativistic phase space, the acceptance under the criteria of eq. F3 is 0.5, again $E_{vis} = m_Z$.

3. Four-Charged or Six-Charged Particle, 0-Photon Events from Charged Particle Production

Here I follow the scheme of the $\tau\tau$ production and decay process. Consider

$$e^+ + e^- \to Z^0 \to f^+ + f^- \tag{F6}$$

with the decay modes

$$f^- \to \ell^- + \nu + \bar{\nu} \tag{F7a}$$

$$f^- \to \ell^- + \ell^+ + \ell^- \tag{F7b}$$

Here, as before, ℓ means e, μ or the one-charged-particle, 0-photon τ decay modes; ν means ν_e, ν_μ or ν_τ.

Such events will be distinctive, particularly the six-charged-particle events with $E_{vis} = m_Z$.

ACKNOWLEDGEMENTS

I am greatly indebted to Timothy Barklow for many valuable conversations and insights on small multiplicity of events, to David Stoker for conversations and calculations on close-mass pairs, to Alfred Peterson for his development of, and knowledge of, Monte Carlo programs for e^+e^- annihilation, and to Bruce LeClaire for his construction of general computer programs for executing analysis of real and simulated data. I am also indebted to my other colleagues in the new Mark II collaboration who are preparing our experiment on the SLC.

REFERENCES

1. Proposal for the Mark II at SLC, SLAC–PUB–3561 (1985).
2. $\cos\theta_{acol} = -\vec{n}_1 \cdot \vec{n}_2$ and $\cos\theta_{acop} = -(\vec{n}_1 \times \vec{n}_+) \cdot (\vec{n}_2 \times \vec{n}_+)/|\vec{n}_1 \times \vec{n}_+||\vec{n}_2 \times \vec{n}_+|$ where $\vec{n}_1, \vec{n}_2$ and $\vec{n}_+$ are the unit momentum vectors for observed particle 1, observed particle 2 and the incident e^+, respectively.
3. R. P. Johnson, Ph. D. Thesis, Stanford Univ., SLAC–REP–294 (1986).
4. M. L. Perl et al., Phys. Rev. **34D**, 3321 (1986).
5. F. A. Berands, P. H. Daverveldt and R. Kleiss, Nucl. Phys. **B253**, 441 (1985).
6. M. L. Perl et al., Phys. Rev. **32D**, 2859 (1985).
7. For discussions of search methods for sequential, heavy, charged leptons from Z^0 decay, see D. Stoker, Mark II/SLC–Physics Working Group Note 0–2 (1986), p. 475; Physics at LEP, CERN 86–02 (1986), edited by J. Ellis and R. Peccei, Vol. 1, p. 424.
8. A. Honma, Proc. 23rd Conf. High–Energy Physics (Berkeley, 1986), to be published; also issued as CERN–EP/86–153 (1986).
9. Mark II/SLC–Physics Working Group Note 0–2 (1986), unpublished.

10. Physics at LEP, CERN 86–02 (1986), edited by J. Ellis and R. Peccei.

11. M.L. Perl, Proc. 23rd Int. Conf. High–Energy Physics (Berkeley, 1986), to be published. Also issued as SLAC–PUB–4092 (1986).

12. S. Raby and G. B. West, Los Alamos Preprint LA–UR–86–4151 (1986). This work proposes a close–mass lepton pair in connection with solar neutrino and dark matter questions.

13. F. J. Gilman, Comments Nuclear and Particle Physics, **16**, 231 (1986).

14. S. Komamiya, Proc. Int. Symp. Lepton and Photon Interactions at High Energies (Kyoto, 1985), edited by M. Konuma and K. Takahashi, p. 612.

15. W. Bartel *et al.*, Z. Phys. **C30**, 545 (1980).

16. H. J. Behrend *et al.*, DESY 84–103 (1984).

COMMENTS ON C10
"Rotor Electrometer: New Instrument for Bulk Matter Quark Search Experiments"

In the middle 1980's, working with John Price and Walter Innes, I began to think about the question of whether quarks could exist in isolation, as separate particles with electric charge $\pm 1/3$ or $\pm 2/3$. Of course, by the middle 1980's this was already an old question and there had been many searches for isolated quarks.

We were not impelled to begin a search for isolated quarks by new theoretical ideas of ourselves or others. Rather we thought up a new method to do the search. Our original idea was to move a small piece of matter periodically back and forth between two Faraday cups, thus producing an oscillating current proportional to the net electric charge on the piece of matter.

As described in this paper, it turned out to be better to leave the piece of matter at rest and to move the Faraday cups past the piece of matter using a rotating cylinder supporting a ring of Faraday cups. We were able to measure the charge with a resolution of 0.3 of an electron charge. We needed better accuracy, about 0.05 of an electron charge, and we believe we could have achieved that accuracy.

But we ran into a problem we could not solve. We had to be able to move the piece of matter away from the Faraday cups without the charge changing on the piece of matter. This was necessary so that the actual size of the charge could be measured. Our plan was to have a small hollow cylinder, a sample holder, always close to the Faraday cups and to move the piece of matter into and out of the sample holder. As described in the paper, we could not see how to do this.

When I think back to this experiment, I ruminate as to what lessons this experiment contains for the experimenter. Should the scientists always have a complete plan for an experiment with all problems solved before the experiment is built? That's too stiff a rule because it is often necessary to be working physically on an apparatus in order to solve a problem with that apparatus.

Our rotor electrometer experiment was not continued because there were actually two problems: how to move the piece of matter in and out of the sample holder without changing the charge, and how to improve the charge accuracy to 0.05 of an electron charge. Solving both problems was too much work. So the rule is, avoid experiments which have *two* serious problems.

I returned to the search for isolated quarks last year with a new version of the old Millikan liquid drop method; this is described in Reprint C14.

Rotor electrometer: New instrument for bulk matter quark search experiments

John C. Price,[a] Walter Innes, Spencer Klein, and Martin Perl

Stanford Linear Accelerator Center, Stanford University, Stanford, California 94305

(Received 6 June 1986; accepted for publication 18 July 1986)

The rotor electrometer is a new instrument which we hope will make possible searches for rare fractionally charged impurities in very large quantities of matter. The ultimate goal of the project is to be able to measure the net charge of 10 mg samples of any material to an accuracy of 0.05 q_e in a few minutes (q_e is the electron's charge). This paper reports the achievement of subelectron (0.3 q_e) charge resolution with the new device. We discuss effects which limit the resolution and consider prospects for improving the performance to the point where a fractional charge search may be attempted.

INTRODUCTION

The strongly interacting elementary particles are described as bound states of constituents called "quarks," which carry fractional charge of $\pm 1/3$ or $\pm 2/3$ q_e, where q_e is the charge of the electron. It was originally assumed that if quarks are the elementary constituents of hadrons, then objects with net fractional charge must exist. Today, because of the negative results of many searches for fractional charge,[1] it is believed that quarks are "confined" inside hadrons. "Free quarks," and combinations of quarks with net fractional charge, are thought not to exist.

However, it may be that fractionally charged particles do exist but are very rare. In 1977, and again in 1979 and 1981, Fairbank, Hebard, La Rue, and Phillips reported observations of fractional charge on 100 μg niobium spheres.[2] This positive result, still unconfirmed, has stimulated a new generation of fractional charge search experiments.[3] Their goal is to search as large a quantity of material as possible, and to search many different substances. Present techniques are limited to, at most, a few milligrams, and usually to materials with special properties.

This paper describes a new charge measuring instrument, the rotor electrometer, that we have developed with the goal of pushing the search for fractional charge beyond the milligram level. We hope to be able to measure a 10 mg sample of any composition in a few minutes, so that as much as a gram of material could be searched in a day. The new instrument differs from others used previously in that it does not rely on the detection of a small force. Instead, a small alternating voltage is generated by rapidly varying the capacitance between a small conducting sample container and a high impedance amplifier. A high speed rotor with an active magnetic bearing is used to implement the varying capacitance. A somewhat similar device has recently been proposed by Willams and Gillies.[4]

The rotor electrometer described is the third in a series of similar instruments built by our group. The first two instruments achieved charge sensitivities of 100 q_e and 5 q_e, respectively, and have been described briefly elsewhere.[5] The present instrument has a resolution of 0.3 q_e for measurements of changes in the charge of a single sample. The calibration is obtained by observation of single electron changes in the sample charge. The sensitivity is presently limited by amplifier noise, sample motion, and an unidentified noise source.

Some improvement in the charge resolution must still be made before a fractional charge search can be attempted, and also the technique must be extended to absolute measurements of the sample charge, or at least to absolute measurements of the difference between the charge of two samples. As will be explained below, because the instrument is very sensitive to small motions of the sample (or sample container), extending the method to absolute charge measurements presents some difficulties, but may be possible with an improved device.

A more complete description of the rotor electrometer is available in the first author's thesis.[6]

I. THE METHOD

The basic principle, but not the actual geometry, of the rotor electrometer is illustrated in Fig. 1. A conducting sample with charge Q is moved periodically back and forth between two Faraday cups. The "signal cup" is connected to the input of a high impedance voltage measuring amplifier. When the sample is inside the signal cup a voltage Q/C appears at the amplifier input. The capacitance C is made as small as possible, limited by the amplifier's input capacitance. A second Faraday cup, the "ground cup," is required

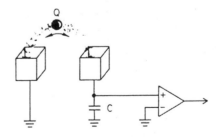

FIG. 1. Principle of the rotor electrometer. Q is the sample charge and C is the total capacitance to ground.

so that half a cycle later the sample charge will be shielded and the voltage at the amplifier input will drop to zero. The rms voltage generated per electron charge on the sample will be

$$v_e = (1/2\sqrt{2})(q_e/C).$$

Since the signal is periodic a lock-in amplifier can be used to limit the bandwidth through which noise may pass. If the total amplifier noise is represented as a voltage spectral density S_n (V²/Hz) at the amplifier input then the rms fluctuations due to the amplifier noise will be

$$v_n = (S_n/2T)^{1/2},$$

so the resolution in units of the electron's charge is

$$\sigma_e = v_n/v_e = (2C/q_e)(S_n/T)^{1/2},$$

where T is the time over which the lock-in's output is averaged. The parameters for the rotor electrometer are $C = 4.3$ pF and $\sqrt{S_n} = 8.8 \text{nV}/\sqrt{\text{Hz}}$ (at the 7.2 kHz signal frequency) which give

$$\sigma_e = 0.47 \times (1 \text{ s}/T)^{1/2}.$$

In practice the resolution is not this good. We lose a factor of 3 in signal due to electrostatic inefficiencies, the total noise is a factor of 2 above the amplifier noise, and also the averaging time is limited to $T = 200$ s by low-frequency drifts.

II. DESCRIPTION OF THE INSTRUMENT

The mechanical arrangement of the actual device is shown in Fig. 2. Instead of moving the sample back and forth between the Faraday cups, the sample is held fixed and the cups are moved. In earlier devices the cups were three sided, but in the present device they have been reduced to one-sided pads in order to reduce the stray capacitance from cup to cup. Eight signal pads and eight ground pads, each 2.5 mm wide and 5 mm high, are spaced around the perimeter of the high speed rotor. As the rotor spins, the sample charge is alternately coupled to the amplifier by a signal pad, and then shielded from the amplifier by a ground pad. Neglecting inefficiencies, a peak-to-peak voltage Q/C appears at the amplifier input, where Q is the sample charge, and $C = 4.3$ pF is the total capacitance of the signal pads to ground, due in part to the total stray capacitance from each signal pad to the adjacent ground pads (2.8 pF), and in part to the amplifier input capacitance (1.5 pF).

The rotor is 18 cm long and 3 cm in diameter. It is made mainly of a single piece of polystyrene, chosen for its high stiffness, low dielectric constant, and small dielectric loss angle. The rotor spins in high vacuum on an active magnetic bearing at 900 Hz. The charge measuring signal is at 8×900 Hz = 7.2 kHz. A 4-in. oil diffusion pump with a liquid nitrogen cold trap keeps the chamber pressure below 10^{-6} Torr. The drag on the rotor is such that the deceleration is only 0.3 Hz/h when the drive motor is turned off. This allows data collection with the rotor coasting, so that noise generated by the motor is not a problem.

The ground pads are capacitively coupled to ground and the signal pads are capacitively coupled to the low noise amplifier. The ground pads and signal pads, the rotating halves of the signal and ground coupling capacitors, and the four reflective tachometer pads are all plated in copper onto the surface of the polystyrene rotor body. The signal coupling

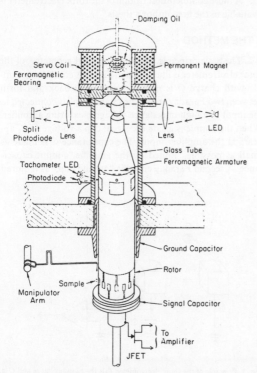

FIG. 2. Cut-away view of the instrument.

FIG. 3. Signal generating part of the device. The front cover of the vacuum chamber has been removed for this photograph.

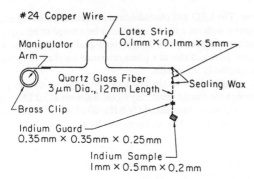

FIG. 4. Test sample and sample mount. The sample is a small piece of indium foil pinched onto a quartz glass fiber. The guard reduces the instrument's sensitivity to motions (see Sec. IV).

capacitor gap is 1.0 mm and the capacitance is 11.4 pF. The lower half of the signal coupling capacitor is made of copper-plated aluminum and is mounted on a quartz glass insulator. The 18.5 pF cylindrical ground coupling capacitor has a 2.2-mm gap and the outer half is also made of copper-plated aluminum. It is bolted to the ceiling of the vacuum chamber.

Figure 3 is a photograph of the signal generating part of the apparatus with a test sample installed. The n-channel junction field-effect transistor (JFET) which forms the first stage of the low noise amplifier is visible below the signal coupling capacitor.

The sample is hung from a quartz glass fiber along the left edge of the rotor next to the pads. Figure 4 shows details of a typical sample used to study the instrument. The quartz glass fiber is chosen for its excellent insulating properties, and indium has been used for test samples because it sticks to the glass fiber. A thin strip of latex rubber damps the motions of the sample. The sample manipulator arm exits the chamber through a flexible bellows, and is attached to a precision manipulator outside the vacuum so that calibrated motions of the sample may be made. Crude adjustments of the sample charge are made with a fine wire probe which is connected to an adjustable potential. An ultraviolet light is used to make precise adjustments of the charge by photoemission.

The upper part of Fig. 2 shows the active magnetic bearing which allows the rotor to spin at high speeds with low vibration. Earlier versions of the instrument used ball bearings, but these were not satisfactory because of vibrations which generated microphonic signals, and because of electrical noise due to the metal-to-metal contacts in the bearings. The active magnetic bearing now employed is based on designs developed by Beams and co-workers.[7]

A samarium cobalt permanent magnet generates a solenoidal magnetic field which supports the weight of the rotor. This arrangement is stable horizontally, so that the rotation axis is drawn to the magnetic axis, but the vertical equilibrium point where the upwards magnetic force equals the downwards force of gravity is unstable, so the vertical motions must be controlled by a servo system. The LED, lenses, and split photodiode are used together with the suspension electronics (see Fig. 5) to measure the vertical position of the rotor. The electronics supply a correction current to the servo coil, which generates a trim magnetic field that pushes on the rotor to make its vertical motions about the equilibrium position stable and damped.

To provide damping of the horizontal motions of the

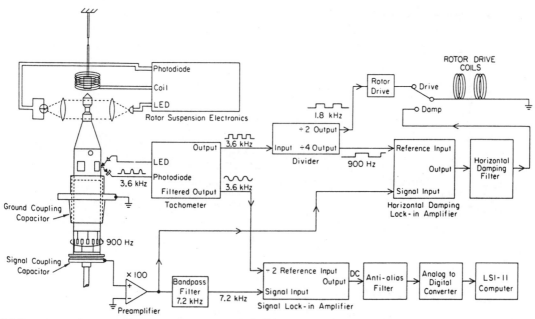

FIG. 5. Block diagram. The signal lock-in is set to lock at twice the reference signal frequency. The reference signal is bandpass filtered to reduce phase noise.

rotor, the permanent magnet is free to move horizontally in a viscous oil bath. The magnet is hung as a pendulum from a point 12 cm above the top of the rotor. This arrangement was not sufficient to damp the slow (6-s period) gravitational precession of the rotor, so an active horizontal damping circuit was added, as shown in Fig. 5. A separate lock-in amplifier demodulates the 900-Hz fundamental signal, which is sensitive to the horizontal position of the rotor. After spin up, the (near dc) output of the damping lock-in is differentiated and applied to the rotor drive coils. The magnetic field generated by the coils applies a small torque to the magnetic moment of the rotor's ferromagnetic bearing. If the lock-in phase is set correctly, this torque can be used to damp the precession.

The rotor is driven by a simple reluctance motor. The ferromagnetic armature, made of alloy 4750 nickel iron, is shown in Fig. 2. The rotor drive coils (not shown in Fig. 2) are mounted outside the glass tube which houses the upper part of the rotor. The two coils are each 70 turns of #18 Cu wire on 3-cm-diam forms. The rotor drive electronics (see Fig. 5) supplies an 8-A current at a 50% duty cycle to the drive coils, resulting in a peak torque of about 1400 dyn cm and an average acceleration of about 1 Hz/s, so that the rotor may be spun up to 900 Hz in 17 min. A reversing switch is included so that the rotor may be spun-down in the same amount of time.

Four reflective copper pads on the rotor are illuminated by an LED and viewed by a photodiode to provide a tachometer. The LED and photodiode are mounted on a ring concentric with the rotor axis that may be rotated to adjust the tachometer phase. The tachometer signal is required for the rotor drive and also as a phase reference for the two lock-in amplifiers in the system.

Figure 6 shows the low noise preamplifier circuit. Although the charge measuring signal is always at 7.2 kHz, the preamplifier is broadband (500 Hz–20 kHz). This has been useful for diagnostic purposes, and because the 900-Hz signal is used for the horizontal damping. No gate bias resistor is needed at the preamplifier input. The JFET will automatically bias with the gate-source voltage near zero, which is its most quiet operating point.

Care has been taken in the preamplifier design to ensure that there are no significant noise sources outside the first stage JFET. The total contribution of the power supplies, the following stages, and the first stage bias resistors is 0.7 nV$\sqrt{\text{Hz}}$ at 7.2 kHz when referred to the input.

With a 2.8 pF dummy load connected to the preamplifier to simulate the capacitance of the apparatus, the total noise referred to the preamplifier input is 8.8 nV/$\sqrt{\text{Hz}}$ at 7.2 kHz. The measured preamplifier input capacitance C_{in} is 1.5 pF. The 2N3686 first stage JFET was selected from 25 tested devices, of types 2N3686, 2N4416, 2N4117, and 2N4220. The device chosen had the smallest value of $C\sqrt{S_n}$, where C is the total capacitance, 2.8 pF + C_{in}, and $\sqrt{S_n}$ is the total noise referred to the input. The forward transconductance of the chosen device is 2 mS.

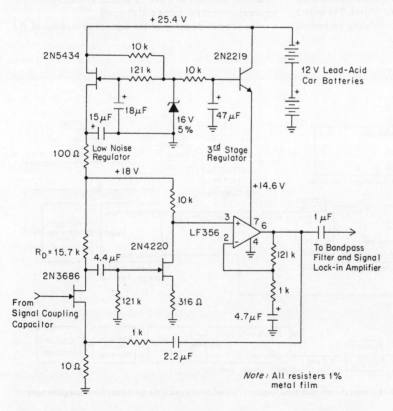

FIG. 6. Preamplifier schematic.

The signal processing chain which follows the preamplifier is shown in Fig. 5. The typical signal amplitude at 7.2 kHz is 1 to 10 μV rms (referred to the preamplifier input), while the signal at 900 Hz is typically 100 times larger. Because of this a bandpass filter at 7.2 kHz is needed to remove the 900 Hz signal so that it will not saturate the lock-in input. The main lock-in amplifier[8] is followed by an antialias filter, an analog to digital converter, and the data logging computer. The computer is also interfaced to the ultraviolet light so that the data logging software can adjust the sample charge.

The instrument as described so far cannot be used for a fractional charge search, since an unknown background signal will be generated by charges on the quartz fiber, and also by other mechanisms. While the sensitivity to charges on the conducting sample may be directly measured (by ejecting photoelectrons, for example), charges on the fiber will be coupled to the instrument with an unknown efficiency and can mimic fractional charge.

Our proposed solution to this problem is shown in Fig. 7. A small conducting sample container is suspended by quartz fibers and the sample itself is placed deep inside the container. The charge is measured, then the sample is removed (perhaps by a small hook or tweezers), and the background signal generated by the sample container and fibers alone is measured. If the sample container is deep enough so that it forms a very good Faraday cup, then the sample-in signal minus the sample-out signal gives a measure of the net sample charge. One could alternatively replace the first sample with a second similar sample. Then the difference of the two readings would measure the difference of the net charges of the two samples, which is sufficient for a fractional charge search.

We have not yet tried to use a sample container in the instrument. So far we have only studied the noise and resolution of the instrument using samples like the one shown in Fig. 4.

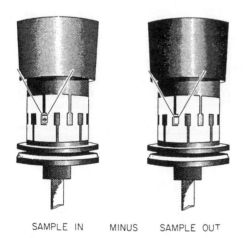

FIG. 7. Use of sample container for background subtraction.

III. DESIGN CONSIDERATIONS

The basic features of the present design follow from the choice of sample size, together with the amplifier noise properties, and some aspects of the mechanics of high-speed rotors.

Since we wish to measure 10-mg samples, the sample dimensions must be about $1\times1\times1$ mm. To provide good shielding, the sample container should be several times deeper than it is wide, for example a small can 1.5 mm in diameter and 4 mm high might be adequate. The pads should be about the same size as the sample container. If they were much smaller the coupling efficiency would suffer, while if they were much larger there would be unnecessary stray capacitance from pad to pad. Some preliminary electrostatics studies with a $\times 10$ scale model have led us to the present dimensions. They are reasonable but perhaps not strictly optimal.

Once the pad geometry is chosen the rotor diameter follows from a choice of the number of pads, or equivalently the signal harmonic number. One pair of pads would seem to be the optimal choice from the point of view of stray capacitance. Unfortunately, there is a very large signal generated at the fundamental (100–1000 μV rms, at the amplifier input), apparently due to the surface potentials in the coupling capacitor gaps. This signal changes by about 1 μV per 1 μm change in the rotor position, so that if the fundamental were also the signal frequency it would be necessary to hold the rotor fixed to about 0.25 nm during the measurement period. (The actual calibration of the instrument is 4.7 nV/q_e, so for a resolution of 0.05 q_e systematic effects must be held below 0.25 nV.) Instead we use the eighth harmonic for the signal frequency, where the background signal with no sample present is only 1 μV rms, and the signal only changes by 0.5 nV per 1 μm of motion of the rotor relative to the coupling capacitors. Thus the rotor must be held stationary only to 0.5 μm, which is required anyway because of the sensitivity to motions of the rotor relative to the sample.

The required signal frequency follows from the amplifier noise properties. Because of the high impedance of the capacitive signal source, the total noise rises very rapidly at low frequencies. The amplifier noise limited S/N doubles between 4 and 8 kHz, but only increases by about a factor of 1.5 in going from 8 to 16 kHz. While it is always better to spin faster, because of mechanical considerations to be discussed below, 7.2 kHz was used as the signal frequency.

Two difficulties arose in attempts to reach the required spin speed. First, it was found to be impossible to pass through flexural critical speeds. These are the rotation frequencies equal to the frequencies of the transverse flexural normal modes of the rotor. At these speeds prototype rotors deformed so much that they hit the sides of the vacuum chamber. Although many high-speed ball bearing systems are able to pass quickly through shaft critical speeds without damage, it seems to be difficult to apply sufficient torque to a high compliance magnetic bearing system. The solution we have adopted is simply to make the rotor subcritical. It has been possible to make the rotor stiff enough so that the first transverse flexural mode is at 1220 Hz, well above the 900-Hz spin speed. To obtain a high critical speed it is helpful to make the rotor hollow. The walls of the present design are

2.5 mm thick, except in the pad region where they are thinned to 1.2 mm to help reduce the stray capacitance from pad to pad.

The second major difficulty in reaching high speeds is an instability of the "forward whirl" which appears in this type of bearing. The basic mechanism was apparently first understood by MacHattie,[9] and a detailed discussion is given in Ref. 6. The forward whirl is a gyrodynamic normal mode of the system, which at very high speeds is identical to the precession of a free symmetric top. The frequency of the whirl mode is proportional to the spin speed (for our rotor the whirl is at 42 Hz when the spin speed is 900 Hz), and because of the mass of the suspension magnet, the damping of the mode due to the oil bath decreases as the rotor spins faster. When there are magnetic losses in the bearing a force arises, increasing with the spin speed, which can drive the forward whirl. As the spin speed is increased eventually the driving force exceeds the damping, and the whirl mode becomes unstable.

Our first prototype of essentially the present design had a bearing made of 1018 cold rolled steel, and had a whirl instability threshold at 400 Hz. The next rotor had a bearing made of unannealed alloy 4750 nickel iron, a lower loss material. It began to whirl at 600 Hz. The present bearing is of annealed alloy 4750. The whirl threshold is now at 1000 Hz, allowing routine operation at 900 Hz with no observable whirl.

IV. PERFORMANCE

The instrument has so far been studied in three stages. First with no sample, then with a sample but without induced charge changes, and then finally with a sample and with periodic charge changes induced by photoemission.

With no sample the total noise observed is simply the 8.8 $nV/\sqrt{Hz}$ noise of the preamplifier, and the spectrum seen at the output of the lock-in amplifier is flat down to the lowest frequencies of interest (about 1 mHz). This measurement ensures that there is no extra noise contribution from the filters, the lock-in amplifier, the sampling, or the digitization.

When a sample is installed the instrument becomes very sensitive to small motions of the sample relative to the rotor. It thus becomes important to fully characterize the motions that are present in the system. By using the available information—the 7.2-kHz signal, the 900-Hz signal, and signals in the bearing servo electronics—it has been possible to obtain spectra of the motions in three orthogonal directions. The sensitivity to motions may be measured by using the sample manipulator to make calibrated translations of the sample. By combining the measured motion sensitivity of a given sample with the motion spectra, it has been possible to calculate the contributions that motions make to the total noise.

Many attempts were made to reduce the motion sensitivities of samples. Flowing ionized N_2 gas over the quartz fiber to discharge it was helpful, as were changes in the sample geometry. With a sample of the sort shown in Fig. 4, but without the guard, the best motion sensitivities achieved were about 20 $nV/\mu m$. The motion amplitudes are 1 to 2 $\mu m/\sqrt{Hz}$, at the important frequencies of a few mHz, so the motional noise was 20 to 40 $nV/\sqrt{Hz}$, well above the amplifier noise. The situation was finally improved by adding the guard. It is a small additional piece of indium, $0.35 \times 0.35 \times 0.25$ mm, pinched onto the quartz fiber 4.3 mm above the sample, as shown in Fig. 4. The guard is placed well above the sample where the coupling to the pads is reduced, but where the coupling has a large dependence on the vertical position. By carefully adjusting the charge of both the sample and the guard it is possible to nearly null the motion sensitivity in both the vertical and the radial directions. In the third direction the motions are small enough so that the total motional noise can be made less than or at least comparable to the amplifier noise. All of the data presented below were recorded with the most recent sample, which is the only guarded sample studied so far. The sample dimensions are as in Fig. 4, and the sample weighs 1 mg.

To make a measurement in a fractional charge search, one would first average the lock-in output for a time T, then change the contents of the sample container, and then average for a time T again. The difference of the two averages gives a measure of the sample charge. To simulate such a measurement we have made data runs where we average the output for a time T, then flash ultraviolet light on the sample to eject a few photoelectrons, and then average again for a time T. This is repeated many times, and the difference of consecutive averages is histogrammed. Peaks appear in the histogram where the sample charge has changed by one, two, or more electrons. The spacing of the peaks gives a calibration of the instrument, and the widths give a measure of the resolution. To more simply study the noise we have also taken data and generated histograms without flashing the ultraviolet light between averaging periods.

Figure 8 shows two typical histograms obtained without flashing the light. Each contains about 7 h of data. The averaging time was 200 s, which is the longest time that can be profitably used because the noise spectrum rises rapidly at

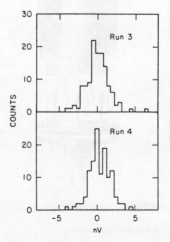

FIG. 8. Histograms of the differences between consecutive 200-s averages of the output. In an actual measurement the sample would be changed between averaging periods. The calibration is 1 $q_e = 4.7$ nV.

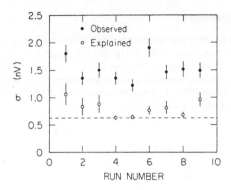

FIG. 9. Observed noises (solid) are the rms widths of histograms like those in Fig. 8. The explained noises (open) are the sum of the amplifier noise contributions and the motional noise contributions. The dotted line shows the amplifier noise level.

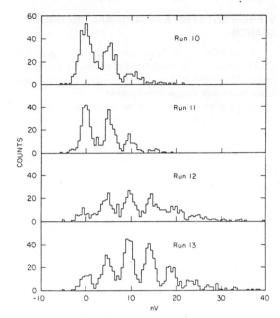

FIG. 11. Histograms with charge changes. These histograms were generated in the same way as those in Fig. 8, except that the sample was exposed to a burst of ultraviolet light between each averaging period.

low frequencies. It has been necessary to subtract a linear fit from each data run because of a nearly linear drift in the data. The drift rate varies from run to run, and is usually about 1 nV/200 s, but was as high as 37 nV/200 s in one run. The slope of the fit is set equal to the average of the slopes of the 200-s segments. This information will still be available when there are unknown changes in the signal between each 200-s period.

Figure 9 shows the measured rms widths of all the runs made without flashing the light. The solid circles give the observed widths, while the open circles give the contribution to the widths which can be accounted for by the sum of the amplifier noise and the motion noise. The dotted line gives the amplifier noise contribution alone:

$$v_n = \frac{\sqrt{2} \cdot 8.8 \text{ nV}/\sqrt{\text{Hz}}}{\sqrt{2 \cdot 200 \text{ s}}} = 0.62 \text{ nV}.$$

The factor of $\sqrt{2}$ in the numerator of this expression occurs because there are two averaging periods contributing to each data point in the histograms. The motional contribution is different for each run because of differences in the sample motion sensitivity. The motional noise contribution is always smaller than the amplifier noise contribution.

There is an important contribution to the total noise that remains unexplained. The average unexplained contribution to the total widths is 1.2 nV, unchanging from run to run within the statistical uncertainties. Figure 10 shows a spectrum of the total noise for runs 3 and 4. The unexplained contribution is dominant in both spectra below about 3 mHz, and it rises rapidly at lower frequencies. The unexplained noise may be due in part to spontaneous charge changes caused by cosmic rays or radioactivity, but a search for steps in the data shows that less than half of the extra noise can be accounted for in this way. We hope that future tests with different samples will provide a clue as to the source of the unexplained noise.

Figure 11 shows histograms for data runs in which the light was flashed between the averaging periods. The light intensity was changed from run to run, but kept constant during any one run. Run 12 seems to be somewhat noisier than the others. It may be that a piece of dust was transferred to the sample when it was reneutralized part way through the run.

In run 13 peaks due to the emission of one, two, three, and four photoelectrons are visible. The peaks are spaced by 4.7 nV, so the instrument's calibration is 4.7 nV/q_e. Since (from the data in Fig.9) the mean observed noise is 1.5 nV, the measured resolution is 0.3 q_e.

From the measured total capacitance of 4.3 pF one should actually expect the calibration to be 13 nV/q_e. Since it is only 4.7 nV/q_e, the instrument is inefficient by a factor of 2.8. Some of the inefficiency is due to the finite size of the coupling capacitances and to imperfect sample to pad coupling.

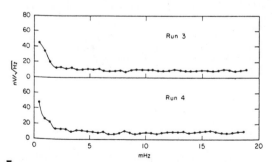

FIG. 10. Spectra of the total noise referred to the input.

V. PROSPECTS FOR A FRACTIONAL CHARGE SEARCH

Before a fractional charge search can be attempted we must first improve the resolution (as measured above) by a factor of 6 to reach 0.05 q_e, and then must extend the technique to absolute charge measurement.

To improve the resolution we must reduce the unexplained part of the noise, the motional noise contribution, and the amplifier noise contribution. The unexplained noise must be reduced by a factor of 5. We do not know if this can be done, but several tests could be made to help us understand the problem. In particular we would like to see how changes in the fiber diameter and sample size affect the unexplained noise.

The motional noise usually contributes about 0.15 q_e to the resolution, so it must be reduced by a factor of 3. This could be done either by reducing the motion sensitivities, or by reducing the motions themselves. The motion sensitivities could be reduced by the addition of a second guard or perhaps by changes in the pad geometry. With a properly placed second guard one could null the motion sensitivity in all three directions, because one would have the charge of three objects available for adjustment. The motions could be reduced by improving the magnetic suspension electronics and by using low thermal expansion materials, or by more careful temperature control.

The amplifier noise contribution also must be reduced by a factor of 3. Improvements could be made by cooling the JFET,[10] by increasing the spin speed, by improving the electrostatic efficiency, or by increasing the averaging time. Some combination of these steps will probably be necessary.

If it should prove difficult to decrease each of the noise contributions then one could still improve the resolution by making multiple measurements of the same sample.

Once the resolution is improved we must then install a sample container in the instrument so that absolute charge measurements can be made. The difficulty is that the sample container position must not be changed when the sample is changed. Since the best motion sensitivities achieved so far are 1 nV/μm = 0.2 q_e/μm, we must maintain the sample container's position to about 250 nm to keep the systematic effects below 0.05 q_e. It will certainly be helpful to swap two identical samples, rather than to go from a full sample container to an empty one, in order to keep the weight constant.

The 250-nm requirement could be relaxed if the motion sensitivity could be reduced, perhaps by the addition of a second guard. Unfortunately, it will still be necessary to keep the relative sample to guard positions fixed to high accuracy. In an improved device it would seem very worthwhile to have some way of constantly monitoring the important motions.

VI. SUMMARY

The rotor electrometer is a new electronic instrument for fractional charge search experiments. A high-speed rotor with an active magnetic bearing is used to rapidly vary the capacitance between a conducting sample and a high impedance preamplifier. In principle, a resolution of 0.05 q_e may be obtained in a few minutes.

The instrument has so far reached a resolution of 0.3 q_e for measurements of changes in the sample charge. The resolution is limited by amplifier noise, sensitivity to motions, and an unexplained effect. Besides the need to improve the resolution, a major outstanding problem is to extend the method to absolute charge measurement. This is made difficult by the instrument's sensitivity to motions of the sample container relative to the rotor.

ACKNOWLEDGMENTS

The authors would like to thank Ronald Baggs, Kenneth Hughes, Robert Leonard, Thomas Nakashima, and Ronald Stickley for their assistance. Work supported by the Department of Energy Contract No. DE-AC03-76SF00515.

---

[a] Present address: Department of Physics, Stanford University, Stanford, CA 94305.

[1] L. W. Jones, Rev. Mod. Phys. **49**, 717 (1977); L. Lyons, Prog. Particle Nucl. Phys. **7**, 157 (1981); G. Susinno, in *Proceedings of the International Conference on Physics in Collision: High Energy ee/ep/pp Interactions, Blacksburg, Virginia, 1981*, edited by W.P. Trower and G. P. Bellini (Plenum, New York, 1982), Vol. I, p. 33.

[2] G. S. La Rue, W. M. Fairbank, and A. F. Hebard, Phys. Rev. Lett. **38**, 1011 (1977); G. S. La Rue, W. M. Fairbank, and J. D. Phillips, *ibid.* **42**, 142 (1979); G. S. La Rue, J. D. Phillips, and W. M. Fairbank, *ibid.* **46**, 967 (1981).

[3] D. Joyce *et al.*, Phys. Rev. Lett. **51**, 731 (1983); M. A. Lindgren *et al.*, *ibid.* **51**, 1621 (1983), and references therein; R. G. Milner, B. H. Cooper, K. H. Chang, K. Wilson, J. Labrenz, and R. D. Mc Keown, *ibid.* **54**, 1472 (1985); P. F. Smith, G. J. Homer, J. D. Lewin, H. E. Walford, and W. G. Jones, Phys. Lett. B **153**, 188 (1985); G. Hirsch, R. Hagstrom, and C. Hendriks, Lawrence Berkeley Laboratory preprint No. LBL-9350 (1979) (unpublished).

[4] E. R. Williams and G. T. Gillies, Lett. Nuovo Cimento **37**, 520 (1983).

[5] W. Innes, S. Klein, M. Perl, and J. C. Price, preprint SLAC-PUB-2938 (1982) (unpublished).

[6] J. C. Price, Ph.D. thesis (Stanford University, 1985) (printed as SLAC-Report-288, Stanford Linear Accelerator Center, 1985) (unpublished).

[7] J. W. Beams, J. D. Ross, and J. F. Dillon, Rev. Sci. Instrum. **22**, 77 (1951). Many references to the Beams work are given by P. J. Geary, *Magnetic and Electric Suspensions* (British Scientific Instrument Research Association, Kent, 1984).

[8] Princeton Applied Research, model #5402.

[9] L. E. MacHattie, Rev. Sci. Instrum. **12**, 429 (1941).

[10] F. Bordoni, G. Maggi, A. Ottaviano, and G. V. Pallottino, Rev. Sci. Instrum. **52**, 1079 (1981); S. Klein, W. Innes, and J. C. Price, *ibid.* **56**, 1941 (1985).

COMMENTS ON C11
"Electron-Positron Collision Physics: 1 MeV to 2 TeV"

In 1988 I reviewed all of electron-positron collision physics for a meeting of nuclear and particle physicists. While preparing this review I realized that there was no experimental data on e^+e^- collisions in the center-of-mass energy region between several MeV and about 400 MeV. There still is no data in this region in 1995. At any energy the reactions

$$e^+ + e^- \rightarrow e^+ + e^-$$

$$e^+ + e^- \rightarrow \gamma + \gamma$$

will occur. Above 200 MeV

$$e^+ + e^- \rightarrow \mu^+ + \mu^-$$

$$ee^+ + e^- \rightarrow \mu^+ + \mu^- + \gamma$$

will also occur.

Does the theory of quantum electrodynamics perfectly predict the dynamics of these processes or is there some unknown new physics hidden in these processes? The accepted answer to this question is that there is no new physics because (a) quantum electrodynamics works perfectly in atomic physics which bounds this energy range from below, and (b) to the best of our knowledge there are no experimental deviations from the predictions of quantum electrodynamics above this energy range. I have not found a model for new physics which would allow deviations from quantum electrodynamics to be found in the several MeV to several hundred MeV center-of-mass energy range. Such a model must not predict deviations which would already have been discovered at lower or higher energy.

But I haven't looked hard for such a new physics model. Soon after I wrote this paper I took up the cause of the tau-charm factory as described in Reprint B2. I didn't have time to push this idea along although I saw how to do the experiment. Use the 50 GeV electron beam at SLAC to make a positron beam of variable energy. Build a neat and small apparatus, probably composed completely of silicon detectors, which can detect electrons, muons, and photons. I don't think a magnetic field is necessary. It would be a sweet experiment.

Of course proposing such an experiment would mean going against the consensus of fruitful ways to look for new elementary particle phenomena and it would be hard to convince a program committee to allot accelerator running time to the experiment. I haven't the time or resources now to push this experiment, I'm too busy with 10 GeV electron-positron collision physics working in the CLEO collaboration and with a new search for isolated quarks (Reprint C14). Perhaps some younger physicist will take up this idea, but given the difficulty of getting permanent jobs today in high-energy physics they would have to be very brave. This kind of experiment is probably only for the old and contrary and tenured.

The second energy region of unknown physics discussed in this 1988 review was above 50 GeV, in particular the region of Z^0 production and decay. LEP and the SLC were still being constructed. Now in 1995 we have a tremendous amount of beautiful experimental information about Z^0 production and decay. We are much wiser but we are disappointed; new particles have not been found, new physics has not been found.

ELECTRON–POSITRON COLLISION PHYSICS: 1 MeV TO 2 TeV*

Martin L. Perl
Stanford Linear Accelerator Center
Stanford University, Stanford, California 94309

ABSTRACT

An overview of electron-positron collision physics is presented. It begins at 1 MeV, the energy region of positronium formation, and extends to 2 TeV, the energy region which requires an electron-positron linear collider. In addition, the concept of searching for a lepton-specific force is discussed.

TABLE OF CONTENTS

	Page
A. Introduction	1185
B. 1 MeV to 500 MeV and Positronium	1186
C. 500 MeV to 3 GeV and Hadron Production	1188
D. 3 GeV to 10 GeV: the ψ, the τ, the c–Quark and Hadron Jets	1189
E. 10 GeV to 56 GeV: b–Quark Physics and Other Successes	1190
F. Overview of Cross Section for $e^+e^- \to \ell^+\ell^-$	1193
G. Overview of Cross Section for $e^+e^- \to$ hadrons	1194
H. 56 GeV to 70 GeV	1195
I. The $Z°$ Region: SLC and LEP	1197
J. 100 GeV to 200 GeV: LEP	1200
K. 200 GeV to 2 TeV: e^+e^- Linear Colliders	1201
L. Speculations on a Lepton–Specific Force	1202
Acknowledgments	1204
References	1204

A. Introduction

The initial purpose of this talk was to tell an audience of nuclear physicists and particle physicists about electron-positron collision physics: what we have learned and what we are doing. As I wrote the talk, I found that it provided me with the opportunity to give a broader view of electron-positron collision physics, to point out the areas where future research might be most fruitful. Some of these areas are well-known: electron-positron annihilation physics at the $Z°$ and studying the properties of hadrons containing the b-quark. Less recognized fruitful areas are: understanding the decay modes of the tau lepton and studying hadrons containing the c-quark. There are two areas which I have never seen discussed: electron–positron collision physics below 500 MeV in the barycentric system and the search for what I call a lepton-specific force. In the written version, I have shifted the emphasis to pointing out these less recognized and unrecognized areas of electron-positron collision physics.

---

*Work supported by the Department of Energy, contract DE–AC03–76SF00515.

© 1988 American Institute of Physics

Reprinted from *Proc. Intersections between Particle and Nuclear Physics*, ed. G. M. Bunce, pp. 1185, Copyright 1988, with kind permission from American Institute of Physics.

Figure 1 is the route map for this talk. I start at 1 MeV and go up in energy, dividing the discussion into energy regions according to the physics, the experimental method and the accelerator technology.

I have restricted the references to a few special areas because I am covering a broad subject.

Unless otherwise noted, all energies are total energies in the barycentric system denoted by $\sqrt{s}$ or $E_{c.m.}$.

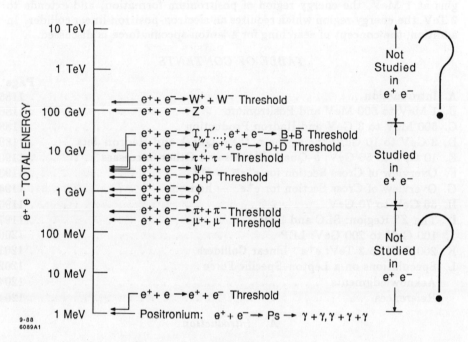

Figure 1

B. 1 MeV to 500 MeV and Positronium

I begin with the regions of 1 MeV to 500 MeV total energy. At the 1 MeV boundary of this region lies a vast area of e^+e^- physics: the formation of positronium

$$e^+ + e^- \to Ps, \quad (1a)$$

and its decay

$$Ps \to \gamma + \gamma, \quad \gamma + \gamma + \gamma. \quad (1b)$$

A minor subject in the region is the study of Bhabha scattering

$$e^+ + e^- \to e^+ + e^-. \quad (2)$$

Three decades ago, Bhabha scattering was studied using a positron beam on a fixed target, then work stopped in most of the energy region. Research continued only at the low-energy end, the area of atomic physics. In the last few years, a special interest in Bhabha scattering at an $E_{c.m.}$ of 1.6 to 1.8 MeV has developed, as I discuss later.

There are two reasons for there being no interest in general studies of Bhabha scattering or photon pair production,
$$e^+ + e^- \to \gamma + \gamma , \qquad (3)$$
in the 1 MeV to 500 MeV region. First, the consensus is, or at least was, that quantum electrodynamics explains all e^+e^- physics in this region. Second, an e^+e^- circular collider of traditional design has low luminosity in this energy region, and no e^+e^- collider has been built in this energy region. The two reasons reinforce each other. If there is no physics interest, there is no incentive to build a collider. If a collider is very hard to build, why get interested in the physics. One set of perhaps anomalous measurements has revived interest in studying Bhabha scattering in the 1 MeV to several MeV region. Some experiments[1] show e^+e^- pairs produced in collisions of high Z ions, these pairs having masses of 1.6 to 1.8 MeV/c^2. The data on these pairs is confusing; the production mechanism is unknown. Electron beam dump experiments[2] have excluded a production mechanism involving a conventional, unstable elementary particle: $\phi \to e^+ + e^-$. Bhabha scattering at an $E_{c.m.}$ of 1.6 to 1.8 MeV might show a peak in the cross section if the unknown production mechanism can occur in a pure e^+e^- system; that is, if the presence of high Z ions is not required. Numerous measurements of Bhabha scattering in the energy region have been made in the past few years. An e^+ beam and a fixed target, usually of low Z, are used. At present, the measurements are contradictory and there is no confirmed observation of a peak.[3]

Interest in the several MeV region has also been stimulated by possible problems in precisely understanding the properties of positronium itself; the lifetime of orthopositronium, for example.[4]

I have been thinking about the best strategy for exploring the 1 MeV to 500 MeV region; measuring with precision the cross sections for
$$e^+ + e^- \to e^+ + e^- , \qquad (4a)$$
$$e^+ + e^- \to \mu^+ + \mu^- , \qquad (4b)$$
$$e^+ + e^- \to n\gamma , \quad n \geq 2 ; \qquad (4c)$$
looking for anomalous differential cross sections and resonances in the total cross section. We do not know how to design an e^+e^- circular collider which can operate from several MeV to 500 MeV, because known design principles limit the dynamic range to three or four. Therefore, I am considering fixed target experiments. The SLAC linear accelerator can produce e^+ beams with a maximum energy of 50 GeV, corresponding to $E_{c.m.} = 220$ MeV. Above about 500 MeV, the VEPP–2M e^+e^- circular (Sec. C) can take up studies of the reactions in Eq. 4; but there would still be a gap in energy coverage.

In Sec. L on lepton-specific forces, I will again take up the 1 MeV to 500 MeV region. There are indirect ways to explore this region, using processes

such as
$$e^+ + e^- \to e^+ + e^- + e^+ + e^-,$$
but these indirect methods have less sensitivity.

C. 500 MeV to 3 GeV and Hadron Production

The history of high-energy, e^+e^- collision physics began in this energy region: ADONE in Italy, the CEA collider in the United States, the DCI in France and the VEPP colliders in the Soviet Union. Today, there is only one e^+e^- collider in this region, VEPP–2M.

In this energy region, the production of hadrons becomes important
$$e^+ + e^- \to \text{hadrons},$$
and it was in this region that two main processes for hadron production were elucidated. Resonances such as the ρ, ω, ϕ, and at higher energies, the ψ and Υ, are produced through the process in Fig. 2a. The continuum production of hadrons,
$$e^+ + e^- \to \text{many hadrons},$$
occurs through the process in Fig. 2b, with the cross section per quark type and color,
$$\sigma(e^+e^- \to q\bar{q} \to \text{hadrons}) = \frac{4\pi\alpha^2 Q_q^2}{3s}. \tag{5}$$
The recognized research to be done in this region concerns more precise studies of hadron production. I also see valuable research to be done on the reactions in Eq. 4, searching for anomalous behavior. Looking ahead, a high luminosity collider producing
$$e^+ + e^- \to \phi \to K^\circ + \bar{K}^\circ$$
can extend our knowledge of the CP violating mechanism in the $K^\circ \bar{K}^\circ$ system.

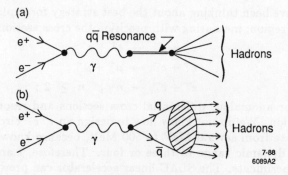

Figure 2

D. 3 GeV to 10 GeV: the ψ, the τ, the c-Quark and Hadron Jets

This is the energy region where four great discoveries were made in the 1970's:

(i) The ψ, ψ', ψ'' family of $c\bar{c}$–hadrons was discovered.

(ii) The τ heavy lepton was found through

$$e^+ + e^- \to \tau^+ + \tau^- . \qquad (6)$$

(iii) The D–hadrons containing a single c–quark were found.

(iv) Hadron jets produced by the process

$$e^+ + e^- \to q + \bar{q} , \quad q \to \text{hadron jet} , \quad \bar{q} \to \text{hadron jet} ,$$

were identified.

One e^+e^- collider is now operating in this region, SPEAR in the United States. A new, higher luminosity e^+e^- collider, BEPC, is now being built in the Peoples Republic of China to operate in this region.

There are a number of recognized research areas in this region. There is much more to be done in studies of the ψ, D and F particle families. The complicated energy dependence of $\sigma(e^+e^- \to \text{hadrons})$ from $E_{c.m.} = 3.5\ GeV$ to about 5 GeV, Fig. 3, is not understood. The τ decay mode puzzle[5,6] needs to be unraveled. Some properties[7] of the τ, such as the mass of ν_τ, can be measured precisely using data from this region. I add to this list more precise studies of the lepton-photon vertices in Eqs. (4) and (6) and the search for a lepton-specific force.

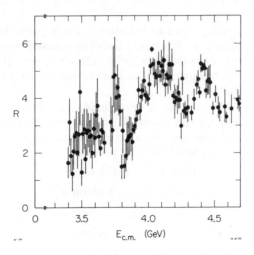

Figure 3

The need for large amounts of data to further the research listed in the previous paragraph has stimulated a proposal for a very high luminosity e^+e^- circular collider in this energy region.[8,9]

E. 10 GeV to 56 GeV: b–Quark Physics and Other Successes

At the lower end of the energy region lies the threshold for the production of the upsilon family of particles, $\Upsilon, \Upsilon', \Upsilon'', \ldots$, made up of a $b\bar{b}$–quark pair, and the threshold for the production of hadrons containing a single b–quark. Three e^+e^- circular colliders operate in the 10 GeV region: CESR in the United States, DORIS in Germany and VEPP–4 in the Soviet Union.

At higher energies, there is the PEP collider in the United States, 20 to 30 GeV, and the new TRISTAN collider in Japan. The TRISTAN collider sets the upper end of this region, 56 GeV, the highest e^+e^- collision energy at which there is data. In the next few years, TRISTAN's energy will move into the 60–70 GeV range.

There have been four successes in the 10–56 GeV range. One triumphant success is the comprehensive research on the properties of hadrons containing the b–quark. The most recent research indicates the possible existence of substantial mixing of B° and $\overline{B}^\circ$ mesons. This in turn allows the possibility of searching for CP violation in the B°–$\overline{B}^\circ$ system and, if it exists, studying CP violation in this new system.

Future detailed studies of b–quark physics and searching for CP violation requires e^+e^- collider luminosities in the range of 5×10^{32} to 10^{34} cm^{-2} s^{-1}. Existing e^+e^- colliders have luminosities in the range of 10^{31} to 10^{32} cm^{-2} s^{-1}. Therefore, a great amount of discussion and design work is being devoted to higher luminosities. Plans and proposals[10] include: increasing the luminosities of the existing single-ring, circular colliders; building new double-ring, circular colliders; building a linear collider; and building a mixed linear-circular collider. These e^+e^-–collider proposals have to be compared with proposals to use the large number of B mesons produced in hadron-hadron collisions, in fixed target or collider experiments.

The second great success is the elucidation of the theory of quantum chromodynamics through the study of hadron production in the continuum. An important part of this elucidation came from the discovery and study of events with three hadron jets from the process

$$e^+ + e^- \to q + \bar{q} + g ,$$
$$q \to \text{quark} - \text{hadron jet} ,$$
$$\bar{q} \to \text{quark} - \text{hadron jet} ,$$
$$g \to \text{gluon} - \text{hadron jet} .$$

The third success is the study of hadron production and related physics in the two-virtual photon process,

$$e^+ + e^- \to e^+ + e^- + \gamma_v + \gamma_v$$
$$\gamma_v + \gamma_v \to \text{hadrons}$$

in Fig. 4. Here γ_v is a virtual photon which in some kinematic conditions is almost a real photon.

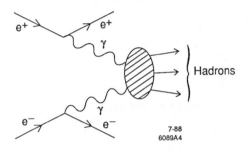

Figure 4

The fourth success is the measurement of the interference of the electromagnetic and weak amplitudes, Fig. 5, as the energy advances up the lower tail of the Z°. A taste of greater things to come.

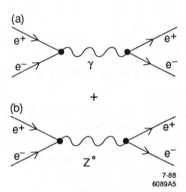

Figure 5

The PETRA e^+e^- collider in Germany contributed data to all of the mass, along with the other colliders listed in this section. PETRA which reached a maximum energy of about 46 GeV is not operated at present.

Out of the large amount of data collected in this 10 to 56 GeV region has also come a strange result. The many measurements of the decay modes of the τ lepton have resulted in a problem in understanding the decay modes with one-charged particle, Table I. There is a discrepancy[5,6] between the inclusive

Rows	Symbol	Decay Mode of τ^-	Branching Fraction (%)
1	B_1	1–charged particle inclusive	86.6 ± 0.3
2	B_e	$\nu_\tau + e^- + \bar{\nu}_e$	17.6 ± 0.4
	B_μ	$\nu_\tau + \mu^- + \bar{\nu}_\mu$	17.7 ± 0.4
	B_π	$\nu_\tau + \pi^-$	10.8 ± 0.6
	B_ρ	$\nu_\tau + \rho^-$	22.5 ± 0.9
		Sum for modes in Rows 2	68.6 ± 1.2
3	$B_{\pi 2\pi^0}$	$\nu_\tau + \pi^- + 2\pi^0$	7.6 ± 0.8
		$\nu_\tau + mK + n\pi^0$ $\longrightarrow$ 1–charged particle $m \geq 1,\ n \geq 0,\ K = K^0$ or K^-	1.8 ± 0.3
		Sum for modes in Rows 3	9.4 ± 0.9
4		$\nu_\tau + \pi^- + n\pi^0$ $n \geq 3$ $\nu_\tau + \pi^- + m\eta + n\pi^0$ $\longrightarrow$ 1–charged particle $m \geq 1,\ n \geq 0,$	
		Sum for modes in Rows 4	≤ 2.7
5		Sum for modes in Rows 2, 3 and 4	80.7 ± 1.5

Table 1. Summary of present knowledge of 1–charged particle branching fraction in percent from Refs. 5 and 6. The numbers in Rows 1, 2, and 3 are the average of measured values and the associated standard deviation. The sum in Row 4 is the 95% upper limit obtained from other data and accepted theory. Note that Refs. 5 and 6 used Gaussian error distributions.

one-charged particle branching fraction and the sum of the known exclusive one-charged particle branching fractions.

The 10 to 56 GeV region has brought great disappointment, as well as great success. The top quark has not been found; a fourth generation quark has not been found; a fourth generation charged lepton has not been found; nor have additional neutral leptons been found—but the neutral leptons searches are not definitive.[11] Further e^+e^- searches for these particles requires higher energy: TRISTAN above 56 GeV, the SLAC Linear Collider (SLC) and LEP in the Z° region of 70 to 110 GeV, LEP above the Z° to about 200 GeV and, ultimately, e^+e^- linear colliders into the TeV region.

Before moving on from what we have measured below 56 GeV to what we hope to measure above 56 GeV, I will summarize our knowledge and expectations about the total cross sections for

$$e^+ + e^- \to \ell^+ + \ell^- \; ; \quad \ell = e, \mu, \tau \,,$$

and

$$e^+ + e^- \to \text{hadrons} \,.$$

These reactions make up the explored land of known e^+e^- collision physics. We have to look beyond that land for new particles or new phenomena.

F. Overview of Cross Section for $e^+e^- \to \ell^+\ell^-$

The total cross section for

$$e^+ + e^- \to \ell^+ + \ell^- \; ; \quad \ell = e, \mu, \text{ or } \tau \,, \tag{7}$$

follows from electroweak interaction theory. In writing down these formulas, I keep in mind that the formulas have only been confirmed directly in the e^+e^- collisions below 56 GeV, and indirectly in the decays of Z°'s produced in $p\bar{p}$ collisions. There may be surprises. Consider three energy regions, Fig. 6: the region centered on the Z° mass of about 93 GeV; the region below the Z°; and the region far above the Z°, above 200 GeV. In these formulas I ignore threshold effects and radiative corrections. The latter may be substantive, changing cross sections by tens of percent.

Below the Z°, the electromagnetic process, Fig. 5a, dominates with

$$\sigma(e^+e^- \to \gamma \to \ell^+\ell^-) = \frac{4\pi\alpha^2}{3s} \approx \frac{87}{s} \text{ nb} \,, \tag{8a}$$

where s is in GeV^2. The weak process itself gives, Fig. 5b,

$$\sigma(e^+e^- \to Z^\circ \to \ell^+\ell^-) = \frac{G^2 s}{96\pi} = 1.8 \times 10^{-7} s \text{ nb} \,, \tag{8b}$$

where s is again in GeV^2. Thus, when $\sqrt{s} \lesssim 50$ GeV, the weak process is detected through its interference with the electromagnetic.

In the vicinity of the Z°, the weak process dominates as a real Z° is produced. Here

$$\sigma(e^+e^- \to Z^\circ \to \ell^+\ell^-) = \frac{G^2 s}{96\pi} \left[\frac{m_Z^4}{(s-m_Z^2)^2 + \Gamma_Z^2 m_Z^2} \right] \,, \tag{9}$$

This is about 1.6 nb at $E_{c.m.} = m_Z$. Radiative corrections reduce this to about 1.2 nb.

Above 200 GeV, if there are no surprises, the electromagnetic process contributes according to Eq. (8a), but it is convenient to put s in TeV^2 and use picobarns instead of nanobarns. Then,

$$\sigma(e^+e^- \to \gamma \to \ell^+\ell^-) \approx \frac{0.087}{s} \text{ pb} \approx \frac{0.1}{s} \text{ pb} \,, \tag{10a}$$

a very small cross section. The weak process, ignoring interference, gives an even smaller cross section:

$$\sigma(e^+e^- \to Z^\circ \to \ell^+\ell^-) \approx \frac{G^2 m_Z^4}{96\pi s} = \frac{0.013}{s} \text{ pb} , \qquad (10b)$$

with s in TeV^2.

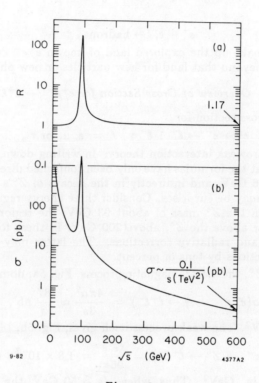

Figure 6

Figure 6 also gives the cross section in terms of R, where

$$R = \frac{\sigma}{\sigma_0}, \qquad \sigma_0 = \frac{4\pi\alpha^2}{3s} . \qquad (11)$$

Thus far above this Z° region, $R(e^+e^- \to \ell^+\ell^-)$ is again about one.

G. Overview of Cross Section for $e^+e^- \to$ hadrons

Again, I consider three energy regions: below the Z°, in the vicinity of the Z° and far above the Z°.

Below the Z°, hadron production is dominated by the electromagnetic process

$$e^+ + e^- \to \gamma \to \text{hadrons} , \qquad (12)$$

operating through the two mechanisms described in Fig. 2: resonance production and continuum production. The cross section for continuum production is obtained from Eq. (5) by summing over the number of quarks multiplied by three, for the three colors. Ignoring quantum chromodynamic corrections

$$\sigma(e^+e^- \to \gamma \to \text{hadrons, continuum}) \approx \frac{4\pi\alpha^2}{s} \sum_q Q_q^2 . \qquad (13a)$$

Usually, as in Fig. 7,[12] R is displayed. Measurements confirm the expectation

$$R(e^+e^- \to \gamma \to \text{hadrons, continuum}) \approx 3 \sum_q Q_q^2 . \qquad (13b)$$

Once quantum chromodynamics is taken into account, as $E_{c.m.}$ rises to the lower tail of the Z°, the weak process contribution

$$e^+ + e^- \to Z^\circ \to \text{hadrons} , \qquad (14)$$

becomes obvious, Fig. 7b.

In the vicinity of the Z°, we expect an enormous increase in $\sigma(e^+ + e^- \to Z^\circ \to$ hadrons), Fig. 8. At the Z°, $\sigma(e^+ + e^- \to Z^\circ \to \text{hadrons}) \approx 40$ nb, including the radiative corrections.

Far above the Z°(above, say, 200 GeV), the production of hadrons occurs through the combined amplitudes for the reactions in Eqs. (12) and (14). In analogy to Eq. (10)

$$\sigma(e^+e^- \to q\bar{q} \to \text{hadrons}) = \frac{0.087\, r_q\, T_q\, (m_q, s)}{s} \text{ pb} . \qquad (15)$$

Here s is in TeV^2, T_q is a threshold factor with $T_q = 1$ for $s \gg m_q^2$, m_q is the mass of the q–quark and

$$r_q = 1.9 , \quad \text{charge } \frac{2}{3} \text{ quarks}: u, c ,$$

$$r_q = 1.1 , \quad \text{charge } \frac{1}{3} \text{ quarks}: d, s, b ,$$

These five known quarks will give

$$\sigma\left(e^+e^- \to \text{hadrons}\right) = \frac{0.6}{s} \text{ pb} . \qquad (16)$$

With these completed overviews of charged lepton and hadron production, I move above the 56 GeV boundary to energies yet to be explored in e^+e^- collision physics.

H. 56 GeV to 70 GeV

The new TRISTAN collider will soon explore this region. The great interest is to search for new particles, particularly the top quark, and to search for new phenomena. Other interests are the study of the interference between electromagnetic and weak amplitudes and the study of two-virtual photon physics.

I have though of a speculative possibility in this region where the cross sections for lepton and quark production are at a minimum before the rise of

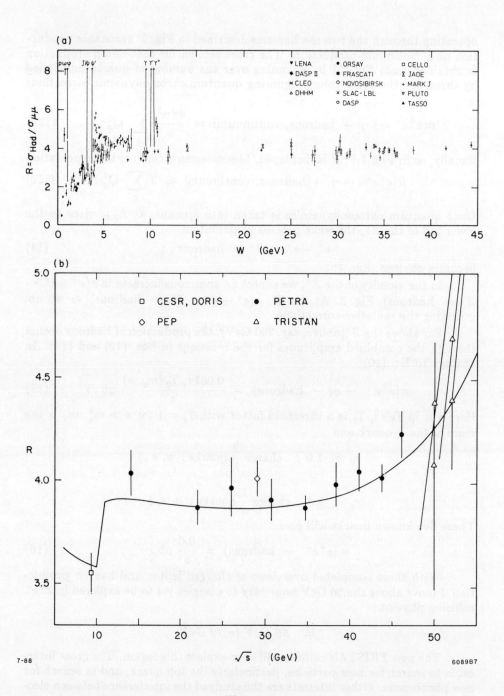

Figure 7

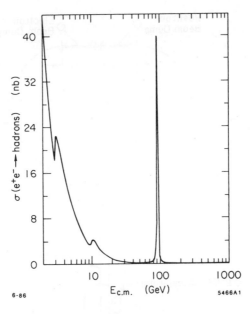

Figure 8

the $Z°$. Suppose there is a new phenomenon in e^+e^- collision physics which occurs at much higher energy. At lower energies, the cross section for effects due to this phenomenon might be proportional to s in analogy to Eq. (8b). The minimum in the cross section for lepton and quark production is an ideal place to look for such effects.

I. The $Z°$ Region: SLC and LEP

We are entering this energy region with two new e^+e^- colliders. The SLAC Linear Collider (SLC) in the United States, Fig. 9, is starting operation at the $Z°$. Its energy range is 70 to 110 GeV. It is the first collider to use the linear collider principle,[13] pointing the way for this new collider technology.

In about a year, the LEP e^+e^- circular collider, Fig. 10, in Switzerland will begin operation; the largest diameter accelerator or collider in the world. It will begin in the $Z°$ energy region. As the supply of radio frequency power to the circulating beams is increased, the energy range will be extended to about 180 GeV.

So much has been dreamed and written about e^+e^- collision physics at the $Z°$; too much to summarize. I will mention a few main points.[14]

The $Z°$ resonance, Figs. 6 and 8, provides an enormous cross section for lepton and quark production at high energy. Neutral elementary fermions, such as neutrinos, are produce as copiously as charged fermions. At $E_{c.m.} = m_z$,

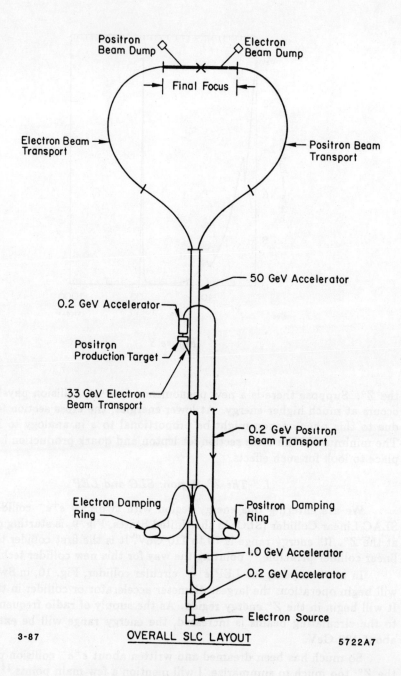

Figure 9

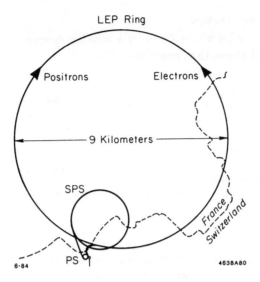

Figure 10

ignoring radiative corrections:

$$\sigma(e^+e^- \to Z^\circ \to f\bar{f}) = \frac{G^2 m_z^4 r_f T_f(m_f, m_z)}{96\pi \Gamma_Z^2}. \quad (17)$$

Here T_f is a threshold factor and

$$r_f = 1.0, \quad f = \ell^- = e^-, \mu^- \text{ or } \tau^-$$
$$r_f = 2.0, \quad f = \nu_\ell,$$
$$r_f = 1.2, \quad f = u \text{ or } c \text{ quark},$$
$$r_f = 1.5, \quad f = d, s, \text{ or } b \text{ quark},$$

Searches for new particles can be direct or indirect. If a new particle is charged, stable or unstable, or if a new particle is neutral and unstable, the search can be direct—unexplained events from the Z° decay.

The production of new stable neutral particles—massive neutrinos, for example—can be detected indirectly through additions to the predicted width of the Z°:

$$\Gamma_Z = \sum_f \Gamma_{Zf},$$

$$\Gamma_{Zf} = \frac{G m_Z^3 r_f T_f}{24\sqrt{2}\,\pi} \quad (18)$$

This, again, ignores radiative corrections which must be precisely calculated. If the mass of the new, neutral particle is of the order of a GeV/c² or less, the

cross section for the reaction
$$e^+ + e^- \to Z^\circ \to \gamma + \text{missing energy} , \quad (19a)$$
will be augmented above that contributed by
$$e^+ + e^- \to Z^\circ \to \gamma + \nu_\ell + \bar{\nu}_\ell , \quad \ell = e, \mu, \tau . \quad (19b)$$

If you believe in the existence of a physical Higgs particle, the elusive H° can be sought through
$$e^+ + e^- \to Z^\circ \to Z^\circ + H^\circ .$$

Returning to known physics, production and decay of the Z° at LEP and the SLC is a source of c–quarks, b–quarks and τ leptons, allowing extension of the knowledge gained in the 3 GeV to 56 GeV regions. In addition, electroweak theory and quantum chromodynamics can be tested in more detail.

J. 100 GeV to 200 GeV: LEP

Moving above the Z°, the energy region up to about 200 GeV will be explored by the LEP collider.[15] New searches will be made for more massive quarks and leptons, for the Higgs particle (if not yet found), and for speculative particles of all sorts.

The reaction
$$e^+ + e^- \to W^+ + W^- , \quad (20)$$
is of great interest because one of its amplitudes, Fig. 11, has a W–Z°–W vertex. Electroweak theory predicts the cross section in Fig. 12. At 200 GeV, $\sigma(e^+e^- \to W^+W^-) = 20$ pb compared to $\sigma(e^+e^- \to \mu^+\mu^-) = 2.5$ pb from Eq. (10). Thus, the cross section for $e^+e^- \to W^+W^-$ is relatively substantial at the upper end of this energy range. Once W's are directly produced, their properties and their decay products can be studied in detail.

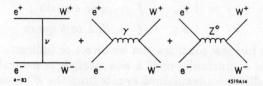

Figure 11

For most of us, the 100 to 200 GeV region is both fascinating and frightening. We expect to understand e^+e^- collision physics thorough the Z° region, although this may be hubris; but the 100 to 200 GeV region is the entrance to *terra incognita*. What if no new particles are found below 200 GeV, not even the top quark? What if $\sigma(e^+e^- \to f\bar{f})$ continues to decrease as $1/s$, beautifully simple, but ever smaller?

It will be eight to ten years before experimenters at LEP have concluded a through exploration of this region. We are too impatient to wait those years

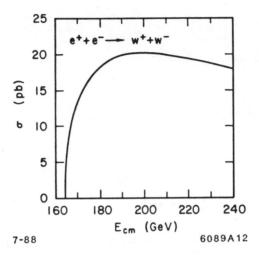

Figure 12

before deciding to push above 200 GeV in e^+e^- collision physics. Hence, our great interest in e^+e^- linear colliders[13] and e^+e^- collision physics from 200 GeV to 2 TeV.

K. 200 GeV to 2 TeV: e^+e^- Linear Colliders

We are excited about with e^+e^- linear colliders, Fig. 13, because the traditional e^+e^- circular collider costs too much to build and operate above 200 or 300 GeV. We don't know yet how to design a 500 GeV or 2 TeV linear collider, but we know the general principles; we know what we have to do. Cross sections of the order of

$$\sigma \sim \frac{.1}{s} \text{ pb},$$

with s in TeV^2, require luminosities in the range of 10^{33} to 10^{34} cm^{-2} s^{-1}. This requires the transverse dimensions of the colliding e^+ and e^- bunches to be 10^{-3} to 10^{-1} μm, putting large demands on controlling the size and position of the bunches as they move through the linear accelerator. Large amounts of microwave power must be produced efficiently. Bunch halos and backgrounds from passage of the bunches through the magnets and collimators must be eliminated so as not to overwhelm the detector.

Linear collider research and development work[13] is going on in Europe, Japan, the Soviet Union and the United States. There is a great deal to learn and invent. It may not be wise to try to jump from the 100 GeV, pioneering SLAC Linear Collider to a 1 or 2 TeV linear collider. The best course may be to build an intermediate energy facility of, say, 500 GeV and then use that experience to go on.

As I have already said, the 200 GeV to 2 TeV region is *terra incognita*. Even conventional processes become strange and wonderful.[16,17] Suppose there

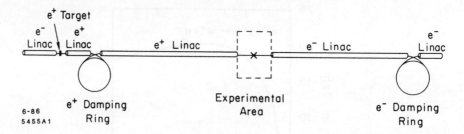

Figure 13

is a massive lepton pair L^-, L° with

$$m_- - m_0 > m_W \,,$$

where m_-, m_0 and m_W are the masses of the L^-, L° and W. Then the major decay mode is

$$L^- \to L^\circ + W^- \,,$$

where the W^- is real and itself decays. The W^- here is like the π^- or ρ^- in

$$\tau^- \to \nu_\tau + \pi^- \,, \quad \nu_\tau + \rho^- \,.$$

Similarly, the major decay mode of a massive quark pair q, q' with

$$m_q - m_{q'} > m_W \,,$$

is

$$q \to q' + W^- \,.$$

Another wonderful effect of very high energy on conventional processes is that the "two virtual-photon" process of Fig. 14a is joined by the "two virtual-W" process of Fig. 14b. A variation of the process in Fig. 14b provides a neat way to produce a physical Higgs particle,

$$e^+ + e^- \to \bar\nu_e + \nu_e + H^\circ \,,$$

via the process in Fig. 14c.

By now we are far in the future; it is almost time to conclude this journey from 1 MeV to 2 TeV. Before concluding, I want to speculate on something we can search for now—a lepton-specific force.

L. *Speculations on a Lepton-Specific Force*

It is strange to me that it is conventional to hope or expect that (a) lepton generations mix just as quark generations mix, and (b) neutrinos have nonzero mass. It is strange because experiment leads to opposite conclusions. There is no evidence for $e - \mu$, $e - \tau$ or $\mu - \tau$ generation mixing. Indeed, one of the most precise measurements in particle physics is the upper limit

$$\frac{\Gamma(\mu^- \to e^- \gamma)}{\Gamma(\mu^- \to e^- \bar\nu_e \nu_\mu)} < 10^{-10} \,.$$

There is no confirmed evidence for neutrino masses being other than zero.

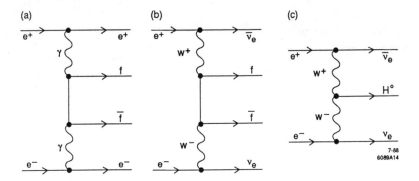

Figure 14

I see the leptons as very different from the quarks, and this has led me to speculate as to the existence of a force only exerted by and on leptons—a lepton-specific force.

I think of the force as being carried by a neutral particle λ, with mass m_λ and an L–λ–L coupling of strength $\sim \sqrt{\alpha_\lambda}$. Here L is a charged lepton or neutrino. This speculation overlaps axion and Higgs particle ideas, but the λ doesn't couple to nonleptons. The precise strength of the coupling may depend on the type of lepton or generation, but I don't think of $\sim \sqrt{\alpha_\lambda}$ as depending, like the Higgs particle, on the lepton mass.

Christopher Hawkins and I[18] have considered what ranges of m_λ and α_λ are ruled out by: atomic measurements such as $g_e - 2$, by previous particle physics measurements and searches, such as axion searches, and by deductions from astrophysics. We find that when

$$m_\lambda \gtrsim 10 \text{ to } 100 \text{ MeV}/c^2 \;.$$

There are few experimental limits on the existence of a particle λ carrying a lepton-specific force.

Three hypothetical processes in e^+e^- collision physics, Fig. 15, allow searches for the λ. The process in Fig. 15a, with a λ in the t–channel, would affect small angle Bhabha scattering—a reaction which has never been tested to better than several percent in e^+e^- collisions physics.

The process in Fig. 15b, annihilation through the λ, would show up as an energy resonance at $E_{c.m.} = m_\lambda$ in

$$e^+ + e^- \to \ell^+ + \ell^- \;, \quad \ell = e, \mu, \tau \;. \tag{21}$$

In Sec. B, I noted that there are precise experiments searching for a resonance in $e^+e^- \to e^+e^-$ at $E_{c.m.} \sim 1.8$ MeV; but many energy regions in $E_{c.m.}$ collision physics have not been explored for an energy resonance in the reactions in Eq. 21.

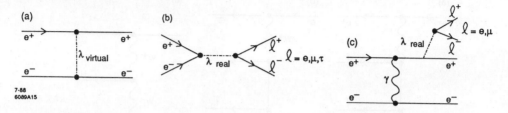

Figure 15

Hawkins and I are looking for the process in Fig. 15c, using 29 GeV data acquired with the Mark II detector at PEP:

$$e^+ + e^- \to e^+ + e^- + \lambda , \qquad (22a)$$

$$\lambda \to e^+ + e^- \quad \text{or} \quad \lambda \to \mu^+ + \mu^- . \qquad (22b)$$

One final e^+ or e^- in Eq. (22a) will usually have an angle close to the beam line; then the detected event is

$$e^+ + e^- \to e^{\pm} + \ell^{\pm} + \ell^{\mp} + \text{missing energy} ,$$

the missing energy being carried off approximately along the beam line. The background is "two virtual-photon" processes.

ACKNOWLEDGMENTS

I have had valuable discussions with my colleagues T. L. Barklow, D. L. Burke, J. M. Dorfan, G. J. Feldman, K. K. Gan, C. A. Hawkins, K. G. Hayes, J. A. Jaros, M. E. Peskin and Y. S. Tsai.

REFERENCES

1. An excellent set of papers is in Physics of Strong Fields (Plenum Press, N.Y., 1987), W. Greiner, ed.
2. M. Davier, Proc. XXII Int. Conf. on High Energy Physics (World Scientific, Singapore, 1987), S. C. Loken, ed.; E. M. Riordan et al., Phys. Rev. Lett. 59, 755 (1987).
3. W. Koenig, paper in this proceeding.
4. C. I. Westbrook et al., Phys. Rev. Lett. 58, 1328 (1987).
5. M. L. Perl, SLAC–PUB–4632 (1988).
6. K. G. Hayes and M. L. Perl, Phys. Rev. (to be published); also issued as SLAC–PUB–4471.
7. R. Stroynowski, CALT–68–1431 (1988).
8. J. Kirkby, CERN–EP/87–210 (1987).
9. J. M. Jowett, CERN LEP–TH/87–56 (1987).
10. See, for example, E. D. Bloom, Proc. 8th Int. Conf. on High Energy Physics (Nashville, 1987); also issued as SLAC–PUB–4604(1988).

11. K. K. Gan and M. L. Perl, Int. J. Mod. Phys. **A3**, 531 (1988).
12. Figure 7 is adapted from S. L. Wu, Phys. Rep. **107**, 60 (1984); S. L. Wu in Proc. 1987 Int. Sym. Lepton and Photon Interactions at High Energies (Hamburg, 1987).
13. See, for example, papers in Proc. 1987 ICFA Seminar on Future Perspectives in High Energy Physics, BNL52114 (1987).
14. J. M. Dorfan in New Frontiers in Particle Physics, (World Scientific, Singapore, 1986), J. M. Cameron, B. A. Campbell, A. N. Kamal, and F. C. Khanna, eds.
15. G. Barbiellini *et al.*, in Physics at LEP, Vol. 2. (CERN 86–02, 1986), J. Ellis and R. Peccei, eds.
16. C. Ahn *et al.*, SLAC–329 (1988).
17. Proc. Workshop on Physics at Future Accelerators, CERN 87–07 (1987).
18. C. A. Hawkins and M. L. Perl, to be published.

COMMENTS ON C12
"Exploration of the Limits on Charged-Lepton-Specific Forces"

This research has not progressed beyond this Physical Review paper by Christopher Hawkins and myself. Like the study of electron-positron collisions in the several MeV to several hundred MeV energy range discussed in Reprint C11, this study was stopped by my work on the tau-charm factory. However now that I am working on electron-positron physics with the CLEO experiment I will be able to take up this study again.

The speculation is that just as the strong force only involves quarks and gluons, there might be a force which only involves charged leptons, a charged-lepton-specific force. My hope is that this force would explain the large, perhaps infinite, ratio of the mass of a charged lepton to the mass of its associated neutrino.

When I began to think about searching for this force using data from the CLEO experiment, I realized again how important it is to me, and I think for most expermental scientists, to be carrying out experiments in order to get new ideas. Some colleagues in the CLEO experiment have already studied the rare decay modes of the tau

$$\tau^- \to e^- + e^+ + e^- + \nu_\tau + \bar{\nu}_e$$
$$\tau^- \to \mu^- + e^+ + e^- + \nu_\tau + \bar{\nu}_\mu$$

and so forth. In hearing about this I thought that perhaps this is a good way to look for a charged-lepton-specific force. I don't know the answer but for me it is an interesting and energizing question.

When I wrote this comment and the previous comment about some of my uncompleted and speculative research, I thought how pleasant it is to be able to write about future research. But I know that research time is always limited, that most research takes longer than expected, that there are often mistakes and delays, that one may have to spend extra time learning an experimental technique or calculational method. Therefore as in the past I will have to continue to make research choices. This has led me to reflect on how I make research choices.

First, as I have written before in these comments, I like to go contrary to the crowd. Second, since the days of my thesis experiment in atomic beams, I like new experimental techniques particularly if they have a mechanical flavor.

Third, as I have learned through the years, it is good to work with colleagues who are one's equal in experimental ability, and it is even better to work with colleagues who are one's superior in experimental ability. This has all sorts of advantages. If there are two ways to do something, you don't have to argue about it, since their way will be as good or better than yours. And if the experiment runs into trouble, perhaps your colleagues can find a solution. Also with smart colleagues you may be able to keep on with the experiment without endless discussions and meetings and reports. I write "may" not "will" because the rituals of endless meetings and discussions and reports seem to have become imbedded in modern science.

My fourth criterion for research choices is to stay away from the colossal, marathon experiments which dominate modern high energy physics: the experiments which take ten years to build after five years of design proposals. This may seem selfish, these experiments are crucial and someone has to prepare for the future. Well it is selfish, but I feel I have done my part in the last forty years.

My final criterion is a result of having grown up in Brooklyn, New York in a family and community of hard-working, ambitious, and practical people. I won't begin an experiment if I don't see how to find funds for it or how to get accelerator time for it. Of course, sometimes as with the tau-charm factory I seriously underestimate the problem of getting financial support. But I won't make that mistake again; it's hard to find two hundred million dollars these days.

Exploration of the limits on charged-lepton-specific forces

Christopher A. Hawkins and Martin L. Perl
Stanford Linear Accelerator Center, Stanford University, Stanford, California 94309
(Received 17 February 1989)

> This paper explores the present experimental limits on the existence of a hypothetical force which would only couple to charged leptons; the neutral particle carrying the force having a mass greater than several MeV/c^2. We consider limits from data on g_e-2, $g_\mu-2$, electron beam-dump experiments, $e^++e^-\to e^++e^-$, $e^++e^-\to \mu^++\mu^-$, and $e^++e^-\to \tau^++\tau^-$. Our purpose is to provide a basis for design of future experiments which would be more sensitive to the existence of a charged-lepton-specific force or other unknown phenomena connected to charged leptons.

I. INTRODUCTION

The known forces which act on leptons (electroweak and gravitational) also act on quarks and other particles. Similarly, proposed interactions involving leptons, the Higgs-particle interaction for example, are also proposed for quarks. Therefore, most searches for new forces have depended upon quarks partaking in that force, even if a final-lepton signature is required. Such searches are irrelevant for a new force which acts only on leptons: a lepton-specific force. In this paper we describe some present experimental limits on the existence and properties of lepton-specific forces *which couple only to charge leptons*. We show that the limits are least imposing when the mass of the particle carrying the force is larger than about 20 MeV/c^2. This leads us to describe possible experiments which could probe further into the question of the existence of a force coupling to charged leptons. We have two interests in such experiments.

One interest comes from puzzling over the peculiar properties of the known lepton compared to the known quarks. Unlike the quarks, the two masses in a lepton doublet are very different; indeed, the neutrino mass may be zero. Unlike the quarks, there is no evidence for generation mixing: μ-lepton-number conservation holds to at least 10^{-10}, τ-lepton-number conservation holds to at least 10^{-4}–10^{-5}. Might another peculiarity of the leptons be that there is a force associated only with charged leptons? The latter might be related to the disparate masses problem.

Our second interest comes from a desire to carry out precise and sensitive measurements at high energy which do not involve complicated or poorly understood properties of quarks. Such measurements must either not involve hadrons or only involve hadrons in a well-understood way. Some electron-positron collision reactions meet these criteria and have been carefully studied:

$$e^++e^-\to e^++e^-,\ e^++e^-\to l^++l^-,\ l=\mu,\tau\ .$$

Other reactions which meet these criteria are

$$\gamma+p\to l^++l^-+p,\ l=e,\mu\ ,$$
$$e+p\to l^++l^-+e+p,\ l=e,\mu\ .$$

The comparison of the precision and sensitivity of different measurements requires a hypothesis as to the unknown physical phenomenon which might be revealed by increased precision or sensitivity. We use the hypothesis of a force coupling only to charged leptons, and a model described next.

In this model the force is carried by a particle called λ of mass m_λ; λ is neutral and does not change lepton number (Sec. II). To get a feeling for the extent of present limits, λ is allowed to be a pseudoscalar or a vector particle.

In describing current limits on a lepton-specific force we will sometimes make use of results from axion and Higgs-particle search experiments. This is done in Sec. III where the limits are recounted from the comparison of the measurements of g_e-2 and $g_\mu-2$ with theory. In Sec. IV we describe additional limits when m_λ is less than about 20 MeV/c^2; these limits are obtained from electron beam-dump experiments. Additional limits for larger values of m_λ are obtained in Sec. V from measurements on the reactions

$$e^++e^-\to e^++e^-,\ \mu^++\mu^-,\ \tau^++\tau^-\ .$$

Our interest is in direct searches for a force carried by a λ with a mass greater than about 20 MeV/c^2. Astrophysical considerations[1] are not of use in this case, although very restrictive limits can be obtained for smaller values of m_λ (less than about 1 MeV/c^2). Therefore, we do not discuss limits coming from astrophysical observations or calculations.

We conclude in Sec. VI with a discussion of possible future experiments on the existence of a charged-lepton-specific force. The emphasis is on the region of large m_λ because this is the region where the limits discussed in this paper exercise the least constraints.

II. MODEL, LIMIT PHILOSOPHY, LIFETIME

A. Model and limit philosophy

We take the λ to be either a pseudoscalar or vector particle which couples only to charged leptons. Using the subscript l to represent a charged lepton, the λ-lepton vertex has one of the following forms:

pseudoscalar: $-ig_{\lambda l}\bar{v}_l \gamma_5 u_l$, (1a)

vector: $-ig_{\lambda l}\bar{v}_l \gamma_\mu u_l$. (1b)

We define $\alpha_{\lambda l} = g_{\lambda l}^2 / 4\pi$.

We do not have a fixed idea as to the dependence of the coupling constants $g_{\lambda l}$ on the nature or properties of the lepton l. Unlike the Higgs-particle hypothesis we do not connect $g_{\lambda l}$ with the lepton mass m_l. We do not assume relationships inside a set of $g_{\lambda l}$'s. Each limit is considered separately and presented on a graph of the type of Fig. 1.

Our philosophy in this paper is to sketch out the approximate pseudoscalar and vector limits on $\alpha_{\lambda l}$ for various ranges of m_λ. We can use approximate limits, usually the 90%-C.L. limit, because we are not testing a specific theory. Our purpose is to find regions where limits on a lepton-specific force are least constrictive; our goal is to carry out search experiments in some of those regions.

There are two other spin and coupling possibilities: scalar and axial vector. We have not reported on all possibilities because it would make too long and repetitive a paper. With these other possibilities the limits are either less restrictive or about as restrictive as the cases we discuss. A further simplification in our considerations is that we assume there is only one λ particle which couples to a specific lepton. We ignore the possibility that two different λ's couple to the same lepton; hence, we avoid the complication that effects from the two λ's weaken or cancel each other.

B. Lifetime when $m_\lambda > 2m_l$

The $\alpha_{\lambda l}$-m_l region of sensitivity of a particular search method usually depends upon the lifetime of the λ, τ_λ. The simplest case is when λ couples to just one lepton l and $m_\lambda > 2m_l$. Then[2]

pseudoscalar: $\tau_\lambda = \dfrac{2\hbar}{\alpha_\lambda m_\lambda} \left[1 - \dfrac{4m_l^2}{m_\lambda^2} \right]^{-1/2}$, (2a)

vector: $\tau_\lambda = \dfrac{3\hbar}{\alpha_\lambda m_\lambda} \left[1 - \dfrac{4m_l^2}{m_\lambda^2} \right]^{-1/2} \left[1 + \dfrac{2m_l^2}{m_\lambda^2} \right]^{-1}$, (2b)

where $\hbar = 6.6 \times 10^{-22}$ s MeV and m_λ is in MeV.

In most searches the crucial parameter is not τ_λ, but the decay length, $L_d = c\gamma_\lambda \tau_\lambda$. Electron beam-dump experiments in which the λ is directly detected require L_d larger than tens or hundreds of meters. Experiments which require the λ to leave a production target before decaying require L_d larger than millimeters or centimeters.

C. Lifetime when $m_\lambda < 2m_l$

Suppose λ couples to only one lepton, the charged lepton l, and $m_\lambda < 2m_l$. Then the dominant decay mode for the pseudoscalar is

$\lambda \to \gamma + \gamma$ (3a)

through a virtual l loop. The lifetime is[3]

$\tau_\lambda \approx \dfrac{16\pi^2 \hbar}{\alpha^2 \alpha_{\lambda l} m_\lambda} \left[\dfrac{m_l}{m_\lambda} \right]^2$. (3b)

Comparing Eq. (3b) to Eq. (2a), the lifetime is much larger because of the factor $\alpha^{-2}(m_l/m_\lambda)^2$.

If λ is a vector it cannot decay to two γ's. The decay mode $\lambda \to 3\gamma$ will have a lifetime longer than that in Eq. (3b) by a factor of about 1000.

III. LIMITS FROM $g_e - 2$ AND $g_\mu - 2$

A classic activity in atomic and particle physics is to search for new physical phenomena by comparing measurements of $g_e - 2$ and $g_\mu - 2$ with calculations. We need only copy the very useful formulas from Ref. 2. In this section we give $\alpha_{\lambda l}$ vs m_λ limits for the cases of λ scalar or axial vector as well as the cases we use throughout the paper of λ pseudoscalar or vector. Using $a_l = (g_l - 2)/2$ with $l = e$ or μ, we define

$\Delta a_l = a_l(\text{measured}) - a_l(\text{calculated})$ (4a)

and

$\Delta a_l = \alpha_{\lambda l} K(r)/2\pi$. (4b)

Here $r = (m_\lambda/m_l)^2$. The function $K(r)$ depends on the nature of λ. The values of $K(r)$ at small and large values of r are instructive: limit as $r = m_\lambda^2/m_l^2 \to 0$,

pseudoscalar: $K \to \tfrac{1}{2}$, vector: $K \to 1$,

scalar: $K \to -\tfrac{3}{2}$, axial vector $K \to 4\ln(r)$; (4c)

limit as $r = m_\lambda^2/m_l^2 \to \infty$,

pseudoscalar: $K \to (1/r)\ln(r)$,

vector: $K \to 2/(3r)$, (4d)

scalar: $K \to -(1/r)\ln(r)$,

axial vector: $K \to -10/(3r)$.

The limits on Δa_l are not symmetric. From Ref. 4,

$\Delta a_e = (-1.11 \pm 1.28) \times 10^{-10}$,

$\Delta a_\mu = (3.9 \pm 8.7) \times 10^{-9}$.

The 90%-C.L. limits are

$\Delta a_e < +0.53 \times 10^{-10}$, $\Delta a_e > -2.75 \times 10^{-10}$,

$\Delta a_\mu < +1.50 \times 10^{-8}$, $\Delta a_\mu > -0.72 \times 10^{-8}$. (5)

The limits on Δa_l in Eq. (5) lead to the excluded regions in Fig. 1. When $m_\lambda \lesssim m_l$, the upper limit on α_λ is of the order of $2\pi \Delta a_l$, a drastic constraint on $\alpha_{\lambda l}$. This constraint weakens when $m_\lambda \gg m_l$, the upper limit increases approximately as m_λ^2. As stated in Sec. II A, we assume only one λ couples to a lepton.

The limits on α_λ provided by Δa_l are a foundation on which we erect other limits from other data and searches (Secs. IV and V). Note that although Δa_μ is about 100

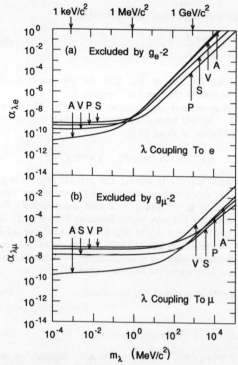

FIG. 1. Upper limits on $\alpha_{\lambda l}$ set by g_l-2 measurements for (a) l =electron and (b) l =muon. A =axial vector, V =vector, P =pseudoscalar, S =scalar.

times larger than Δa_e, at large values of m_λ there is a stronger constraint on $\alpha_{\lambda\mu}$ compared to $\alpha_{\lambda e}$ due to the effect of the muon mass.

IV. LIMITS FROM ELECTRON BEAM-DUMP EXPERIMENTS

A. $0 \leq m_\lambda < 2m_e$

Two electron beam-dump experiments,[5,6] schematically described by Fig. 2 have been carried out at SLAC. A beam of 20-GeV electrons is dumped into a target containing at least several radiation lengths. Directly downstream of the target, a distance D, is a track-detecting and electromagnetic-shower-detecting, thick plate chamber. The distance D is partially filled with shielding. In the experiment of Rothenberg,[5] D was about 60 m, the detector consisted of four optical spark chambers with thick aluminum plates, and the total number of effective radiation lengths in the detector was about 9.4. In the experiment of Bjorken et al.,[6] D was about 400 m, the detector consisted of aluminum or iron plates interleaved with multiwire proportional chambers, and the total number of effective radiation lengths was about 4. In the former experiment, the physicists looked for events which might be neutrino interactions, these events consisting of one or more charged particles. Electromagnetic showers of sufficient energy would also have been detect-

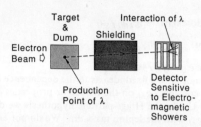

FIG. 2. Schematic of electron beam-dump experiments which set limits on $\alpha_{\lambda e}$ and m_λ when $m_\lambda < 2m_e$.

ed. In the latter experiment, the physicists specifically looked for electromagnetic showers. Neither experiment reported any unexplained source of electromagnetic showers.

We can interpret the null results of these experiments for our purposes by noting that the λ could be produced by the process, Fig. 3(a),

$$e^- + \text{nucleus} \rightarrow e^- + \lambda + \text{nucleus or nucleons} , \quad (6a)$$

analogous to electron bremsstrahlung. Some λ's which reach the detector and have sufficient energy will interact with the material in the detector through the process, Fig. 3(b),

$$\lambda + \text{nucleus} \rightarrow e^+ + e^- + \text{nucleus or nucleons} , \quad (6b)$$

analogous to photoproduction of e^+e^- pairs.

An order-of-magnitude calculation shows that this is a sensitive search method for λ coupling to an e when $m_\lambda < 2m_e$. This mass restriction combined with the $g_e - 2$ constraint in Fig. 1(a) means that the sensitivity of the beam-dump experiment need only be investigated for $\alpha_{\lambda e} < 3 \times 10^{-9}$. The lifetime for a pseudoscalar λ, Eq. 3(b), is

$$\tau_\lambda(m_\lambda < 2m_e) > 9 \times 10^{-8} \text{ s} . \quad (7a)$$

We will only consider λ's with energy greater than 2 GeV; hence, the decay length is

$$D_\lambda(m_\lambda < 2m_e) > 5 \times 10^4 \text{ m} . \quad (7b)$$

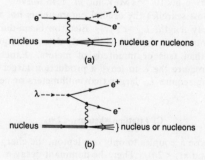

FIG. 3. Diagrams for (a) $e^- + \text{nucleus} \rightarrow e^- + \lambda + \text{nucleus or nucleons}$ and (b) $\lambda + \text{nucleus} \rightarrow e^+ + e^- + \text{nucleus or nucleons}$.

This decay length is much larger than the target to detector distance in either experiment. Thus both experiments are applicable. Although the experiment of Rothenberg[5] and Donnelly et al.[5] is in principle more sensitive than that of Bjorken et al.,[6] we will analyze only the experiment of Bjorken et al.,[6] because we have been able to discuss the detector sensitivity with one of the experimenters.[7]

No electromagnetic showers of greater than 2-GeV energy were found in 4 radiation lengths of the detector when a total of 30 C of 20-GeV electrons were used. We show here the calculation of the limit on $\alpha_{\lambda e}$ for λ being massless and a vector. The pseudoscalar case and mass dependence are related to the massless vector case using relations given by Tsai.[8]

An EGS (Ref. 9) shower simulation gives the number of λ's produced in the dump that are within the detector acceptance to be $3.76\,\alpha_{\lambda e}/\alpha$ λ's per incident electron. For C coulombs, the number of produced λ's is

$$N_\lambda \sim 2.3 \times 10^{19} C \alpha_{\lambda e}/\alpha \;.$$

The shielding contains about 1300 radiation lengths, but since we are only concerned with $\alpha_{\lambda e} < 10^{-8}$ because of the $g_e - 2$ constraint, there is negligible attenuation of λ's in the shielding. The probability that a λ of 2 GeV or more energy produces a shower in an N_{rad}-radiation-length detector is $\tfrac{7}{9} N_{\text{rad}} (\alpha_{\lambda e}/\alpha)$. Therefore, the number of showers expected is

$$N_{\text{shower}} = 2.3 \times 10^{19} C \tfrac{7}{9} N_{\text{rad}} (\alpha_{\lambda e}/\alpha)^2 \;. \quad (8a)$$

Taking the upper limit on N_{shower} as 2.3, with $C=30$ and $N_{\text{rad}}=4$,

$$\alpha_{\lambda e} \lesssim 2.4 \times 10^{-13} \;. \quad (8b)$$

Thus for $m_\lambda < 2m_e$, our interpretation of these beam-dumped experiments decreases the upper limit on $\alpha_{\lambda e}$ from about 10^{-8} to about 2×10^{-13}. These limits as a function of m_λ are shown later in Fig. 5.

Once $m_\lambda > 2m_e$ the λ lifetime becomes too short to use our interpretation of this experiment. But at this boundary another set of search experiments can be used, those connected with the possibility of the production of anomalous e^+e^- pairs in heavy-ion collisions.

B. $2m_e < m_\lambda \lesssim 15 \text{ MeV}/c^2$

In the past decade there has been continuing but confusing evidence[10] that there is anomalous production of e^+e^- pairs when heavy ions such as Th and U collide. When the kinetic energy of the incident ion is about the energy required to overcome the Coulomb barrier between the nuclei, there appear to be peaks in the e^+e^- mass spectrum between about 1.5 and 1.8 MeV/c^2. A great deal of theoretical and experimental research has involved the hypothesis that the e^+e^- pair are the decay products of a neutral particle produced in the heavy-ion collision, which we call λ.

One area of experimental research has looked for the λ through the sequence

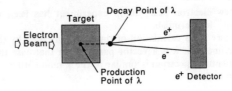

FIG. 4. Schematic of electron beam-dump experiments which set limits on $\alpha_{\lambda e}$ and m_λ when $m_\lambda > 2m_e$.

$$e^- + \text{nucleus} \rightarrow e^- + \lambda + \text{nucleus or nucleons} \;, \quad (9a)$$
$$\lambda \rightarrow e^+ + e^- \;. \quad (9b)$$

The reaction in Eq. (9a) is the same as that in Eq. (6a) and Fig. 3(a). Tsai,[8] Olsen,[11] and Holvik and Olsen[12] give the theory and cross sections for this reaction. The decay process, Eq. (9b), must take place outside the production target, Fig. 4, setting a lower limit on the λ lifetime. As shown in Fig. 4, in these experiments the existence of the λ would be demonstrated by an excess of positrons in the forward direction.

Most of the searches[13–15] have used electron beams. One search[16] analyzed data from a proton beam-dump experiment; high-energy electrons are produced in the dump through the sequence

$$p + \text{nucleon} \rightarrow \pi^0 + \cdots, \quad \pi^0 \rightarrow \gamma + \gamma \;,$$
$$\gamma + \text{nucleus} \rightarrow e^+ + e^+ + \cdots \;. \quad (10)$$

The results of all these searches[13–16] were null.

Davier[17] has combined the limits from Refs. 13–16 and from an earlier electron beam-dump search for axions.[18] We apply the same limits to a pseudoscalar λ in Fig. 5. We also show in Fig. 5 the excluded region from the beam-dump experiments discussed in Sec. IV A. Thus the excluded region from Sec. IV A is extended to larger values of m_λ, to about 15 MeV/c^2. In a narrow range of m_λ, the upper limit on $\alpha_{\lambda e}$ is reduced to 10^{-14}.

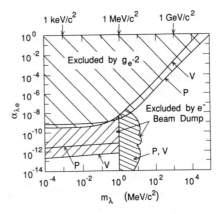

FIG. 5. Excluded regions of $\alpha_{\lambda e}$ vs m_λ from considerations in Secs. III and IV. $V=$ vector, $P=$ pseudoscalar.

A new electron beam-dump experiment has been proposed[19] for Fermilab.

There is no calculation of the limits imposed by these experiments if λ is a vector particle. In that case the λ momentum spectrum is similar to the γ momentum spectrum from

$$e^- + \text{nucleus} \rightarrow e^- + \gamma + \text{nucleus or nucleons} \quad (11a)$$

and the background from

$$\gamma + \text{nucleus} \rightarrow e^- + e^+ + \text{nucleus} \quad (11b)$$

is more serious. However Riordan[20] pointed out to us that the production cross section for the reaction in Eq. (9a) is larger for a vector λ compared to a pseudoscalar λ. Riordan concludes[20] that the $\alpha_{\lambda e}$-m_λ excluded region is about the same for the two types of λ. Hence, we use these beam-dump limits, Fig. 5, for a vector λ.

The type of electron beam-dump search discussed in this section becomes less sensitive as m_λ increases above 15 MeV/c^2. The production cross section decreases.[8] Furthermore, the decay length is proportional to m_λ^{-2} for fixed $\alpha_{\lambda e}$ and fixed energy; hence, a thinner target must be used. Therefore, other search methods must be used for large values of m_λ.

V. LIMITS FROM e^+e^- COLLISION DATA

A. $e^+ + e^- \rightarrow e^+ + e^-$

1. Vector λ

The Feynman diagrams in Fig. 6 are, for Bhabha scattering,

$$e^+ + e^- \rightarrow e^+ + e^-, \quad (12)$$

through both γ and λ exchange. If λ is a vector particle the cross section is given by the formula for γ and Z^0 exchange[21] with the Z^0's axial-vector coupling parameter set to 0.

In the barycentric system

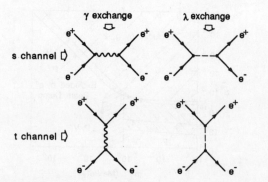

FIG. 6. Feynman diagrams for the process $e^+ + e^- \rightarrow e^+ + e^-$ taking place through γ and λ exchange.

$$\frac{d\sigma(e^+e^- \rightarrow e^+e^-)}{d\cos\theta} = \frac{\pi\alpha^2}{s}(f_{\gamma\gamma} + xf_{\gamma\lambda} + x^2 f_{\lambda\lambda}),$$

$$x = \frac{\alpha_{\lambda e}}{\alpha_\gamma}, \quad (13a)$$

where $f_{\gamma\gamma}$, $f_{\gamma\lambda}$, and $f_{\lambda\lambda}$ are the contributions from the product of γ-γ, γ-λ, and λ-λ amplitudes. Specifically,

$$f_{\gamma\gamma} = \left[\frac{s^2+u^2}{t^2} + \frac{2u^2}{st} + \frac{t^2+u^2}{s^2}\right], \quad (13b)$$

$$f_{\gamma\lambda} = 2\left[\left[\frac{s^2+u^2}{t^2}\right](R_t)_{\text{real}} + \frac{u^2}{st}(R_t + R_s)_{\text{real}}\right.$$
$$\left. + \left[\frac{t^2+u^2}{s^2}\right](R_s)_{\text{real}}\right], \quad (13c)$$

$$f_{\lambda\lambda} = \left[\left[\frac{s^2+u^2}{t^2}\right]|R_t|^2 + \frac{2u^2}{st}(R_s R_t)_{\text{real}}\right.$$
$$\left. + \left[\frac{t^2+u^2}{s^2}\right]|R_s|^2\right], \quad (13d)$$

where

$$t = -s(1-\cos\theta)/2, \quad u = -s(1+\cos\theta)/2,$$
$$(13e)$$

$$R_t = t/(t - m_\lambda^2 + i\Gamma_\lambda m_\lambda), \quad R_s = s/(s - m_\lambda^2 + i\Gamma_\lambda m_\lambda).$$

We found that the most sensitive search for a nonzero x uses high-energy e^+e^- storage-ring data on the partial total cross section

$$\sigma'(e^+e^- \rightarrow e^+e^-) = \int_{-c}^{c} \left[\frac{d\sigma(e^+e^- \rightarrow e^+e^-)}{d\cos\theta}\right] d\cos\theta,$$
$$(14a)$$

where the limits of integration $\pm c$ depend on the experiment. The prime indicates the cross section is for part of the $\cos\theta$ range. From Eqs. (13),

$$\sigma'(e^+e^- \rightarrow e^+e^-) = \sigma'_{\gamma\gamma} + x\sigma'_{\gamma\lambda} + x^2\sigma'_{\lambda\lambda}. \quad (14b)$$

Suppose an experiment reports the upper limit

$$\frac{\sigma'_{\text{meas}}(e^+e^- \rightarrow e^+e^-) - \sigma'_{\text{QED}}(e^+e^- \rightarrow e^+e^-)}{\sigma'_{\text{QED}}(e^+e^- \rightarrow e^+e^-)} < \epsilon.$$
$$(15a)$$

Then in our model,

$$x^2\left[\frac{\sigma'_{\lambda\lambda}}{\sigma'_{\gamma\gamma}}\right] + x\left[\frac{\sigma'_{\gamma\lambda}}{\sigma'_{\gamma\gamma}}\right] < \epsilon \quad (15b)$$

gives the limit on x. Using

$$r_2 = \sigma'_{\lambda\lambda}/\sigma'_{\gamma\gamma}, \quad r_1 = \sigma'_{\gamma\lambda}/\sigma'_{\gamma\gamma}, \quad r_2 x^2_{\text{lim}} + r_1 x_{\text{lim}} = \epsilon. \quad (16)$$

The sizes of r_1 and r_2 and hence the relation of the x_{lim}'s to ϵ depends on m_λ. When $|s - m_\lambda^2| \sim m_\lambda \Gamma_\lambda$ the s-channel resonance in R_s dominates σ', and the deviation from pure photon exchange will depend on x^2. Otherwise the $\gamma-\lambda$ interference term, $r_1 x$ in Eq. (16), is most important and the deviation depends on x.

In looking for a deviation from pure photon exchange it is crucial to examine how the luminosity was determined in a measurement of $\sigma'(e^+e^- \rightarrow e^+e^-)$. The use of large-angle $e^+e^- \rightarrow e^+e^-$ scattering to determine the luminosity negates the search for a deviation. We have used the comparison of $\sigma'(e^+e^- \rightarrow e^+e^-)$ and $\sigma'(e^+e^- \rightarrow \gamma\gamma)$ of Derrick et al.[22] Using data from the High Resolution Spectrometer (HRS) detector at the SLAC e^+e^- storage ring PEP they give

$$\left[\frac{\sigma'(e^+e^- \rightarrow e^+e^-)}{\sigma'(e^+e^- \rightarrow \gamma\gamma)}\right]_{\text{meas}} \Big/ \left[\frac{\sigma'(e^+e^- \rightarrow e^+e^-)}{\sigma'(e^+e^- \rightarrow \gamma\gamma)}\right]_{\text{QED}}$$
$$= 0.993 \pm 0.009 \pm 0.008 , \quad (17)$$

where c in Eq. (14a) is 0.55, and QED means the theory is pure quantum electrodynamics calculated to third order in α. In our model λ does not enter into the reaction $e^+ + e^- \rightarrow \gamma + \gamma$ in lowest order; hence, we use Eq. (17) to set the deviations allowed in $e^+ + e^- \rightarrow e^+ + e^-$ due to the presence of the λ. Following our philosophy of giving approximate, exploratory limits we add quadratically the statistical and systematic errors to give a measure of the allowed deviation. The 90%-C.L. limit is

$$\epsilon = 0.008 . \quad (18)$$

Using Eqs. (13), (14), and (16) we obtain the limits on $\alpha_{\lambda e}$ in Fig. 7.

The width of resonance at $\sqrt{s} = 29$ GeV is set mostly by the variation in the beam energy of PEP over the course of several years' data acquisition. We took the variation in the $\sqrt{s}$ to be $\pm 0.002\sqrt{s}$. Values of $\alpha_{\lambda e}$ as small as 10^{-3}–4×10^{-5} are excluded by the measurement of Derrick et al.[22] At 29 GeV the $\alpha_{\lambda e}$ limit reaches below 10^{-5}, but these values are dependent on our uncertain estimation of the experimental resonance width.

The limits on $\alpha_{\lambda e}$ for $m_\lambda > 29$ GeV might be further examined through the use of $e^+e^- \rightarrow e^+e^-$ data from the DESY PETRA or KEK TRISTAN storage rings, but we have not found published data that we could directly use; uncertainties in the luminosity determination negate the advantage of the higher energy. This can certainly be overcome by experimenters who have their own data from these storage rings.

2. Pseudoscalar λ

When λ is pseudoscalar or scalar the Bhabha-scattering differential cross section in the barycentric system is

$$\frac{d\sigma(e^+e^- \rightarrow e^+e^-)}{d\cos\theta} = \frac{\pi\alpha^2}{s}(f_{\gamma\gamma} + x f_{\gamma\lambda} + x^2 f_{\lambda\lambda}) .$$
(19a)

Here

$$f_{\gamma\gamma} = \left[\frac{s^2+u^2}{t^2} + \frac{2u^2}{st} + \frac{t^2+u^2}{s^2}\right] , \quad (19b)$$

$$f_{\gamma\lambda} = -\left[-\frac{u}{t}(R_s)_{\text{real}} + \frac{u}{s}(R_t)_{\text{real}}\right] , \quad (19c)$$

$$f_{\lambda\lambda} = \tfrac{1}{2}[|R_t|^2 + (R_s R_t)_{\text{real}} + |R_s|^2] , \quad (19d)$$

where the notation is described in Eq. (13e).

We again use the limit from Derrick et al.,[22] Eq. (18), and the analysis described in Sec. V A 1. The excluded regions of $\alpha_{\lambda e}$ are given in Fig. 8.

The excluded regions are smaller than the vector case, Fig. 7, because r_1 and r_2 are smaller in the pseudoscalar case compared to the vector case. For example, set $m_\lambda = 0$, then

pseudoscalar: $r_1 = -0.107$, $r_2 = 0.214$,

vector: $r_1 = 2.000$, $r_2 = 1.000$.

The limits in Fig. 8 also apply to a scalar λ.

B. $e^+ + e^- \rightarrow \mu^+ + \mu^-, \tau^+ + \tau^-$

1. Vector λ

The s-channel reaction

$$e^+ + e^- \rightarrow l^+ + l^-, \quad l = \mu, \tau \quad (20)$$

provides limits on

$$\alpha_{\lambda el} = g_{\lambda e} g_{\lambda l}/4\pi$$

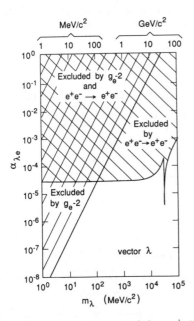

FIG. 7. Limits on $\alpha_{\lambda e}$ for a vector λ from $e^+e^- \rightarrow e^+e^-$ at 29 GeV and $g_e - 2$.

through the diagrams in Fig. 9. The barycentric differential cross section has the simple form

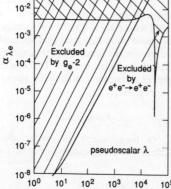

FIG. 8. Limits on $\alpha_{\lambda e}$ for a pseudoscalar λ from $e^+e^- \rightarrow e^+e^-$ at 29 GeV and $g_e - 2$.

$$\frac{d\sigma}{d\cos\theta} = \frac{\pi\alpha^2\beta}{s}(2-\beta^2+\beta^2\cos^2\theta)$$
$$\times [1+2x(R_s)_{\text{real}}+x^2|R_s|^2], \quad (21a)$$

where

$$x = \alpha_{\lambda el}/\alpha_\gamma, \quad R_s = s/(s-m_\lambda^2+i\Gamma_\lambda m_\lambda) \quad (21b)$$

and β is v_l/c.

The magnitude of $d\sigma/d\cos\theta$ is determined in part by the luminosity, which in turn depends upon large-angle Bhabha scattering. As in Sec. V A we define a partial total cross section σ' obtained by integrating $d\sigma/d\cos\theta$ within the range $-c < \cos\theta < c$. The limits on x are obtained from the ratio

$$\rho(e^+e^- \rightarrow \gamma\gamma)$$
$$= \frac{\sigma'_{\text{meas}}(e^+e^- \rightarrow l^+l^-)/\sigma'_{\text{meas}}(e^+e^- \rightarrow e^+e^-)}{\sigma'_{\text{QED}}(e^+e^- \rightarrow l^+l^-)/\sigma'_{\text{QED}}(e^+e^- \rightarrow e^+e^-)}.$$

An exact treatment of this ratio requires recognition that $\sigma'_{\text{meas}}(e^+e^- \rightarrow e^+e^-)$ was used to set limits on $\alpha_{\lambda e} = g_{\lambda e}^2/4\pi$. It is sufficient for our purpose to quadratically add the errors in $\sigma'_{\text{meas}}(e^+e^- \rightarrow e^+e^-)$ to the larger er-

rors in $\sigma'_{\text{meas}}(e^+e^- \rightarrow l^+l^-)$ and use the combined error ϵ, where

$$\frac{\sigma'_{\text{meas}}(e^+e^- \rightarrow l^+l^-)}{\sigma_{\text{QED}}(e^+e^- \rightarrow l^+l^-)} = 1+\epsilon. \quad (22)$$

From Eqs. (21) and (22),

$$x_{\text{lim}}^2|R_s|^2 + 2x_{\text{lim}}(R_s)_{\text{real}} = \epsilon, \quad (23)$$

$$\mu: \epsilon_\mu = 0.038, \quad \tau: \epsilon_\tau = 0.064. \quad (24)$$

The ϵ_μ value comes from the 29-GeV results in Refs. 22 and 23. The ϵ_τ value comes from the 29-GeV results in Ref. 24 and an additional uncertainty due to the problem in understanding the τ decay modes.[25] When ϵ_μ or ϵ_τ are inserted in Eq. (23) we obtain the limits in Figs. 10 and 11.

2. Pseudoscalar λ

If λ is pseudoscalar there is no interference between the γ-exchange and λ-exchange amplitudes:

$$\frac{d\sigma}{d\cos\theta} = \frac{\pi\alpha^2\beta}{s}[(2-\beta^2+\beta^2\cos^2\theta)+x^2|R_s|^2], \quad (25)$$

where the notation is given in Eq. (21b). The limit on $\alpha_{\lambda e\mu}$ is given by

$$x_{\text{lim}}r = \sqrt{\epsilon}, \quad (26a)$$

where

$$r = |R_s|/(2-\beta^2+\beta^2c^2/3)^{1/2}. \quad (26b)$$

This is for the partial cross section for the range $-c < \cos\theta < c$.

The ϵ values given in Eq. (24) lead to the upper limits on $\alpha_{\lambda e\mu}$ in Fig. 10 and on $\alpha_{\lambda e\tau}$ in Fig. 11.

FIG. 9. Feynman diagrams for $e^+ + e^- \rightarrow \mu^+ + \mu^-, \tau^+ + \tau^-$.

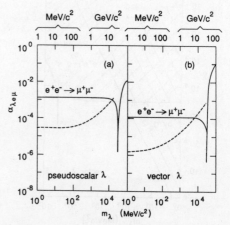

FIG. 10. In (a) λ pseudoscalar and (b) λ vector the solid curve gives the upper limit on $\alpha_{\lambda e\mu}$ from $e^+e^- \rightarrow \mu^+\mu^-$. The dashed curve gives the upper limit on $\sqrt{\alpha_{\lambda e}\alpha_{\lambda\mu}}$.

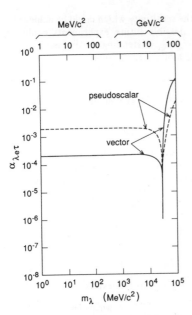

FIG. 11. The curves marked pseudoscalar and vector give the upper limit on $\alpha_{\lambda e\tau}$ from $e^+e^- \to \tau^+\tau^-$ for λ pseudoscalar and λ vector.

3. Discussion of $\alpha_{\lambda e\mu}, \alpha_{\lambda e\tau}$

The upper limits in Figs. 10 and 11 on

$$\alpha_{\lambda el} = g_{\lambda e} g_{\lambda l}/4\pi \qquad (27)$$

do not set limits on $g_{\lambda l}$ unless we know a connection between $g_{\lambda e}$ and $g_{\lambda l}$. In the special case of $g_{\lambda e} = 0$, the upper limit on $\alpha_{\lambda el}$ tells us nothing about $g_{\lambda l}$. In a model which copies the Higgs-particle hypothesis with $g_{\lambda l} = (m_l/m_e)g_{\lambda e}$, the individual upper limits are given by $\alpha_{\lambda l} = g_{\lambda l}^2/4\pi = (m_l/m_e)\alpha_{\lambda el}$ and $\alpha_{\lambda e} = g_{\lambda e}^2/4\pi = (m_e/m_l)\alpha_{\lambda el}$.

In Fig. 10 we compare the $\alpha_{\lambda e\mu}$ upper limit (solid curve) with the upper limit on $\sqrt{\alpha_{\lambda e}\alpha_{\lambda \mu}}$ (dashed curve). The $\alpha_{\lambda e}$ limit is obtained from $e^+e^- \to e^+e^-$ (Sec. IV A); the $\alpha_{\lambda \mu}$ limit is obtained from $g_{\mu-2}$ (Sec. III). For much of the range of $m_\lambda < 29$ GeV, the $\sqrt{\alpha_{\lambda e}\alpha_{\lambda \mu}}$ upper limits are smaller than the $\alpha_{\lambda e\mu}$ upper limit.

VI. SUMMARY AND DISCUSSION

In the spirit of our model we discuss separately the e, μ, and τ. We remark on the limits given in this paper, we point out other existing data that can be examined, and discuss possible future experiments.

A. Electron-specific forces

1. Remarks on limits and use of other data

Figures 1(a), 7, and 8 summarize the limits on the λ-e system. When m_λ is greater than about 200 MeV/c^2, the smallest upper bound on $\alpha_{\lambda e}$ comes from $e^+e^- \to e^+e^-$.

Depending on the properties assumed for λ, the upper bound lies between 10^{-2} and 10^{-5} for most of the m_λ range. Smaller upper bounds occur, of course, at the resonant mass for the data we used, $m_\lambda = 29$ GeV/c^2. But such a bound has little use because it only applies to an m_λ mass range about 0.1 GeV/c^2 wide at 29 GeV/c^2.

The increased sensitivity of the $e^+e^- \to e^+e^-$ cross section at the resonance $m_\lambda = E_{c.m.}$ might be used over a broader mass range by analyzing data required during energy scans. For example, the energy range from about 3 to 6 GeV was scanned at the SLAC storage ring SPEAR (Ref. 26), and from about 30 to 46 GeV was scanned at PETRA (Ref. 21). To use such scan data close attention must be paid to how the large-angle Bhabha scattering was normalized. We have not made such a study.

We are studying[27] data from the SLAC storage ring PEP on the reaction

$$e^+ + e^- \to e^+ + e^- + e^+ + e^-, \qquad (28a)$$

looking for the process

$$e^+ + e^- \to e^+ + e^- + \lambda, \quad \lambda \to e^+ + e^- \qquad (28b)$$

through detection of an e^+e^- mass peak at m_λ. Figure 12(a) shows one of the Feynman diagrams for this hypothetical process.

2. Possible future experiments

A λ search method analogous to that in Eq. (28b) uses electroproduction on a proton:

$$e^- + p \to e^- + p + \lambda, \quad \lambda \to e^+ + e^-. \qquad (29)$$

Figure 12(b) shows one of the Feynman diagrams. Again the λ would be detected by an e^+e^- mass peak. In an ep fixed-target search using an e^- beam of energy E_{beam}, the mass range is limited by $m_\lambda < \sqrt{2E_{\text{beam}}m_{\text{proton}}}$. This limit is smaller than the $m_\lambda < E_{c.m.}$ limit for the $e^+e^- \to e^+e^-\lambda$ process in a storage ring. However the ep fixed-target search experiment can be designed[28] for a higher effective interaction rate and hence greater sensitivity. The search can also be carried out at the HERA ep collider now under construction.

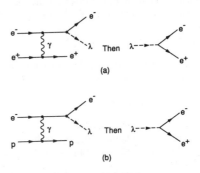

FIG. 12. Examples of Feynman diagrams for (a) $e^+ + e^- \to e + e + \lambda$, $\lambda \to e^+ + e^-$, and (b) $e^- + p \to e^- + p + \lambda$, $\lambda \to e^+ + e^+$.

In thinking about possible methods to search for a λ coupled to an e, one can consider a deliberate energy scan of the $e^+e^-\to e^+e^-$ cross section, looking for the resonance at $E_{\text{c.m.}}=m_\lambda$. The scanning could be done in an e^+e^- collider or in a fixed-target experiment. Unfortunately, such a scanning search at existing e^+e^- storage rings would be a long experiment and could not be justified at this time. The mass range in a fixed-target scanning search is limited to $m_\lambda < \sqrt{2E_{\text{beam}}m_e}$, about 220 MeV/$c^2$ for the 50-GeV e^- beam at SLAC. We have not investigated whether such a search could extend into the unexplored regions in Figs. 7 and 8.

B. Muon-specific forces

1. Remarks on limits and use of other data

If the λ couples only to the μ, our only limits on $\alpha_{\lambda\mu}$ come from $g_\mu-2$, Fig. 1(b). As discussed in Sec. III, the larger size of m_μ compared to m_e leads to the limits imposed by $g_\mu-2$ extending to larger values of m_λ. Comparing Fig. 7 for $\alpha_{\lambda e}$ with Fig. 1(b) for $\alpha_{\lambda\mu}$, one sees that most of the $\alpha_{\lambda e}$-m_λ region excluded by $e^+e^-\to e^+e^-$ is excluded for $\alpha_{\lambda\mu}$-m_λ by $g_\mu-2$.

The study of the $\mu^+\mu^-$ mass spectrum in muon trident production,

$$\mu^\pm + N \to \mu^\pm + N' + \mu^+ + \mu^- , \quad (30\text{a})$$

also provides a way to search for a muon-specific force. One would look for the process

$$\mu^\pm + N \to \mu^\pm + N' + \lambda, \quad \lambda \to \mu^+ + \mu^- , \quad (30\text{b})$$

which would occur through a diagram similar to that in Fig. 12(b) with all e's replaced by μ's. Sloan[29] has brought to our attention a study[30] of the $\mu^+\mu^-$ mass spectrum from the reaction in Eq. (30a), the data having been obtained by the European Muon Collaboration.[30] There are no unexplained peaks in the $\mu^+\mu^-$ mass spectrum. The upper limits which this null result imposes on $\alpha_{\lambda\mu}$ have not been calculated.

We have already noted in Sec. V B 3 that we learn little new from $e^+e^-\to \mu^+\mu^-$ compared to the joint limits from $e^+e^-\to e^+e^-$ and $g_\mu-2$. This assumes our model in which $|\alpha_{\lambda e\mu}|=|\sqrt{\alpha_{\lambda e}\alpha_{\lambda\mu}}|$. There may be more complex models which do not have this equivalence.

We are studying[27] data from PEP looking for the process

$$e^+ + e^- \to e^+ + e^- + \lambda, \quad \lambda \to \mu^+ + \mu^- . \quad (31)$$

This could take place through a Feynman diagram analogous to that in Fig. 12(a).

2. Possible future experiments

If the precision of the $g_\mu-2$ measurement is improved, then the unexplored region of $\alpha_{\lambda\mu}$-m_λ in Fig. 1(b) can be entered. The initial work leading to such an improvement has begun.[31]

Another way to extend the search for a λ which couples only to the μ is to study the process in Eq. (30a) with increased statistics compared to Ref. 30. We have not studied the sensitivity which could be achieved.

Other possible future experiments could explore the product $\alpha_{\lambda e}\alpha_{\lambda\mu}$. Extending the discussion in Sec. VI A 2, one could look for the process

$$e^- + p \to e^- + p + \lambda, \quad \lambda \to \mu^+ + \mu^- . \quad (32)$$

Or a deliberate energy scanning search could be made for a resonance in

$$e^+ + e^- \to \mu^+ + \mu^- . \quad (33)$$

From a broader viewpoint, there are still unresolved experimental questions concerning the production of muons when a high-energy e^- dissipates in a thick target. We refer to the work of Nelson and Kase[32] and of Nelson, Kase, and Svenson.[33] We do not know if the experimental results of these authors[33] have anything to do with the speculations in this paper. However a new high-energy study of

$$e^- + p \to \mu^+ + \mu^- + \cdots$$

would help clarify those results

C. τ specific forces

1. Remarks on limits and use of other data

The upper limit on $\alpha_{\lambda e\tau}$ from $e^+e^-\to \tau^+\tau^-$, Fig. 11, gives an upper limit on the $\alpha_{\lambda\tau}$ only if one assumes a relation between $g_{\lambda e}$ and $g_{\lambda\tau}$. If the λ couples only to the τ, there are two ways a λ-τ coupling could affect existing τ data: (a) The λ-τ coupling would add to the τ-γ-τ vertex a correction term proportional to $\alpha_{\lambda\tau}$. The measurements of $\sigma(e^+e^-\to\tau^+\tau^-)$ then limit the size of $\alpha_{\lambda\tau}$. (b) If $m_\lambda < m_\tau$ there would be an effect on τ decays proportional to $\alpha_{\lambda\tau}$. These two types of limits require some discussion of τ physics and data and will be presented elsewhere.

2. Possible future experiments

We have no suggestions on how to better explore the limits on $\alpha_{\lambda\tau}$ if the λ couples only to the τ. Indeed little more can be done even if the λ also couples to the e. One cannot get much sensitivity from a search using

$$e^+ + e^- \to e^+ + e^- + \lambda, \quad \lambda \to \tau^+ + \tau^- ; \quad (34)$$

there will not be a $\tau^+\tau^-$ mass peak even if $m_\lambda > 2m_\tau$. The τ remains a challenge to experimenters.

ACKNOWLEDGMENTS

We appreciate discussions with E. M. Riordan and his very useful reading of our first draft. As always, Y. S. Tsai has been very helpful in providing theoretical insights and methods for calculation. We appreciate the discussions with, and support from, the members of the PEGASYS Collaboration. We appreciate discussions with W. R. Nelson on the production of muons by electron beams. We are grateful to T. Sloan for the information he provided on muon trident production. We thank H. A. Olsen for correcting an estimate in this paper. This work was supported by the Department of Energy, Contract No. DE-AC03-76SF00515.

[1] H.-Y. Cheng, Phys. Rev. D **36**, 1649 (1987).
[2] J. Reinhardt, A. Schäfer, B. Müller, and W. Greiner, Phys. Rev. C **33**, 194 (1986).
[3] J. Steinberger, Phys. Rev. **76**, 1180 (1949).
[4] V. W. Hughes, Nucl. Phys. **A463**, 3C (1987).
[5] A. F. Rothenberg, Ph.D. thesis, Stanford University, Report No. SLAC-147, 1972. The experiment is briefly described in T. W. Donnelly et al., Phys. Rev. D **18**, 1607 (1978).
[6] J. D. Bjorken et al., Phys. Rev. D **38**, 3375 (1988).
[7] L. W. Mo (private communiction).
[8] Y. S. Tsai, Phys. Rev. D **34**, 1326 (1986).
[9] W. R. Nelson, H. Hirayama, and D. W. O. Rogers, Report No. SLAC-PUB-4721, 1985 (unpublished).
[10] The anomalous production of e^+e^- pairs by heavy ions is discussed in many papers in *Physics of Strong Fields*, edited by W. Greiner (Plenum, New York, 1980).
[11] H. A. Olsen, Phys. Rev. D **36**, 959 (1987).
[12] E. Holvik and H. A. Olsen, Phys. Scr. **38**, 324 (1988).
[13] E. M. Riordan et al., Phys. Rev. Lett. **59**, 755 (1987).
[14] A. Konaka et al., Phys. Rev. Lett. **57**, 659 (1980).
[15] M. Davier et al., Phys. Lett. B **180**, 295 (1986).
[16] C. Brown et al., Phys. Rev. Lett. **57**, 2101 (1986).
[17] M. Davier, in *Proceedings of the XXIII International Conference on High Energy Physics,* Berkeley, California, 1986, edited by S. C. Loken (World Scientific, Singapore, 1987).
[18] D. J. Bechis et al., Phys. Rev. Lett. **42**, 1511 (1979).
[19] M. Crisler et al., Fermilab Proposal No. P774, 1986 (unpublished).
[20] E. M. Riordan (private communication).
[21] See, for example, S. L. Wu, Phys. Rep. **107**, 220 (1984).
[22] M. Derrick et al., Phys. Rev. D **34**, 3286 (1986).
[23] M. Derrick et al., Phys. Rev. D **31**, 2352 (1985).
[24] K. K. Gan, Ph.D. thesis, Purdue University, 1985.
[25] K. G. Hayes and M. L. Perl, Phys. Rev. D **38**, 3351 (1988).
[26] A. Boyarski et al., Phys. Rev. Lett. **34**, 762 (1975).
[27] C. A. Hawkins and M. L. Perl (unpublished).
[28] An $e+p \rightarrow e+p+\lambda$ search is part of the PEGASYS proposal for studying general electroproduction of hadrons. The PEGASYS experiment would use the e^- beam in the PEP storage ring and a gas-jet target.
[29] T. Sloan (private communication).
[30] N. Dyce, Ph.D. thesis, University of Lancaster, 1988.
[31] V. W. Hughes and G. T. Danby, in *Intersections between Particles and Nuclear Physics—1984,* proceedings of the Conference, Steamboat Springs, Colorado, edited by R. E. Mischke (AIP Conf. Proc. No. 123) (AIP, New York, 1984).
[32] W. R. Nelson and K. R. Kase, Nucl. Instrum. Methods **120**, 401 (1974).
[33] W. R. Nelson, K. R. Kase, and G. K. Svenson, Nucl. Instrum. Methods **120**, 413 (1974).

COMMENTS ON C13
"Notes on the Landau, Pomeranchuk, Migdal Effect: Experiment and Theory"

This was a rare experiment for the 1990's world of high energy physics: it was built in six months, mostly out of parts of previous experiments; data was acquired in a month; and a handful of physicists carried it all out. In the 1950's the Russian physicists Landau and Pomeranchuk predicted that the rate of emission of photons by ultra relativistic electrons would be affected in a peculiar way by the density of matter. The prediction was made more quantitative by Migdal, hence the name Landau, Pomeranchuk, Migdal effect; LPM effect for brevity.

Until our experiment there was no quantitative test of the theory of the LPM effect. The theory used quantum mechanics and semi-classical electrodynamics to predict that the multiple scattering of the electron as it passed through dense matter would suppress the emission of low energy photons. Since no one today doubts the correctiveness of quantum mechanics or of semi-classical electrodynamics, why bother to carry out a quantitative test of the theory? Why did we do the experiment?

We did it because Spencer Klein thought of a neat way to do the experiment, we did it because we could use the SLAC linear accelerator in a parasitic way, not using expensive main beam time, we did it because it was a sweet experiment. As my paper reports, we have confirmed the prediction of Landau, Pomeranchuk, and Migdal, and incidentally we have pointed out that their formulation is imprecise and lacks generality. This in turn has stimulated new theoretical work by my colleagues, Richard Blankenbecler and Sidney Drell.

Still we found no new physics and it will be sometime before I work again on an experiment that is only neat and sweet. I enjoyed the experimental work on the LPM effect and I was pleased when I finally understood Migdal's tortuous mathematics in simple terms. I even had a vague hope that this experiment would stimulate me to think of ways to turn the experimental technique into something deeper. But back that hope melted away and the data analysis matured, my thoughts turned hard to my other deeper research work. Much of that research work is with the CLEO detector at the Cornell electron-positron collider CESR. With the large amount of data on tau lepton physics being collected with the CLEO detector, I am going to try again to find new physics through the tau lepton.

SLAC-PUB-6514
May 1994
T/E

NOTES ON THE LANDAU, POMERANCHUK, MIGDAL EFFECT: EXPERIMENT AND THEORY*

Martin L. Perl
Stanford Linear Accelerator Center,
Stanford University,
Stanford, California 94309

Abstract

The status of the Landau, Pomeranchuk, Migdal Effect is briefly reviewed. A recent experiment at the Stanford Linear Accelerator Center substantially agrees with the existing theoretical formulation. However, that formulation suffers from an imprecise foundation and a lack of generality. The difficulty of finding a simple, explanatory picture of the $1/\sqrt{k}$ behavior of the Effect is also noted.

Talk presented at:
LES RENCONTRES DE PHYSIQUE DE LA VALLEE D'AOSTE,
La Thuile, Italy – March 6–12, 1994

*This work was supported by the Department of Energy, contract DE–AC03–76SF00515.

I. Introduction and Bethe-Heitler Bremsstrahlung Theory

In 1993 a group of colleagues and myself[1,2,3,4] carried out the first precise experiment on the Landau, Pomeranchuk, Migdal Effect (LPM Effect).[5,6,7,8] This effect occurs when an ultrarelativistic particle emits low energy bremsstrahlung photons as the particle passes through dense matter; fewer photons are emitted than predicted by bremsstrahlung theory for isolated atoms. Our measurements are in substantial agreement with existing LPM Effect theory as developed by Migdal[6]

However, our use of this theory has accentuated the limitations of this theory. In Sec. II I give a qualitative description of the theory, its predictions, and its limitations. I also note the problem of finding a simple, semi-quantitative picture of the effect; a picture which could be useful in thinking about the underlying physics. In the final section, Sec. III, I summarize our first analyzed experimental results.[2,4]

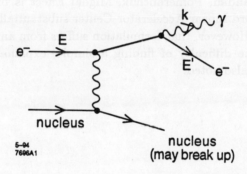

Fig. 1. Bremsstrahlung process on an isolated atom.

I begin by considering the cross section for bremsstrahlung by an ultrarelativistic electron of mass m and energy E on an isolated atom with nuclear charge Z (Fig. 1). The criteria for ultrarelativistic in this case is not only $E/m = \gamma \gg 1$, but also the criteria given in Eq. B4 of Tsai.[9] Next, let k be the photon energy and use the complete screening approximation[9] so that

$$k \ll E \tag{1}$$

is required.

Then, the differential cross section is[9]

$$\frac{d\sigma}{dk} = 4\alpha r_e^2 \frac{1}{k} \times \left\{ \left[\frac{4}{3} - \frac{4}{3}y + y^2\right] [Z^2 F_{el} + Z F_{inel}] + \frac{1}{9}[1-y][Z^2 + Z] \right\} \quad (2)$$

Here $y = k/E$, α is the fine structure constant, and r_e is the classical radius of the electron. In terms of the electron charge e and mass m.

$$r_e = \frac{e^2}{mc^2} = 2.82 \times 10^{-13} \text{ cm} \quad (3)$$

In Eq. (2) there has already been integration over the other kinematic variables of the photon, scattered electron, and produced hadrons. F_{el} and F_{inel} are the results of this integration over the elastic and inelastic atomic from factors[9], they do not depend on y or E, and are given approximately by[9]

$$F_{el} = \ell n\left(\frac{184}{Z^{1/3}}\right), \quad F_{inel} = \ell n\left(\frac{1194}{Z^{2/3}}\right) \quad (4)$$

Therefore when $y \ll 1$, Eq. (2) becomes the well known.

$$\left(\frac{d\sigma}{dk}\right)_{y \ll 1} \approx \frac{\text{constant}}{k} \quad (5)$$

I define for later use the probability per unit length of emitting a photon with energy between k and $k + dk$

$$P_{BH} = n\frac{d\sigma}{dk} = 4n\alpha r_e^2 \frac{1}{k} \times \left\{ \left[\frac{4}{3} - \frac{4}{3}y + y^2\right] \left[Z^2 \ell n\left(\frac{184}{Z^{1/3}}\right) + Z \ell n\left(\frac{1194}{Z^{2/3}}\right)\right] + \frac{1}{9}[1-y][Z^2 + Z] \right\} \quad (6a)$$

The subscript BH denotes the Bethe-Heitler bremsstrahlung theory and n is the number of atoms per unit volume. For future use I note that if we ignore the $Z\ell n\left(\frac{1194}{Z^{2/3}}\right)$ term and the $Z^2 + Z$ terms in Eq. (6a) and set $y \ll 1$

$$P_{BH} = 4n\alpha r_e^2 Z^2 \ell n\left(\frac{184}{Z^{1/3}}\right)\frac{1}{k}$$

Using the radiation length X_0 defined approximately by

$$X_0^{-1} \approx 4n\alpha r_e^2 Z^2 \ell n\left(\frac{184}{Z^{1/3}}\right) \quad (6b)$$

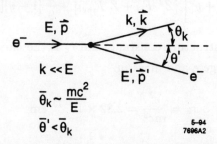

Fig. 2. Kinematic quantities for $k \ll E$.

$$P_{BH} = \frac{1}{X_0 k} \tag{6c}$$

Also I note, although I have not discussed the angular distributions, that the average values of the angle θ_k and θ' (Fig. 2) are

$$\bar{\theta}_k \sim \frac{mc^2}{E}$$
$$\bar{\theta}' \lesssim \bar{\theta}_k \tag{7a}$$

when

$$y \ll 1 \tag{7b}$$

that is, when $k \ll E$. The longitudinal momentum transfer, Fig. 2, is

$$q_\parallel = p - p' \cos \theta' - \frac{k}{c} \cos \theta_k \ ,$$

and for the conditions in Eqs. (7) simplifies to

$$q_\parallel \approx p - p' - k/c \approx \frac{(mc^2)^2 k}{2EE'c} \approx \frac{k}{2\gamma^2 c} \tag{8}$$

where $\gamma = E/mc^2$. The c appears in Eq. (8) because k is an energy.

The uncertainty principle requires that the spatial position of the bremsstrahlung process have a longitudinal spatial uncertainty of

$$\ell_\parallel \approx \frac{\hbar}{q_\parallel} \approx \frac{2\hbar c \gamma^2}{k} \tag{9}$$

If the atom taking part in the process is isolated from all other atoms by distances greater than $\ell_\parallel$, the uncertainty principle has no effect. However, if the atom is not isolated then Eq. (9) can not be ignored, and it leads to the LPM Effect.

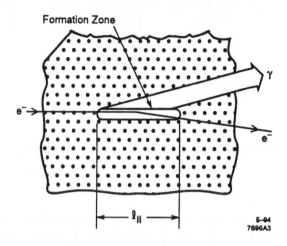

Fig. 3. A qualitative picture of the LPM Effect. The bremsstrahlung γ is produced coherently over the entire length $\ell_\parallel$ of the formation zone. The multiple scattering of the e^- inside the formation zone suppresses the bremsstrahlung probability.

II. The Migdal Formulation of the LPM Effect

Suppose the brehmsstrahlung process just described occurs in a dense medium, Fig. 3, such that

$$\ell_\parallel \gg 1/\sqrt[3]{n} \tag{10}$$

Then the point of occurrence of the brehmsstrahlung process is uncertain within the formation zone shown in Fig. 3.

Any process which changes the path of the electron inside the distance $\ell_\parallel$ will reduce P, the probability per unit length of bremsstrahlung emission. Of course, the process most likely to occur is multiple scattering by the incident or final electron on the atoms in the formation zone.

Consider the simple equation for the mean square multiple scattering angle over a length ℓ,

$$\overline{\theta_s^2} = \left(\frac{E_s}{E}\right)^2 \frac{\ell}{X_0} \tag{11}$$

where X_0 is a radiation length and $E_s = \sqrt{\frac{4\pi}{\alpha}} mc^2 = 21$ MeV. The bremsstrahlung probability per unit length, P, will be reduced when

$$\overline{\theta_s^2} \gtrsim \bar{\theta}_k^2 = \left(\frac{mc^2}{E}\right)^2 \quad (12)$$

Using $\overline{\theta_s^2} = \bar{\theta}_k^2$ define

$$\ell_{LPM} = \left(\frac{mc^2}{E_s}\right)^2 X_0 \quad (13)$$

Then Eq. 12 leads to the condition

$$\ell_\| \gtrsim \ell_{LPM} \quad (14)$$

for reduction of P by multiple scattering. Table 1 gives some examples of ℓ_{LPM} and $\ell_\|$ for our experiment.[2]

Table 1. Values of X_0 and ℓ_{LPM}. Values of k_{LPM} for 25 GeV incident electrons.

Material	Z	X_0	ℓ_{LPM} (μm)	k_{LPM} (MeV)
C	6	18.8	109	8.6
Fe	26	1.76	10.2	92
Au	79	0.33	1.9	490
U	92	0.32	1.9	510

A first problem in working with the LPM effect is to try to develop a physical picture as to how and why the reduction of P occurs under conditions of Eq. (12). Qualitatively the bremsstrahlung photon has to be emitted by a coherent process over the length $\ell_\|$. The repeated changing of electron direction due to multiple scattering inside $\ell_\|$ destructively interferes with that coherence. Galitsky and Gurevich[10] have presented a quantitative picture of the LPM effect which I recommend to the reader.

Indeed other processes which destructively interfere with that coherence also reduce P. One example is the effect of a magnetic field changing the electron's direction. Another example is photon absorption, or scattering in the medium. However, this paper is restricted to the multiple scattering effect.

Migdal[6] has provided a derivation of P_{LPM} which replaces P_{BH} (Eq. 6). But the derivation is much too complicated to summarize here and I know of no simpler quantitative derivation. Therefore, I proceed to his results. Using Eq. (14), the LPM Effect requires

$$\frac{\ell_{LPM}}{\ell_\|} \lesssim 1 \quad (15a)$$

where

$$\frac{\ell_{LPM}}{\ell_{\parallel}} = \frac{\alpha k X_0}{8\pi\hbar c\gamma^2} = y\left(\frac{mc}{\hbar}\right)\left(\frac{mc^2}{E}\right)\frac{\alpha X_0}{8\pi} \quad (15b)$$

From Eqs. (15), the LPM effect requires

$$k \lesssim \frac{8\pi\hbar c\gamma^2}{\alpha X_0} = k_{LPM} \quad (16a)$$

With X_0 in cm

$$k_{LPM} = 6.78 \times 10^{-8}\gamma^2/X_0 \text{ MeV} \quad (16b)$$

Table 1 gives values of k_{LPM} for $E = 25$ GeV. Remember that $k \lesssim k_{LPM}$ is not a sharp criterion. Within a factor of 2 or so, when k falls below k_{LPM} the LPM Effect begins to reduce the bremsstrahlung probability.

Migdal uses a dimensionless variable

$$s = \frac{1}{2}\left[\frac{y}{1-y}\right]^{\frac{1}{2}}\left[\frac{mc}{\hbar}\frac{mc^2}{E}\frac{\alpha X_0}{8\pi\xi(s)}\right]^{1/2} \quad (17a)$$

There are two differences between Eq. (17a) and $\sqrt{\ell_{LPM}/\ell_{\parallel}}$ from Eq. (15b). First, there is the function:

$$\begin{aligned}\xi(s) &= 1 \quad, \quad s > 1 \\ 1 &< \xi(s) < 2 \quad, \quad s < 1\end{aligned} \quad (17b)$$

Second there is the $1 - y$ term in the first bracket, which is unimportant when $y \ll 1$. Thus, for $y \ll 1$, s is proportional to $\sqrt{\ell_{LPM}/\ell_{\parallel}}$ within a factor $\sqrt{2}$, hence

$$\begin{aligned}s &\gg 1 \quad : \quad \text{no LPM Effect} \\ s &\ll 1 \quad : \quad \text{strong LPM Effect}\end{aligned} \quad (17c)$$

Migdal replaces $\alpha X_0/\pi$ by

$$B = 2\pi Z^2 r_e^2 \, n \, \ell n\left(\frac{183}{Z^{\frac{1}{3}}}\right)\xi(s) = \frac{\pi\xi(s)}{2\alpha X_0} \quad (18)$$

and writes

$$s = \frac{1}{8}\left[\frac{y}{1-y}\frac{m^2 c^3}{\hbar E}\frac{1}{B}\right]^{\frac{1}{2}} \quad (19)$$

Having defined s and B and indicated their significance. I now give the Migdal replacement for P_{BH} in Eq. (6).

$$P_{LPM} = \frac{2\alpha B}{3\pi k} \left\{ y^2 G(s) + 2\left[1 + (1-y)^2\right]\phi(s) \right\} \qquad (20)$$

where for

$$\begin{aligned} s &\to 0 & \phi(s) &\to 6s & G(s) &\to 12\pi s^2 \\ s &\to \infty & \phi(s) &\to 1 & G(s) &\to 1 \end{aligned} \qquad (21)$$

Therefore where $s \to \infty$

$$P_{LPM} = 4n\alpha r_e^2 \frac{1}{k} \left[\frac{4}{3} - \frac{4}{3}y + y^2\right]\left[Z^2 \ell n \frac{183}{Z^{\frac{1}{3}}}\right] \qquad (22)$$

Comparing this with P_{BH}, Eq. (6a) we see almost the same formula except two small terms are missing:

$$Z\ell n \left(\frac{1194}{Z^{2/3}}\right) , \quad \frac{1}{9}[1-y][Z^2 + Z] \qquad (23)$$

When $s \ll 1$ and with the approximation $\xi(s) \approx 1$, Eq. (20) becomes

$$\begin{aligned} P_{LPM} &\approx \frac{8\alpha}{\pi k} Bs \left[1 + (1-y)^2\right] \\ P_{LPM} &\approx \frac{\alpha}{\pi} \left[\frac{m^2 c^3}{\hbar} \frac{B}{kE(E-k)}\right]^{1/2} \left[1 + (1-y)^2\right] \end{aligned} \qquad (24)$$

Looking at Eq. (24) we see that for $s \ll 1$ and $y \ll 1$, P_{LPM} is proportional to $1\sqrt{k}$. This is in sharp contrast to P_{BH} which for $y \ll 1$ is proportional to $1/k$. To emphasize this point, I set $y \ll 1$ in Eq. (24), then

$$P_{LPM} \approx \frac{2\alpha}{\pi} \left[\frac{m^2 c^3}{\hbar E^2} \frac{B}{k}\right]^{1/2}$$

and using from Eq. (18), $B \approx \pi/2\alpha X_0$

$$P_{LPM} \approx \sqrt{\frac{2\alpha}{\pi}} \frac{mc^2}{E} \left[\frac{1}{\hbar c X_0 k}\right]^{1/2} , \quad s \ll 1, y \ll 1 \qquad (25a)$$

Contrast this to a simplified form of Eq. (6a) with $y \ll 1$ and the term in Eq. (23) ignored, namely Eq. (6c)

$$P_{BH} \approx \frac{1}{X_0 k} , \quad y \ll 1 \qquad (25b)$$

Comparing Eqs. (25a) and (25b) we note in addition to the $1/\sqrt{k}$ change from $1/k$ other differences. P_{LPM} is proportional to $1/E$ whereas P_{BH} is independent of E in these approximations. P_{LPM} is proportional to $\sqrt{n}$ whereas P_{BH} is proportional to n. Here n is the number of atoms per unit volume.

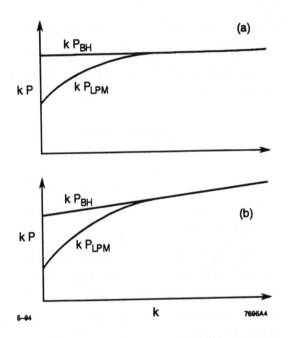

Fig. 4. A comparison of the theoretical curves for kP_{BH} and kP_{LPM} for (a) an ideal experiment, and (b) our experiment in which the e^- may emit more than 1 photon while passing through the target.

Remember that Eq. (25a) is for $s \ll 1$ and is the extreme form of the LPM Effect caused by multiple scattering. The more general form of P_{LPM} is Eq. (20), and for $s \gg 1$ $P_{LPM} \to P_{BH}$. This is pictured graphically in Fig. 4a where kP is plotted against k. The kP_{BH} curve is a horizontal straight line as long as $k \ll E$. The kP_{LPM} curve falls below kP_{LPM} when $k \gtrsim k_{LPM}$ as defined in Eqs. (16).

As I describe in the next section, our experiment[2,3,4] is in substantial agreement with the Migdal formulation of the LPM Effect. Nevertheless, there are a number of problems in this formulation. First, the derivation in Midgal's paper[6] has many approximations, I was not able to estimate their validity. Indeed I am surprised the formulas work so well. Second, Migdal's formulation at large s does not agree perfectly with Bethe-Heitler theory, his P_{LPM} does not have the terms listed in Eq. (23). Third, the Migdal formulation is for a medium of infinite extent along the electron trajectory, it does not possess any direct way of calculating the boundary effect described in the next section. Fourth, there is no physical insight for the

$1/\sqrt{k}$ behavior of P_{LPM} at $s \ll 1$. In terms of dimensions, the $1/\sqrt{k}$ is taken care of by

$$\left(\frac{1}{\hbar c X_0 k}\right)^{1/2} = \frac{1}{\hbar c}\left(\frac{\hbar c/X_0}{k}\right)^{1/2}$$

so that P_{LPM} in Eq. (25a) has the proper dimensions of (energy length)$^{-1}$. But why $k^{1/2}$?

Therefore, more work is needed to provide a more precise foundation, more physical insight, and a more general formulation for the LPM Effect. Meanwhile, as we will show in the next section, the Migdal formulation works quite well.

III. Experiment

The only direct measurement previous to ours[2,3,4] is that of Varfolomeev et al.[11] published in 1976. However, the results of Varfolomeev et al. are limited in the range of k and they are difficult to use for tests of the theory because they are given in terms of ratios of bremsstrahlung spectra for pairs of materials.

Our experiment[3] was carried out in 1993 in End Station A of the Stanford Linear Accelerator Center. We used 25 GeV to 400 MeV electron beams with an average intensity of one electron per pulse and 120 pulses per second. These beams were obtained parasitically from the Stanford Linear Collider (SLC) beams while the SLC was operating for the SLD e^+e^- annihilation experiment at the Z^0 energy.

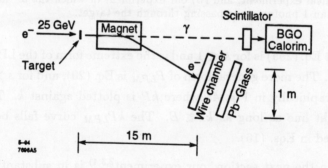

Fig. 5. Schematic picture of the experiment.

Figure 5 shows the experimental arrangement.[2] Targets of 1% to 6% of a radiation length of C, Al, Fe, W, Au, Pb, and U were used, 0.1% of a radiation length was also used for Au. The bremsstrahlung photon was detected in a downstream BGO calorimeter and the scattered e^- was bent in a 3.25 T-m magnet and detected in a wire chamber and a Pb-glass calorimeter. Hence, both k and E' were measured.

In this note I reproduce our results for E=25 GeV electrons on C, Au, and U targets for $5 < k < 500$ MeV. In order to compare our measurements with the Migdal formulation we must note that sometimes an e^- will emit more than 1 photon in passing through the target. For example, if two photons of energies k_1, and k_2 are emitted, the BGO calorimeter measures the sum $k_1 + k_2$ and the scattered e^- has energy $E' = E - (k_1 + k_2)$. This multiple photon emission depletes $d\sigma/dk$ for small k and enhances $d\sigma/dk$ at large k. Hence the ideal kP_{BH} and kP_{LPM} spectra in Fig. 4a are distorted in the data to the curves in Fig. 4b.

The comparison of our measurements for 25 GeV e^- and C, Au, and U targets with Bethe-Heitler theory (BH) and Landau, Pomeranchuk, Migdal theory (LPM) are given in Figs. 6, 7, 9, and 10. The cross section units, kdN/dk, are photons per bin (with 25 bins per decade) per 1000 photons with k in MeV. Normalization and errors are discussed in Ref. 2.

Figures 6, 7a, and 7b clearly show the LPM effect and demonstrate that LPM theory is in much better agreement with the data than BH theory. We see that the Migdal formulation of LPM theory is reasonably well verified. But as is most clear in Figs. 7a and 7b, there is not perfect agreement at the smaller values of k, in that region $(kdN/dk)_{measured} > (kdN/dk)_{LPM}$.

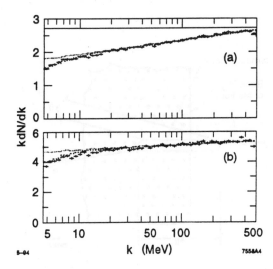

Fig. 6. kdN/dk in number photons per bin per 1000 electrons versus k in MeV for 25 GeV electrons incident on (a) 2% and (b) 6% radiation length carbon targets. Our measurements are denoted by crosses, the Bethe-Heitler theory prediction is denoted by the dotted histogram, upper curve, and the Landau, Pomeranchuk, Migdal theory prediction is denoted by the dashed histogram, lower curve. The latter is a better fit to the measurements.

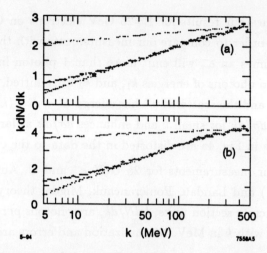

Fig. 7. kdN/dk in number photons per bin per 1000 electrons versus k in MeV for 25 GeV electrons incident on (a) 3% and (b) 5% radiation length uranium targets. Our measurements are denoted by crosses, the Bethe-Heitler theory prediction is denoted by the dotted histogram, upper curve, and the Landau, Pomeranchuk, Migdal theory prediction is denoted by the dashed histogram, lower curve. The latter is a better fit to the measurements.

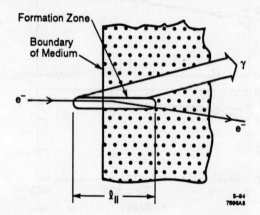

Fig. 8. A qualitative picture of the LPM Effect when the formation zone extends beyond the boundary of the radiating medium.

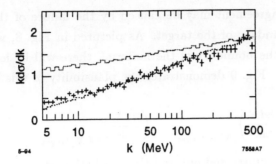

Fig. 9. kdN/dk for a 5% radiation length uranium target minus kdN/dk for a 3% radiation length uranium target. Our measurements are denoted by crosses, the Bethe-Heitler theory prediction is denoted by the dotted histogram, upper curve, and the Landau, Pomeranchuk, Migdal theory prediction is denoted by the dashed histogram, lower curve. The subtraction of the two spectra roughly removes the boundary effect and brings the Landau, Pomeranchuk, Migdal theory prediction closer to the measurements.

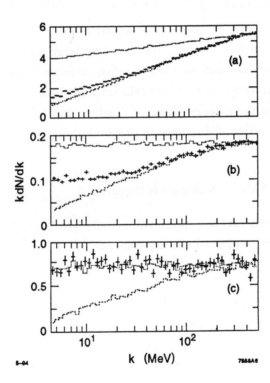

Fig. 10. kdN/dk in number photons per bin per 1000 electrons versus k in MeV for 25 GeV electrons incident on (a) 6%, (b) 1%, and (c) 0.1% radiation length uranium targets. Our measurements are denoted by crosses, the Bethe-Heitler theory prediction is denoted by the dotted histogram, upper curve, and the Landau, Pomeranchuk, Migdal theory prediction is denoted by the dashed histogram, lower curve. As the target becomes thinner the spectrum passes from the Landau, Pomeranchuk, Migdal Effect regime to the isolated atom regime.

We think that this disagreement may be caused by the failure of the Migdal formulation to account for the boundary of the target. As pictured in Fig. 8, when the formation zone of length $\ell_\parallel$ overlaps the boundary of the material, there will be less reduction of the bremsstrahlung probability. Fig. 9 demonstrates the plausibility of this boundary effect by showing for U.

$$\left(\frac{kdN}{dk}\right)_{5\% X_0\ target} - \left(\frac{kdN}{dk}\right)_{3\% X_0\ target}$$

The boundary effect is now subtracted out and there is better agreement between the data and the LPM theory prediction.

Figures 10 for Au with 6% X_0, 1% X_0 and 0.1% X_0 target is a dramatic demonstration of the transition from LPM theory conditions to BH theory conditions. As the target thickness decreases the boundary effect becomes more important. For a very thin target, the boundary effect cancels out the LPM suppression and we are back to the case of an isolated atom.

Thus our experiment has clearly demonstrated the existence of the LPM Effect and has in large part verified the Migdal formulation. We have two large remaining tasks, we have to complete the analysis of our data and we have to refine or improve the theory so that we can make more straightforward comparison of experiment and theory. An important part of improving the theory concerns yet smaller values of k where the LPM Effect must include dielectric suppression, the reduction of bremsstrahlung probability due to interaction of the photon with the medium.

IV. Acknowledgement

I wish to thank my colleagues on the experiment described in Sec. III and particularly to acknowledge Spencer Klein who conceived the experiment.

References

1. The experiment was carried out by S.R. Klein, M. Cavalli-Sforza, L.A. Kelly of the University of California at Santa Cruz; P. Anthony, R. Becker-Szendy, L.P. Keller, G. Niemi, L.S. Rochester, and M.L. Perl of the Stanford Linear Accelerator Center, Stanford University; and P.E. Bosted and J.White of the American University.

2. S.R. Klein *et al.*, *Proc. XVI Int. Symp. Lepton and Photon Interactions at High Energies* (Ithaca, 1993), Eds. P. Drell and D. Rubin, p. 172.

3. R. Becker-Szendy *et al.*, SLAC-PUB-6400 (1993), to be published in *Proc. 21st SLAC Summer Institute on Particle Physics.*

4. M. Cavalli-Sforza *et al.*, SLAC-PUB-6387 (1993), to be published in the *Trans. IEEE 1993 Nucl. Sci. Symp..*

5. L.D. Landau and I.J. Pomeranchuk, Dokl.Akad.Nauk. SSSR **92**, 535 (1953); **92**, 735 (1953). These two papers are available in English in L. Landau, *The Collected Papers of L.D. Landau*, Pergamon Press, 1965.

6. A.B. Migdal, *Phys. Rev.* **103**, 1811 (1956).

7. E.L. Feinberg and I. Pomeranchuk, *Nuovo Cimento*, Supplement to Vol. **3**, 652 (1956).

8. J.S. Bell, *Nuclear Physics* **8**, 613 (1958)

9. Y.-S. Tsai, *Rev. Mod. Phys* **46**, 815 (1974).

10. V.M. Galitsky and I.I. Gurevich, *Il Nuovo Cimento* **32**, 396 (1964).

11. A.A. Varfolomeev *et al.*, *Sov. Phys. JETP* **42**, 218 (1976).

COMMENTS ON C14
"Efficient Bulk Search for Fractional Charge with Multiplexed Millikan Chambers"

As I write this comment we are half way through the first experiment based on this paper and so I have the opportunity to write on an experiment in progress, something rarely done.

Previous general searches for free fractional electric charge have examined at most about 1 milligram of material per experiment. They found no evidence for free fractional electric charge, therefore the upper limit is less than 1 free fractional charge per 10^{21} nucleons. The usual candidate for a free fractional charge is an isolated quark with a charge $\pm\frac{1}{3}e$ or $\pm\frac{2}{3}e$ where e is the magnitude of the electron's charge.

The object of this paper is to describe how to increase the sensitivity of a fractional charge search by 10^4 to 10^5, searching through 10 to 100 grams of matter. But in our first experiment, in which we are developing the technology, we are examining about 1 milligram of matter. The experiment is being carried out mainly by Nancy Mar for her Stanford University PhD. research, by Eric Lee, and by me.

The method derives from the original work of Millikan and from free quark searches carried out at San Francisco State University. Small drops of liquid fall in air under the combined forces of gravity and a vertical electric field; the electric field changes sign periodically. At present we use oil drops with a 7 micrometer diameter which fall with an average velocity of 1.4 millimeters per second. The electric field changes sign every 0.1 seconds and the trajectory of the drops is observed over a vertical distance of about 2 millimeters.

The small diameter of the drops and the fractional resistance of the air leads to the drops falling with a constant velocity, usually called the terminal velocity. If the drop has non-zero electric charge, there are two different values of the terminal velocity, one value when the electric force adds to the gravitational force and the other value when the electric force subtracts from the gravitational force. Applying Stoke's law to these two velocities gives the electric charge on the drop and the mass of the drop.

The trajectory of the drops is measured using a CCD camera whose output is stored on a "frame grabber" board in a desktop computer. The computer calculates the charge and mass of each drop, stores the data, and operates the experiment.

As we are in the midst of this first experiment, my mind is filled with its progress and problems. When we began to design the apparatus I saw three specific design questions and some vague general problems and worries. The design questions were: (1) how to make small drops, 10 micrometers or less in diameter; (2) how to illuminate the drops for the CCD camera; and (3) how to devise a computer program to find with maximum precision the position of a drop. The vague problems and worries included: would there be distortion of the drop trajectory by air currents or non-uniform electric fields, what would be the natural electric charges on the drops, and would a special and very expensive CCD camera be needed?

I will describe in reverse order how all this is turning out. First, the vague problems and worries. As often happens in experimental work the first worries are not the final worries.

We found that with moderate care we could eliminate distortions of the drop trajectory due to air currents or non-uniform electric fields. We operate the experiment now with most drops having zero electric charge, a few with $\pm 1\ e$ or $\pm 2\ e$. We are able to use a commercial CCD camera that costs less than $1000. But unexpectedly we had a great deal of trouble with shaking of the apparatus due to building vibrations coming through the floor and walls. That is mostly solved now and we are able to measure the drop's electric charge with a root mean square error of about $0.025\ e$. I am pleased with this since about half of this error is caused by the Brownian motion of the drops in the air, an error we cannot reduce. Of course the little worries never end. Just last week the rate of bad m! easurements, usually less than 1% , increased during the hours of midnight to 8 a.m. Is some electrical interference being created during those hours, is the room's temperature control malfunctioning during those hours, we still don't know?

Turning to specific question (3), my colleagues did a lot of thinking about algorithms for finding the position of a drop. The solution turned out to be a combination of a simple algorithm and the experimental answer to my specific question (2) about illuminating the drops for the CCD camera. The algorithm was developed by George Fleming, a student from San Francisco State University, who worked with us part time. The algorithm requires uniform backlighting of the drops; this we are able to achieve by carefully adjusting physically the position of a stroboscopic light and by carefully setting with the computer the algorithm's image detection threshold.

The first question on how to make the drops has turned out to be the hardest to answer. Our general approach is to use a liquid-filled tube, several millimeters inner diameter, with a 10 micrometer or so hole in the bottom end and a constriction in the top end. A quick squeezing of the tube then produces a drop using the same principle as the old ink jet printer heads. I began the development of the apparatus with the tube itself consisting of a hollow cylindrical piezoelectric crystal. The crystal contracts when a voltage pulse is applied between the inner and outer lateral surfaces. For the liquid I chose water: easy to get, easy to clean up, and the liquid base of all ink jet printer inks. Water turned out to be a *bad* choice. We struggled for a year with two problems: air dissolved in water interfers with drop production; and as a drop of water falls its mass decreases due to evaporation. To reduce evaporation, I tried 100% humidity air, but then there were electri! cal leakage and corona problems.

At about this time we were joined by Eric Lee, an electronic engineer with a vast amount of laboratory experience in mechanics, optics and electronics. Eric saved the experiment. He designed and built a much simpler drop maker, a brass tube passing through a doughnut shaped piezoelectric crystal; and he replaced the water with a silicone oil.

With the experiment now working at the 1 milligram level, I can think about how to get the several factors of 10 we need in data rate. We must keep the drop diameter at 7 to 8 millimeters, therefore we must increase the rate at which we produce and measure drops. At present our drop production rate is about 1 per second. The paper sets a goal of 10^4 to 10^5 drops per second. The optical system can cope with such a rate, and with faster computers our present drop position algorithm can cope with that rate.

The rate of drop generation is the bottleneck. We need a dropper which can produce parallel streams of drops at a high rate. Eric Lee has some ideas about making the drop producing mechanism out of silicon using microfabrication, that seems to be our best hope. The physics of small drop production is a very complicated area of fluid mechanics, and so I don't expect much help from using fluid mechanics theory. I think we will depend mostly on the Thomas Edison approach, try lots of things.

Meas. Sci. Technol. 5 (1994) 337-347. Printed in the UK

Efficient bulk search for fractional charge with multiplexed Millikan chambers

Charles D Hendricks†, Klaus S Lackner‡, Martin L Perl§ and Gordon S Shaw||

† Livermore National Laboratory, Livermore CA 94551, USA and W J Schafer Asscociates, Livermore, CA 94550, USA
‡ Los Alamos National Laboratory, Los Alamos NM 87545, USA
§ Stanford Linear Accelerator Center, Stanford CA 94305, USA
|| University of California, Irvine CA 92717, USA

Received 1 June 1993, in final form 4 February 1994, accepted for publication 10 February 1994

Abstract. We outline the design of a fully automated Millikan droplet apparatus that could detect a single free fractional charge in several hundred grams of matter even without the use of a prefilter. This would constitute an improvement over current limits by about three orders of magnitude. The experiment achieves high material throughput and high background rejection through on-line processing which allows for a feedback system that can concentrate the measurement effort on anomalous droplets. The task is simplified by generating a monodisperse stream of droplets which will be preprocessed to let only a very narrow range of charges enter the Millikan chamber. Because the droplets can act as carriers of finely dispersed materials it is also possible to search for fractional charge in matter that has not undergone extensive refinement that may have excluded fractionally charged atoms from the sample. In a large refinery style operation many such Millikan chambers could run concurrently to achieve extremely large material throughput.

1. Introduction

The fundamental charge in nature is $\frac{1}{3}e$. This has been well established experimentally and is a basic ingredient of the standard model of particle physics [1]. It is generally believed that the standard model predicts confinement of fractional charge. However, it is difficult to test this aspect of the model experimentally. Small changes to the standard model that are still in agreement with the body of experimental data would introduce stable, fractionally charged particles that can exist in isolation. Neither a slightly broken symmetry of colour SU(3) [2], nor an electrically neutral, heavy member of a colour triplet are ruled out by experimental evidence, and would imply the existence of isolated, fractionally charged particles. These particles would have formed in the early stages of the expanding universe and some would have survived to this day. However, most scenarios predict very low abundances of fractionally charged particles, so low that their discovery in past matter searches would have been unlikely. Whether fractionally charged particles exist in isolation is a purely experimental question that still remains unanswered.

Here we describe an experimental technique for detecting fractional charges in bulk matter that is four to five orders of magnitude more sensitive than past experiments. Finding isolated fractional charges at this level, <1 per 10^{26} nucleons, would not only be of enormous scientific interest, but would also have significant practical applications. For example, one might obtain fractional charges in numbers sufficient for catalysing fusion [3]. Even if this search does not lead to their discovery, it would set a reliable and much more stringent limit on the cosmic abundance of fractionally charged particles. Such a limit would rule out or severely restrict the most natural models that predict free fractional charge.

The technique we describe is conceptually simple and has become possible by recent advances in computing technology. We automate the basic Millikan apparatus utilizing simple, real-time pattern recognition algorithms to measure the charge on a large number of drops simultaneously. Through feedback, the effort is concentrated on those drops that initially appear to have non-integer charge. This allows for extremely high background rejection while keeping the average measurement time short. The Millikan-type charge measurement is simplified by generating streams of monodisperse droplets, i.e. droplets of a single size,

Reprinted from *Measurement Science Technology* 5, pp. 337-346,
Copyright 1994, with kind permission from Institute of Physics Publishing

which are prefiltered in charge so as to allow only a narrow range of charges to enter the Millikan chamber. Because they are fully automated, many such Millikan chambers could run concurrently in a refinery style operation to achieve extremely large material throughput.

The simplicity of this table top experiment is its strongest advantage. By automating simple procedures measuring the charge on very small droplets, we can avoid the pitfalls of other bulk charge measurements. The difficulty of an individual measurement is minimal, interference of separate measurements is rare and detectable. Background rejection is achieved through immediate feedback which allows one to take, if necessary, a second and third look at a drop and even collect drops that pass all tests for fractional charge. This method avoids highly accurate and therefore difficult individual measurements. In a sense, experimental difficulty has been traded for sophistication in on-line data analysis. The latter has become possible in recent years as computer power increased and pattern recognition algorithms have become more refined.

2. The concepts

Matter searches for fractionally charged particles typically proceed by measuring the net charge on samples of bulk material. A fractional net charge would indicate the presence of a fractionally charged particle. The classic example of an experiment determining the fundamental unit of charge in this way is the Millikan droplet experiment [4–6]. The experimental difficulty of matter searches for fractional charge stems from the need of processing large amounts of material.

In the past, experimenters have addressed this difficulty using two complementary approaches. The first, represented by, e.g., Fairbank [7] and Morpurgo [8, 9], attempts to maximize the mass of a sample. This leads to difficult, time-consuming measurements. However, in a single measurement a large amount of matter is processed. As was shown by Smith et al [10, 11] the magnetic levitation technique can be applied to arbitrary materials, for example meteorites, and does not require the high purity required in earlier experiments. This is an important point since fractinonal charges being impurities themselves, are suppressed during the purification process [12, 13].

Instead of maximizing the size of a sample, the second approach minimizes the measurement time for individual samples. Many small samples whose charges are much more easily determined are processed serially. Cylindrical liquid jet break-up technologies fall into this category [14]. The amount of material searched is limited by the speed with which measurements can be performed. The difficulty of this approach is background noise; the tails of the measured charge distributions must not extend to the fractional charges of interest. This accuracy requirement limits the reduction in measurement time. In dropper type experiments, the stringent requirements on the reproducibility in drop generation lead to difficulties which ultimately limited throughput. Nevertheless, the dropper experiment of Bland et al [15, 16] reached one of the highest sensitivities to fractional charges in a bulk matter search (cf table 1). This experimental technique was also used for measurements on unprocessed materials.

Our method advances the second approach. We measure the charge of small drops in basically the same fashion as Millikan. By performing the data analysis on-line we can use the current best estimate of the charge to decide whether a further refinement is necessary. This is only the case if the apparent charge is compatible with a residual of $\frac{1}{3}e$ or $\frac{2}{3}e$. As a result, the average effort spent on a sample can be held low, while background noise is suppressed to very low levels.

The mass of a droplet is chosen so that it can be lifted by the electric field and therefore may be observed for an indefinite time. Thus, a considerable effort can be concentrated on a single drop if it appears to have fractional charge. A hierarchy of more and more complex measurements, first eliminates statistical fluctuations and then detects systematic errors. The introduction of an interactive measurement strategy which requires on-line, real time data analysis is a major improvement because it effectively eliminates background as a problem in bulk matter searches. Another benefit of the automatization of the charge measurement is the ability to perform a large number of measurements in parallel.

The task for the pattern recognition algorithm is simplified by generating monodisperse droplets with small size variations. Before entering the Millikan chamber these droplets are grossly segregated in an electric field according to their charge. Droplets entering the Millikan chamber, fall into a narrow charge range that could completely exclude integer charges. In this case the dominant contribution to the droplet stream is likely to be drops that changed charge during their fall

Table 1. Limits obtained in bulk matter searches. This table has been taken from reference [15]. The positive result of LaRue is generally considered to be spurious. For a detailed discussion see [11].

Group	Material	Mass (mg)	fractional charge/nucleon
LaRue et al [7]	niobium	1.1	2.1×10^{-20}
Morpurgo et al [9]	iron	3.7	1.3×10^{-21}
Ziock et al [17]	iron	0.72	6.9×10^{-21}
Smith et al [10]	niobium	4.87	1.1×10^{-21}
Milner et al [18]	niobium/tungsten		$10^{-19} - 10^{-20}$
Joyce et al [19]	sea water	0.05	9.8×10^{-21}
Savage et al [15]	native mercury	2.0	2.9×10^{-21}

time. Because the mass of the droplets varies very little, such a prefilter could increase the total efficiency of the system by three to five orders of magnitude. If drops have similar mass and charge, they move through the device at similar speeds minimizing the number of close approaches which complicate the pattern recognition problem.

This experimental design improves on the state of the art in two ways. One is the staging which leads to a multiplicative reduction of the experimental error and eliminates a large class of systematic errors. The stages are the prefilter, a series of Millikan oil drop measurements, consistency checks on the density, size, unit charge, and Brownian motion of a drop, followed by the extraction of a fractionally charged droplet from the device. The second is the automated on-line measurement protocol which sets the level of effort invested into a droplet based on the outcome of earlier measurements.

3. The Millikan chamber

To keep the design simple, we consider an air-filled Millikan chamber operating at room temperature and ambient pressure. The electric field is maintained between two horizontal, flat plates. The size of the plates and their spacing follows from other design parameters and is of the order of centimetres. An additional set of plates may be used to move drops horizontally. At appropriate times, a pulsed, ultraviolet light source induces positive charge changes in droplets. The chamber is designed to minimize fluctuations in temperature, to avoid sudden pressure changes, and minimize air flow driven by heat convection or pressure gradients. It is worthwhile pointing out that these are not precision measurements. If gradual changes occur in temperature, in the electric field or even in the gas flow through the chamber, the measurements can account for these because of the effective calibration due to the large number of integer charge measurements that are continuously performed. Each measurement is a relative measurement against many others close in time and position.

Droplets are back-lit and imaged on a charged coupled device (CCD). In order to minimize the rate of light induced charge changes, the light source is strobed, which also provides high timing accuracy. The light entering the chamber should be limited to a narrow spectral bandwidth. A system of lenses is used to obtain optimal resolution and the enlargement may differ in the horizontal and vertical direction. A large array CCD camera may generate an image of about 4000×4000 pixels. The CCD is read out at 3–200 Hz and a small part of the raw data is processed during each cycle to obtain charge measurements. In normal operation, the electric field is switched between E_{max} and $-E_{max}$.

Stoke's law relates the radius R of a droplet, its velocity v relative to the ambient gas, and the viscosity η of the gas to the force F on the droplet,

$$F = 6\pi \eta R v.$$

Therefore, one can calculate a droplet's mass and charge from its trajectory in a time varying electric field. In our case, mass measurements serve mainly as consistency checks but also improve accuracy.

The three most important parameters describing the operation of the Millikan chamber are the maximum electric field, the droplet size and the charge measurement accuracy. Together, these parameters determine the layout of the chamber and its serial throughput. In order to estimate the potential of this approach, we present in the remainder of this section the layout of a nearly optimal design. Clearly, an initial attempt would start with a smaller system.

The electric field in a Millikan chamber is limited by the breakdown characteristics of the gas and by electrohydrodynamic droplet emission from liquid coated surfaces in the chamber. At high field strengths, liquid surfaces distort and form points from which small, highly charged droplets are emitted into the chamber. An electric field $E_{max} = 10000$ V cm^{-1} has been demonstrated [15] and can be maintained if care is taken to minimize liquid layers on electrode surfaces.

If one measures the droplet charge to 0.5% accuracy, a nominal charge of $20e$ is optimal. In this case, a residual charge of $\frac{1}{3}e$ or $\frac{2}{3}e$ differs by three standard deviations from the nearest integer. Because droplets in the Millikan device may occasionally change charge, the size of the droplet is chosen such that a somewhat smaller charge can be levitated. Charge changes may also be induced artificially in order to obtain a consistency check on a measurement. Since induced charge changes are obtained through the emission of photoelectrons, they are biased in favor of positive ones which suggests the choice of a positive droplet charge. Monte Carlo simulations show that even if the ratio of positive to negative charge changes is only 65 : 35, no more than two per cent of all droplets will ever fall below $15e$ (cf figure 1). In the following we will assume that the electric field can just levitate a charge of $15e$. This limits the droplet mass to $M = 2.4 \times 10^{-10}$ g or 1.5×10^{14} nucleons.

A typical density of the droplet fluid is $\rho = 1$ g cm^{-3} which implies a radius of $R = 3.9$ μ. In air which under ambient conditions has a viscosity of $\eta = 182$ μ poise (1.82×10^{-4} g cm^{-1} s^{-1}), the fall velocity is

$$v_{fall} = \frac{2\rho g R^2}{9\eta} = 1.8 \text{ mm s}^{-1}$$

where $g = 980$ cm s^{-2} is the gravitational acceleration.

Brownian motion of the droplet limits the accuracy of velocity measurements. The mean thermal contribution to the vertical droplet velocity is

$$v_{brownian} = \sqrt{\frac{kT}{M}} = 0.13 \text{ mm s}^{-1}.$$

The relaxation time δt follows from Stoke's law

$$\dot{v} = \frac{9\eta}{2\rho R^2} v.$$

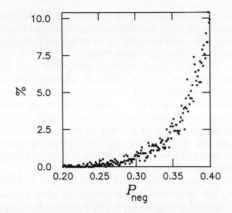

Figure 1. Monte Carlo estimate of the percentage of droplets that will eventually fall below a charge of $15e$ as function of the probability of an individual charge change to be negative. Random charge changes were applied to a droplet of an initial charge of $20e$. Droplets were followed until they either dropped below $15e$, exceeded $60e$ or had not left this interval after 1000 charge changes.

δt also applies to transients from changes in the electric field and is given by

$$\delta t = \frac{2\rho R^2}{9\eta} = 1.8 \times 10^{-4} \text{ s}.$$

The droplet performs a random walk around its average position with a mean stepsize of $v_{\text{brownian}}\delta t$ and a mean time step of δt [20]. For measurement times $\Delta t \gg \delta t$

$$\Delta v = v_{\text{brownian}}\sqrt{\frac{\delta t}{\Delta t}}.$$

If Δt is of order δt or less, the uncertainty becomes the Brownian velocity. A measurement accuracy of 0.5% of the fall velocity requires a minimum measurement time of

$$\Delta t = \left(\frac{\Delta v}{v_{\text{fall}}}\right)^{-2}\left(\frac{v_{\text{brownian}}}{v_{\text{fall}}}\right)^{2}\delta t = 0.038 \text{ s}.$$

The corresponding fall distance is 0.07 mm or only 9 drop diameters. Accordingly, a positional accuracy of $0.3\,\mu$ is required. This is close to the diffraction limit but not impossible to achieve. A frame rate of 30 Hz would be barely sufficient. However, with parallel processing it is possible to reduce the cycle rate and thus the required magnification without sacrificing throughput.

The measurement accuracy is not only limited by Brownian motion. In particular, the effects due to the dipole moments induced by the electric field on the droplets must be carefully controlled [21]. The dipole moment of a dielectric sphere in a constant electric field is given by

$$\boldsymbol{P} = 4\pi\epsilon_0\left(\frac{\epsilon-1}{\epsilon+2}\right)R^3 \boldsymbol{E}.$$

For our choice of parameters and a typical value of the dielectric constant ϵ not too close to one, the induced dipole charge is much larger than the net charge on the droplet. It is given by

$$Q_D = 3\pi\epsilon_0\left(\frac{\epsilon-1}{\epsilon+2}\right)R^2 E.$$

For materials considered here, the factor $(\epsilon-1)/(\epsilon+2)$ is on the order $1/4$ to 1, the latter corresponding to a conductor with a dipole charge of $8000e$.

Since the induced dipole of a droplet interacts with electric field gradients, gradients for example due to the droplet entrance hole must be minimized in the region of measurement. In our case however, an additional 0.5% error would require a relatively large gradient of 250 (V cm^{-1}) cm^{-1}. For this experiment, the more likely error source are droplet-droplet interactions whose potential is given by

$$\frac{1}{4\pi\epsilon_0}\left(\frac{\boldsymbol{P}\cdot\boldsymbol{P}}{r^3}-3\frac{(\boldsymbol{P}\cdot\boldsymbol{r})^2}{r^5}\right).$$

At 30 radii, the dipole–dipole correction to the total force is about 0.5%. In the vertical direction the force is attractive. Thus, to avoid collisions between drops one must maintain a minimum distance between them. To maintain droplets separated for at least 100 s requires a spacing of 50 radii which could be dominated by lateral rather than vertical spacing.

If a charge measurement is compatible with a residual charge of $\frac{1}{3}e$ or $\frac{2}{3}e$, the measurement must be repeated. One out of a few thousand drops has to be measured twice simply because of statistical fluctuations. A charge change during a measurement would also necessitate additional measurements. With a vertical viewing distance of 4000 pixels, ten consecutive pairs of measurements can be taken on the same drop without changing the voltage protocol. The number of pixels could be reduced, if on average the electric field pulls droplets upwards. Ten consecutive charge measurements allow one to establish the residual charge on a droplet with high confidence.

Even though the likelihood of spontaneous charge changes should be minimized, charge changes may be induced intentionally in order to determine the apparent size of the charge quantum. In this way one can rule out a simple distortion of the charge scale (e.g. due to an anomalous density) as the cause of an apparently fractional charge. By inducing charge changes with a pulsed light source every few measurement cycles one can determine the charge scale during the time a droplet moves through the field of view.

Only if the charge on the droplet still appears fractional after it has fallen to the bottom of the field of view is the standard measurement protocol interrupted. Further measurements deal exclusively with this anomalous drop and apply an electric field of different magnitude and time variation in order to rule out other causes of an apparently fractional charge. For an optimized design this exclusive time should be minimized. One way to achieve this would be to extract these special drops from the chamber and process them elsewhere.

We now consider the ultimate limit to which a single automatic Millikan device of this type could be pushed. The obvious limitations arise from the number of droplets that can be observed simultaneously. Vertical spacing between droplets should be large enough to avoid complicated, intertwined and intersecting trajectories. For pattern recognition, a spacing of maybe 10 diameters would be sufficient. In this case more than half of the average distance between droplets required for controlling dipole–dipole interactions would have to result from horizontal spacing between droplets. For a very large CCD, 4000 by 4000 pixels, this would lead to a limit on the order of 20000 drops simultaneously in the field of view. With a fall time of the order of one second this would correspond to 20000 drops s^{-1}.

The next limit stems from the data processing requirement, at 20000 drops s^{-1} the average drop can only get 5×10^{-5} s of attention from the data processing system. A top of the line workstation with parallel processing capabilities may be rated at 2000 MIPS. This would leave 10^5 machine instructions for the average drop. We show in section 6 that this is more than sufficient for the typical drop that only requires a single charge measurement. Droplets that initially appear to be fractionally charged require a much larger computational effort, but they are rare and only change the average effort by a small factor.

Finally, the probability that a droplet will pass all preliminary tests and thereby become a candidate for exclusive attention by the processing system sets another limit on the total rate of drops that can be processed. If the device is blocked for a time t_{excl} by such a candidate, it can be shown that the throughput through the device is optimized if the average time between candidates is equal to the exclusion time t_{excl}. If the probability of a droplet to be such a candidate is P_{cand} then the optimal drop rate D for the device is given by

$$D = \frac{1}{P_{cand} t_{excl}}.$$

If one in 10^5 droplets needs special treatment, $P_{cand} = 10^{-5}$, and if it takes on average 5 s to either process such a droplet or to move it out of the Millikan chamber, we obtain again a limiting drop rate of 20000 drops s^{-1}. P_{cand} is currently unknown; it depends on a number of variables concerning the reproducibility of the droplet generation, the suppression of dust particles and other causes for background. However, our estimates are reasonably close to what has been achieved in previous dropper experiments [15, 16].

If the system operates in this limit, the cycle frequency of the measurements does not affect the throughput. It is only affected by the size of the field of view, the processing power of the computer hardware, the effort required for each droplet and the frequency of droplets that passed all tests for fractional charge. Longer measuring times for individual droplets are compensated for by a larger number of droplets that are simultaneously under observation. Therefore, it may be possible to slightly relax the high rate of data taking without compromising throughput. If the observation time of the average droplet becomes too long, the large fall distances between measurements complicate drop identification and therefore increase the effort in pattern recognition.

The parameters governing an optimal device are listed in table 2. Running continuously, it could handle 10000 drops or 1.5×10^{18} nucleons per second, or 1.3×10^{23} nucleons in a day. An additional increase in effective throughput can be obtained with the prefilter we discuss in the next section. Since the experiment is fully automated, it is also possible to run many such Millikan chambers in parallel.

Table 2. Parameters of Millikan Chamber.

Parameter	Value
Plate Spacing	1–2 cm
Plate Radius	10–20 cm
Observed Fall Distance	0.14 cm
Vertical Image Resolution	0.35 μ/pixel
Vertical Image Size	4000 pixel
Frame Rate	20–200 Hz
Electric field strength	10000 V cm^{-1}
Switching frequency	20–30 Hz
Drop radius	3.9 μ
Drop density	1 g cm^{-3}
Fall velocity	0.18 cm s^{-1}
Transient response time	1.8×10^{-4} s
Maximum number of tracks	$\sim$1000
Horizontal Image size	4000 pixel
Maximum throughput	20 drops/track/s

4. Droplet generation and prefiltering

The technology for generating a monodisperse stream of droplets is well established [22–27]. Hendricks [27] describes a single jet device that generates 10^6 droplets of $R = 5\ \mu$ per second. Their uniformity is excellent at a fractional standard deviation in radius of about 10^{-3}. The mass rate of a single jet is $\sim$1 mg s^{-1}. This exceeds the processing rate in the Millikan chamber by four orders of magnitude. Even higher rates, grams per second, were achieved with multiple jets. The charge of individual droplets in a jet can be controlled within narrow margins. Control is limited by thermal fluctuations:

$$\left| \frac{q\, \Delta q_{therm}}{4\pi \epsilon_0 R_0} \right| = kT \quad \text{or} \quad \Delta q_{therm} = 3.5 e.$$

In principle, it is possible to include in the liquid a fine dispersion of other materials as long as the grain size is significantly smaller than the droplet radius. Methods for including such grains have been studied in the past [28, 29], and others are currently under investigation. If droplets can carry solid materials that may have undergone only minimal processing once can test a wide variety of materials. An upper limit of the fractional charge occurring in such samples can be translated into a meaningful upper limit on their cosmic abundance.

A drop generator consists of a linear array of jets which break up inside a charging electrode that sets the mean charge to approximately $20e$. Each jet corresponds to one of the tracks observed in the Millikan chamber. After the jets break up into droplets, these droplets fall through a transverse electric field that deflects their trajectory by an angle proportional to the charge on the drop. Since the mass of the droplets is known very accurately it is possible to separate the different integer charges. The entry hole into the Millikan chamber can then restrict the droplet charge to a fraction of a charge and the deflecting electric fields can be tuned via feedback to exclude integer charges. For example, only droplets with an effective charge between $20.2e$ to $20.8e$ may enter the Millikan chamber. The majority of droplets that satisfy this criterion have changed charge in flight. The probability of a charge change can be kept low. As a result, the prefilter can greatly enhance the effective throughput, although it is difficult to estimate its efficacy without having actually built one. The mass accuracy would easily allow for a rejection factor of 10^3, as should dipole–dipole interactions and charge changes. For a fractional charge search this charge separation method would have far too high a rate of false positives, but as a prefilter, it provides a large multiplicative factor in the total material throughput. However, the fraction of false positives in the Millikan chamber will also increase and a reasonable compromise must be obtained.

The droplets enter the Millikan chamber in well defined points, each marking the beginning of an individual track. The timing of the droplets is by necessity random because only a very small fraction of the initially formed droplets will pass the prefilter stage of the droplet generator. Droplets are highly uniform in mass and vary only little in their initial charge. This uniformity makes the pattern recognition task manageable. It also allows for a strong suppression of background from the accidental introduction of other materials into the droplet stream.

5. Pattern recognition

The optical system generates a raster image of the droplets in the Millikan chamber. For the design considered here, a droplet image covers approximately 20 raster points. This complicates pattern recognition on the one hand, because the position of the droplet must be deduced from its extended image. On the other hand, it greatly decreases the chance of two droplet images completely overlapping within the time of a measurement. It therefore reduces the effort in the drop identification algorithm. Furthermore, the apparent droplet size and image intensity, which vary in a predetermined manner across the diameter of the droplet, provide a consistency check. At a frame rate of 30 Hz, a droplet moves about 10 diameters or 200 pixels in a single update.

The computational task can be broken up into several steps. First is the analogue-to-digital conversion, which takes the CCD signal and converts it into a digital representation of the image intensity. This step, typically in conjunction with a simple filter operation to clean up the raw image, is performed in commercially available, dedicated equipment well matched to the characteristics of the imaging system. The result is a continuously updated data array that resides in special purpose memory to which the data processing machine also has access. It is not necessary for the CPU to process megabytes of image data on every time step. Instead the processing is focused on small subareas that contain the images of the droplets under observation. The next steps involve the extraction of a location and the calculation of charge and mass.

From the raw image one obtains the size, the brightness and the vertical position of the droplet. Size and brightness are expected to be constant within the accuracy of the device, they also have to be within the established margins for the average drop. If these parameters fall outside this regime or fluctuate, the droplet is considered anomalous and measurements are rejected. Unlike the position measurement which is used to update charge and mass estimates for a droplet, such consistency checks are not performed on all drops and not for every single measurement cycle.

For a single droplet the measurement is quite simple. Applying a predictor corrector algorithm, we use the current position to predict the droplet's position in the next frame. If the deviation from this position exceeds a prespecified amount, and the droplet is not found at a predicted position within one or two more frames, it is considered lost. The precise number of frames one is willing to wait will depend on the accuracy to which its parameters were already known and on the background noise which must be determined empirically.

After the actual droplet position has been determined it is used to update the current best estimate of charge and mass. We assume that both are known from the beginning with an experimental error that is Gaussian and is described by an error ellipse in (q, R) space. Initially we assume $q = 20 \pm 0.5e$, $R = 3.9 \pm 0.01$ μ. In general the probability that $x = (q, R)$ is the location of a droplet in the (q, R)-plane is given by

$$P_0(x) = \frac{1}{Norm} \exp\bigl(-(x - x_0)\Sigma_0(x - x_0)\bigr)$$

where Σ_0 is the error matrix whose eigenvalues are the inverse squares of the major axes of the error ellipse and x_0 the best estimate of the droplet's charge and radius. A single measurement puts a linear constraint on q and R and requires that x satisfy another Gaussian probability distribution

$$P_m(x) = \frac{1}{Norm} \exp\bigl(-(x - x_m)\Sigma_m(x - x_m)\bigr).$$

x_m is an arbitrary point in the (q, R)-plane that satisfies the linear constraint. Σ_m can be obtained from the partial derivatives of the fall distance D with respect to q and R and is singular with a zero eigenvalue whose eigenvector points in the direction in which x_m is unconstrained. Thus, the particular choice of x_m does not affect P_m. The probability distribution that combines the constraints from the latest measurement P_m with the prior knowledge P_0 is again Gaussian. It is characterized by Σ_c and x_c and is proportional to the product of P_0 and P_m. One can show that

$$\Sigma_c = \Sigma_0 + \Sigma_m$$

$$x_c = \Sigma_c^{-1}(\Sigma_0 x_0 + \Sigma_m x_m).$$

Although Σ_m is singular, Σ_c is not and its inverse is well defined. Based on these equations, it is possible to update q, R and the associated error matrix after each measurement. Furthermore, x_c should be compatible with both P_0 and P_m thus providing a consistency check that could be used to detect charge changes which would require the repetition of the measurement.

Once the error ellipse has shrunk to the point that a residual charge of $\frac{1}{3}e$ and $\frac{2}{3}e$ can be ruled out, the droplet can be abandoned. Normally this will only require two measurements at different values of the applied electric field.

Track crossings in which droplet images overlap only pose a problem if at least one of the droplets has been measured multiple times, i.e. is a candidate for fractional charge. One can use its charge and mass to predict its trajectory through the encounter. One may attempt to include corrections for dipole–dipole interactions, but they are difficult to estimate because only the projection of the droplet's motion into the plane of view is measured. If two droplets stay close for more than the predicted time or afterwards follow unexpected trajectories they are considered lost. Note, that partial occultation still allows the tracking of one of the two droplet edges which reduces the time without measurements. The point at which a drop is considered lost depends on the precision to which its parameters were known beforehand. The physical merger of droplets is quite unlikely, if it were to happen it would be recognized from the large change in fall velocity.

Track crossings, particularly in the later stages are further complicated by the fact that not all droplets in the field of view are tracked. It is therefore necessary to define a local protected area around a tracked droplet which is considered its own 'air space'. If another droplet enters this space, it will also be tracked in order to prevent losing the first droplet in a close encounter. This protected area is only important for droplets which have undergone many measurements. In the case that normal measurements are suspended and a single drop is tracked exclusively, a large fraction of the available processing power is used in tracking potentially close encounters. It may even be possible to avoid such collisions by appropriately tuning the electric field.

The purpose of the pattern recognition software is to measure the charge on a large number of droplets simultaneously. With idle processing power, the system searches for currently untracked droplets near the start of the vertical tracks. Droplets with unusual fall velocities, charges outside the expected range or unusual mass are immediately discarded. Such outliers may be handed to a special subprocessor dealing with tracks likely to intersect others. All drops whose charge measurement result within one standard deviation of an integer charge are discarded. The remaining ones are tracked through a second stage and possibly more stages until their charge is considered to be integer. Charge changes are induced and recorded, and if $\Delta q/q$ is integer, the droplet is again discarded. If a droplet is followed to the bottom of the field of view and still appears to have a fractional charge, the electric field is changed to levitate this drop back into the center of the apparatus.

Such a droplet may then be subjected to a number of further tests. First one validates Stokes law for various strengths and directions of the electric field. Then one checks the size of the Brownian motion and finally one ascertains that the apparent charge change is independent of the measured charge. The apparent brightness and steadiness of the droplet image provide additional consistency checks. If a droplet passes all tests it is considered fractionally charged and is maneuvered into a special holding tank.

An important issue in the automatic processing is the self-calibration of the device. The nearly continuous set of measurements together with the constraint that charge distributions must peak at integer values provides a calibration method. The correction may be a function of both position and time. If the corrections vary too rapidly in time or are larger than a predetermined limit, the system is taken off line.

6. Processing power requirements

A rough estimate indicates that even an optimal system with 20000 drops s^{-1} could be controlled by a single, high powered workstation. We express our results in terms of machine cycles required for the pattern recognition algorithms. These in turn are measured from timing simple unoptimized code fragments on two RISC workstations with 25 MIPS and 40 MIPS respectively. The faster station whose result we quote here required noticeably (30%) less machine cycles to perform operations dominated by double precision arithmetic which is not likely to be implemented in the final form of the pattern recognition code. All operations were performed one hundred thousand times during normal multitasking of the workstation. Thus, the timings take into account the small overhead of the operating system as well as idle cycles due to delays in memory access.

Our discussion is based on the typical drop for which we expect to perform the following tasks. First find the droplet in a one-dimensional stretch of about 200 pixels (this operation is performed once), then determine the drop location four times within a window of 40 pixels to the accuracy of one pixel and finally calculate twice, from the travel distance and elapsed time, the charge, the mass and the associated error matrix of the droplet. 200 pixels is the average spacing between droplets therefore it should be sufficient to search a stretch of this length to find a new drop. A window of 40 pixels around the predicted position of the droplet in the next measurement cycle should contain the droplet even if it changed charge. Complicated processing dealing with charge changes and intersecting tracks is not considered for the typical drop. It is only relevant for droplets that are already candidates for fractional charge.

The first step of finding a drop requires on average a scan through 100 pixels using a stepsize of two. Searching for four out of five consecutive points to be above some threshold takes on the average 440 machine instructions.

To determine the accurate droplet position within a window of 40 pixels should require even less instructions. Simple edge detection using four consecutive points to find the left and right edge, and to compute the droplet position as the arithmetic mean of the two, uses on average 280 cycles. More costly (1600 cycles) is the determination of the center of the intensity distribution by calculating the sum and the first moment of all intensities. Under the most pessimistic assumptions concerning the noise in the data, one would have to resort to a convolution of the measured intensity function with a predefined kernel,

$$C(y) = \sum_{x=1}^{40} M(x) P(x - y)$$

where $M(x)$ is the measured intensity as function of pixel position, and $P(x - y)$ the intensity distribution of a noiseless droplet signal centered at y. The correlation function $C(y)$ has its maximum at the location y of the observed droplet. Since it is possible to evaluate $C(y)$ only for a small subset of all y in order to find its maximum and because 40 is not a simple power of 2, the use of FFTs does not appear to be favorable. Instead we have implement a direct search for the maximum of $C(y)$ which requires an average of 5720 machine cycles.

Finally we implemented a specific algorithm using 1400 machine cycles to calculate q, R and the associated error matrix based on the change in position, and time between consecutive measurements.

In summary, using simple test code written in high level languages like C and C++ without any serious effort in optimization shows that the processing required for the typical drop is well below 30000 machine cycles. This falls comfortably within our budget of 10^5 cycles available on a top of the line multiprocessor machine with 2000 MIPS measuring 20000 drops s^{-1}. For the typical drop, we consider the estimate of 30000 machine cycles to be conservative. Any steps omitted in this simplified estimate (e.g. the looping over different tracks when searching for a new drop) are likely to be offset by increased efficiency in an optimized implementation). In particular, the determination of the drop position which dominates the estimate is likely to be achieved with much less effort. The computational effort for anomalous drops is obviously much larger, however, their number is several orders of magnitude smaller and therefore their inclusion should not increase the average effort by more than a factor of three.

7. Discussion

We have outlined the design of an automated Millikan apparatus and discussed the issues surrounding its optimization. A sketch of our design is presented in figure 2. The choice of droplet radius and charge and the rate of droplet flow represents a compromise between errors due to Brownian motion and dipole–dipole interactions. It is likely that for a specific implementation with its own particular constraints the optimization will vary somewhat from the choices given here.

To estimate the sensitivity of an automated Millikan apparatus we can combine the sensitivity of the different stages of the system. The first stage, which prepares uniform drops within a narrow charge change is relatively inefficient. Only a small number of drops will be in the proper charge range. For a continuous charge distribution no more than 1 : 5 could fall into the allowed range of $20.2e$ to $20.8e$. This ratio is based on the thermodynamic limit, other effects may widen the charge distribution and therefore lower the ratio. To be specific we will assume that 1 : 20 may fall into the allowed range. However, since charge is quantized and the allowed range is less than one unit of charge, the effect on integer and fractional charges differs and varies with the position of the window. If random fluctuations in the drop deflection can be held to 0.5% of the total, fractionally charged droplets could be enriched in the sample by better than a factor of 1000. Added to this is another component of a non-gaussian background stemming from in-flight charge changes. If, in total, one achieves an enrichment by a factor of 1000, the total rate of droplet production must exceed the throughput through the Millikan apparatus by a factor of 20000. Hence the rate of droplet generation for a very large automated Millikan apparatus must exceed 4×10^8 s^{-1} or about 0.1 g s^{-1}. This would be achievable by multiplexing about 400 of the current droplet generator designs [27]. Note, however, that we already assume multiplexing by a factor of 1000 in order to produce the parallel streams entering the Millikan chamber.

Under these assumptions, of the total number of fractionally charged droplets approximately one in twenty would reach the Millikan chamber where its detection probability is high, its theoretical maximum for a fully occupied Millikan chamber being 50%. At this rate the effective mass probed becomes 2.4×10^{-3} g s^{-1} or 1.5×10^{21} nucleons s^{-1}. With a year of running time such a device could establish an upper limit on the order of one part in 5×10^{28} on the abundance of fractionally charged particles. If the oil

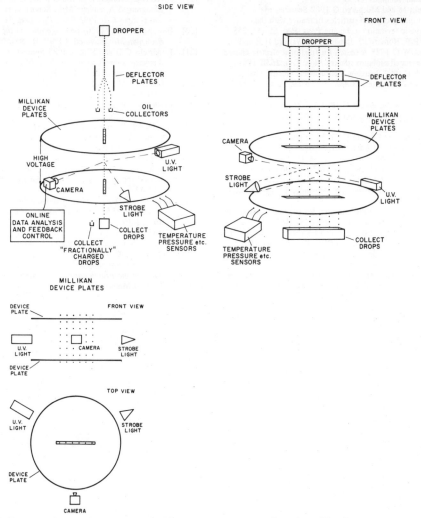

Figure 2. Sketch of the automated Millikan apparatus as seen from three different orientations.

is used as a carrier for a finely dispersed material then this number is reduced by the mass fraction of the suspension. In comparison to other methods of searching for fractionally charged particles the potential of this method stands out by orders of magnitude. Even an initial experiment using a much smaller start-up version of the apparatus could reach a respectable material throughput and could also demonstrate the feasibility of testing unprocessed materials finely dispersed in the oil for fractional charges.

Acknowledgments

We would like to thank J D Bjorken, Roger Bland, John Dressler, Ed Garwin, Howard Matis, Gary Niemi and Cherrill Spencer for helpful discussions.

References

[1] Cvetic M and Langacker P (eds) 1991 *Testing the Standard Model* (Singapore: World Scientific)
[2] Slansky R, Goldman T and Shaw G L 1981 Observable fractional electric charge in broken quantum chromodynamics *Phys. Rev. Lett.* **47** 887
[3] Zweig G 1978 Quark catalysis of exothermal nuclear reactions *Science* **201** 973–9
[4] Millikan R A 1910 *Phil. Mag.* **19** 209
[5] Millikan R A 1911 The isolation of an ion, a precision measurement of its charge, and the correction of Stoke's law *Phys. Rev.* **32** 349–97
[6] Millikan R A 1917 *The Electron* (Chicago: University of Chicago)
[7] LaRue G S, Phillips J D and Fairbank W M 1981 *Phys. Rev. Lett.* **46** 967
[8] Morpurgo G, Gallinaro G and Palmieri G 1970 *Nucl.*

[9] Marinelli M and Morpurgo G 1982 Searches of fractionally charged particles in matter with the magnetic levitation technique *Phys. Rep.* **85** 161–258
[10] Smith P F, Homer G J, Lewin H D, Walford H E and Jones W G 1985 A search for fractional electric charge on levitated niobium spheres *Phys. Lett.* **153B** 188–94
[11] Smith P F 1989 Searches for fractional electric charge in terrestrial materials *Ann. Rev. Nucl. Part. Sci.* **39** 73–111
[12] Lackner K S and Zweig G 1982 The chemistry of fractionally charged atoms *Lett. Nuovo Cimento* **33** 65–73
[13] Lackner K S and Zweig G 1983 Introduction to the chemistry of fractionally charged atoms; electronegativity *Phys. Rev. D* **28** 1671
[14] Hendricks C D and Zweig G 1982 *Detection and Enrichment of Fractionally Charged Particles in Matter* Proposal to the Division of Advanced Energy Projects, DOE
[15] Savage M L *et al* 1986 A search for fractional charges in native mercury *Phys. Lett.* **167B** 481–484
[16] Hodges C L *et al* 1981 Results of a search for fractional charges on mercury drops *Phys. Rev. Lett.* **47** 1651
[17] Liebowitz D, Binder M and Ziock K O H 1983 *Phys. Rev. Lett.* **50** 1640
[18] Milner R G, Cooper B H, Chang K H, Wilson K, Labrenz J and McKeown R D 1985 *Phys. Rev. Lett.* **54** 1472
[19] Joyce D C, Abrams P C, Bland R W, Johnson R T, Lindgren M A, Savage M L, Scholz M H, Young B A and Hodges C L 1983 *Phys. Rev. Lett.* **51** 731
[20] Einstein A 1956 (1926) *Investigations on the Theory of the Brownian Movement* ed R Fürth R (New York: Dover)
[21] Hendricks C D and Kim K 1985 Interaction of a stream of dielectric spheres in an electric field in a high vacuum *IEEE Transactions on Industry Applications* **21** 705–8
[22] Schneider J M and Hendricks C D 1964 *Rev. Sci. Instrum.* **35** 1349
[23] Lindblad N R and Schneider J M 1965 *J. Sci. Instrum.* **42** 635
[24] Lindblad N R, Schneider J M and Hendricks C D 1965 *J. Colloid Sci.* **20** 610
[25] Erin T and Hendricks C D 1968 *Rev. Sci. Instrum.* **39** 1269
[26] Berglund R N and Liu B YH 1973 Generation of monodisperse aerosol standards *Environ. Sci. Technol.* **7** 147–53
[27] Hendricks C D 1974 Micron and submicron particle production *IEEE Transactions on Industry Applications* **10** 508–10
[28] Garwin E, Schwyn S and Schmidt-Ott A 1988 Aerosol generation by spark discharge *J. Aerosol Sci.* **19** 639–42
[29] Hendricks C D November 1985 *Means and Methods of Adding Normally Solid Materials to the Working Fluid in the Fractional Charge Experiment* (Livermore: Internal Memo)

Part D
Essays in Physics

COMMENTS ON D1
"Popular and Unpopular Ideas in Particle Physics"

I have little to add to this 1986 article which was in part a plea for greater professional and academic rewards for those who develop new techniques in accelerators and detectors for elementary particle physics research. In the past decade the rewards have increased, the notable example being the 1992 Nobel Prize in physics awarded to Georges Charpak for his particle detector discoveries.

But overall there is still too much attention given to fashionable theories in our journals and meetings and in the popular scientific press. There has been a shift in popularity to speculative particle astrophysics and cosmology, more about "dark matter" these days. Still some of the old favorites persist; without anymore evidence than we had in 1986, superstring theory still gets a headline every once in awhile.

Popular and unpopular ideas in particle physics

As particle physics becomes a deeper but more difficult field, there is too much emphasis on fashionable ideas and the search for anomalies, and too little reward for improving accelerators and detectors.

Martin L. Perl

I finished writing this article while the 23rd International Conference on High-Energy Physics was taking place in Berkeley, California. It was a well-organized conference with a variety of theoretical talks, comprehensive plenary lectures and many reports on new experiments. Yet the atmosphere of the conference could be summed up in one word: waiting. The participants knew that no major experimental or theoretical advance in particle physics was to be announced at this conference. They listened to the experimental and theoretical results, but they were prepared to wait until the next conference for a breakthrough.

Much of particle physics is in a period of expectant waiting. We are waiting for the first results from the new accelerators—TRISTAN, LEP, the Tevatron Collider, the SLAC Linear Collider. We are waiting to see if new and quantitative predictions can be deduced from superstring theory, predictions that present experimental techniques can test. We are waiting to see if anomalous effects reported in the last few years will stand up to further experimentation.

The world's particle-physics community is industrious and adventurous in this waiting period. We are constructing new accelerators and their attendant detectors as fast as technology and funds will allow. We are improving experimental methods to get greater precision and to probe for anomalies in our standard ways of understanding elementary particles and the forces between them—the "standard model." We are extending theoretical calculation and speculation in all directions, connecting with astrophysics and cosmology. We are working hard to extend the revolution that occurred over the last two decades in particle physics. However, there are some small changes we can make to increase our chances of continuing that revolution; that is the subject of this article.

Deeper and narrower

It has become harder to make substantial progress in particle physics because we have redefined particle physics to be a deeper but narrower field. We concentrate on the deepest questions: What sets the masses of the particles? How can the four forces be unified? We attribute little importance to discoveries that we don't believe are central to the field. For example, 20 years ago the discovery of an additional hadronic resonance was an important event in our world; now such a discovery gains no recognition beyond a new entry in the particle data tables. There is nothing we can do to reverse this. The discoveries of the last 20 years have taught us that some measurements are central to particle physics but others are too peripheral or too complex to be of direct use.

This narrowing of the field has been harder on the experimenter than on the theorist. The theorist can extend his or her interests into general relativity or cosmology or the borderline between nuclear and particle physics. The experimenter, faced with long, complex and expensive experiments, must always ask, "Will this experiment probe into the heart of matter?" The narrowing of the field and the harsher measures of progress have led to two deleterious tendencies in particle physics. One of these tendencies, acceptable and understandable at first glance, but damaging on closer analysis, is the emphasis in particle physics on popular and fashionable ideas rather than unpopular and unfashionable ideas. (As figure 1 indicates, this problem is not unique to particle physics.) The other tendency is the addiction to the search for experimental anomalies.

The fashionable and the popular

A theoretical concept or an idea for a model in particle physics may be popular and fashionable for good reasons. The concept may promise to explain a large set of data; it may point to new

---

Martin Perl is a professor at the Stanford Linear Accelerator Center, in Stanford, California.

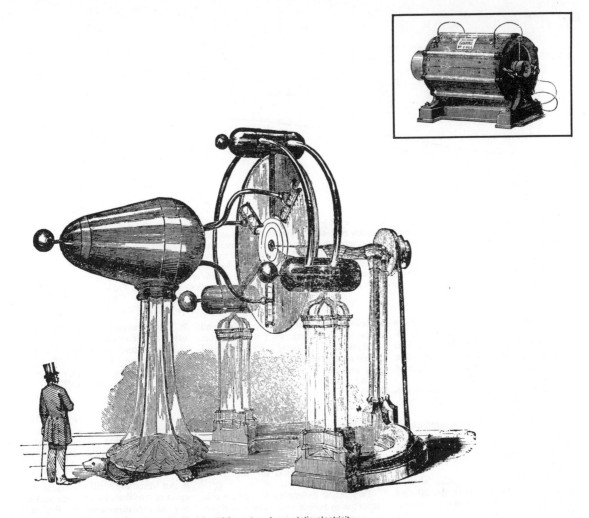

Fashion and the future. In the second half of the 19th century, large static-electricity generators, such as this machine with its 10-foot-diameter glass plate, made popular and dramatic displays. But the future of electromagnetic technology lay with the ungainly electric-current generator (inset). Paradoxically, once static-electricity machines had sunk into obscurity, the idea was reborn in the Van de Graaff generator. Figure 1

directions for experimental exploration; it may connect particle physics with other areas of physics. However, a concept may also become popular and fashionable simply because it gives the experimenter something to search for and the theorist something to calculate.

An example is the question of whether there are neutral leptons other than the neutrinos that are associated with the three generations of leptons—the electron, muon and tau—and, more particularly, whether there are neutral leptons that have mass. The known neutrinos present a simple experimental picture in which each neutrino is associated with its own charged lepton and each has a mass close to zero compared with the mass of the charged lepton. At present there is no confirmed evidence that the masses of the neutrinos are other than zero. Nor is there evidence that any coupling exists between a neutrino in one generation and the charged leptons in the other two generations.

In thinking about the possibility of additional neutral leptons, the simplest extension of the present picture is to consider the existence of additional generations of neutrino and charged-lepton pairs, with the neutrino having zero mass. In this minimum extension there would be no coupling between the new neutrino and any of the known charged leptons.

If you look through recent literature[1] on massive neutrinos, whether experimental or theoretical, you find that this model is not popular. On the contrary, popular models assume, or at least

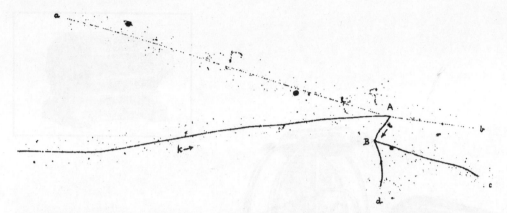

Kaon decay in an emulsion. The decay $K^\pm \to \pi^\pm + \pi^+ + \pi^-$ illustrates the importance of anomalies. The decay of a hadron into nothing but hadrons on a time scale many orders of magnitude slower than the usual time for hadronic interactions was an anomaly. It was eventually explained as the effect of a new, approximately conserved quantum number, strangeness. Track k is the incident K meson, which decayed at point A after nearing or reaching the end of its range. The particle going from A to B is identified as a π^- meson, which produced a nuclear disintegration at B. The other two tracks coming from A are taken to be π mesons, but the possibility that one or both are μ mesons could not be excluded by measurements in the emulsion. This picture is magnified; track k is actually about 1/2 mm long. (From reference 3.)

Figure 2

hope, that the new neutrino will have nonzero mass. Most models also assume nonzero coupling between this new neutrino and the electron, muon or tau generation. Three ideas suggest the second assumption:
▶ The present standard model of elementary particles allows such coupling.
▶ Such coupling occurs between quark generations.
▶ A new symmetry principle would be required to prevent coupling between the lepton generations.
Furthermore, the assumption of nonzero coupling gives experimental and theoretical particle physicists something to work on. The experimenter can try to detect the new neutrino through rare decays of mesons or through its production and subsequent decay in electron–positron annihilation. The theorist can construct models that draw on the analogy with the coupling between quark generations.

Yet these fashionable nonzero-coupling models for massive neutrinos are in contradiction to other basic ways of thinking in physics. Usually we look for the simple solution, and if nature presents us with a simple and regular pattern, we try to understand that pattern, not complicate it. The introduction of nonzero coupling between lepton generations is certainly a complication. To first order there is nothing wrong with our working, experimentally or theoretically, with such nonzero-coupling models. However, to second order these models may be popular not because they are obvious or elegant or solve a mystery, but merely because they give us something to work on.

For the zero-coupling model, by contrast, there appears to be nothing new theoretically that can be said, and experimental work with it is difficult and indirect. Current experimental work is restricted to setting a limit on the number of small-mass neutrinos using the resonance width of the Z^0 or the cross section for the reaction

$$e^+ + e^- \to \nu + \bar\nu + \gamma$$

Our experimental reach will improve drastically when LEP and SLC begin operation. Then we can make much better measurements of the Z^0 width and the cross section for the above reaction.

Two other examples of fashionable concepts in recent years are supersymmetry and axions. One or both ideas could be right, but in view of the continued lack of evidence for their validity, it is surprising how much experimental and theoretical work was, and still is, done on these concepts.

The first remedy for the overemphasis on the fashionable and the popular in particle physics is for the old hands in the field to regain their balance and judgment. Those of us who watched the rise and fall of Regge poles in the 1960s should be cautious about supersymmetry and axions in the 1980s. The second remedy is for the new hands to be more skeptical: "If the old hands know so much, why don't they know the answer?"

Anomalies

In the past decade the particle-physics community has become addicted to the search for experimental anomalies. This addiction is useful to a certain extent but is sometimes injurious. Anomalies have played an honorable and important part in the history of particle physics:

▶ The strange particles were discovered as anomalous events in cloud chambers and emulsions. George D. Rochester and Clifford C. Butler's V particles turned out[2] to be the strange particle that we now call Λ:

$$\Lambda \to p + \pi^-$$

▶ The four-track events[3] of R. Brown and her coworkers were anomalous, as figure 2 explains. These events are now known to be the decay

$$K^\pm \to \pi^\pm + \pi^+ + \pi^-$$

▶ An anomaly still unexplained in spite of two decades of more and more precise measurements is CP violation in K-meson decays. The decay of the long-lived neutral K meson to two pions, for example, violates CP conservation.

Searches for experimental anomalies can be injurious in several ways. First, too much emphasis on searching for anomalies may pull us away from another path to progress in physics— the steady accumulation of data that gradually forces a change in how we view the physical world. Second, too much emphasis on the importance of experimental anomalies may have bad effects on the young women and men entering particle physics. They become distracted by a rush of talks and papers on possible evidence for a new axion, or for a new heavy lepton, or for a fifth force. Excitement in the field is fine, but it should not distract young physicists from learning the craft of physics. In the learning of any craft, one needs time to learn the methods, to develop one's own thoughts, to build confidence in one's skill.

I am not advocating that the experimenters suppress evidence for experi-

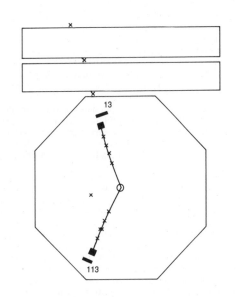

Detector and particle tracks. a: The Mark I detector, sketched in cross section. This was one of the first large-solid-angle, multiple-purpose particle detectors. **b:** An electron–muon event resulting from the decay of a pair of tau leptons. The track at 7 o'clock is due to an electron, as indicated by the large amount of electromagnetic energy deposited in the shower counter. The energies are given in arbitrary units. The track at 11 o'clock is due to a muon, as indicated by its passage through the iron flux return and the two blocks of concrete.
Figure 3

mental anomalies. On the contrary, it is the duty of an experimenter to talk about or publish an experiment; this is crucial to the life of science. The injurious secondary effect is the positive feedback that occurs between experimenters and theorists. This feedback occurs because theorists, being human, are eager to explain new experimental results and to explain them first. Experimenters, also being human, are equally eager to interact with the theorists and provide them with the data they need for their theories. The cure here—and we need only a second-order cure—lies within the power of the journal editors and conference organizers: Damp the positive feedback by limiting the number of published explanations of new and untested data, by spacing the talks on new and untested speculations based on preliminary data and by waiting before organizing conferences on new areas.

Fashion in theories can have an effect similar to that of addiction to the search for anomalies. Many people bemoan our increasingly mathematical and speculative particle theories, and their distance from present experiments. I don't much like the separation either, being a poor mathematician and having been an engineer before becoming a physicist. But there is nothing to be done. Nature dictates the correct theory; if it is mathematical and remote we still have to live with it. My concern is with the effect on experimental directions of fashionable, but unproven, speculative theories. As I have discussed[4] at greater length elsewhere, we need experiments that cover broad areas, experiments that simultaneously acquire data on known phenomena and make wide searches for new phenomena. Experiments based on speculative theories and with narrow goals teach us little if the answer is no—only that the theory is wrong or, more likely, that the parameters in the theory need adjustment. Again it is the time and energy of the young experimenter that are most at risk. One may argue that the young have infinite time and energy, but half-decade or longer experiments use up time, even for the young.[5]

Large detectors, large accelerators

Many people also bemoan the great increase in the cost, complexity and size of accelerators and particle-physics experiments. We have no choice with respect to accelerators. Almost all recent progress in particle physics has

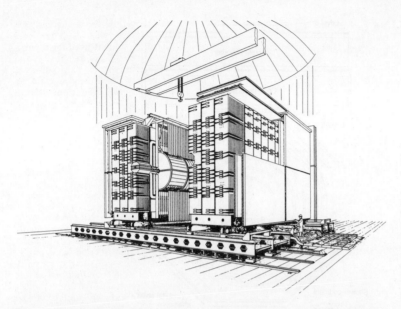

The UA1 detector at CERN, which produced the Z^0 event shown in figure 5. Note the large size of the detector as compared with the person shown standing at lower right in this sketch, or even as compared with the Mark I detector, shown in figure 3a. Figure 4

been made by building higher-energy and more diverse accelerators and colliders. In contrast, in recent times we have learned much less about elementary particles from nonaccelerator experiments, and it is not for lack of trying.

Sometimes there appears to be a choice in experiment: for example, many small particle detectors at an accelerator instead of a few, large-solid-angle, multiple-purpose detectors. However, large-solid-angle, multiple-purpose detectors have two great advantages. First, there is the obvious increase in the efficient use of accelerator time and apparatus when one detector collects data for many experiments at once. Second, a multiple-purpose detector is more powerful than the sum of its components. One can study reactions that cannot be studied by small experiments, and one can do measurements and searches beyond those conceived by the builders of the detector.

For example, we discovered[6] the tau lepton with one of the first large-solid-angle, multiple-purpose particle detectors, the Mark I at SPEAR. We used the reaction sequence

$$e^+ + e^- \rightarrow \tau^+ + \tau^-$$
$$\tau^+ \rightarrow e^+ + \nu_e + \bar{\nu}_\tau$$
$$\tau^- \rightarrow \mu^- + \bar{\nu}_\mu + \nu_\tau$$

The Mark I, shown in figure 3, could identify both electrons and muons. If we had used two separate detectors in separate experiments—one detector sensitive to electrons, the other to muons—it would have been much harder to find the tau.

Modern large-solid-angle, multiple-purpose detectors are bigger and more complex than the Mark I detector, but they also have much greater ability to acquire and analyze data. The UA1 detector, used for the discovery[7] of the W and Z^0 particles, is an example (see figures 4 and 5). Such detectors are the only way we have to study most of the processes that occur in very-high-energy particle collisions. Whether or not we like the cost and sociology associated with the building and use of such detectors, we have no choice as a community: These detectors are essential for progress in particle physics.

The necessity for large accelerators and large detectors impels us intellectually and practically to find ways to improve these devices, increase their experimental reach and minimize their cost, size and complexity. What can we do to stimulate technical progress and invention in accelerators and detectors? I'll address this question for detectors; the answer for accelerators is analogous.

Publish what or perish?

Our reward and recognition system for the experimenter—the PhD, tenure, grants—is based mostly on the publication of experimental results. A graduate student's most useful achievement may be building an efficient trigger system for a multiple-purpose detector, but the faculty will award the PhD for one more measurement of the lifetime of a D meson, B meson or tau lepton. An assistant professor in a collaboration may be valued for his or her ability to design drift chambers, but this physicist's tenure request must contain a substantial list of the collaboration's published results along with a complicated explanation of who did what. In considering a grant request from an experimenter, the physics staff at a funding agency is most interested in the experimental strength of the applicant's group, but a list of publications of experimental results is still needed for the files.

We need more powerful detectors, more compact detectors, more efficient detectors and cheaper detectors, yet the reward system gives little direct recognition for contributions to such goals. This is perverse because it is the physics community itself that sets up the reward system by setting the standards for experimental achievement in particle physics. It is also perverse historically, because the traditional view of an experimental physicist is someone whose work designing and building apparatus is inextricably mixed with the collection and analysis of data.

Our recognition through publication also is counterproductive. As soon as graduate students begin their work in experimental particle physics, they learn that the most prestige comes from publishing in *Physics Letters B* or in *Physical Review Letters*. When did either of those journals publish a paper on particle detectors or accelerator

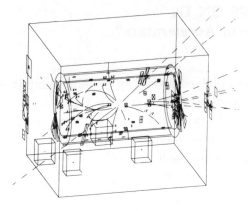

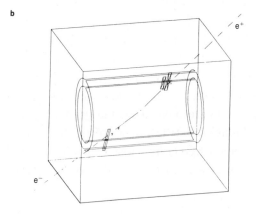

Event containing a Z⁰ particle, found by the UA1 detector at the CERN proton–antiproton collider. The computer drawing in **a** shows the tracks of all the particles produced in the event. In **b** the tracks of the electron and positron produced in the event are shown by themselves. This electron–positron pair comes from the decay of the Z⁰. Figure 5

observing the strange particles and their interactions.... There was therefore a great need for a particle detector of high density and large volume—tens to hundreds of liters—in which tracks could be photographed and scanned at a glance, and in which precision measurements of track geometry could be made.

Today we could well use inventions in particle detection as far-reaching as was the bubble chamber in the 1950s.

Similarly, invention is required to build a new generation of ultra-high-energy accelerators. Burton Richter summarizes[9] the requirements:

These machines will have to have much higher energy than is available today and will have to be built at a cost that the taxpayers of the country (or perhaps the world) will be willing to bear. In the past the scientific community has come up with new techniques of acceleration when the progress of science required it and when the cost of the old techniques, extrapolated to higher energy, became prohibitive.

The more popular and rewarding it is to work on new techniques for acceleration and detection, the sooner we will have the needed inventions.

References

1. For a review of all models see F. J. Gilman, SLAC Publication No. 3898, Stanford Linear Accelerator Center, Stanford, Calif. (1986), submitted to Commun. Nucl. Part. Phys.
2. G. D. Rochester, C. C. Butler, Nature **160**, 855 (1947).
3. R. Brown, U. Camerini, P. H. Fowler, H. Muirhead, C. F. Powell, D. M. Ritson, Nature **163**, 82 (1949).
4. M. L. Perl, New Scientist, 2 January 1986, p. 24.
5. For a discussion of the sociological aspects of the increase in the size of particle-physics experiments, see A. R. Pickering, W. P. Trower, Nature **318**, 243 (1985).
6. M. L. Perl, G. S. Abrams, A. M. Boyarski, M. Breidenbach, D. D. Briggs, F. Bulos, W. Chinowsky, J. T. Dakin, G. J. Feldman, C. E. Friedberg, D. Fryberger, G. Goldhaber, G. Hanson, F. B. Heile, B. Jean-Marie, J. A. Kadyk, R. R. Larsen, A. M. Litke, D. Lüke, B. A. Lulu, V. Lüth, D. Lyon, C. C. Morehouse, J. M. Paterson, F. M. Pierre, T. P. Pun, P. A. Rapidis, B. Richter, B. Sadoulet, R. F. Schwitters, W. Tanenbaum, G. H. Trilling, F. Vannucci, J. S. Whitaker, F. C. Winkelmann, J. E. Wiss, Phys. Rev. Lett. **35**, 1489 (1975).
7. C. Sutton, *The Particle Connection*, Simon and Schuster, New York (1984), chap. 10.
8. *Nobel Lectures: Physics, 1942–1962*, Elsevier, Amsterdam (1964), p. 530.
9. B. Richter, in *Laser Acceleration of Particles*, C. Joshi, T. Katsouleas, eds., AIP, New York (1985), p. 8. □

---

research?

We are dependent on governments, large institutions and universities for places to work and for money to do our work. However, our reward and recognition system operates within our own small institutions: our departments, our laboratories, our journals. We can certainly change the reward system.

Inventions

I have made suggestions for the practice of experimental particle physics that can increase the rate of progress in the field. While these are suggestions for second-order changes, there is the chance for a first-order change. A scientist can make major advances when deeply immersed in research that is supported and rewarded. Immersion in a subject—even obsession with it—is almost always a necessary condition for a major discovery. I would like our community to give full support to experimenters absorbed in accelerator and detector research, to support them on an equal basis with those absorbed in exploring the standard model of particle physics or looking for new particles. This will stimulate inventions in the apparatus of particle physics, inventions that may be necessary to continue the revolution in the field.

I conclude with two quotations. Donald Glaser, deliberately setting out to improve particle detection, invented the bubble chamber, which revolutionized particle physics in the 1950s. In his words:[8]

I became interested in trying to devise new experimental methods for investigating the physics of elementary particles in 1950, not long after the new "strange particles" had been discovered in cosmic rays.... Greatly stimulated by these developments, I began to wonder whether it would be possible somehow to speed up the rate of

COMMENTS ON D2
"Science in the Age of Accelerators"

With the final reprint in this volume I have indulged myself by combining my interest in physics with my interest in the history of mechanical and construction technology. I have a strong interest in the nineteenth and early twentieth centuries when mechanical invention and innovation reached its height. I collect mechanical antiques: farm machinery, old typewriters and sewing machines, acoustic phonographs. Finding old Meccano and Erector sets is a special pleasure. (I stay away from old toy electric trains, there's too big a crowd there.)

I had hoped to write more about the connections I see of physics with popular science and public technology, but I have not had the time, or not chosen to make the time. Perhaps I will make time in the future. For example, the best known and most popular toy construction sets used nuts and bolts to connect the parts: Erector in the United States, Meccano in France and the British Empire, Marklin in Germany. Yet toy inventors have known for a long time that nuts and bolts are hard to use, are easily lost, and discourage many children from playing with construction toys. There have been many attempts to devise boltless construction sets, some such as the Lionel construction set of the late 1940's have been manufactured. The only success has been LEGO, but LEGO copies a brick technology, not a steel girder technology. Here then is a subject that deserves a paper: mechanics and toy construction sets.

SLAC – PUB – 4363
June 1987
(E)

Science in the Age of Accelerators[*][†]

MARTIN L. PERL

Stanford Linear Accelerator Center
Stanford, California 94305

Private Science, Public Science

Accelerators have brought the particle physicist to work and live in three worlds: the private world of science, the public world of science, and the world of large accelerators. Our private world is our apparatus, our data, our theories, our colleagues, our journals, our meetings, and above all our understanding of elementary particles. There are more intimate areas in that private world, the childhood toys and dreams that led us

Talk presented at the American Physical Society Meeting:
Arlington, Virginia, April 20–23, 1987 and
also contributed to Proc. Joint US-CERN School on Particle Accelerators:
South Padre Island, Texas, October 23–29, 1986

---

[*] This work was supported by the Department of Energy, contract DE-AC03-76SF00515.
[†] Figures in this paper without attribution are from SLAC archives or the private collection of the author.

Reprinted from *Proc. Physics of Particle Accelerators*, eds. M. Month and M. Dienes, pp. 2098, Copyright 1988, with kind permission of American Institute of Physics.

into physics. There is a connection between building huge accelerators and Erector sets and Meccanos and ham radios. There is a connection, sometimes a painful one, between childhood reading about lone science heros: Pasteur, Madame Curie, Einstein; and then growing up to be part of a group building or using a huge accelerator.

The public world of science is how society sees us, how we want to be seen in newspapers and on TV, how we interact with governments, and most important, how governments support science. Since the 1940's most of us in basic research have not been able to avoid the public world, even if we wanted to stay in our private world. Public money is needed to study agriculture as well as atoms, libidos as well as leptons. If the apparatus is table size, if the laboratory is room size, with a little obliqueness the dependence on public money can be ignored. The builders and users of large accelerators, of large telescopes, of space rockets and satellites cannot ignore their dependence.

At the Spring, 1987 Meeting[1] of the American Physical Society I used many slides and two screens to visually trace the intertwining of the private and public worlds of science with the coming of the age of particle accelerators. I was not trying to do the history or sociology or politics of accelerators. Rather I was illustrating some of the themes laid out historically in Fig. 1. (During my talk, Fig. 1 was always projected, here the reader will have to refer back to it).

There is not space here for all the pictures I used; I retain the unfamiliar images. The reader knows the familiar ones: Rutherford in the Cavendish Laboratory standing under a sign reading "TALK SOFTLY PLEASE"[2] or Livingston and Lawrence in front of the 37-inch cyclotron.[3] These and other familiar images I used came from Refs. 2, 3, and 4, of which the most entrancing is *The Particle Explosion* by Close, Marten and Sutton.[2]

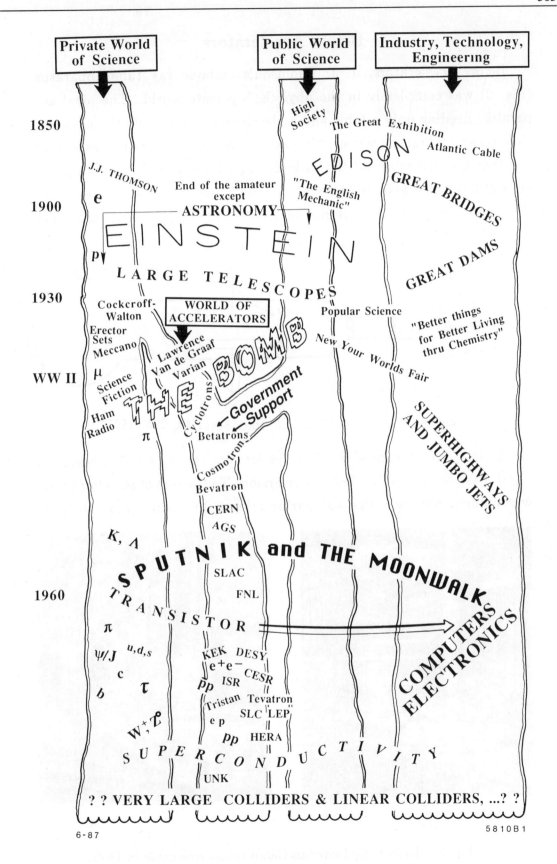

Before Accelerators

Before accelerators, J. J. Thomson's cathode ray tube apparatus (Fig. 2) was completely in the physicist's private world. The ideal apparatus, needing only a table and a glassblower, to identify the electron. Not so easy. Thomson writes, "It was only when the vacuum was a good one that the deflection [of the cathode rays] took place." Vacuum problems ninety years ago. In the same article Thomson asks "...what are these particles? Are they atoms, or molecules, or matter in a still finer state of subdivision?"

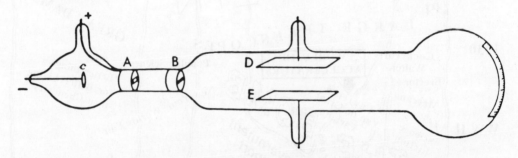

Fig. 2. From J. J. Thomson, Phil. Mag. **44**, 293 (1897).

Society with a capital S (High Society in Fig. 1) was interested in physics as culture and intellectual diversion. To the rest of society physics was hidden, remote. The submarine telegraphic cable (Fig. 3) is my

Fig. 3. Laying the Dover-to-Calais submarine cable in 1850.

metaphor. The public interest is in the enterprise and danger in laying the cable; it is in the wonder of connecting islands and continents. A great engineering feat. Hidden in all this is our physicist hero, Kelvin, and his theory of telegraphic signaling[5] and his idea of a stranded cable.

It is not pure science, but it is great engineering feats which catch the interest and enthusiasm of masses of people in the nineteenth and twentieth centuries: the Atlantic Cable, railroads, large steamships, great bridges (Figs. 4 and 5), great dams. Scientific apparatus can also be great engineering structures. First came the large telescopes, then space rockets and satellites, now huge particle colliders. I will return to this idea later because particle colliders as engineering feats can have special affection from the public and special support from governments. This brings benefits and dangers to particle physics.

Fig. 4. The Forth Bridge, near Edinburgh, Scotland, completed in 1890. The first large bridge using the cantilever and central girder principle.

Fig. 5. The Brooklyn Bridge, New York, U.S.A., completed in 1883. One of the first large suspension bridges.

It is not pure science, but it is new and visible technology which catches the interest and enthusiasm of masses of people. From the Great Exhibition of 1851 — the Crystal Palace — in London (Fig. 6) to the Trylon and Perisphere of the 1939 New York Worlds Fair (Fig. 6), new and future technology has brought the crowds. The Great Exhibition was arranged in four departments: Raw Materials, Machinery, Manufacturers, and Fine Arts. Science is buried in technology.[6] Worlds Fairs fail these days because we are so immediately immersed in new technology.

In North America, Edison was and still is the great symbol of new and visible technology. During the reign of Edison, our private physics world moved on with Maxwell and Hertz and Lorentz and Planck, but Edison was a thousand times more famous. Only Einstein crossed the fame barrier out of our private world. His name stretched across the private and public worlds of physics (Fig. 1).

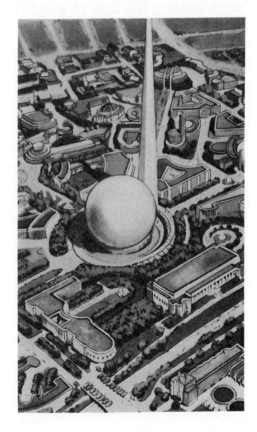

Fig. 6. Two Worlds Fairs: *Above*: The Great Exhibition of 1851 in London usually called the Crystal Palace. *Right*: The Trylon and Perisphere of the 1939 New York Worlds Fair. From a colored postcard.

The End of the Amateur

In the last decades of the nineteenth century, a new gulf appeared between the private and public worlds of science. As sciences developed amateurs could no longer contribute or even fully understand. The usual example is the ninth edition of the Encyclopedia Britannica (1889), the last edition whose physics articles were useful to the professional and to the amateur. After that we have our Handbuch der Physik, and the encyclopedias stay with the public world.

An example I like is *The English Mechanic*, a combined do-it-yourself and amateur scientist magazine (Fig. 7). Building a steam car is an impressive hobby, but the inside contents of *The English Mechanic* are more impressive. In this issue there is a summary of a lecture by Dewar on liquid and solid hydrogen; a note on the Curie's work on induced radioactivity; the positions of two new variable stars are given; and there are dozens of queries from readers on subjects ranging from using ammonia for renovating felt hats to using the formula

$$\int_0^{2\pi} \int_{r=0}^{r=6} AB d\theta \, r dr$$

There are no amateur science magazines or amateurs like that today. Except in astronomy. That lucky science has its subject in full view, still has crucial contributions from amateurs, and has apparatus which are also engineering feats. A hundred years ago telescopes were already impressive structures (Fig. 8).

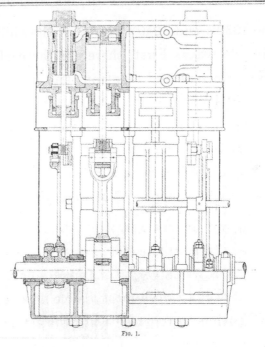

Fig. 7.

Fig. 8. *Left*: The 1864, 80 cm diameter, reflecting telescope of Foucault at Marseilles incorporating the innovations of a silver-on-glass surface and a parabolic figure. *Above*: The last of the great refracting telescopes, the 40 inch diameter at the Yerkes Observatory, installed in 1897.

The Thirties and Forties

The 1930's and 1940's represent the childhood of the age of accelerators in two ways. First, there are the early accelerators and their inventors: cyclotrons and betatrons and linear accelerators; Cockcroft and Walton and Lawrence and Van de Graff and Wideroe. Familiar names and images. Second, the thirties and forties were the childhood years of the physicists who have since dominated the building and use of large accelerators. That generation is retiring, or will soon retire, from the private world of science. What were our images of physics?

I think our images were quite different from the childhood images of physicists born after the Bomb or Sputnik or the Moonwalk. Before the bomb, physics was a very, very small and private world. In the thirties in the United States the most visible new technology and the public science was chemistry represented by the slogan of the Dupont Company — "Better Things for Better Living Through Chemistry". We, at least the accelerator builders and experimenters, came to physics mostly indirectly through Erector sets and Meccanos (Figs. 9 and 10) and ham radio. Our reading was the science and hobby magazines (Fig. 11) which were compounded of futuristic technology, science projects usually too complicated for our skill or pocket money (Fig. 12), and occasional perpetual motion (Fig. 13). Popular science magazines have degenerated since the nineteenth century, there was no perpetual motion in the *English Mechanic* because the editors knew the first law of thermodynamics.

We knew about a few physics greats: Kepler, Newton, Madame Curie if we went to the movies, and, of course, Einstein. But not Bohr or Schrodinger or Fermi or Michelson or Hahn and Strassman. These were great engineering projects going on in the thirties. Boulder Dam (Fig. 14), huge battleships, ocean liners, the China Clipper, and the Twentieth Century Limited entranced us. But we knew that wasn't science. Physics and chemistry and biology and astronomy were science; the problem was, "Could you make a living doing science?"

Fig. 9. Cover of 1929 Erector set manual.

Fig. 10. Page of instructions from a 1930's Meccano manual.

Fig. 11. *Left*: Cover of December, 1930 Everyday Mechanics featuring the rotor force idea which was popular for futuristic ships and airplanes in the 1930's. The editor Gernsbach pioneered hobby electronics magazines and science fiction magazines in the 1920's. *Right*: Cover of February, 1934 Modern Mechanix.

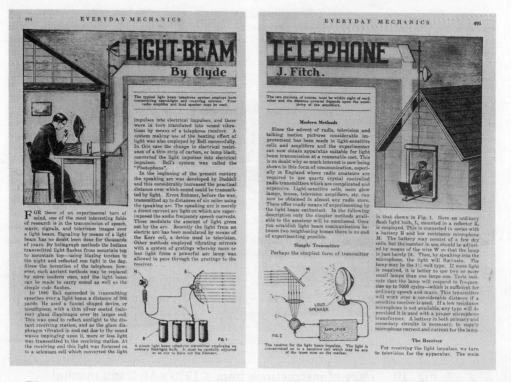

Fig. 12. A project from the December, 1930 issue of Everday Mechanics.

New Rail Car Runs on Air-Electric Perpetual Drive

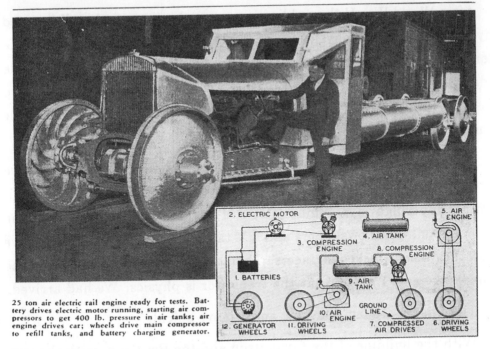

25 ton air electric rail engine ready for tests. Battery drives electric motor running, starting air compressors to get 400 lb. pressure in air tanks; air engine drives car; wheels drive main compressor to refill tanks, and battery charging generator.

FROM coast to coast by rail in 24 hours, traveling literally on air—that is what W. E. Boyette of Atlanta, Georgia, claims for his invention, a railroad engine that runs almost entirely on air.

Air for fuel—speeds of up to 125 miles an hour on rails—low transportation costs—these are possibilities conjured by Boyette's air electric car. After being started by batteries, the car needs only air to keep it running—a close approach to perpetual motion.

Fig. 13. A perpetual motion proposal from the February, 1934 issue of Modern Mechanix.

Fig. 14. *Right*: Boulder Dam across the Colorado River in the U.S.A. completed in 1935. *Left*: A tunnel used during the construction of Boulder Dam.

Physics Leaves Its Private World

After World War II the public discovered physics or rather discovered physicists. Not many physicists and not much physics, mostly it was the atom bomb and the hydrogen bomb and nuclear reactors and a little radar that caught the public. But this was enough for all of physics to cross the Rubicon into the public world of science: public interest, public scrutiny, public money.

There was only modest public interest or public scrutiny because we were still not associated with the new technologies and feats of engineering: superhighways, jumbo jets, Sputnik, and an astronaut walking on the moon. The newspapers and TV talked of rocket scientists but we knew they meant rocket engineers. Then the transistor appeared, and the newspapers and TV said that the transistor is physics. We had arrived.

The images are familiar now and I move quickly.

With the building of the Cosmotron and the Bevatron, with the construction of large alternating gradient synchrotrons, with the establishment of CERN, DESY, the Rutherford Laboratory, SLAC, (Fig. 15), Fermilab (Fig. 16), government support begins to flow steadily into the world of accelerators (Fig. 1). The builders and users of accelerators now live in three worlds.

We particle physicists are not alone in the necessity of living in three worlds. The same thing happens to space science, to plasma physics, and eventually to material science with its need for high intensity neutron and photon sources. Only the astronomers stay lucky — still able to get some of their telescopes from the world of private wealth. Although not the Hubble Space Telescope.

With public money, following the accelerator pioneers of the thirties and forties, following the dreams started with Erector sets and *Modern Mechanix*, we found there were two kinds of neutrinos (now three) and the proton was made of quarks. We found the ψ/J particles, the τ heavy lepton, the heavy b quark, the gluon that carries the strong force, and the $W^{\pm}$ and Z^0 that carry the weak force. It has been a splendid, an amazing, twenty-five years. I'm sorry that these discoveries have been given the awful and dull name "standard model". We have come so far in answering Thomson's question " ...or matter in a still finer form of subdivision?", our work deserves a better name.

Fig. 15. A view of SLAC showing the new Collider Hall for the SLC in the lower left corner. Photograph by Joe Faust.

Fig. 16. A view of Fermilab showing the experimental areas and Accelerator Complex, from the 1986 Annual Report of the Fermi National Accelerator Laboratory.

517

Large Particle Colliders:
New Physics, New Technology, Great Engineering Feats

As we plunged forward in our private world of quarks and leptons and intermediate bosons, our accelerator world moved closer and closer to the public world of science, and began to spill over into the industrial world (Fig. 1). In the past, connections between the accelerator world and the industrial world were fitful. We hungrily used some of their new technology — solid state electronics and computers. The passing from electron-positron storage rings to synchrotron light sources has begun to provide important applied research tools to industry. But mostly we kept to ourselves, except for the civil construction involved in building accelerators and accelerator laboratories: SLAC (Fig. 15), Fermilab (Fig. 16), and LEP (Fig. 17).

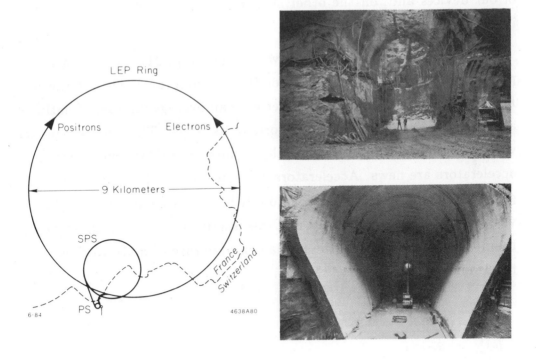

Fig. 17. *Left*: The LEP electron-positron collider under construction at CERN. *Upper Right*: The cave for the DELPHI experiment. *Lower Right*: The cave for the L3 experiment. It is interesting to observe the similarities between these photographic images and the Boulder Dam tunnel in Fig. 14. The photographs are from the CERN Courier, December 1986.

17

We can no longer keep to ourselves, as accelerators get bigger the civil construction to be done by the industrial world grows bigger. We might still build our own accelerator technical components, but we now have obligations to use industry for the sake of industrial development and for the sale of national economies.

We can no longer keep to ourselves. We need the spiritual and cultural and financial support of the public world of science. The public can provide that support because they continue to be interested and excited by great engineering feats, and that is what our accelerators have become. New and visible technology also interest the public. Our new technology is still esoteric: superconducting magnets (Fig. 18) and linear colliders (Figs. 19 and 20), but it is futuristic technology, and that is interesting to lots of people. There is also a fascination with the contrast between, on one side, the very small objects we study and the precision of some of our devices and, on the other side, the huge size of our tunnels and interaction regions and detectors.

As a demonstration my final two images (Fig. 21) are from a newspaper. Not the New York Times or the Washington Post, but the San Francisco Examiner[7] — a newspaper with an average mixture of national affairs, crime, local politics, and sports in its pages. The science reporter was given the space to do this article because the editors knew that large accelerators are news. Accelerators are news primarily because they are great engineering achievements, secondly because they incorporate highly visible new technology, thirdly, and this is a distant third, because we use them to learn more about the fundamental nature of matter. This is my thesis and my conclusion.

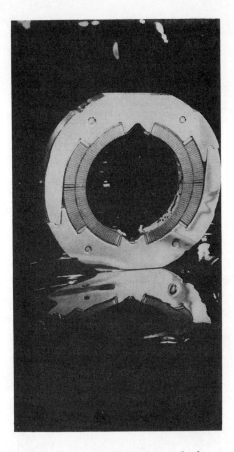

Fig. 18. A cross section of the superconducting magnet coil used in the Tevatron proton-antiproton collider. From the 1986 Annual Report of the Fermi National Accelerator Laboratory.

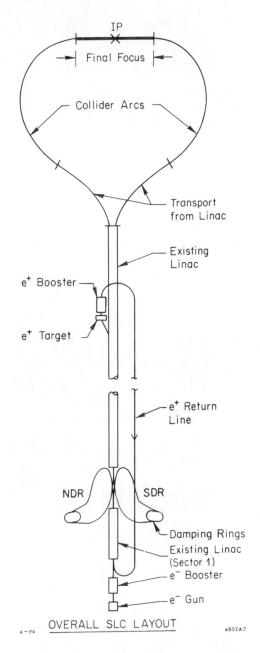

Fig. 19. Schematic of the principles of operation of the SLAC Linear Collider now being commissioned.

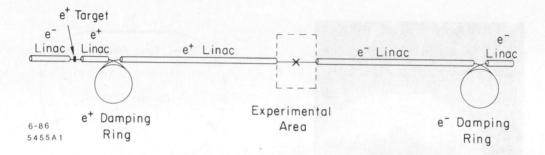

Fig. 20. Diagram for a future linear collider from J. Rees, SLAC-PUB-4037 (1986).

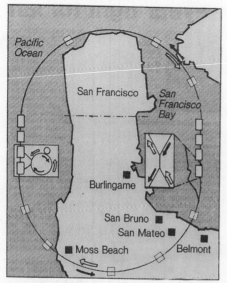

The Superconducting Super Collider won't be built in the Bay Area. But to visualize its size, imagine 52 miles of tube circling the northern tip of San Francisco, out into the Pacific, down below Moss Beach and San Mateo and over to the shores of Oakland. Price: more than $5 billion. It will work by launching two beams of protons in opposite directions at close to the speed of light. They will collide head-on in chambers, one enlarged at right, to create even smaller particles.

Fig. 21. Figures from articles in the San Francisco Examiner of April 19, 1987. The captions are from the articles.

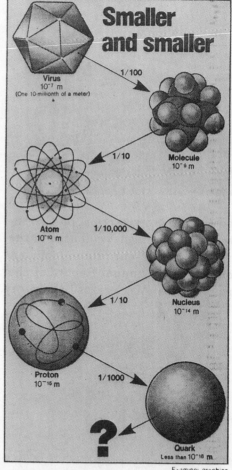

Stages in the structure of matter, from virus to quark. The characteristic sizes are given in meters. It is unknown if quarks have a measurable extent, with some internal structure, or if they are truly elementary, pointlike objects.

I end with two warnings. The engineering images in this talk were of successful buildings and machines, there were no pictures of the Tacoma Narrows bridge, of the Challenger, of Chernobyl. If we are to build and use successfully the huge new accelerators, we must follow the principles of good engineering as well as good physics. We must know our technology well, we must design carefully — better we must overdesign, we must construct for strength and reliability and durability. If we can't get the public support to build our accelerators truly and well, we had better be honest with the public and tell them we can't do it. We must not fail with the huge accelerators we propose.

My second warning comes from the private world of science — in that dark country where we cannot know what is ahead in the physics of elementary particles. When Roebling designed the Brooklyn Bridge in the 1870's, he could promise that the bridge would take people from Brooklyn to Manhattan and back. It still does. We cannot promise that the next accelerator will take us to the Higgs particle or to the theory of everything or to the next heavy lepton, or even to the top quark.

It is difficult to avoid promises when science gets discussed and displayed in newspapers, on TV, and in government hearings, Witness the new high-temperature superconductors (Fig. 1). However if the promises of these superconductors are not kept, the public world will soon forget. There will be little harm to material sciences or solid state physics. If our huge accelerators fail, our promises will not be so easily forgotten.

Acknowledgements

I am greatly indebted to Melvin Month for suggesting this topic and encouraging me to prepare this paper. I wish to thank Lydia Beers for preparing and assembling this manuscript and Sylvia McBride for drawing Fig. 1.

I apologize to my colleagues who are not from the United States or Great Britain for mostly using historical images from those two regions, those are the images I grew up with, feel closest to, and know best.

References

1. Martin L. Perl, Bull. Am. Phys. Soc. **32**, 1100 (1987).

2. F. Close, M. Marten, and C. Sutton, *The Particle Explosion* (Oxford Univ. Press, Oxford, 1987); the photograph of Rutherford is on page 40.

3. E. Segrè, *From X-Rays to Quarks*, W. H. Freeman, San Francisco, 1980); the photograph of Livingston and Lawrence is on page 232.

4. D. Varian, *The Inventor and the Pilot*, (Pacific Books, Palo Alto, 1983).

5. E. Whittaker, *A History of the Theories of Aether and Electricity* (Harper, New York, 1960), Vol. 1, page 228.

6. Facsimile of *The Art-Journal Illustrated Catalogue of the Industries of All Nations* (Crown Publishers, 1970), section entitled "The Science of the Exhibition".

7. San Francisco Examiner, April 19, 1987.

Part E

Reflections on Experimental Science

REFLECTIONS ON EXPERIMENTAL SCIENCE

1. Choosing The Experiment: We Want to Know

Why does an experimenter choose to do a particular experiment or set of experiments. The *standard answer* is that the experimenter wants to know something not already known, wants to understand something not already understood. This standard answer is sometimes the correct answer. The experiments whose papers are reprinted in Part A, the experiments which led to the discovery of the tau lepton, were chosen to try to learn something new about leptons and about the electron-muon problem.

Another example in my research where the standard answer is the correct answer is our present experiment (Reprint C14) searching for free quarks. We want to know, if we look harder than anybody else has looked, can we find free quarks? These would be quarks made in the early universe which escaped combining with other quarks, thus avoiding being forever buried in ordinary mesons and baryons.

However the standard answer is often not correct and, even when it is correct, it leads to deeper questions about the choice of experiments. If an experimenter hopes to learn something new from experiment A and also hopes to learn something new from experiment B, why do the experimenters choose A not B. Why are we looking for free quarks, why not look for dark matter? Dark matter is the undetected material which is supposed by many astronomers and astrophysicists to make up most of the mass of the universe. The reader can probably detect my residual skepticism about the existence of dark matter. But I am only slightly less skeptical about the existence of free quarks. More about this later.

In the next several sections of these Reflections I write about the reasons and emotions which lead to a choice of experiment A over experiment B. I will write about experiments which are not done for the conventional reason, that is, are not done because the experimenters expect to learn something new or measure something new.

2. Choosing The Experiment: Theory Predicts

Most physicists and chemists who do basic research, as opposed to applied research, specialize as experimenters or specialize as theorists. I don't know how prevalent this specialization is in the other physical sciences or in the social sciences. The reason for the split is straightforward. In highly developed scientific fields such as physics and chemistry, few people have the time and the talent to do the mathematics and calculations required for theoretical work while carrying out complicated experiments.

Experimenters often choose to do an experiment to test the validity of a theory or measure quantities important in a theory, the theory having been devised by someone else. In many cases there is nothing wrong with this, it is simply a division of labor in complicated sciences.

There is a dangerous aspect to the "theory predicts" reason for choosing to do an experiment. "Theory predicts" can start out as a helpful companion for experimenters, helping them choose an experiment, helping them to defend the doing of the experiment before committees and funding agencies. But "theory predicts' can end up as a tyrannical master over experimenters, distorting their research. This change from companion to master is obvious in my own field of elementary particle physics where theoretical work is generally put on a higher plane than experimental work. I wrote a little about this hierarchy in the comments on Reprint D1.

In my youth, and that is my excuse, I succumbed to some "theory predicts" domination of my research. In choosing to do experiments on pion-proton elastic scattering, Reprint C4 for example, I was certainly entranced by the claims that the Regge pole model of strong interactions was fundamental. I remember reading papers and monographs on that model and thinking, if I could only learn more about the mathematics of complex variables then I could understand the fundamental nature of the strong interaction. Thirty years later we do have a basic theory of the strong interaction, quantum chromodynamics, but it did not come out of the Regge pole model or out of pion-proton elastic scattering experiments.

I succumbed again to "theory predicts" about ten years later when we carried out a very arduous optical spark chamber experiment on the production of resonances in pion-proton collisions. I began to hate the experiment soon after we started it and it was only through the great efforts of two very competent graduate students, William Kaune and John Pratt, that the experiment was completed and the results published. But the effort drove them out of elementary particle physics research. I certainly have not reprinted those publications in this book, if the reader wants the references they will have to send me electronic mail.

These two experiences with "theory predicts" and my contrary taste in physics have prevented me from succumbing to some modern tyrannical theories of elementary particle physics. I won't name these theories here, send me electronic mail if you want my list.

Certainly when experimenters are preparing to explore a new area in physics it is a comfort for them to have some theory about what they may find, the theory may even be useful in designing the apparatus. But the theory must be their servant not their master.

3. Choosing The Experiment: This is What We Do

Most scientific papers on the results of an experiment begin with the equivalent of "We report a new measurement of ..." or "We report the first study of ..." This sounds like the conventional reason for doing an experiment, Sec. 1, to learn or measure something new. But often this masks a deeper motivation: "We have carried out this experiment because this is the kind of experiment we do."

The experimenters have done several experiments in a particular subject. They have an apparatus or technique that works well and their results have been accepted by others and confirmed. They understand the theory of the subject and they know the scientific literature. They even know how to improve the apparatus or technique. And so they do more experiments in the subject using the same apparatus, an improved apparatus, or a similar apparatus.

This sounds so critical that I better take the example from my own research. As described in the Memoir A1 on the discovery of the tau, the existence of the tau and its main properties had been established by 1982. Yet I continued for five more years to study the physics of the tau lepton using electron-positron colliders. I then stopped doing experiments involving tau leptons because there was no longer a copious source of tau leptons at my laboratory, the Stanford Linear Accelerator Center (SLAC). In 1995 I am about to start again studying the physics of tau leptons using the Cornell electron-positron collider CESR. Why? The answer is a paraphrase of the previous paragraph. I have made many measurements on the tau lepton using an electron-positron collider and a large-solid-angle detector. I know the technique well, the theory well, and the literature well. The CESR electron-positron collider and the associated large-solid-angle detector called CLEO are much improved over the equipment I used in the past t o study the tau lepton. And so I am going to participate in the CLEO Collaboration in more studies of the tau lepton.

There are many advantages to the "this is what we do" motivation for choosing experiments. It is a good way for experimenters to build a sound body of knowledge about a subject. The research is steady and pleasurable, there are few worries and setbacks. The close study of a subject may reveal unexpected discoveries. Indeed this seemed to be happening for a time in tau physics in the early 1990's when the tau seemed to decay in mysterious ways a few per cent of the time. Unfortunately the mystery has dissipated with more precise measurements.

In most basic science research the underlying criterion for selecting a string of experiments is "this is what we do". Individual experiments will have a specific objective to learn something new or measure something new, but each of the experiments will fit the "this is what we do" criterion.

As I wrote above, this is a productive and pleasant and broadly employed way to select experiments. But there are hazards, particularly for old experimenters, old collaborations, old laboratories. The experiments may become repetitive with less and less knowledge added with each succeeding experiment, as the experimenters come up against technological and theoretical limits. The subject may become overworked with articles and meetings devoted to trivia.

This happened to research on the multiplicity of particles produced in high energy inelastic collisions of hadrons, a fashionable research subject in the 1960's and 1970's. I never worked on that topic, but it interested me and I devoted a lot of space to it in my 1970's book *High Energy Hadron Physics*. The experiments were easy, the data plentiful, and the theories clever. At first some nice empirical regularities were found in the data, regularities which are still useful today in designing high energy experiments. But within a decade the subject became a morass of over-abundant data, too detailed measurements, and superficial theories.

The rule about "this is what we do" experiments is enjoy them but be wary of them.

4. Choosing The Experiment: Sweet Experiments

Sometimes we do an experiment because it is a neat idea, it uses a clever apparatus, it is a sweet experiment. The recent experiment by my colleagues and myself on the Landau, Pomeranchuk, Migdal (LPM) effect (Reprint C13) was just such an experiment. We knew that Landau being a very smart physicist had very probably made the right prediction about how high energy electrons would behave in passing through dense matter. His prediction was based on straightforward quantum mechanics, and we didn't expect quantum mechanics to go wrong in dealing with the gentle particle collisions that would cause the LPM effect.

But we had a neat, quick, and easy way to do the experiment, and there was no previous precise experiment on the LPM effect. It was a sweet experiment and Landau turned out to be right, as we expected.

Sweet experiments are most prevalent in the experimental sciences when their costs are small. They are rare in my field of high energy physics. I have a few I'd like to do but I don't think I'll get to do them. For example, in the past half century there has been a tremendous amount of research using positronium, the beautifully simple system consisting of an electron and a positron bound together by electrical attraction. However the beautiful system consisting of a negative muon and a positive muon bound together by electrical attraction has never been made; there are no published experiments on this system. It would be a sweet experiment to make such a system and study it; and it would be an easy experiment. But there seems to be nothing *new* to be learned from the electrical binding of a negative muon to a positive muon. The experiment would take two years to prepare and carry out, and I don't think that I can talk an accelerator program committee into giving me the accelerator time, so I haven't ch osen to do that sweet experiment.

5. Choosing The Experiment: Personality and Personal Taste

In addition to the various motives for choosing an experiment which I have just described, there are the hidden and often unconscious motives deriving from the experimenter's personality and personal scientific taste. In Sec. 1 of these Reflections I remarked that I was roughly equally skeptical about the existence of dark matter and the existence of free quarks, yet I have chosen to do an experiment, actually my second experiment, searching for free quarks. This decision comes out of my personality and personal scientific taste.

Experimental and theoretical research on the possible existence of dark matter occupies lots of physicists at present. On the other hand, I think my colleagues and I are the only physicists searching for free quarks in bulk matter at present. It was precisely this imbalance of numbers of researchers that attracted me to the quark search. I like the feeling of being part of a small band of explorers starting out on a very uncertain search, the feeling of being free to set our own pace and make our own mistakes. I even like not having to keep up with new articles on free quark searches, not having to attend conferences devoted to free quark searches. The next article will be ours, and we are our own conference. This is not rationality, it is personality and personal scientific taste.

Having worked with hundreds of other experimenters in the past forty years and having met and talked with many hundreds more, I know that personality and scientific taste plays a large role in experimental choices. Some experimenters are very social, they enjoy working in an area where there are lots of fellow researchers and lots of conferences. They also enjoy working in large experimental groups where there is lots of interaction between collaborators. Other researchers are very competent in the administration and management required in carrying out large, long-duration experiments. And so their personality takes them into large group physics.

But I know that many researchers share my taste for small and lonely bands of scientific explorers. After all, the popular image of the scientist is Pasteur proposing his germ theory of disease while his colleagues laughed at him, the popular image of the scientist is Einstein working in the Patent Office. These images sent many of us into science. Unfortunately the desire to be part of a small, lonely band of researchers can not always be realized for the reason discussed at the beginning of Sec. 8.

6. Doing Experiments: Old Technology

Having written down my reflections on choosing an experiment, I now turn to doing the experiment. It is easy and comfortable for experimenters to use well-understood and well-developed technology — old technology. There is absolutely nothing wrong with this. I used optical spark chambers for ten years for all sorts of experiments. I enjoyed building them and felt secure in their use. I preach the rule, "If you can use an old technology for a new experiment, use it, enjoy it, and sleep well at night."

But an old technology gradually loses its power. I wrote about the obvious examples of accelerators and telescopes in Reprint D2. New technology is eventually needed.

7. Doing Experiments: New Technology

It is new technology which provides so much of the challenge and pleasure and pain in an experiment. I have written about these things in many of my comments in Part C of this book, I will not repeat those reflections here.

It is very strange that the scientific paper which repeats an experiment or describes a new technology rarely reports on the experimenter's pleasures and worries and intermediate failures. I can't recall an experimental paper in a scientific journal in which the authors write, "We made a mistake in the mechanical design and had to rebuild the apparatus," or "The superconducting magnet was a lemon when purchased and fixing it delayed us two years," or "We didn't really understand the chemistry behind our new technique and we kept getting wrong results for a year." At best the problems in a new technology are buried in formal discussions of experimental errors.

There is a macho flavor about the reluctance in scientific papers to describe the ups and downs of experiments. This is doubly strange because these ups and downs are the main topic of conversation among researchers, particularly the "downs" of other people's experiments.

This leads to two rules about new experimental technologies. First rule, if you are pioneering a new technology be prepared for trouble, worry and sleepless nights; nature is hostile to innovation. Second rule, if you are copying another experimenter's new technology don't just read the published description, talk to the graduate student, talk to the technician.

8. Doing Experiments: Support and Supportive

Perhaps the most significant difference between experimenters and theorists is that experimenters need money to buy equipment, to build an apparatus, to pay for operating the experiment, to pay for computer time, to hire helpers of all sorts — engineers, technicians, programmers; theorists only need money for a library and computers. Although these days theorists need a bit more money for graduate students and travel. But the expensive people in experimental science are the experimenters.

Since the primary support required to do experiments is money, usually called by the euphemism funding, it is obvious that the choice of an experiment can be influenced by the availability of funding and by fashions in funding. It is easiest to get funding when your experiment is like several other experiments immersed in a fashionable field. It is hardest to get funding when you are an isolated experimenter. This is the reason for the observation at the end of Sec. 5 that more researchers would like to be in small, lonely bands than are in such bands. Some experimenters, no names here, get around this problem by developing a dual research personality. One personality works on "this is what we do" or "theory predicts' fashionable experiments. This brings in funding for students and technicians and a laboratory. The other personality steals a little from this cornucopia to do an unfashionable experiment, perhaps a "sweet" experiment.

An experimenter needs more than funding support and helpers and laboratory space, she or he needs a supportive environment in their institution and in their scientific community. I've always been fortunate in this way, at my two institutions, first the Physics Department of the University of Michigan and then the Stanford Linear Accelerator Center. In a supportive environment there are colleagues and technicians who know all sorts of things, who can lend an experimenter all sorts of odd equipment. When I began the quark search experiment which uses very small drops of oil I needed to use a high-power optical microscope to look at the tiny orifice in the mechanism that produces the drops. I had never used a high-power microscope and my experimental group didn't own any microscope. From one colleague, Clive Field, I received some basic lessons in the use and limitations of high-power optical microscopes; from another colleague, Edward Garwin, I was able to borrow an old but very high-quality microscope which w e still use.

A different and subtle aspect of a supportive surrounding for an experimenter is the attitude of one's colleagues toward an experimenter's mistakes and bad luck. Particularly important is the attitude of the senior colleagues. Experimenters make mistakes and adventurous experimenters make more mistakes. It is important to be tolerant about mistakes, but it is also important to warn the experimenter when he or she is being foolish. If a colleague is about to publish an experimental result which in a subtle way violates Newton's Law, $f = ma$, it is important to slow down that publication until the colleague finally realizes the chance they are taking.

What about my long shot free quark search? The attitude of my colleagues is one of tolerant amusement.

9. The Results of an Experiment: Success

Most experiments are successful: the experiment worked, the results were published, contemporary or subsequent experiments by others agreed with the result. There are all types and degrees of success.

Of course, there is the astonishing success, a discovery or measurement that changes a science. The discovery of the tau lepton is the example from my research. This type of a success is rare in any experimental science. I can give you some rules about what *not to do* if you seek such a success. Don't spend too much time trying to confirm someone else's speculative theory. If you prove them right your experimental work will be called beautiful, their theoretical work will be called astonishing. Don't work on a specific exploration with too many colleagues or when there are too many similar experiments. The degree of astonishment awarded to a discovery by the scientific world is proportional to the ratio of the significance of the discovery to the number of involved experimenters.

I have already used the term beautiful applied to an experiment's results. A beautiful success need not be the confirmation of a theory, it can be attached to an experimental result which clears up a confusion in a field. Sometimes beautiful results can be applied to an entire set of experiments, for example the experiments carried out at the LEP and SLC electron-positron colliders in the past five years which established the properties of the Z^0.

Then there are the great bulk of experiments which are simply successful. This means the experiment was carried out professionally, the results are right and there is even some interest in the results. There is more pleasure and satisfaction for the experimenter in a simply successful experiment than is appreciated or understood by someone who has not done experiments. In some of my comments in Part C, I have discussed the pleasure and satisfaction.

The carrying out of an experiment is like a difficult trek through hazardous country. The apparatus may break down and have to be fixed. For a long time the results may look wrong, full of distortions and mysteries. The experimenters may be running out of time and energy. There are the middle-of-the-night panics when the experimenter wakes up with a new worry about the experiment. If the experiment is large and operated by people on shift, she or he may telephone seeking reassurance. Or, the worried experimenter may go into the laboratory in the middle of the night. There is nothing useful to do but it's better than insomnia.

The completion of a successful experiment is like the return home of the traveler at the end of the difficult trek. The hazards are passed, the traveler is once more safe and warm and well-fed. The contentment does not last very long, and like the compulsive traveler the experimenter soon sets out again.

The completion of a successful experiment brings special pleasure to many experimenters if competing experimenters are bested by publishing the results first or publishing slightly better results. This is a great deal of irrationality in this pervasive completion in science.

First the rational, or at least understandable, aspects of scientific competition. If one makes a discovery first, if one's results are much better than the results of the competing experimenters, then there is certainly the basic human pleasure in winning. This is the applause of one's colleagues; there may be the tangible reward of more funding; there are promotions and prizes.

But why compete to be the first to confirm someone else's theory? If the theory is confirmed, the credit goes, and should go, to the originator of the theory. Or why compete to knock down a theory? If a theory is wrong, what does it matter which experimental result first showed the theory to be wrong? The only situation I understand is when experimenters fear that being late and last with a duplicative result, the scientific journals may refuse to publish their articles. The final proof of the success of an experiment is publication. But in my science, well-established experimenters and groups of experimenters still compete over every experiment, even though by tradition they are guaranteed publication.

As I completed this section I thought that I was sounding above it all and I began to think, "Am I competitive in this irrational way." Of course I am competitive this way, though not as competitive as when I was younger. Looking into my experimenter's psyche I think I understand this irrational competitiveness. An experiment becomes ephemeral once a subfield is established and the accumulated experimental knowledge is summarized in a review article or a monograph. The experiment is a number in the bibliography. All that work and worry becomes just a bibliographic entry.

The experiment is even more of a ghost if the subfield can be summarized in a simple law. Thus all the hundreds of experiments on how the tau lepton decays can be replaced by stating the conclusion: to the best of our knowledge the tau decays according to the well-known theory of the weak interaction with the constraint that lepton type is conserved. If one has published first, perhaps the ghost is not as wispy.

10. The Results of an Experiment: Failure

Experimenters don't like to talk about their own failed experiments. Of course they love to talk about the failures of others, that is part of the competitiveness about which I just wrote. Therefore to explore the significance to an experimenter of his or her failed experiment I must rely on my experience and those of a few close colleagues.

Failure in an experiment is painful, there is the time and work and money wasted, there is the worry that future funding will be reduced, there is the worry about the effect on a career. There is usually something to learn from a failure, but the pain outweighs the utility.

Therefore experimenters move the boundary between the definition of a successful experiment and the definition of a failed experiment. They move the boundary so that the class of failed experiment is as small as possible.

The moving of the boundary between success and failure is done in various ways. When I decided to put Reprint C2 into this book I thought of it as an example of what I did using bubble chambers early in my career, I thought of it as a primitive but successful experiment. But now that I have almost completed this book I think it is an example of a failed experiment. After all we learned little of significance about the behavior of K^+ mesons. Yet by the definition of a successful experiment given in the previous section it was a success.

In my research life the only experiment I considered a failure at the time was the search for free quarks using a rotor electrometer, Reprint C10. We were not able to carry out the search. Yet I am proud of the idea and the skill of my colleagues, I don't regret the attempt.

In particle physics there is a large class of search experiments which are called successful, sometimes even called beautiful, yet which never found the particle or phenomenon they were designed to find. I'll give an example. The theory of elementary particles called supersymmetry says that for every type of known quark or lepton there is a supersymmetric partner with somewhat similar properties, but with a different quantum spin and a different mass. The theory does not predict the mass of these supersymmetric partner particles and so for twenty years there have been searches for partner particles. The searches have extended to larger and larger particle masses as new high energy accelerators and colliders has been introduced. But no supersymmetric partner particle has been found. All the searches have *failed*. Are these experiments failures or successes? The world of particle physics has defined these searches as successful.

In the end experimenters in all sciences protect themselves and their field by defining failed experiments as narrowly as possible. An experiment fails when the experimenter gets no answer or gets the wrong answer. All other experiments are called successful.

11. Discovery and Invention

In all of experimental science the great goals are to discover something new about nature or to invent a new experimental technology. Discovery and invention. This is what the student dreams of doing, this is what the experimenter hopes for and works for, this is what brings fame and prizes, this is what gets the experimenter's name into the histories of science.

A tremendous amount has been written about the arts of discovery and invention, and about creativity, that mysterious property of the mind which is supposed to lead to discovery and invention. There have been an endless number of books and courses and even video tapes on how to be creative. How many times have we read the story of how Kekulé in a dream discovered the cyclic structure of benzene?

I have read books on creativity, I have known scientists who have made wonderful discoveries and surprising inventions, I have myself tried various prescriptions for creative thinking. In the end I have realized that I know very little about creativity. I don't know if creativity in the artist is the same as creativity in the scientist. I am not a mathematician or a theoretical physicist, so I don't know how the mind of a Newton or an Einstein or a Feynmann works. I do know enough to mistrust what others write about those minds.

However, over the years I have developed some personal maxims and observations related to discovery and invention. A few years ago I began to collect these maxims and observations when I found them sprinkled through my conversations with graduate students. One of the pleasures of doing research is working with graduate students; they have helped me to learn UNIX and I have been able to tell them about the experimenter's way. The writing of this book solidified my collection of maxims and observations.

My first observation is that nature is hard and unyielding. This may seem depressing advice to give a novice experimenter, but it is critical advice. The experimenter does not have the luxury of the theorist, the luxury to spin out some speculations or to build a mathematical tower and then hope to be right. The theorist gets credit for a clever theory until proven wrong. If the theory cannot be tested, the credit can go on for a long time. But the experimenter gets no credit for an inconclusive experiment.

The experimenter comes up against hard, unyielding nature in several ways. If an apparatus gives measurements with intrinsic errors of 10%, it is hopeless to try to use the apparatus to look for a 1% effect. These are measurements one would like to make but nature resists. For example, for sixty years we have wanted to learn if neutrinos have masses greater than zero. Present technology allows some limited precision, direct, measurements of a neutrino's mass, and there is one indirect method which depends upon the speculation that neutrinos can change their type. A young experimenter may try to find a highly precise method to determine a neutrino's mass. But she or he must know that no one has succeeded in doing this in sixty years. I do not tell the young experimenter that a highly precise method is impossible, but I do tell them that nature is hard and unyielding.

The hardness of nature must be taken into account in the choice of an experiment, the subject of the first section of these Reflections. How unyielding will nature be if the experimenters try something new? The usual way to try to answer this question is to begin with detailed designs and calculations about the proposed experiment. Sometimes prototypes of parts of the proposed apparatus are built and tested. This procedure is certainly necessary for a large apparatus which takes years and many millions of dollars to build. But this procedure can prevent the carrying out of a potentially successful experiment. Therefore for small experiments I sometimes start building the apparatus and even using it before I see how to solve all the problems. In doing this I am partly depending on previous experience that the actuality of building or operating stimulates solutions to problems. I am also partly depending on my colleagues to pull me out of difficulties, and I am always hoping for good luck.

The hardness of nature is most obvious to me during the starting up and the operation of an experiment: parts break, errors are larger than expected, mysterious effects contaminate the data. This produces in me a tension between solving each problem as it arises and pushing ahead to complete the experiment. The argument for solving each problem is obvious. The experiment may fail or produce unreliable data if problems are not solved. There is also a not obvious argument, sometimes solving a minor problem will give the experimenter a deeper insight into the experiment, leading to a major improvement or even a new experiment.

Well then, why push ahead ignoring problems? Sometimes there is no choice because the experiment uses a facility with limited access time. An accelerator, a reactor, or a satellite for example. Sometimes there is a more delicate reason for pushing on. The experimenter judges the problem to be minor and decides to live with it rather than delaying the data acquisition. When I was young I was more for pushing on to combat my anxiety that the experiment would never be completed. Now my anxieties are more centered on the experiment's problems than on the experiment's completion. I don't know if this is wisdom or tenure.

One more observation about nature being hard and unyielding. The solving of the problems that arise in the operation and completion of an experiment is a personal battle between me and nature. When the major problems are solved and when half the data is safely recorded I began to relax and to believe that nature has yielded a little. This is the situation with our present search for free quarks, by the end of last month, we had recorded on tape half of the data we wanted for the first experiment. Then the experiment was interrupted by another part of nature — people. The roof above our laboratory was taken apart for a five-month-long renovation of the building air conditioning system. The experiment was plagued by scheduled and unscheduled interruptions of our laboratory's electrical power and air conditioning. We finally had to move the experiment to another location. I told Nancy Mar, the graduate student on the experiment, that we must be winning our battle with nature since nature had to resort to roof renovation to slow down this free quark search.

With the hard and unyielding qualities of nature as the reality, the experimenter looks for new ideas for discovery and invention. Where have the new ideas come from in my research? New ideas have come from being obsessed with a subject, they have been given to me by colleagues, they have come from discussions with colleagues and working with colleagues, they have come from the need to simply use a new set of data or a new technique. Before elaborating on these sources of new ideas, I want to write about some sources which are sterile for me.

I have found over the years that I cannot get a new idea by following the prescription (a) that the scientist frees his or her mind of everyday research careers, and (b) then lets one's mind speculate and dream about new directions. I have tried keeping an idea notebook, it soon ends up in a corner. I have tried drawing speculative pictures of apparatus and writing down stream-of-consciousness research thoughts, they have always been for me gibberish. I have tried taking a few hours in the science library and scanning the journals. Sometimes I find interesting things, but I don't get new ideas.

Another often discussed source of new scientific ideas is the general intellectual commerce of science: going to scientific meetings, attending lectures and colloquia, serving on scientific committees, reviewing proposals and grant applications. This doesn't work for me. Perhaps the main reason that I easily get bored and restless at meetings and lectures is just that.

Enough negative thoughts, I turn to the sources of new ideas that have worked for me. First there is obsession with a specific limited subject: not all of physics, not even all of elementary particle physics, but a few specific subjects in elementary particle physics. Early in my research life it was the strong interaction, elastic scattering of elementary particles. This obsession brought me the pleasures of introducing with my colleagues new experimental methods and of carrying out good experiments. The obsession was slowly worn away by the intractable nature of the theory of strong interaction elastic scattering.

As I described in Part A of this volume, I have been obsessed since the early 1960's with the nature of leptons. I don't expect to have the luck to find another tau, but my lepton obsession brings me new ideas. If I hear or read about a new technology in high energy physics, I ask myself whether it can be applied to leptons with any profit. I follow all the big and little, conventional and odd, experiments on leptons. For example, experimenters at the GSI nuclear physics laboratory in Germany have been finding some unexplained regularities in the production of electron-positron pairs by the collision of heavy nuclei. I'm not a nuclear physicist and I can't repeat their experiments, but I often wonder what those regularities signify about leptons and I look for related experiments that I might do.

In addition to my lepton obsession I have my more recent obsession about searching for free quarks. This obsession began in the middle 1980's with the electrometer apparatus described in Reprint C8. After the failure of that experiment I kept looking for other search methods, but nothing appealed to me until my colleagues Charles Hendricks, Klaus Lackner, and Gordon Shaw approached me with their idea described in Reprint C14. This illustrates the second source of new ideas for me, ideas which come from colleagues. I prefer collaborating with such colleagues for two reasons. First, I strongly feel that if the idea succeeds they should get a substantial part of the credit, not just an acknowledgment in a paper. Second, since they conceived the idea and have probably thought a lot about the experiment, they will be strong research companions when the experiment comes up against the hardness of nature. Sometimes a colleague simply wants to give the idea away and not work on it. I will accept that, too.

The third source of new ideas for me is discussion with colleagues and working with colleagues; ideas which we develop together and usually carry out together. This requires a special research relationship where we respect each other's research skills and where we have similar research interests and styles. These special research relationships have usually been with experimenters or engineers: at Michigan with Lawrence Jones, Donald Meyer and Michael Longo, at SLAC with Gary Feldman, Walter Innes, John Jaros, Rafe Schindler, Eric Lee and quite a few others. In the wider world there are experimenters with whom I have found new ideas through working together on a previous experiment. The working together is essential for me, discussion by itself is not enough.

Therefore it is to be expected that it is rare for me to have a new idea producing, research relationship with theoretical physicists. It's not the mathematics or the formal theory, I'm competent in mathematics. But I have a deep resistance to working on ideas which come surrounded by mathematics and formal theory. However there are some theoretical physicists with whom I have simulating, long term, research relationships, they include Paul Tsai, Fred Gilman, Haim Harari, James Bjorken, Sidney Drell, and Richard Blanckenbecler.

The fourth source of new ideas for me may sound peculiar and I don't know if other experimental scientists have experienced this fourth source. If they have experienced it, they don't talk about it or write about it. This fourth source is the need to use a new set of data or to use a new experimental technique. I'll begin with a very recent example.

As I have written elsewhere in this volume, a few colleagues from SLAC and myself were fortunate in being able to join in the past year a very successful, ongoing experiment in electron-positron annihilation physics at Cornell University — the CLEO experiment using the 10 GeV CESR electron-positron collider. We are collaborating in upgrading and operating the experiment, and we are beginning to work with some of the data already acquired. It is a tremendous job to learn how to use the data acquired by a modern, large high-energy physics experiment. The computer programs are immense and there is much to learn about the properties and limitations of the apparatus. I wanted to learn how to work with CLEO data and was looking for a simple way to start. An Ohio State University colleague, K.K.Gan, also a member of the CLEO experiment, visited SLAC a few weeks ago and asked me if I was interested in working with him in studying in the CLEO data some properties of the eta meson. I have never been interested in the eta, but his idea involved looking at pairs of charged particles, something relatively easy to do and an excellent way to learn how to analyze CLEO data. Also, his proposal revived in me my old speculations about lepton pairs, Reprint C9, and so we have begun.

Thus this fourth source of new ideas reverses the conventional view of experimental science. In the conventional view the experimenter builds an apparatus or acquires data to answer an initial question. In this reversal, the experimenter has an apparatus or a new technology or a set of data and then looks for an idea to make use of the apparatus or technology or data. This source of new ideas is important to me.

A final reflection on new ideas. I find it remarkable that a new idea in experimental science is often obvious after someone else gets the new idea. Why is that? If the new idea is obvious why didn't I think of it? This phenomenon of the non-obvious suddenly being obvious comes out of the mysterious heart of the process of discovery and invention in experimental science. The mysteries of nature and the mysteries of creativity are mixed together in my research life; the mixture heightens the pleasures.

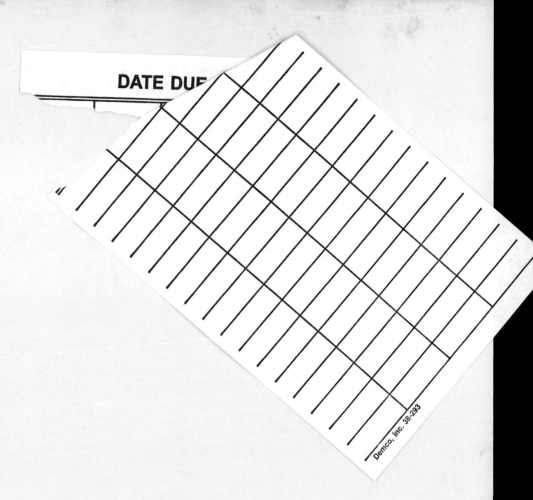